AIR QUALITY IN THE 21ST CENTURY

Air Quality in the 21st Century

Gaia C. Romano

and

Alice G. Conti

Editors

Nova Science Publishers, Inc.

New York

Library of Congress Cataloging-in-Publication Data

Air quality in the 21st century / Gaia C. Romano and Alice G. Conti, editors.
p. cm.
ISBN 978-1-60456-793-9 (hardcover)
1. Air quality management. 2. Air--Pollution. I. Romano, Gaia C. II. Conti, Alice G.
TD883.A5766 2008
628.5'3--dc22 2008023351

Published by Nova Science Publishers, Inc. ✣New York

CONTENTS

PREFACE

This book presents the latest research on the crucial issues and the standards necessary to assess, monitor and increase air quality. Particular emphasis is paid to problems related to urban air quality.

Chapter 1 – The chapter presents an integrate effort in a medium sized European city located in south Europe to measure and simulate the emissions of benzene in the ambient air. The campaign included ambient air monitoring (through passive and active samplers) in 9 sites of the city and personal exposures of special groups of the population (taxi drivers, policemen, filling station employees), as well as, the general population in several microenvironments. Additionally meteorological and traffic data were measured.

The analysis of the collected data was attempted with the aid of emission and dispersion models, as well as, with the aid of traditional statistical and contemporary techniques (neural networks). The limited size of the city, the roads radial network, the old vehicles fleet, the meteorological conditions and the geometry of the roads affect significantly the ambient benzene concentrations in the city, that found increased when specific circumstances occur (calm wind conditions, proximity to filling stations, traffic congestion, street canyons etc).

The average mean concentrations of benzene were found significantly higher of the limit threshold imposed by the European Union at hot spots near the street canyons but lower in urban background sites. The values are found higher during summer. The personal exposure values were found significantly varied for the different groups of the population presenting their highest values in the filling stations employees group following by the policemen. Active sampling results revealed that driving in a traffic congested road is an activity leading to elevated exposure levels of benzene. Almost similar to car refueling procedure, in terms of benzene exposure is the walking in the roadside of a congested road.

Risk evaluation due to exposure to benzene revealed that general population runs a risk equal to $5.3*10^{-5}$, which increases to 10-40% for the examined occupational groups. Filling stations employees refueling cars, run the higher risk ($7.4*10^{-5}$).

Finally several "what if" scenarios presenting different composition of the vehicle fleet were attempted to examined their contribution to a possible reduction of the emitted benzene in the air. It was found that the replacement of the older technology cars by newer (from EURO1 to EURO5) as well as the replacement of the non catalytic with catalytic cars will decrease the released benzene up to 77%.

Chapter 2 – This paper provides a survey on the role of the atmospheric boundary layer and its role on air quality. Based on the evaluation of a weather forecast model it is found that the nighttime or stable boundary layer is difficult to forecast. We briefly summarize the relevant processes in this boundary layer prototype, and we discuss their interactions. Next, a climatology of flow meandering and intermittency of the atmospheric turbulence, and its spatial extent under these conditions, is presented based on field observations. Finally, a new formula for the stable boundary layer height is derived.

Chapter 3 – Using ground-based spectral solar extinction data taken in the greater urban Athens atmosphere during a field survey, experimental and modeled aerosol optical depths have been retrieved. The Angstrom's turbidity parameters (**α, β**) were derived via three widely used techniques: the Volz, the logarithmic and the direct fitting methods to the experimental aerosol optical depths. This chapter investigates the ability of the different methods to determine similar Angstrom's turbidity parameters further addressing their dependence on the spectral ranges considered for their derivation. The results obtained reveal that the various techniques result in different pairs of the Angstrom's parameters mainly at the shorter wavelengths. It also was found that the Angstrom's turbidity parameters obtained by any of the three methods at shorter wavebands are not representative of the whole spectrum exhibiting large uncertainties especially under low turbidity levels. From the overall analysis it is established that the logarithmic method predict almost identical results as those obtained via the direct fitting method, also possessing the least wavelength dependence.

Chapter 4 – Odor episodes and environmental air quality are topics of worldwide concern, mainly due to the fact that industrial facilities are often located very close to inhabited areas. A newly developed methodology based on the simultaneous application of meteorological analysis techniques, social control, chemical control and numerical modeling makes it possible to identify and evaluate air quality at different points in urban and industrial areas, as well as to identify the emission source of the odor episodes. Social control makes it possible to build databases of odor episodes, obtain sensory measurements (by determining the annoyance index) and obtain samples during odor episodes. This is done by switching on air samplers at the beginning and end of odor episodes. When databases of odor episodes are treated statistically, a valuable tool is obtained: annoyance index maps, which can be used to calibrate the dispersion modeling results. The chemical compounds are identified by means of chemical control. A validated analytical method, based on thermal desorption (TD) coupled with gas chromatography (GC) and mass spectrometry (MS), is used to determine a wide range of volatile organic compounds (VOCs) that cause odor nuisances and affect air quality in indoor and outdoor air, including alkanes, aromatic hydrocarbons, aldehydes, alcohols, chlorides, esters, ketones, terpenes, amides, carbon disulfide and isocyanates. If the meteorological conditions and main VOC emission sources of a given location are known, numerical modeling can be used to create impact prediction maps that closely resemble the impact maps created after analysis of the VOCs. These impact maps are obtained with a Gaussian-type model developed at the Environmental Center Laboratory (LCMA) of the Department of Chemical Engineering at the Polytechnical University of Catalonia (UPC). They make it possible to study the dispersion of odor compounds in the atmosphere surrounding emission sources. These maps are a three-dimensional representation of time-averaged accumulative concentrations for a particular period of time around the emission sources. In addition, meteorological models can be used to track, backwards in time, the source of the air mass responsible for the discomfort, mainly to find possible VOC sources outside the urban area. The procedure combines an analytical approach based on the acquisition of samples, which requires the participation of the affected population (i.e. social participation used as a scientific tool), with a modeling approach.

Chapter 5 – Lichens have been widely used as biomonitoring of atmospheric pollution, because these organisms respond to phytotoxic gases (especially SO_2 and NO_x) at cellular, individual and community level. Nevertheless, due to the biological and ecological variability of the organisms, scientists often find difficult to discern the effects of pollution against a natural background noise. On the other hand, statistically supported responses are needed if we aim to include these techniques in decisional processes of environmental management. In this chapter we try to investigate this complex trade-off, by focusing the attention on bioindication techniques, based on the observation of a decreased richness of lichen communities, related to increasing concentration of atmospheric pollutants. In particular we present worked examples of biomonitoring applications in heterogeneous environments, to provide an overview of main sampling, analytical and statistical procedures, which can lead scientists and stakeholders to a better knowledge of the relationship between lichen diversity and air pollution. We take into account three main factors of variability for lichen communities across spatial and temporal scales.

In the first case, we investigate the different responses of lichens to atmospheric pollutants in relation to macroclimatic predictors. Because of close dependence on the atmosphere for their metabolic processes, lichens are strongly influenced by climate-related variables (e.g. rainfall and temperature). It was shown how, under the same level of air pollution, lichen diversity is higher in areas with higher rainfall. We present results from dry vs. humid bioclimatic areas in Italy and we discuss how synergistic and antagonistic relations among pollutants, lichen and climate may significantly vary, together with the predictive power of biomonitoring methods.

Modern approach to environmental assessments includes an evaluation of the pollution impact not only on human health, but also on natural heritage. Hence, lichen biomonitoring was used to detect the effects of gaseous pollutants on natural ecosystems. In the second example, we show that, in the framework of lichen biomonitoring in natural forests, habitat-related variables may play a major role on lichen diversity and more caution should be taken when interpreting these data in terms of direct effects of pollution.

At a more detailed spatial scale, the variability of lichen diversity can be very high and consequently the level of uncertainty in the interpretation of data may arise considerably. This should be taken into account when higher scale surveys are planned, by considering an adequate sampling intensity at within-site scale. In the third working example we discuss this issue, showing data from intensive experiments and comparing the variability unrelated to atmospheric pollution at local scale in natural environments with that observed at higher scale in anthropized ones.

Chapter 6 – Health, ecological, and socio-economic impact of hydrocarbon contaminants (crude oils, natural gas, lubricating oils, and so on) in soil, water, and air have been reviewed for purposes of self-risk recognition and assessment. Over 56 hydrocarbons are involved in environmental contamination. Daily exposure doses of 0.1 to 5.0 milligrams per kilogram (oral ingestion) and 0.1 to 6.3 milligrams per kilogram (inhalation) of gasoline to kerosene range hydrocarbons (C_6 to C_{16}) can cause kidney, liver and hematological malfunctions. Leaded gasoline is the single largest source of lead exposure in urban areas. Urban congestion, worn-out vehicles, low quality fuels (having sulfur greater than 0.1 percent), and odor-emitting wastes, which provide 'maternity wards' and 'free lunch counters' for flies and rats, are some of the despairs of poverty in developing nations. A careful husbandry of the scientific, political and socio-economic measures suggested herein might enhance the quality and stability of the environment.

Chapter 7 – In this paper, we consider recently introduced bivariate stochastic volatility models commonly used to analyse financial time series, to study problems related to air pollution data. Such models are used here to estimate the volatility of weekly averaged ozone measurements taking into account two different sets of data provided by the monitoring network of Mexico City. A Bayesian analysis is developed using Markov Chain Monte Carlo (MCMC) methods to simulate samples from the joint posterior distributions and perform the estimates of interest.

Chapter 8 – Air pollution sampling site selection is one of the most important and yet most vexing of the problems faced by those responsible for regional and urban air quality management and for the attainment and maintenance of national ambient air quality standards. This chapter will present different optimization techniques for the design of optimal Air Quality Monitoring Networks (AQMN). The objective of the optimization is to provide maximum information about the presence and level of atmospheric contaminants in a given area and with a limited budget. A criteria for assessing the allocation of monitoring stations is developed by applying a utility function that can describe the spatial coverage of the network and its ability to detect violations of standards for multiple pollutants. The use of mathematical models based on the Multiple Cell Approach (MCA) to create spatial distributions for the concentrations of the pollutants emitted from different emission sources will be illustrated. The optimization techniques presented in the chapter are illustrated on a number of case studies and the number of monitoring stations and their locations for each case are obtained. In addition, the effect of the spatial correlation coefficient on total area coverage are discussed.

Chapter 9 – This paper examines the sensitivity of predicted concentrations for regulatory applications using air quality models for urban areas. The issue is explored in detail using the latest air quality model AERMOD promulgated by the US Environmental Protection Agency (EPA) and the meteorological and source data available from Toledo, Ohio for the years 1990, 1991, and 1992. The analysis suggests using the twelve sectors' land use parameter values around three kilometer radius of the meteorological site and the population around the major sources to obtain predicted concentrations. The use of this method resulted in lower predicted concentrations than the concentrations predicted from using the area method. The study also found that the AERMOD model was highly sensitive to variation in urban population for shorter averaging time periods and almost had negligible effect for the longer averaging time periods. This study demonstrates that the input of land use parameters and population affect the ground level concentrations for an urban area with multiple sources.

Chapter 10 – The combination of chemical substances and specific meteorological parameters cause the formation of the ground-level ozone layer. High temperature, sunlight intensity and increased surface pressure are the "obvious" conditions that mostly help ozone in its formation. In addition, local circulations (e.g. sea breeze, valley winds) with light or stronger winds may assist in ozone formation by creating adequate dilution conditions that accelerate photochemical reactions. Wind direction has also been considered as an important factor in the formation of ozone.

The strong spatial and temporal variability of traffic-related air pollution detected at roadside locations in large or medium-sized cities has raised the question of how representative the site and time period of air quality measurements actually can be. In general, measurements follow symmetric patterns and show their dependence on meteorological conditions. A CFD urban street canyon modeling approximation in local scale is presented in order to finalise a few issues that are observed in relation to ozone production, temperature and solar radiation.

In: Air Quality in the 21st Century
Editors: G. C. Romano and A. G. Conti, pp. 1-65
ISBN 978-1-60456-793-9

Chapter 1

PARAMETERS CONTROLLING AMBIENT AIR BENZENE CONCENTRATIONS AND HUMAN EXPOSURE IN A MEDIUM SIZED SOUTHEASTERN EUROPEAN CITY

Pavlos A. Kassomenos[1,*], Georgios A. Pilidis[2], Costas L. Papaloukas[3] and Spyros P. Karakitsios[2]

[1] University of Ioannina, Department of Physics, Laboratory of Meteorology, GR-45110, Ioannina, Greece

[2] University of Ioannina, Department of Biological Applications and Technologies, Laboratory of Environmental Chemistry, GR-45110, Ioannina, Greece

[2] University of Ioannina, Department of Biological Applications and Technologies, Laboratory of Bioinformatics, GR-45110, Ioannina, Greece

ABSTRACT

The chapter presents an integrate effort in a medium sized European city located in south Europe to measure and simulate the emissions of benzene in the ambient air.

The campaign included ambient air monitoring (through passive and active samplers) in 9 sites of the city and personal exposures of special groups of the population (taxi drivers, policemen, filling station employees), as well as, the general population in several microenvironments. Additionally meteorological and traffic data were measured.

The analysis of the collected data was attempted with the aid of emission and dispersion models, as well as, with the aid of traditional statistical and contemporary techniques (neural networks). The limited size of the city, the roads radial network, the old vehicles fleet, the meteorological conditions and the geometry of the roads affect significantly the ambient benzene concentrations in the city, that found increased when specific circumstances occur (calm wind conditions, proximity to filling stations, traffic congestion, street canyons etc).

The average mean concentrations of benzene were found significantly higher of the limit threshold imposed by the European Union at hot spots near the street canyons but lower in

[*] Corresponding author: Professor P.A. Kassomenos. E-mail address: pkassom@uoi.gr Fax: +30-26510-98671, Tel: +30-26510-98470.

urban background sites. The values are found higher during summer. The personal exposure values were found significantly varied for the different groups of the population presenting their highest values in the filling stations employees group following by the policemen. Active sampling results revealed that driving in a traffic congested road is an activity leading to elevated exposure levels of benzene. Almost similar to car refueling procedure, in terms of benzene exposure is the walking in the roadside of a congested road.
Risk evaluation due to exposure to benzene revealed that general population runs a risk equal to $5.3*10^{-5}$, which increases to 10-40% for the examined occupational groups. Filling stations employees refueling cars, run the higher risk ($7.4*10^{-5}$).
Finally several "what if" scenarios presenting different composition of the vehicle fleet were attempted to examined their contribution to a possible reduction of the emitted benzene in the air. It was found that the replacement of the older technology cars by newer (from EURO1 to EURO5) as well as the replacement of the non catalytic with catalytic cars will decrease the released benzene up to 77%.

Keywords: Benzene, modeling, human exposure, risk assessment

1. INTRODUCTION

Benzene (C_6H_6) is a Volatile Organic Compound (VOC) with a very stable chemical ring structure that constitutes the base of the aromatic hydrocarbon family. It is a natural component of crude oil (less than 1% by weight) and is thus found in certain refined products, such as gasoline (Schnatter, 2000). As lead-containing antiknock additives have been reduced and eliminated, more aromatics (including benzene) are blended into gasoline for antiknock purposes and, therefore, benzene contented in gasoline varies from less than 1% up to 5%, depending on fuel quality, countries legislation and season (Medinsky et. al., 1995; Verma and Tombe, 2002).

Benzene is rapidly but incompletely absorbed by humans and animals following inhalation exposure (ATSDR, 1997). It is classified by the International Association on the Risks of Cancer (IARC, 1987) as a Class 1 carcinogen. By the 1970s, the first attempts to quantify the risk of leukemia due to benzene were made (McMichael et. al., 1975; Infante 2001). While these studies were rather crude by contemporary standards, they strongly suggested that the risk of leukemia was increased in workers exposed to high concentrations of benzene. Shortly thereafter, more robust methods of cohort enumeration and exposure estimation allowed refined estimates of the dose response relationship between benzene and aggregate leukemias (OSHA, 1977; ACGIH, 2001). More recently, several case-control studies, cohort studies and updates have been reported, some of which contain quantified benzene exposure estimates and more robust diagnostic procedures for leukemia subtypes. Furthermore, although benzene exposure in humans has been associated with an increased risk of acute myeloid leukemia (AML), it has also been shown in a literature review of epidemiological studies, that the link with other leukemia cell types is no less persuasive (Savitz and Andrews, 1997).

Due to the above mentioned health effects (Guerra et al, 1995 ; USEPA, 2000), usage of benzene is in general restricted· nowadays is used only in industries where it is produced or applied as a raw material in oil refineries, in petrochemical industries, and in coke-oven industries (WHO, 2000). Progressive reduction of the use of benzene have ensured that exposure to high levels of benzene in the workplaces does not constitute any long a serious problem (Capleton and Levy, 2005). Nowadays, the main interest of studies on benzene is focused on continuous and prolonged exposure to relatively low levels of benzene, both in occupational and urban environmental settings (Duarte-Davidson et al, 2001).

Urban environment air quality is mostly dominated by traffic emissions (Fenger, 1999; Colvile et al., 2001) and despite the significant improvements in fuel and engine technologies, ambient concentrations of traffic-related pollutants like CO, NO_x, hydrocarbons, and particles still constitute a considerable health impact. Among the hydrocarbons, benzene is the compound of greatest environmental impact. Concern about benzene leaded to the exigency of monitoring benzene ambient air concentrations, as much as personal exposure of several population groups, occupational or not. In order to reduce benzene exposure levels in urban environments, European Union has set through 2000/69 directive (Official Journal of the European Communities, 2000) a maximum limit of 5 $\mu g/m^3$ to be reached in 2010.

According to the above, since the past few years, there is an increased interest in determining the ambient concentrations of volatile organic compounds (VOCs) and (consequently benzene) due to their significant adverse effects on human health. Several papers have been published so far, concerning VOCs measurements in Greek cities as well as other cities European or not. For example, in the area of Athens, Greece, Moschonas and Glavas, (1996), used electropolished canisters to measure VOCs for three sort periods during summer, while Kourtidis et.al, (1999), presented measurements of NO_X, CO and C_5-C_{12} hydrocarbons during the MEDCAPHOT-TRACE experiment in a street canyon with heavy traffic load. Petrakis et al, (2003) determined BTX concentrations using a Differential Optical Absorption Spectroscopy system (DOAS). Using a similar device Kourtidis et al, (2002) performed BTX measurements in the city of Thessaloniki, but they also presented measurements of O_3, NO_2, and SO_2. Polycyclic aromatic hydrocarbons in the ambient air of several Greek towns in relation to the atmospheric pollutants or PM_{10} concentrations have been the subject of various works in the past (Papageorgopoulou et al, 1999, Mantis et al, 2005). Finally, Pilidis et al, (2005) conducted a BTX measurements campaign including passive and active sampling measurements in Ioannina, Greece. Brocco et al, (1997), made a series of measurements in the city of Rome using DOAS, also deriving high benzene levels, while Ferrari et al, (1998) utilized classical analytical methods to determine the levels of aromatic hydrocarbons and aldehydes in the atmosphere of Grenoble. A similar interest on VOCs has also been manifested in the rural Southeast United States of America by Hagerman et al, (1997), as well as in Korea (Na et al, 2001).

Research on personal exposures to pollutants has shown that the measurements at fixed point monitoring stations at urban background locations may not well represent the exposures of individual members of the general population (Cocheo et al., 2000). A few studies showed a good correlation between the estimation of personal exposure directly by attaching personal samplers to people, or by fixed point monitors in individual microenvironments combined with activity data defining the time spent in each of the microenvironments. Studies on the subject have been conducted by Leung and Harrison (1998), Maitre et al. (2002) and Bono et al. (2003).

In recent years, a significant number of exposure studies related to benzene (or VOCs in general) were conducted. Four related exposure studies, EXPOLIS (Edwards and Jantunen, 2001; Edwards et al., 2001a, b), MACBETH (Cocheo et al., 2000), AIRMEX (Kotzias, 2005) and PEOPLE (Ballesta et al., 2006) and have assessed the European situation in various cities. The EXPOLIS study measured a wide range of VOC with participants measuring their exposure over a 48 h period. The MACBETH study measured aromatic hydrocarbons with participants measuring their exposure over a 108 h period. The EXPOLIS study had a greater focus upon indoor measurements and the importance of indoor emission sources, in particular cleaning products. The MACBETH study aimed to contrast exposures of particularly exposed groups, for example traffic wardens, with non-exposed groups, as school teachers (Skov et al., 2001; Gonzalez-Flesca et al., 2000). In the frame of AIRMEX, measuring campaigns in various cities in Southern and Central Europe were carried out to estimate indoor/outdoor relationships and personal exposure concentrations for selected volatile organic compounds. The PEOPLE study had a much shorter sampling time (12 h) compared to the above studies and a distribution of samplers that is representative of urban commuting populations. The analyses of the results indicated some remarkable relations among the exposure levels and the urban background concentrations.

Beside these international studies, several other remarkable studies were conducted on national scale in Europe (Laia et al., 2004; Gonzalez-Flesca et al., 2007), USA (Heavner et al., 1995; Fruin, et al. 2001), Asia (Chan et al, 2003; Son et al, 2003; Ohura et al, 2006) and Australia (Horton et al, 2006).

In Greece, the first extended exposure study to benzene was performed in Athens by Chatzis et al., (2005), including several population groups and a large amount of ambient air samples. The importance of the parameters that constitute the exposure patterns was accessed by the measurements results and the questionnaires via statistical methods.

This chapter presents the results of an extended measuring campaign performed in a medium sized Southeastern European City. The measurements conducted are:

1. Measurements of ambient benzene concentrations (through both passive and active tubes) in several city locations (urban background, streets, street canyons, filling stations etc.).
2. Measurements of traffic parameters (traffic flow, vehicles speed, density, fleet contribution etc.).
3. Meteorological observations (wind speed/directions, ambient temperature/humidity).
4. Exposure patterns related to some characteristic population groups (policemen, filling station employees, taxi drivers, general population).

The variety of the performed measurements gained an in depth analysis and quantification of the parameters affecting ambient benzene concentrations in several microenvironments.

Further to the experimental study, numerical models were also used. Specifically the STREET dispersion model (Johnson et al, 1973), Neural networks and other statistical methods, CALINE 4 dispersion model were applied to compute the dispersion of benzene in specific microenvironments.

For the exposure studies beside the passive sampling measurements related to the exposed subjects, active sampling measurements where also performed under several conditions and activities in order to determine their contribution to the total amount of exposure.

Finally different "what if" scenarios were tested to understand the contribution of each kind of vehicle to the overall benzene concentration and to examine and especially propose environmental policies.

2. Materials and Methods

2.1. General Information

The campaign has taken place in Ioannina (Figure 2), a medium sized (100,000 inhabitants) city in North Western Greece (Figure 1) covering an area of approximately 12 km^2.

While, seemingly, a city of that size would not have been expected to present raised levels of air pollution, ambient air benzene levels seem to be elevated (Pilidis et al , 2005), as a result of the combination of a number of factors, namely:

- The high ratio of vehicles per inhabitant, which according to the local transport authority is about 0,5.

Figure 1. Map of Greece.

- The composition of the vehicles fleet. A large percentage of the vehicles in circulation are not even equipped with a catalytic converter, and from those having one very few have a last generation catalytic converter. There are also a large number of motorbikes not subject to any emission reducing directive, while the same also applies to heavy duty vehicles. Furthermore the prohibition against using diesel engines in private cars (it is only allowed for taxis and all vehicles of this category use diesel as fuel) results in high emission levels for some pollutants like benzene.
- The star-shaped morphology of the city road network, with the main arteries of the city converging towards the city centre and crossing each other, results in a rapid increase of traffic density (in combination to the large amount of vehicles for the size of the city) approaching to the city centre.

- The lack of sufficient parking space that adds to the traffic problems.
- The geomorphology of the greater area. Ioannina is built on the shore of a mountain lake located in a basin. This favors the development of stable atmospheric conditions, as well as prolonged periods with little or no wind, conditions contributing to the accumulation of atmospheric pollutants.

It is worth mentioning that the area has minimal industrial activity, therefore the dominant source of air pollution over the city is vehicle circulation, aided by central heating during the cold period of the year.

2.2. Measurements Campaign Design

The measurements campaign, consisted of two sampling periods (February and August 2006), in order to investigate the seasonal variation. Both sampling periods included measurements of benzene in ambient air while also personal exposure of the same pollutant to sixty volunteers.

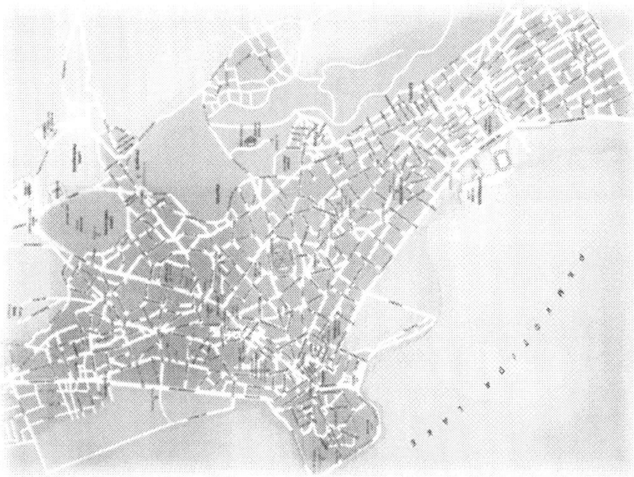

Figure 2. Map of Ioannina.

2.2.1. Ambient air Measurements

Measurements of ambient benzene concentrations (performed mostly by active sampling) were conducted in several microenvironments (indoor, outdoor, in vehicles, etc) in order to determine the configuration of the benzene concentrations under different environmental parameters (traffic volume, meteorological conditions, proximity to filling stations or street canyons).

The first indicator for urban air benzene was given by placing passive samplers at 12 points in the city. The sampling points were chosen in the pavements of the city main streets in a way to represent variant prevailing traffic conditions in combination to the street plan diversity. Beside them, one more sampler was placed in the Historical Castle of the city, which is a monument located in the city limits, but in a place not directly affected by traffic emissions, in order to determine the urban background concentration due to the procedure of dispersion. In all sampling points passive samplers were exposed for one week, placed 1,8 m above and 3 m away from the center of the nearest traffic lane.

Extended measurements of active sampling, traffic volume/composition, and wind speed/direction where conducted in the street canyon of the city in which the highest levels of benzene concentrations where usually recorded. The detailed study of this road gave to the research team the feasibility to develop, evaluate and compare two predictive models (Kassomenos et. al, 2004, Karakitsios et. al, 2006) enabling traffic canyon characteristics in combination to traffic density. Thus, additive parallel active sampling measurements were conducted in streets with similar traffic characteristics but different street layout (canyon-non canyon) and vice versa, in order to testify the significance of these parameters in ambient air benzene levels.

Finally, passive and active sampling measurements were performed around five filling stations, an urban, three suburban and one rural. The position of the urban filling station, excluded a significant amount of contribution due to other emission sources. The only additional emission source was a main road in front of the filling station. Therefore the urban background concentration had a minimized and easy calculated effect on the ambient benzene concentrations. The second filling station was chosen to be a rural one, in order to observe the effect of the emissions from the filling station alone, due to the non significant contribution of the road and the absence of background concentrations. In the third case, a suburban road of 2 km length was chosen to be studied. This road has an excess interest because of the existence of three filling stations. In each area, the average weekly benzene concentration was determined in 16 points. At the urban and the rural filling stations two of the sixteen samplers were placed in the filling station near to the pumps, in order to observe the levels of benzene in this area. The rest 14 samplers were placed in different peripheral places covering a wide range of directions ($0\text{-}360^0$) and distances (up to 100m) (Figures 1, 2) at a height of 1.8m. In the suburban location, all the passive samplers (16) were placed equidistant along the street. At the adjacent roads of the filling stations, benzene active sampling measurements were also performed, in order to model the contribution of the road to the ambient benzene concentrations (Karakitsios et al, 2007).

A major issue of the whole concept was the accurate determination of urban background concentration. The urban background concentration can be defined in several ways. The commonest (and more reliable) modelling practice is to use background concentrations obtained from measurements at urban locations that are not directly affected by local sources (Vardoulakis et al, 2003). Based on this, two passive samplers were placed in two different urban locations (not affected by traffic or other known benzene sources) and their values were averaged in order to exclude the urban background concentration. All passive samplers were placed for one week. The procedure was conducted in the same way whenever it was necessary.

2.2.2. Personal Exposure Measurements

The volunteers group included twenty people which constituted the control group, twenty policemen, ten filling station employees and ten taxi drivers. The passive samplers were put during the working hours (8:00-16:00) every day for one week. All participants kept a personal daily questionnaire, in which they referred to the kind and the duration of the performed activities during sampling time.

The policemen group consisted of officers with several kind of activities including patrol on foot or in vehicles (car or motorcycle), traffic regulation at intersections of central roads in the city, guarding outside the police station and policemen working in offices.

People working in filling stations were doing three major activities; cars refuelling, miscellaneous activities (like car washing or oil checking) and cashiers. Most of the employees were doing a combination of these activities.

Taxi drivers were driving in a variety of roads of the city, near to the centre as much to the suburban roads. None of the taxi drivers used to smoke during the shift, in order to avoid increasing the in-car benzene concentrations by other sources than traffic emissions.

Control group was necessary in order to compare risk assessment results with a group with no specific occupational exposure to the studied pollutants. The aim was to test the hypothesis of the increased cancer risk of occupationally exposed population. People of the control group were taking several kinds of activities during sampling time, in order to represent the exposure due to everyday life habits, like staying at home, walking outdoor, shopping or visiting several entertainment places.

Active sampling measurements were also performed in personal exposure. In this case, the exposed subjects were carrying the pumps with the samplers for 30', performing several activities common for the policemen, the taxi drivers and the control group, like driving, walking in the pavements of the streets, staying in indoor environment etc. The samples were taken under several conditions, in order to formulate a clear opinion about the contribution to each activity under specific circumstances to the total amount of exposure.

Similar measurements were also conducted to the filling station employees. In this case, the diversity of the activities is expressed through the three different kind of activities mentioned above performed by the employees. The amount of fuel traded, in combination with meteorological conditions were measured in parallel, in order to develop an exposure model (Karakitsios et al, 2007)

2.3. Instruments and Apparatus

The equipment used during the measurement campaign and analyses is listed below:

1. Gas chromatograph Varian 3900 GC with flame ionization detector (FID). SBM^{TM}-5 capillary column 30 m x 0,53 mm x 0,5 μm film thickness, by Supelco, Italy.
2. Thermal desorption system MARKES Thermal Desorption Cold Trap Injector.
3. 3 SKC model 222 low volume pumps.
4. Flow meter with a measurement range of 0,010 to 12 l/min, DryCal CD-Lite (Bios International, USA).
5. MARKES CARBOGRAPH 1TD sampling tubes (proper for both active and passive sampling) filled with 400 mg of sorbent.
6. Cup anemometer and wind vane Qualimetrics, model 2020 and 2032, respectively.

2.4. Quality Assurance and Quality Control

The quality assurance and quality control (QA/QC) procedure included laboratory and field blanks, parallel samples and duplicate measurements of samples. Before each sampling, blank sample (zero air) was analyzed to ensure that the concentration of all compounds inside was below 0.02 ppbv. Parallel samples and duplicate measurements of samples were analyzed to test the precision of the sampling and analytical techniques, respectively. The mean relative standard deviations (RSD) for all the compounds were less than 8%. A new calibration curve was determined each time. The detection limit of each compound was calculated from the data of seven replicate measurements of low concentration samples and observed from their standard deviation. Correlation coefficient for benzene was 0.9983. Response linearity tests showed that the response signal was proportional to injection volume for benzene concentrations (0.5–220 $\mu g/m^3$).

3. Ambient Air Concentrations

3.1. Environmental Data

3.1.1 Traffic Data

Passive samplers were placed in the main streets of the city where (except sampler 13 that it was not near a street) traffic measurements including traffic flow, fleet composition and speed were performed. The average values for the above streets are illustrated in Table 1.

Detailed description of the vehicles fleet composition is presented in Table 2. The fleet pattern circulating in the city is dominated by passenger cars equipped with a catalytic converter (61.28% of the total fleet). The percentage of passenger cars not equipped with a catalytic converter is also significant (12.42%), followed by motorcycles (8.84%). Both of them present strong benzene emission potential. Buses and trucks do not represent a significant part of the fleet (1.71 & 1.78% respectively), but they are responsible for traffic delays and increased congestion. It is easily observed that diesel passenger vehicles constitute a small amount of the traffic fleet (7.78%) and are only taxis.

Traffic flow in the city presents maximum between 8:00-9:00 and 2:00-3:00, in some days also between 17:30-18:30 and 20:30-21:30, related to commuting to working places. Traffic speed is in general irreversibly correlated to traffic flow (especially under congestion), but in most of the cases has an average of 25 km/h, with a deviation of ± 10 km/h.

Table 1. Traffic flow in main streets of the city

	1	2	3	4	5	6	7	8	9	10	11	12
Catalytic (veh/h)	1060	1160	1340	1150	1100	1000	1000	1100	1060	1030	1260	980
Non Catalytic (veh/h)	240	280	340	210	180	180	170	190	220	220	270	160
Diesel vehicles (veh/h)	160	180	130	170	160	180	130	110	150	130	140	140
Light duty vehicles (veh/h)	100	140	150	120	110	130	100	90	80	110	90	120
Heavy duty vehicles (veh/h)	10	40	50	40	40	40	50	35	10	20	35	15
Buses (veh/h)	25	50	60	30	25	30	30	20	24	20	30	25
Motorcycles (veh/h)	210	150	170	170	150	170	120	150	130	170	150	170
Average speed (km/h)	15	20	25	32	30	40	35	28	25	30	30	30

Table 2. Detailed traffic fleet composition (in percentages %/)

Percentage %	Catalytic	Non Catalytic	Diesel vehicles	Light duty vehicles	Heavy duty vehicles	Buses	Motorcycles
Pre ECE		1					
ECE 15-		3					
ECE 15-02		5					
ECE 15-03		23					
ECE 15-04		39					
Conventional		29	10	9	62	83	100
EuroI	59		24	27	26	13	
EuroII	26		35	39	9	4	
EuroIII	12		29	25	3		
EuroIV	3		2				

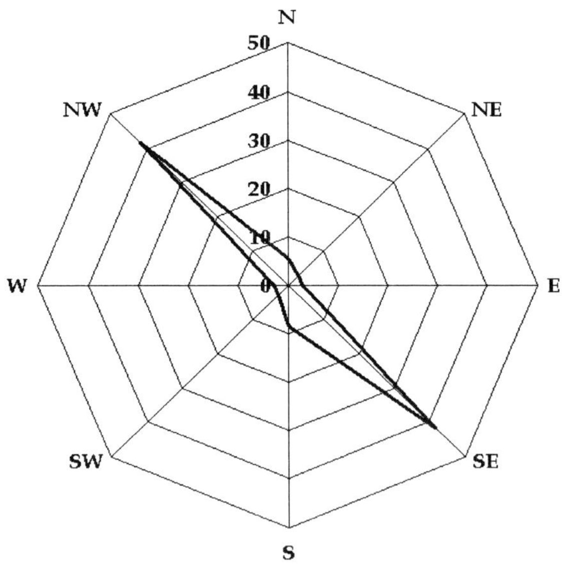

Figure 3. Wind rose in the area of Ioannina.

3.1.2. Meteorological Data

The area in which the measurements campaign was performed presents a rather unfavorable meteorology which affects the accumulation of the released air pollutants. Since the city is located in a basin (470-m mean altitude from the sea level) it is characterized by the frequent formation of temperature inversions which trap the pollutants and in combination with the prevailing light winds does not favor the quick remotion of the air pollution from it.

Wind speed and direction measurements confirmed the historical data on the wind field over the area. The prevailing wind directions, during the sampling periods, were from southeast and northwest directions (Figure 3).

Mean wind speed ranged from 0 to 7 m/sec, for the summer period, while the average wind speed was 2 m/sec. In summer, wind speed tended to increase after the evening hours. In the winter, wind speed ranged from 0 to 5 m/sec, having an average value of 1.1 m/sec. Table 3 presents the mean temperature, humidity, wind speed and rain in the basin per month during the last 40-years.

Ambient temperature due the time of summer measurements ranged from 14 to 28 ^{o}C, having an average value of 23 ^{o}C, while for the winter the corresponding values were -2 to 13 ^{o}C and 7 ^{o}C. Temperature variation through the day was regular.

3.1.3. Urban Ambient Air Benzene Concentrations

The results of the average weekly ambient air concentrations are presented in Figure 4.

At each sampling point, the concentrations of benzene are presented as a couple of values (the upper value is for summer, while the lower one is for winter) expressed in $\mu g/m^3$ (Figure 4).

The results are similar to recent measurements performed in other Greek cities and higher compared to other European cities (at the measurement campaigns referred above), reflecting the high emission levels caused by the relatively old vehicles fleet.

Table 3. Meteorological data for the area of Ioannina (1967-2007)

	Mean T (°C)	Average Max T (°C)	Average Min T (°C)	Mean Atm. Pressure (hPa)	Mean Relative Humidity (%)	Mean Wind speed (knots)	Mean Precipitation (mm)
January	4.7	10.1	0.2	1019.3	76.9	3.1	124.2
February	6.1	11.5	1.0	1017.7	73.7	3.7	111.6
March	8.8	14.4	3.2	1016.4	69.5	4.1	95.4
April	12.4	17.7	5.9	1014.3	67.9	3.5	78.0
May	17.4	23.0	9.6	1014.9	65.9	3.0	69.3
June	21.9	27.6	12.8	1014.4	59.1	3.0	43.5
July	24.8	30.8	14.9	1013.3	52.4	2.8	32.0
August	24.3	30.9	15.0	1013,7	54.4	2.7	31.2
September	20.1	26.7	12.2	1016.8	63.6	2.5	54.0
October	14.9	21.2	8.5	1018.8	70.8	2.9	99.5
November	9.7	15.5	4.7	1019.2	79.8	1.9	167.9
December	5.9	11.1	1.8	1018.7	81.5	2.3	174.9
Annual	14.2	20.0	7.5	1016.5	68.0	3.0	1081,5

Even though all sampling points were located on roads with similar traffic flow and fleet characteristics, there is a spatial variation of benzene concentrations obtained by the passive sampling method that can be attributed to two factors.

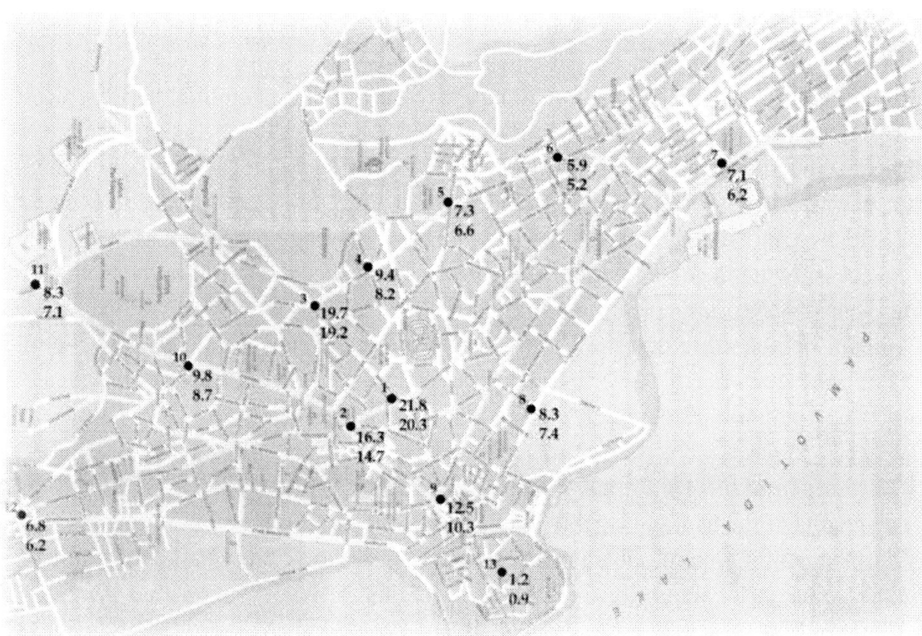

Figure 4. Urban ambient air benzene concentrations (in μg/m³).

The first is the topography of each area. It is known that roads with the same pollutants' emission rate have different pollution concentrations, if the dispersion and diffusion conditions are different (Berkowicz et al, 1997).

As already mentioned some of the sampling points were inside street canyons, while others were not. The measurements showed that in the first case weekly average benzene concentrations were never lower than 14 $\mu g/m^3$, whereas in the later case concentrations did not exceed 7-8 $\mu g/m^3$, with the exception of point 2, where an average value of 15 $\mu g/m^3$ was found. The increased value at point 2, in comparison to the rest of the point with similar topography, is due to the second reason responsible for the observed variation of benzene concentrations, namely the increased traffic density observed throughout the day. The significance of traffic density as a factor affecting air pollution levels is also known (Dirks et al, 2003), and its increased values at point 2 is in direct relation with the fact that this point is located in the city centre, where all major city roads converge as a result of the star-shaped road network of the city, rendering traffic congestion inevitable.

Inferences on the role of traffic density in the formation of the observed benzene concentrations (and for the rest of the measured pollutants) were derived from the active sampling measurements at point 1 and the development of a corresponding model (Kassomenos et al, 2004), discussed excessively in a later paragraph.

Related to the above is the fact that in general benzene concentrations are increased in areas close to the centre of the city, while going outward the concentrations tend to decrease. This is more obvious at sampling points 3-4-5-6, in which the samplers were placed along the same street. As far as we approach to the centre of the city (from point 6 to 3), the concentrations tend to increase in a non linear way. This tendency is expected due to the radial traffic network of the city. Under these conditions, traffic density increases significantly near to the centre of the city, having as a consequence the proportional increase of air pollution (Dirks et al., 2003).

The increase of benzene concentrations in summer is attributed to the significant increase in the number of motorcycles circulating in the city. This fact is in concurrence with the proper weather conditions that favour the use of motorcycles. It is well known that motorcycles (especially the older ones) have a significant potential of benzene emissions (Ntzaichristos and Samaras, 2000), affecting in a similar way the observed benzene concentrations. Furthermore some motorcycle owners use unleaded petrol because of its lower cost, and since these vehicles are not equipped with a catalyst, the benzene emission factor is even larger that that of non catalytic vehicles.

3.2. Parameters Rising Ambient air Benzene Levels in Local Scale

3.2.1. Traffic Density

3.2.1.1. General Information

Although there is quite a lot of work done towards the direction of studying the procedures that describe the distribution of VOCs in a street canyon (Xie et al, 2003 among others), only recently some modeling studies of benzene in urban street canyons have appeared. Specifically, Carr et al, 2002, modeled annual benzene concentrations, in an urban environment using a DET model, while Vardoulakis et al, 2003, published a thorough review of the techniques used for street canyons modeling of several pollutants, among them benzene. Up to now all papers referring to benzene modeling, use traffic flow as the determinative parameter for calculating road emissions.

In our case DET, eventhough constructed according to the basic modeling principles (Xie et al, 2003 among others), uses traffic density a basic parameter. This parameter can describe the size of traffic volume as in the road in a more realistic way. Compared to traffic flow, density allows the estimation of Road Emission Rate (RER) for several traffic conditions, ranging from zero values of traffic density (free flow), up to a maximum theoretical value (traffic jam conditions). Traffic density also represents a better measure of the emissions, because at each value of traffic flow correspond two values of traffic density, with consequently significantly different road emission rates (Kassomenos et al., 2004). Moreover complicated emission modes, requiring complex calculations like the start-stop conditions, can be adequately approached.

On the other hand ANNs have been used to predict atmospheric pollutants concentrations (Vioti et al, 2002). The main reason for this is that they are able to handle non linear systems. The non reactive pollutants are non linear systems (in a much smaller degree than photochemical pollutants) since they are highly depend on atmospheric dispersion conditions and vehicles emissions.

During the past decade, several attempts, to forecast urban air quality via neural network systems, have been presented. These works mainly aimed to predict concentrations of classic pollutants such as NO_x, SO_2, O_3 (Niska et al, 2004).

Recently, non reactive pollutants forecasting via ANNs (Gardner and Dorling, 1998; 1999; Dorling et al, 2003), has gained an increased interest, because ANNs can handle the large amounts of data, now routinely acquired through modern monitoring techniques. If these data (e.g. air quality, meteorological and traffic) are used for training an ANN, this can lead to reliable prediction of atmospheric concentrations of emitted pollutants in a way easier and faster than a deterministic model (Nagendra et al, 2004).

In this work, both an ANN and a DET (Kassomenos et al., 2004) models are developed and evaluated in comparison with measurements which have been recorded in a street canyon.

In the following section the data and the models used, as well as, the methodology adopted and the analysis of the results are described and discussed.

3.2.1.2. Data

Data (traffic flow, type of vehicle, air quality and meteorological) were collected during an experimental campaign. All measurements were made near the side of sampling point 1, which is on of the busiest roads in the city. This road is a typical street canyon having a width of 10m, surrounded by buildings of about 15 m height. Four measurement campaigns were carried out lasting one week each, representing the four seasons of the year. The measurements were recorded for 12.5 hours on a daily basis (09:00- 21:30 Local Time).

The following parameters were measured:

- Traffic flow from two independent observers. Vehicles were registered in seven main categories (catalytic passenger vehicles, non catalytic passenger vehicles, diesel passenger vehicles, light duty vehicles, heavy duty vehicles, buses and motorcycles) every half an hour. It was found that the maximum traffic density value (D_j) in the specific road, is nearly 200 vehicles/km, while the maximum theoretical speed is 50 km/h, equal to the speed limit for the road. In order to estimate the traffic density the vehicles' speed was calculated as the ratio of the distance (a part of the street canyon) to the time needed for the vehicle to cross this part of the road.

- Wind speed (m/sec) and direction (degrees) on an hourly basis on the roof of the same building in front of which the rest of the observations were made. The height in which wind speed and direction were measured is 15-m above ground. This height is the mean height of the buildings in the studied canyon. The threshold of the anemometer was 0.3 m/s.
- Benzene concentration every two hours, at a height of 1.8 m and at a horizontal distance of 3 m from the center of the nearest traffic lane, by the method of active sampling. The sampling procedure (active sampling) included adsorption of benzene on tubes using a personal pump. The sampling flow rate was set to 100 ml/min and the sampling time was set to 30 min. Approximately 3 lt of air were pumped through the tube. The samples were analyzed in a thermal desorption cold trap injector, combined with a gas chromatography system with flame ionization detector (FID).
- Average weekly benzene concentrations, using passive samplers, at an urban background site. The values of the background benzene concentrations were: 3.7 $\mu g/m^3$ for the Spring, 4.3 $\mu g/m^3$ for the Summer, 4 $\mu g/m^3$ for the Autumn and 3.7 $\mu g/m^3$ for the Winter, and were used as the urban background concentrations in the dispersion model.

3.2.1.3. Modelling Techniques

3.2.1.3.1. DET Model

The DET model is extensively described in Kassomenos et al. (2004). It consists of three modules: the traffic, the emissions and the dispersion module. The main objective of this arrangement (i.e. the three traffic modules) was the development of a model in which benzene RER depends on traffic density. Traffic density depends on traffic flow, vehicles speed and the theoretical maximum density of the road traveled.

To calculate RER, traffic flow was divided into seven main vehicle categories based on their type and fuel used (Table 2) and the fleet percentage distribution per measurement season is presented on Table 4.

Table 4. Vehicles fleet composition in the under study street (%)

Percentage per season %	Catalytic	Non Catalytic	Diesel vehicles	Light duty vehicles	Heavy duty vehicles	Buses	Motorcycles
Spring	60	13	9	5	1	1	11
Summer	56	12	8	5	1	1	17
Autumn	59	14	9	6	1	1	10
Winter	62	13	9	6	1	1	8

The correct estimation of the fleet composition in an application is very important, because each vehicle category has different emission factors and RER ($g.km^{-1}.s^{-1}$) calculation results as the sum of the distributed emission rates from the seven vehicle categories mentioned above.

In the last stage of the methodology, the STREET dispersion model was used to predict benzene concentrations (Johnson et al, 1973). In this model, pollutant's concentration is a function of RER, urban background concentration, wind speed and direction and street's canyon height and width.

I. Traffic Model

If we suppose that the vehicles along a road are moving with a steady speed V and are equidistant to each other while moving in a total flow equal to T, then the density D of the traffic flow is given by the equation

$$D = \frac{T}{V} \tag{1}$$

Also, it has been found that the movement speed is related to the traffic density D, the maximum density D_j (jam density) and the maximum speed V_0 that can be observed on a certain road.

$$V = V_0(1 - \frac{D}{D_j}) \tag{2}$$

From the two equations above we obtain that:

$$T = DV_0(1 - \frac{D}{D_j}) \tag{3}$$

Consequently, the density value is estimated for each flow using the following

$$D = (\frac{D_j}{2})(1 \pm \sqrt{1 - \frac{4T}{V_0 D_j}}) \tag{4}$$

We can see from the equation above, that each flow value is characterized by two density values representing different movement speeds. One represents traffic jam (+) (4a) and, consequently, low speed conditions, while the other stands for free flow conditions (-) (4b) and high speeds (Greenshields, 1943).

In order to estimate traffic density from flow data in a proper manner, it was considered necessary to convert the total number of moving vehicles into Units of Passenger Vehicles before calculating the traffic flow. This was done because due to their size, not all vehicles take up the same space on the road; therefore they also contribute differently in the density formation.

All passenger vehicles and light duty trucks have a coefficient equal to one (1), heavy trucks equal to two (2), buses equal to three (3) and motorcycles equal to half a unit (1/2). This coefficient is especially important for calculations in "what if" scenarios concerning changes in the composition of the vehicle-fleet or in possible traffic regulations and prohibitions.

II. Emissions Model

The development of a traffic model describing the road under consideration in an adequate manner is a key factor for the estimation of the Road Emission Rate (RER), or the mass emitted per unit length of road per unit time.

The difficulty in the estimation of the emission rate of a certain pollutant is due to different emission rates from the various vehicle categories given that they use different types of fuel (diesel or gasoline, leaded or unleaded), they have different technology (catalytic or not, various anti-polluting specifications), they have different fuel consumption as well as different level of maintenance. The values of the emission factors for the several vehicle categories are based on the COPERT methodology (Ntzaichristos and Samaras, 2000). The emission factor used for each vehicle category is an average resulting from the analysis of each category in sub-categories. For example, the emission factor for the catalytic passenger vehicles, results from the following equation:

$$Catalytic\ passenger\ vehicles\ Emission\ Factor = \frac{\sum_{i=I}^{IV} (num_i)(EF_i)}{\sum_{i=I}^{IV} num_i}$$

where the num_I is the number of catalytic passenger vehicles circulating in a city that fulfill the emission specifications EuroI adopted by the EU and emit benzene by an emission factor EF_I, num_{II} is the number of catalytic passenger vehicles circulating in a city that fulfill the emission specifications EuroII, num_{III} for EuroIII and finally num_{IV} for EuroIV.

The same applies to the rest of the categories. The estimation of the population of each vehicle category (Table 3) was done by the traffic flow measurements and the associated sub-categories (Table 2) were based on statistical data from the Transportation Authority Department of the region. The data concern the grand total of the moving vehicles, the number of vehicles withdrawn from circulation as well as the sales of new models. Working similarly for the rest of the vehicle categories we obtained the emission factors that appear in Table 5.

The estimation of the emission factors was done for steady movement speeds of 50, 40, 25 and 10 km/h.

Since a vehicle's speed is constantly changing, the estimation of a the real emission rate of a vehicle depends on the traffic characteristics of the road on which the vehicle is moving (Dirks et al, 2003), estimated that the emission factor approaches a minimum value during free flow with the maximum speed, while it tends to infinite under traffic jam conditions.

Table 5. Benzene emission factors for the corresponding vehicles categories in several speeds

Speed (km/h)	Katalytic emission factor (g/km)	Non katalytic emission factor (g/km)	Diesel vehicles emission factor (g/km)	Light duty vehicles emission factor (g/km)	Heavy duty vehicles emission factor (g/km)	Buses emission factor (g/km)	Motorcycles emission factor (g/km)
10	0,048	0,24	0,006	0,0078	0,182	0,224	0,06
25	0,024	0,144	0,0036	0,0065	0,098	0,112	0,036
40	0,012	0,096	0,0024	0,0052	0,07	0,07	0,024
50	0,0096	0,072	0,0018	0,0039	0,056	0,056	0,018

To estimate the real emission factor VER for the given traffic conditions and for each one of the main vehicle categories, the following equation is used (Dirks et al, 2003):

$$VER = \frac{(aD^2 + bD + c)D}{D_j - D} + VER_0 \tag{5}$$

where D is the actual and D_j the jam densities, and VER_0 is the emission factor for the maximum speed V_0 in free flow conditions. The coefficients a, b, c are determined by solving the equation for three speeds, hence densities, with known emission factors in our case 40, 25 and 10km/h (Table 6).

After determining the coefficients a, b and c we can estimate the emission rate Q_{total} (in mg/cusec) of the road. The Q_{total} will be equal to the sum of all the partial Q from the seven main vehicle categories we examined. The Q of each category is the product of the respective emission factor, VER, multiplied by the flow per second, of the respective vehicle category.

$$Q = \frac{VER \cdot T}{3600} \tag{6}$$

$$Q_{total} = \sum Q \tag{7}$$

Table 6. Coefficients (a,b,c) values obtained by the equations

co-efficient	Cataly-tic	Non Catalytic	Diesel vehicles	Light duty vehicles	Heavy duty vehicles	Buses	Motor-cycles
a	-	-	-	$1.355*10^{-}$	$4.86*10^{-7}$	-	$4.3*10^{-6}$
b	2.667*1	-	$7.08*10^{-6}$	$-6.23*10^{-}$	-	$2.72*10^{-4}$	$-9.71*10^{-4}$
c	1.07*10	0,109	$2.7*10^{-3}$	-	0.067	0.048	0.056

From equations (6) and (7) it is obvious that traffic flow values are necessary in order to calculate Q_{total}. The question is which of the two flow values we will accept as the correct one. This will be judged by the observed traffic flow data, from which it is obvious whether the vehicles are free to move or are being jammed. The road constituted by two traffic lanes, which were not treated separately.

III. Dispersion Model

The dispersion model which was used for our calculations is STREET (Johnson et al, 1973), according to which the concentration value of a pollutant C is equal to the sum of the road contribution C_s which is due to the road emissions, and the background contribution C_b, which is due to the pollution of the air that enters from the top of the canyon.

$$C = C_s + C_b \tag{8}$$

The value of C_s is proportional to the emission rate and conversely proportional to the wind speed on the top of the canyon. For winds that blow in a direction forming an angle greater than 30^0 with the axis of the road, there are two equations for the leeward (9) and the windward (10) side respectively:

$$C_s = \frac{k \cdot Q}{(\sqrt{x^2 + z^2} + h_0)(U + u_s)} \tag{9}$$

where k is a constant empirically defined to have a value equal to seven, Q is the road's emission rate, U is the speed of the wind on the top of the canyon, x is the horizontal distance between the receptor and the center of the nearest traffic lane, z is the vertical distance between the receptor and the ground, u_s accounts for the mechanically induced air movement caused by traffic ($u_s = 0.5\ ms^{-1}$) and h_0 accounts for initial mixing of pollutants ($h_0 = 2m$).

For the windward side the equation becomes:

$$C_s = \frac{k \cdot Q \cdot (H - z)}{(U + u_s) \cdot H \cdot W} \tag{10}$$

where H is the height of the canyon and W its width. For winds that their directions are parallel or almost parallel to the axis of the road, the mean value from the two equations (9 & 10) is used.

3.2.1.3.2. ANN Model

The initial concept about artificial intelligence systems was introduced by Robbins and Monro in 1951. The whole idea was the generation of a new method on how to handle statistical questions. As time passed by, neural network models were adopted in more complicated problems, because they have the ability to describe highly non-linear relationships.

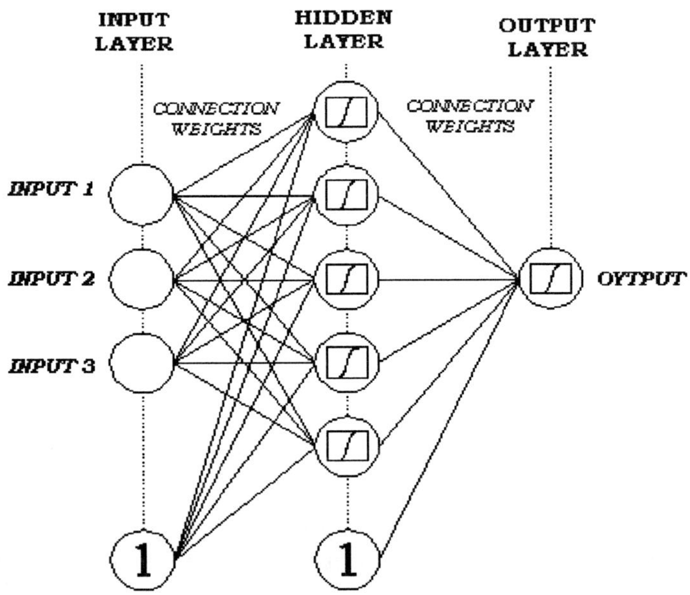

Figure 5. Construction of an ANN.

A neural network is a computer model whose architecture essentially mimics the knowledge acquisition and organizational skills of the human brain. The basic architecture of neural networks has been widely covered (Rumelhart, 1986, Lippmann, 1987).

A neural network consists of a number of interconnected processing elements, commonly referred to as neurons. The neurons are logically arranged into two or more layers as shown in Figure 5, and interact with each other via weighted connections.

These scalar weights determine the nature and strength of the influence between the interconnected neurons. Each neuron is connected to all the neurons in the next layer. There is an input layer where data are presented to the neural network, and an output layer that holds the response of the network to the input. It is the intermediate layers, also known as hidden layers that enable these networks to represent and compute complicated associations between patterns. Neural networks essentially learn through the adaptation of their connection weights.

In this work the data available (196 records) were randomly divided into two groups. The first group (75% of data) was used to train the model. The model developed by the first procedure was used then to predict the benzene concentrations for the second group (25% of the data).

The input parameters were: flow of each vehicle category, vehicles' speed and wind speed. The output parameter was average benzene concentration in both sides of the road (leeward and windward). Since data is not available from both sides of the road, mean benzene concentration was calculated using the basic equations of the STREET model (equations 9 and 10) considering which observed value (leeward or windward) was given by the measurements. The mean benzene concentration was considered equal to the observed value when the wind was parallel to the street canyon or during calm conditions.

The ANN architecture consisted of three layers: input, hidden and output (Figure 6). In order to reduce the number of input neurons and improve the modeling results, diesel passenger vehicles and light duty vehicles were integrated into the same category, named light diesel vehicles, because of their almost similar benzene emission factors. For the same reasons, heavy duty vehicles and buses were integrated into one category, named heavy diesel vehicles. Consequently, the ANN's first layer consists from seven neurons, one for each input, the second layer consists from eight neurons (experimentally found to yield best results) and the third layer includes only one neuron, as much as the output parameters (benzene concentration).

It should also be mentioned that at the hidden layer sigmoid units are employed with hyperbolic tangent as activation function while at the output layer a linear unit are adopted.

It should also be noted that the input parameter values are pre-processed in order to be normalized according to Eq. 11, the inputs and targets will fall in the interval [-1, 1].

$$x_n = 2\frac{x - \min(x)}{\max(x) - \min(x)} - 1 \tag{11}$$

Network training was performed using the Bayesian regularisation approach (MacKay, 1992; Foresee and Hagan, 1997), which is a supervised learning method that determines the optimal regularization (or in other words generalisation) parameters in an automated fashion. According to this approach, the weights and biases of the network are assumed to be random variables having specified distributions. The regularization parameters are related to the unknown variances associated with these distributions. Consequently, using statistical techniques these parameters can then be estimated. Specifically, within this framework the following objective function is minimised:

$$E = a_1 \sum_{i=1}^{N} (t_i - o_i)^2 + a_2 \sum_{i=1}^{M} \mathbf{w}_i^2 \ , \tag{12}$$

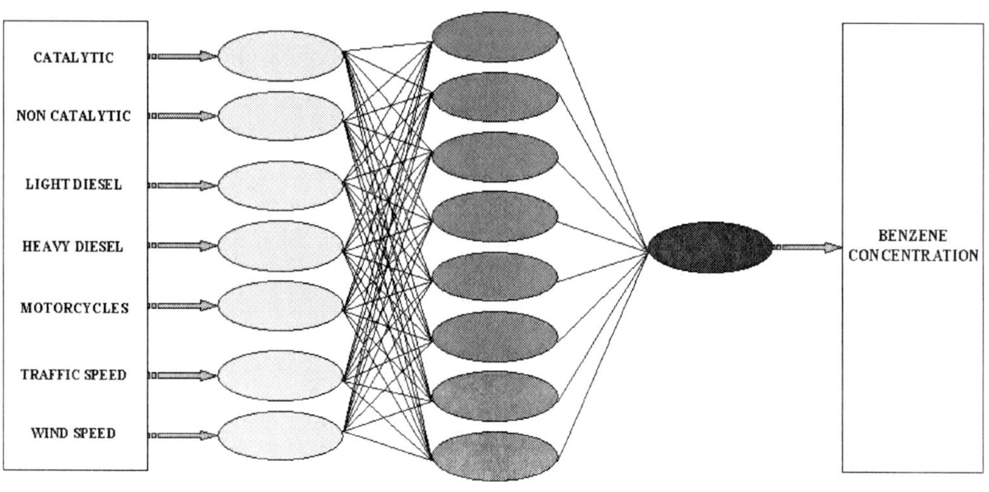

Figure 6. Construction of the developed ANN.

where t_i are the desired network outputs, o_i are the network outputs during training, $\mathbf{w}_i$ are the network parameters (weights and biases), M is the number of those parameters and N is the number of the training patterns. The hyper parameters a_i are estimated at each iteration according to the following formulas:

$$a_1 = \frac{N-\gamma}{2\sum_{i=1}^{N}(t_i - o_i)^2} \quad , \tag{13}$$

$$a_2 = \frac{\gamma}{2\sum_{i=1}^{M}\mathbf{w}_i^2} \quad , \tag{14}$$

where γ (called the number of effective parameters) is given as:

$$\gamma = N - 2a_2 \mathrm{tr}(\mathbf{H})^{-1} \quad , \tag{15}$$

with $\mathbf{H}$ being the Hessian matrix of the objective function and can be approximated using the Jacobian matrix.

The network parameters $\mathbf{w}_i$ are updated according to the Levenberg-Marquardt optimisation technique:

$$\mathbf{w}_{i+1} = \mathbf{w}_i - \left[\mathbf{J}^T\mathbf{J} + \mu\mathbf{I}\right]^{-1}\mathbf{J}^T\mathbf{e} \quad , \tag{16}$$

where $\mathbf{J}$ is the Jacobian matrix, $\mathbf{I}$ is the unit matrix, $\mathbf{e}$ is the vector of network errors and μ is a scalar parameter (Hagan and Menhaj, 1994). The parameters of the network are initialised according to the Nguyen-Widrow method (Nguyen and Widrow, 1990) and a_1 and a_2 are initially set to one and zero, respectively. It should be noted that the objective function E is adapted at each iteration, since the hyperparameters a_i are reestimated. The training procedure is considered to be completed when the effective number of parameters, γ, has converged, indicating that the ANN is characterised by generalisation and has not overfitted the training data.

Other training algorithms, besides Bayesian regularisation, were also studied, like the resilient backpropagation (Riedmiller and Braun, 1993), the scaled conjugate gradient (Moller, 1993), the BFGS quasi-Newton (Dennis and Schnabel, 1983), the one step secant (Battiti, 1992) and the Levenberg-Marquardt (Hagan and Menhaj, 1994). However, in all cases Bayesian regularisation proved to be more consistent and robust concerning the generated results.

Other neural network modelling techniques have been involved by many researchers, dealing with atmospheric pollution. In 1998, Nunnari et al, applied two different neural network techniques to the modelling of time-series of atmospheric pollution data. Two kinds of neural networks are used in this paper, the multilayer perceptrons (MLPs) and the fuzzy neural networks (FNNs). FNNs are special neural networks in which the weights of the connections represent the parameters of a set of fuzzy rules. Fuzzy sets are linguistic terms that are expressed in an exact mathematical way by using the concept of membership function, which represents the degree of truth of an assertion in a range (0, 1). Membership functions can have several shapes, the most common being Gaussian, trapezoidal, triangular and sigmoidal. Lek and Guegan in 1999, introduced two algorithms frequently used, one supervised network, the backpropagation algorithm and one unsupervised network, the Kohonen self-organizing mapping (SOM) algorithm. Kohonen SOM falls into the category of unsupervised learning methodology, in which the relevant multivariate algorithms seek clusters in the data. Conventionally, at least in ecology, reduction of multivariate data is normally carried out using principal components analysis or hierarchical clustering analysis. Unsupervised learning allows the investigator to group objects together on the basis of their perceived closeness in n dimensional hyperspace (where n is the number of variables or observations made on each object). Formally, a Kohonen network consists of two types of units: an input layer and an output layer. The array of input units operates simply as a flow-through layer for the input vectors and has no further significance. In the output layer, SOM often consist of a two dimensional network of neurons arranged in a square (or other geometrical form) grid or lattice. Each neuron is connected to its n nearest neighbours on the grid. The neurons store a set of weights (weight vector) each of which corresponds to one of the inputs in the data. The SOM algorithm can be characterized by several steps. To predict hourly O_3 concentrations 1 day ahead, Balaguer et al (2002), used an autoregressive-moving average with exogenous input (ARMAX) models, multilayer perceptrons (MLP) and finite impulse response (FIR) neural networks. In order to introduce dynamic capabilities in a static neural network, FIR neural networks substitute the static synaptic weights by dynamic connections (digital filters). The FIR network models each weight as a basic FIR linear filter. This filter can be modelled with a tapped delay line. In a FIR filter, the output y (k) corresponds to a weighted sum of past delayed values of the input x. Three modelling techniques where used by Corani (2005) to predict air quality in Milan: feed-forward neural networks (FFNNs), pruned neural networks (PNNs) and lazy learning (LL). In particular, feed-forward neural networks (FFNNs), currently recognized as state-of-the-art approach for statistical prediction of air quality, were compared with two alternative approaches derived from machine learning: pruned neural networks (PNNs) and lazy learning (LL). PNNs constitute a parameter parsimonious approach, based on the removal of redundant parameters from fully connected neural networks. LL, on the other hand, is a local linear prediction algorithm, which performs a local learning procedure each time a prediction is required.

3.2.1.4. Models Evaluation and Discussion

Although both models were evaluated in all seasons separately, for simplicity reasons the evaluation results are presented in a unified graph in Figure 7.

The benzene values predicted by DET model were very close to those measured. RMSE value was 9.248μg/m^3, RRMSE was 1.538 and R^2 was 0.9859 (Table 7).

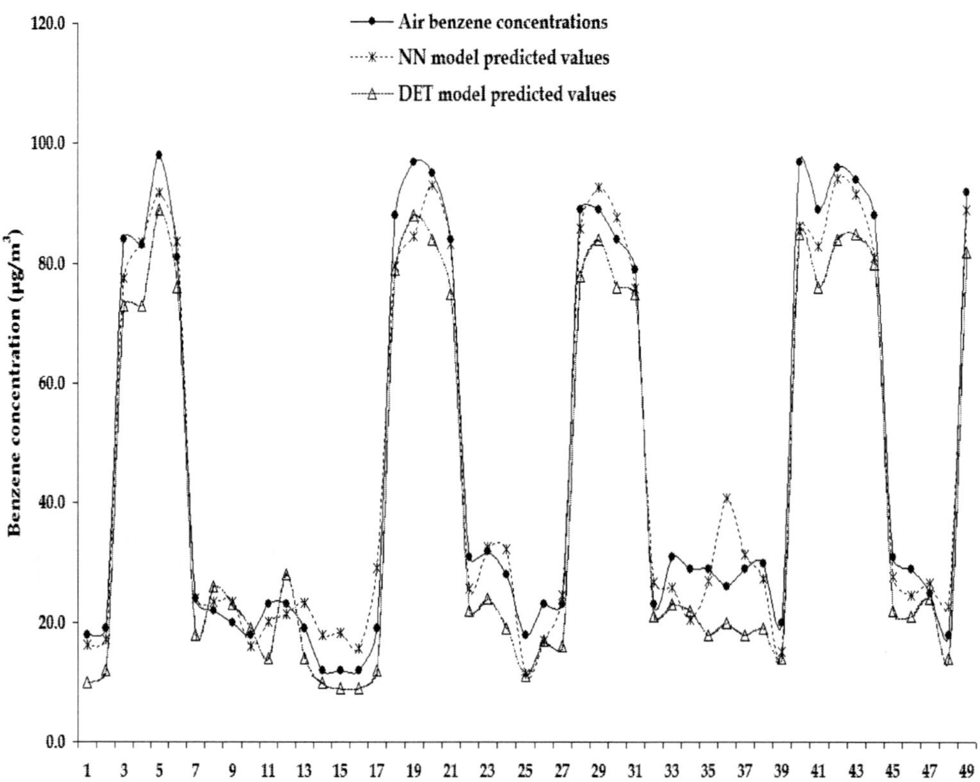

Figure 7. DET and ANN models evaluation results.

It is obvious from the analysis, presented in Figure 7, that DET model slightly underestimates the measurements. This is possibly due to the underestimation of the emission factors compared to the actual ones. There are many reasons that the emission factors are higher than the theoretical factors: degradation of the catalyst, insufficient maintenance of the engine, bad fuel quality, use of unleaded petrol in motorcycles etc. Higher differences were observed for lower concentrations, as well as, when the wind speed was non zero, while under calm conditions and high concentrations the error percentage was significantly smaller.

Also using traffic density as the parameter that determines the emission factors, makes forecasting of the concentrations in conditions with completely different traffic characteristics, feasible. The road we studied appears to have an almost steady flow throughout the day. The flow density though, depends on how fast the vehicles are conveyed to crossing roads. The last is due to the fact that the road under study, leads to the center of the city where many public services and trading stores are situated and given that there is a lack of sufficient parking spaces therefore traffic flow through any road is highly interconnected to that of all roads in a wider area around it. For the road considered herein both solutions of equation (4) are acceptable and directly connected to the total activity of the surrounding area. So even though we did not detect significant changes during normal working days, there are marked variations in flow density during the weekends as well as during Monday and Wednesday afternoon hours when the stores remain closed.

Table 7. DET model evaluation parameters

	Spring	Summer	Autumn	Winter
RMSE	10.452	9.071	10.11	8.1755
RRMSE	1.562	1.407	1.54	1.495
R^2	0.9862	0.9919	0.992	0.9723

The rest of the working days, Tuesday, Thursday, and Friday, present almost steady flows except for midday hours when, again, the stores close for a couple of hours, and the surrounding area gets discharged from traffic. Consequently, for normal working hours, when the market is open, that is all hours apart from Monday, Wednesday and Saturday afternoon and the whole of Sunday, the road is under constant traffic jam conditions and the density is estimated by equation (4a), while for the rest of the hours the density is estimated by equation (4b).

As far as the composition of the vehicle fleet is concerned, important changes are noticed only on a seasonal scale, with main difference the increased number of two wheel vehicles during the Summer period, a fact directly related to the warm weather conditions favoring their use. This can also be seen in Table 4, presenting the fraction of each vehicle category with respect to the total of the vehicles measured during each measurement period. The two wheeled vehicles, due to the fuel type they are using and the lack of a special anti-polluting technology, have a high potential of benzene emission, and in combination with their relatively large number, have an important contribution to the formation of the atmospheric accumulation of benzene. This result in higher observed values the benzene during the Summer compared to Spring and Autumn, a fact also predicted by the methodology and confirmed by the measurements. For summer we examined the possibility of fuel's evaporation and found that is was lower than 2%. The special value of the methodology is its ability to distinguish between traffic conditions that a simple measurement of traffic flow, as this which is done by automatic recorders (pneumatic counters) would have considered as identical, since as illustrated above for the same traffic flow two widely different density conditions can be observed. In addition, given that the density of a road is subject to its rate of decongestion, it is possible to find the interaction between roads that branch. This is so because possible changes in the flow of one road will bring about changes in the flow and the density of all crossing roads, with direct consequences to the benzene emissions and the related atmospheric concentration, something that is now possible to be estimated.

DET model was able to follow the variations of the measured benzene concentrations when fed by real traffic data (Figure 7). Such variations could be caused by changes in the composition of the fleet even, when only a very small change in traffic density was observed. This is the reason of the very high R^2 values computed. In addition, DET model can describe conditions in which traffic flow and fleet composition seems to have insignificant variation but they lead to a completely different RER, due to the fact that for the same traffic flow, two completely different density conditions exist (Kassomenos et al, 2004).

On the other hand, benzene values predicted by the ANN model were closer to the measured ones (Figure 7) and its evaluation parameters were better to those of DET model (Table 8). Unlike to DET model, ANN model does not presents any specific trend and does not overestimate or underestimate the observed values.

Table 8. ANN and DET model evaluation parameters

	ANN model	DET model
Mean Absolute Error	4,3	7,3
Minimum Absolute Error	0	1
Maximum Absolute Error	15	14
RMSE	5.325	7.9154
RRMSE	0.9487	1.2479
R2	0.9760	0.9868

Neural network modelling can also assess the importance of each of the input variables by using the network weights. The method for partitioning the connection weights proposed by Garson (1991) was used. The technique involves partitioning the hidden– output connection weights of each neuron into components associated with each input neuron (Goh, 1995). The results of the calculations are shown in Figure 8, displayed as columns representing the relative importance of the various input variables.

From this figure, it is evident that wind speed is the most significant factor affecting air quality in an area with a relative weight of almost 21%, followed by the speed of the vehicles (~18%) and the not catalytic cars (~17%). Since, air quality models usually consist of an emission and a meteorological part; we grouped our inputs in two main categories: meteorological-input parameters (represented in our case by wind speed) and emission-input parameters (e.g. vehicle speed, vehicle type etc.)

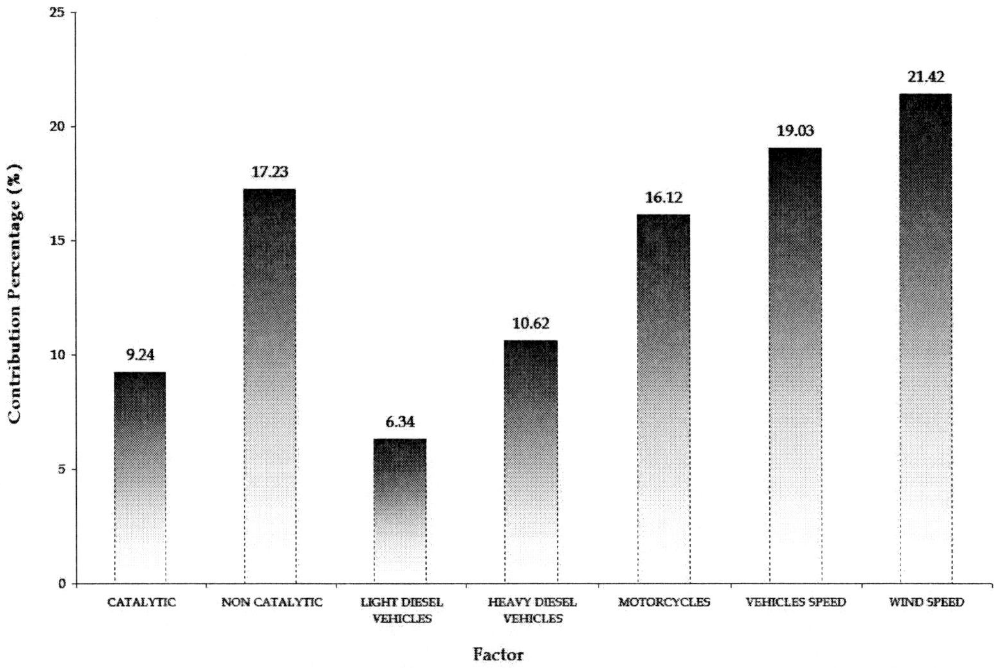

Figure 8. Relative Importance of the parameters controlling ambient air benzene concentrations.

Among the emission parameters, vehicle speed, plays the most significant role. As it was shown above vehicle speed is the key factor in the calculation of traffic density and consequently RER.

The remaining emission parameters refer to the total traffic flow. The relative contribution of each is a combination of the emission factor and the percentage of the total traffic flow, a specific vehicle category represents.

Non catalytic passenger cars have the highest relative importance among the traffic flow components. They account for almost 25% of the traffic flow contribution while represent only 13% of the total traffic flow. Catalytic passenger cars, on the other hand account for a significant lower part of the traffic related contribution (9%), although they represent almost 60% of the total traffic flow.

Heavy diesel vehicles despite their small number (2% of the total fleet) have a contribution factor slightly higher than catalytic passenger vehicles. This is a combined consequence of their high emission factors, lower speed and the fact that they have an extra effect in traffic density, considering that one heavy diesel vehicle corresponds to two (heavy duty vehicle) or three (buses) passenger vehicles in of Mean of Passenger Car Units.

Motorcycles, present almost the same contribution with non catalytic passenger cars (16%). This is due to the fact that they have higher benzene emission factor than other vehicle categories and their fraction (13%) of total traffic composition. The motorcycles determining contribution to benzene concentrations was also observed by the seasonal differences among summer and winter mentioned above.

Finally, light duty vehicles have the lowest contribution factor (6%), a fact that was expected taking in mind the percentage of traffic flow they represent (14%) and the lowest benzene emission factor among all vehicle categories they process.

Although the ANN prediction of an almost non reacting pollutant as benzene does not have the non-linearity degree and consequently the indispensability of predicting ozone for example, a validated ANN model may give valuable information (Reich et al, 1999).

3.2.2. Street Canyon Effect

The term street canyon ideally refers to a relatively narrow street with buildings lined up continuously along both sides (Nicholson, 1975). High pollution levels have been often observed in urban street canyons due to the increased traffic emissions and reduced natural ventilation. The dispersion of gaseous pollutants in a street canyon depends generally on the rate at which the street exchanges air vertically with the above roof-level atmosphere and later all with connecting streets (Riain et al., 1998). Skimming flow, a feature of regular canyons, provides minimal ventilation of the canyon and is relatively ineffective in removing pollutants (Hunter et al., 1992). Field measurements (DePaul and Sheih, 1985; Qin and Kot, 1993) show increased concentrations of traffic-related pollutants on the leeward side of the canyon, and decreasing concentrations along with height above the ground on both sides of the street. From a population exposure point of view, air quality in street canyons is of a major importance, since the highest pollution levels and the larger targets of impact are often concentrated in this kind of streets (Hertel et al., 2001).

In this study, the canyon effect was observed through the ambient air measurements, where streets with almost similar traffic characteristics (flow and density) presented different average weekly concentrations. Therefore, it is not surprising the fact that points 1,3,4,9 and 10 (Figure 4) that are located in street canyons present an average benzene value of 15,9 μg/m^3 which is clearly elevated compared to the average concentration of 8,2 μg/m^3 of the rest of the points.

Active sampling measurements also among these sets of streets revealed that calm wind conditions in combination to heavy traffic, leaded to ambient concentrations to 3-4 times higher in the canyons than in the other streets.

3.2.3. Filling Station Proximity

3.2.3.1. General Information

In urban areas it has been estimated that the amount of air benzene concentrations is mainly emitted by traffic. An additional contribution to this amount is being made in a local scale by filling stations. There are several studies that indicate the effect of refueling procedure and gasoline leaks to the observed elevated benzene levels and the related health effects to the gas station attendants (Carpleton and Levy, 2005; Carrieri et al, 2006). However, there are not as many studies that outline the contribution of gasoline refilling station to the overall presence of benzene concentrations in the wider vicinity.

In the present study, it is attempted to introduce a methodology that includes monitoring in the vicinity and modeling of the observed concentrations by the local emission sources which are the under study filling stations and the approximate roads.

The present study was based in monitoring the ambient benzene levels in five filling stations, an urban, three suburban and one rural. The position of the urban filling station, excludes a significant amount of contribution due to other emission sources. The only additional emission source is a main road in front of the filling station. Therefore the urban background concentration has a minimized and easy calculated effect on the ambient benzene concentrations. The second filling station was chosen to be a rural one, in order to observe the effect of the emissions from the filling station alone, due to the non significant contribution of the road and the absence of background concentrations. In the third case, a suburban road of 2 km length was chosen to be studied. This road has an excess interest because of the existence of three gas stations. The streets orientation according to the north is 15° to the west for the urban, 10°to the east for the suburban and 30° to the west for the rural.

In each area, the average weekly benzene concentration was determined in 16 points. At the urban and the rural filling stations two of the sixteen samplers were placed in the filling station near to the pumps, in order to observe the levels of benzene in this area. The rest 14 samplers were placed in different peripheral places covering a wide range of directions (0-360^0) and distances (up to 100m) (Figures 9, 10) at a height of 1.8m. In the suburban location (Figure 11), all the passive samplers (16) were placed equidistant along the street. A major issue of the whole concept was the accurate determination of urban background concentration. The urban background concentration can be defined in several ways. In the rural area also two samplers was placed to investigate any possible background concentration or the existence of any other benzene source beside the road and the gas station emissions that may affect the measurements results. All passive samplers were placed for one week.

Apart from passive sampling, the following parameters were measured:

- traffic flow from two independent observers, where the vehicles were registered in seven main categories (catalytic passenger vehicles, non catalytic passenger vehicles, diesel passenger vehicles, light duty passenger vehicles, heavy duty passenger vehicles, buses and motorcycles) every half an hour. In the urban and the suburban sites the measurements were continuous from 8:00 to 21:00 and randomly taken after that hour. In the rural site the measurements were in general randomly taken and it is worth mentioning that, the only difference in traffic flow patterns was between weekends and weekdays, since the daily variation was observed. Vehicles speed was calculated by the quotient of the distance (a part of the road) and the time, which was needed to cover that distance.

- wind speed (m/sec) and direction (degrees), on an hourly basis on the site where the observations were made. Meteorological data about the daily variation of ambient temperature, amount of cloudiness and solar radiation were obtained by the local meteorological station.
- 16 samples of benzene concentration by the method of active sampling in the urban filling station (where the passive samplers were placed) in different hours of the day, in order to examine the daily variation of the concentrations, also checking for the possibility of repletion of the passive samplers. The sampling procedure (active sampling) included adsorption of benzene on the same tubes using a personal pump. The sampling flow rate was set to 100 ml/min and the sampling time was set to 30 min.
- 154 samples of benzene concentration by the method of active sampling, 77 from each location, in order to validate the CALINE model.

The equipment used during the measurement campaign and analyses, while also the quality assurance and quality control (QA/QC) procedure where the similar referred above.

3.2.3.2. Measurements Results

3.2.3.2.1. Passive Sampling Results

In the urban gas station (Figure 9) the mean benzene concentration is about 50 $\mu g/m^3$, decaying while the distance is increasing in a way similar to a Gaussian distribution. In the gas station roadside the concentrations are 10 $\mu g/m^3$ higher than the other side of the road, due to the additive contribution of these two emission sources. In the suburban site, the observed curving of the plotted benzene concentrations (Figure 11) due to the presence of the filling stations is noticeable and enough by alone to represent in a significant degree the effect on the observed concentrations. Finally, in the rural site, the rate that the concentrations tend to decrease while the distance from the filling station increases (Figure 10), indicates that in this vicinity the dominant source of benzene is the filling station.

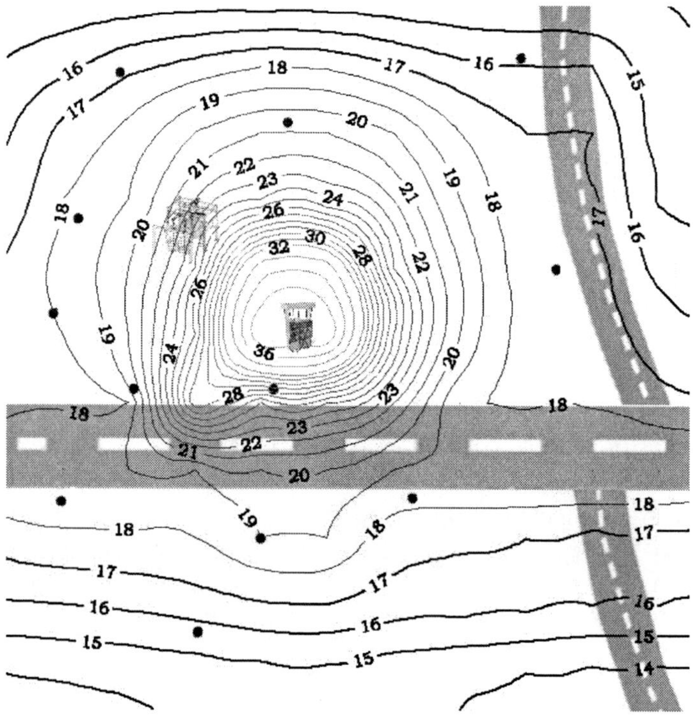

Figure 9. Urban filling station benzene isoconcentration lines (μg/m^3).

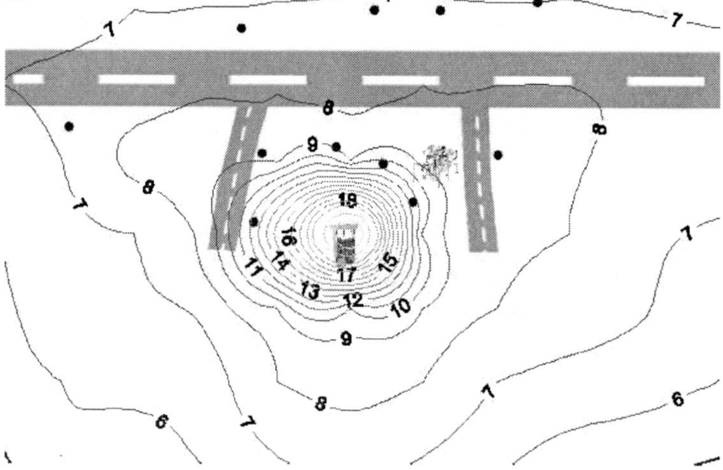

Figure 10. Rural filling station benzene isoconcentration lines (μg/m^3).

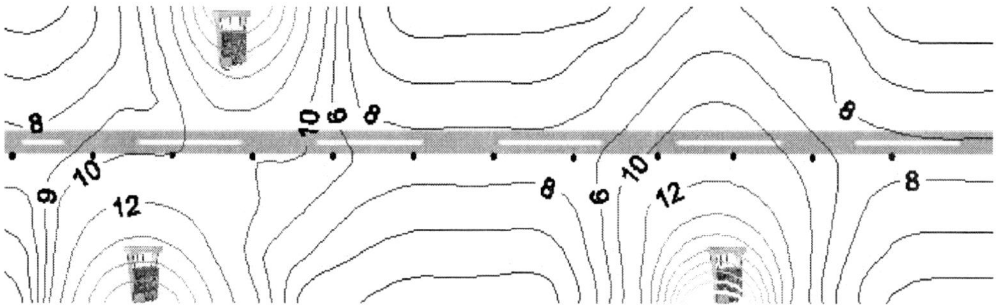

Figure 11. Suburban filling station benzene isoconcentration lines (μg/m^3).

The urban background concentration (affecting also the suburban road) was determined 3,8 and 4,2 μg/m^3 at the two measurement locations, giving the average value of 4 μg/m^3. At the rural sampling point, background concentration was insignificant (0,1 μg/m^3) , excluding the possibility of the existence of another emission source than the road and the filling station.

3.2.3.2.2. Active Sampling Results

As it was mentioned before, in the urban filling station, also 16 samples by the method of active sampling were taken. The results are shown in Figure 12, were also the tendency of wind speed and traffic flow are presented.

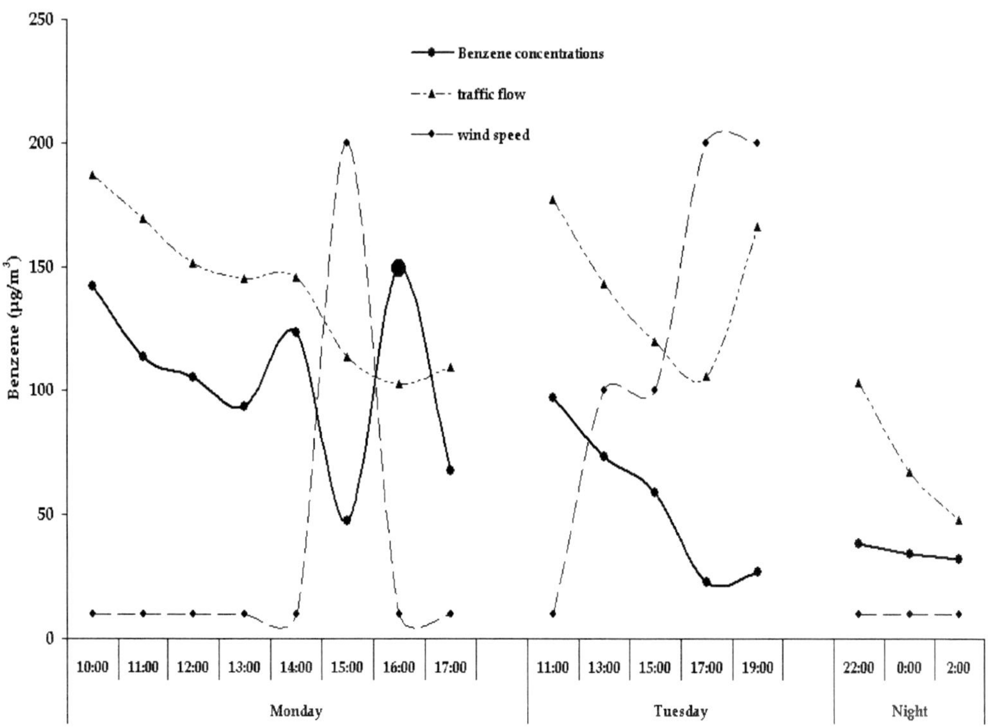

Figure 12. Urban filling station daily variation of ambient benzene concentrations (μg/m^3).

It is interesting that there is a good correlation among the road traffic flow (and consequently to the cars that refuel in the gas station) and the benzene values (0,72), a fact that indicate the correlation among the benzene concentrations and the total amount of fuels being distributed. The fact that the correlation is higher to the wind speed (0,86) indicate that there are also some other strong emission sources rather than the refuel loss, and these are the effect of the urban background concentration in combination with the road emissions and furthermore the leaks from the filling station restoring deposits. The last conjecture was affirmed by the samples taken in the night, where the contribution of the road was insignificant and also no fuel was distributed. The highest value of benzene was 155 μg/m^3, and this value was measured when the deposits of the filling station were refueled (the big dot branded spot in Figure 12).

3.2.3.2.3. Traffic Data

As it was mentioned before, the traffic flow (Figure 13) on each road in front of the gas stations was measured and the traffic speed was calculated. The urban and the suburban roads present similar fleet pattern, dominated by the passenger cars equipped with a catalytic converter. The traffic speed in both cases is equal to the road limit (50 and 70 km/h respectively), having a deviation of ± 10 km/h. In the rural road the fraction of heavy diesel vehicles (trucks and buses) is significantly increased, while the traveling speed is also 70 km/h.

Traffic flow daily variation has a similar behavior in urban and suburban roads (Figure 13), and almost similar in the rural one, where the variation between day and night is not so wide.

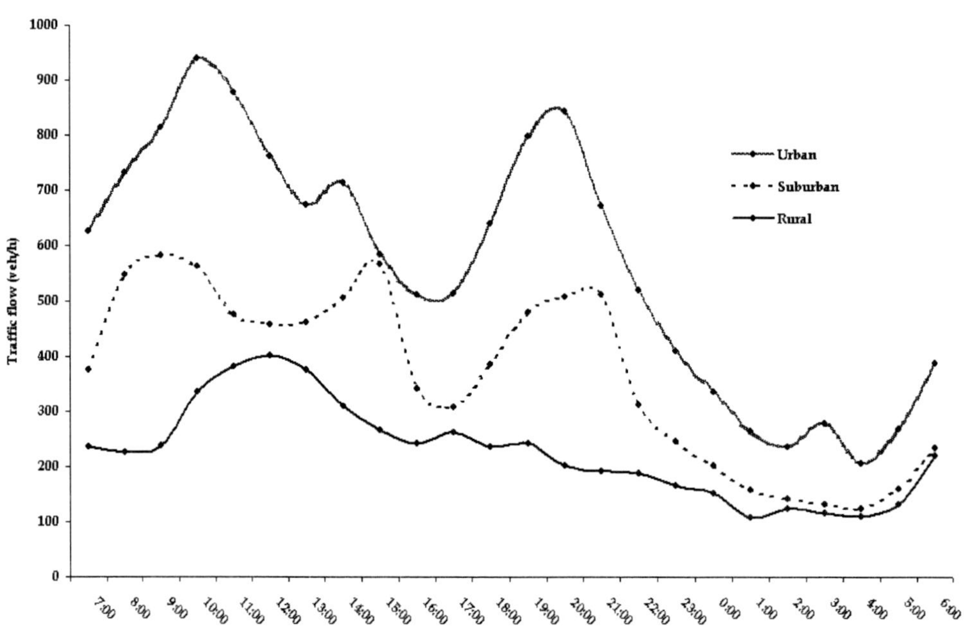

Figure 13. Traffic flow on the adjacent streets of the filling stations.

Traffic data were necessary in order to calculate with preciseness benzene's road emission rate. Based on the data of daily traffic variation, detailed fleet composition, vehicles speed and emission factors for each vehicles subcategory (Table 2) given by COPERT (Ntziachristos and Samaras, 2000), benzene emission rate for urban, suburban and rural road ranges from 1,4 – 0,26 $gr.km^{-1}.sec^{-1}$, 0,6 – 0,12 $gr.km^{-1}.sec^{-1}$ and 0,5 – 0,11 $gr.km^{-1}.sec^{-1}$ respectively.

3.2.3.2.4. Meteorological Data

The variability of atmospheric stability, ambient temperature, wind velocity and the amount of solar radiation was modest during the measurement campaign.

Wind speed and direction measurements were in agreement to previous data acquired from the wind field over the area. The mean wind speed for the measurements period ranged from 0 to 7 m/sec, while the average wind velocity was 1.1 m/sec.

The above parameters were completely necessary in order to derive the stability classes, used in the applied model. Atmospheric stratification varied from stable to sligtly unstable. In terms of the Pasquill classes (denoted by letters from A to G), the fractions of slightly unstable (class C), neutral (class D), slightly stable (class E) and stable (F) and cases were 7%, 32%, 51%, and 10%, respectively.

3.2.3.3. Interpretation Using Models

3.2.3.3.1. Model Formulations

The CALINE model represent roadway links as a series of elements with corresponding finite line sources (FLS). The pollutant concentration at a receptor location is estimated by summing the concentration contribution from each FLS. The CALINE 4 model uses a parameterization scheme quite similar to CALINE 3 for pollutant dispersion, and in general, CALINE 3 and CALINE 4 pollutant estimates are statistically similar. For additional details, interested readers are referred to Benson (Benson, 1979, 1984, 1992).

At the heart of the CALINE model is the concept of a "mixing zone" that exists above the roadway where the intense mechanical turbulence, augmented by buoyancy, results in enhanced mixing of pollutants. The primary use of the mixing zone is to establish initial Gaussian dispersion parameters at a reference distance near the edge of a roadway. This mixing zone is defined as the region over the travelled way plus 3 m (approximately two vehicle widths) on either side. The additional width accounts for the initial horizontal dispersion imparted to pollutants by the vehicle wake. Within the mixing zone, the mechanical turbulence created by moving vehicles and the thermal turbulence created by hot vehicle exhaust are treated as significant dispersive mechanisms (Dabbert et al., 1981).

The CALINE dispersion parameterizations are based in part on roadway geometry and wind direction. Downwind of the roadway edge, the CALINE model determines the Gaussian dispersion parameters with modified Pasquill–Turner curves.

Incremental downwind concentrations are computed using the crosswind Gaussian formulation for a line source of finite length:

$$C(x,y) = \frac{q}{\pi \sigma_z u} \int_{y_1 - y}^{y_2 - y} \exp\left(\frac{-y^2}{2\sigma_y^2} \right) dy \tag{17}$$

where q is the lineal source strength, u is the wind speed, σ_y and σ_z are the horizontal and vertical Gaussian dispersion parameters, and y_1 and y_2 are the FLS endpoint y-coordinates.

A CALINE roadway link is assigned an equivalent line source strength based on the product of a fleet-averaged vehicle emission factor (grams of pollutant per vehicle per mile traveled) and vehicles flow rate (vehicles per hour). The roadway link emissions are distributed to each element's FLS, which is centered on the midpoint of each element and oriented perpendicular to the mean wind direction. The length of each FLS is based on the wind direction, link width, and element length. Since the FLSs are positioned perpendicular to the wind direction, FLSs are rarely parallel, and often times are not physically coincident, with the roadway link they represent.

3.2.3.3.2. Evaluation and Optimization of CALINE 4

CALINE is a widely used and well evaluated dispersion model (Benson, 1992, Vardoulakis et al, 2003,). In its latest form of CALINE 4 software program is easy to be applied, requiring traffic and meteorological data.

The evaluation and optimization of CALINE was critical in order to use the results for abstracting road's contribution from the measured values. The procedure was performed in the three monitoring sites, had a duration of two days and it was made exactly by the same way. The measurements used for CALINE's validation were performed in a distance of 500 m along to the filling stations, in order to exclude any possible effect in the observed values.

In each road, at the distances of 8 and 40 m from the roadside, by the method of active sampling, 2 samples were taken every 3 hours. Background concentration was also determined by the method of active sampling by taking one sample every three hours (simultaneously with the measurements at 10 and 40 m from the roadside) in the place where the one of the two passive samplers for the determination of urban background concentration was placed as mentioned before. Totally for each monitoring site 48 samples were collected (including the urban background concentration determination). The schedule was made in such a way because the model must be validated in several cases of traffic conditions and meteorological parameters (wind speed and direction, amount of cloudiness, temperatures, solar radiation) in order to optimize its performance.

The validation results in all the streets show an over prediction of the measured values in most of the cases. This was expected because it is well known from older studies on CALINE (Benson, 1992; Held, 2003) that it overestimates concentrations in cases of low wind speeds (or winds perpendicular to the road), a situation commonly occurred in the wider region, as mentioned before.

The scatter plots of predicted versus observed benzene concentrations at the distances of 8 and 40 m respectively are presented in Figures 13 (a-c). The bold line is a statistical linear fit to the data points (executed with the standard Excel line-fitting program). The slopes of the linear fits of the scatter plots deviate almost to 20% from the ideal agreement in all cases. Clearly, the slopes of such correlations are sensitive to the modelling of the emission coefficients. For instance, a fractional increase in the vehicular benzene emission coefficient would increase all the predicted benzene concentrations by approximately the same fraction (Levitin et.al, 2005).

To eliminate this artifact, by the observed values the emission factor used as input in the CALINE was derived (model optimization). The correction of the emission factor was equal to the coefficient of the overpredictions. In each monitoring site that coefficient is depended to the angle that the road has with the dominant wind direction. The coefficients of correction were 0.813, 0.832, and 0.847 for urban (30°), suburban (55°) and rural (75°) road respectively. The derived emission factor was then used to calculate the contribution of the road in each monitoring site.

3.2.3.4. Contribution of Filling Stations to the Observed Benzene Concentrations

After the optimization of CALINE, the calculation of the roads contribution to each monitoring site was done. To achieve this, the model was run for every single hour for one week duration and then the results were averaged. An advantage of CALINE 4 is the fact that in every single run, all the sampling positions (receptors) are included, a fact that makes the model very fast to use, especially when the number of the receptors is large (16 in our case).

In Figures 14-16 the benzene concentrations in each area without the contribution of background concentration and adjacent road emissions are presented.

It is obvious that the greatest amount of effect is in the vicinity of the rural area (Figure 15). This fact was expected, since there were no other emission sources except the road, in which the benzene emissions are kept in a lower level because of the small traffic volume and the relatively high speed. In the urban area (Figure 14), the absolute contribution is higher in terms of μg/m^3, but relatively a little bit lower. This is due to the fact that the road affecting the concentrations is heavily trafficked and the vehicles speed is the lowest among the three locations. The combination of these factors leads to increased benzene emissions. Furthermore, the contribution of the urban background concentration is also significant (4 μg/m^3) affecting in a determinant way the observed concentrations.

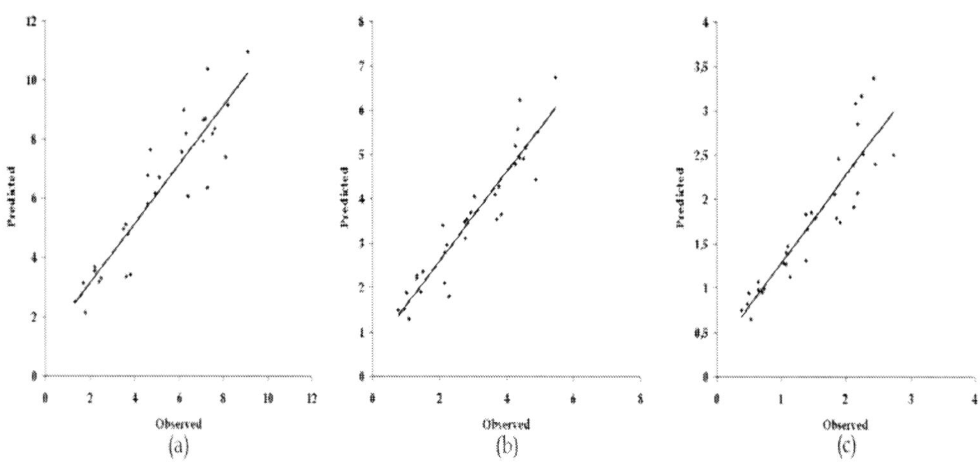

Figure 13. CALINE 4 evaluation (Predicted vs Observed values).

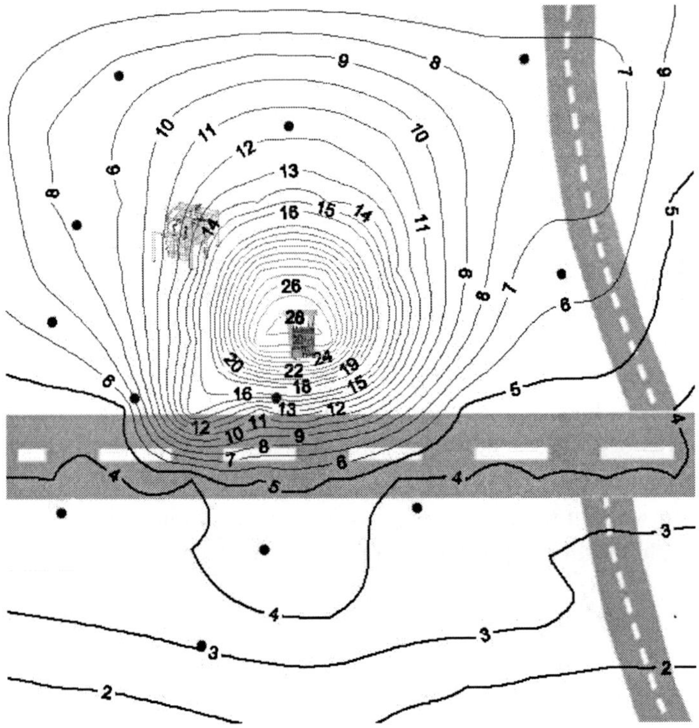

Figure 14. Urban filling station benzene isoconcentration lines ($\mu g/m^3$).

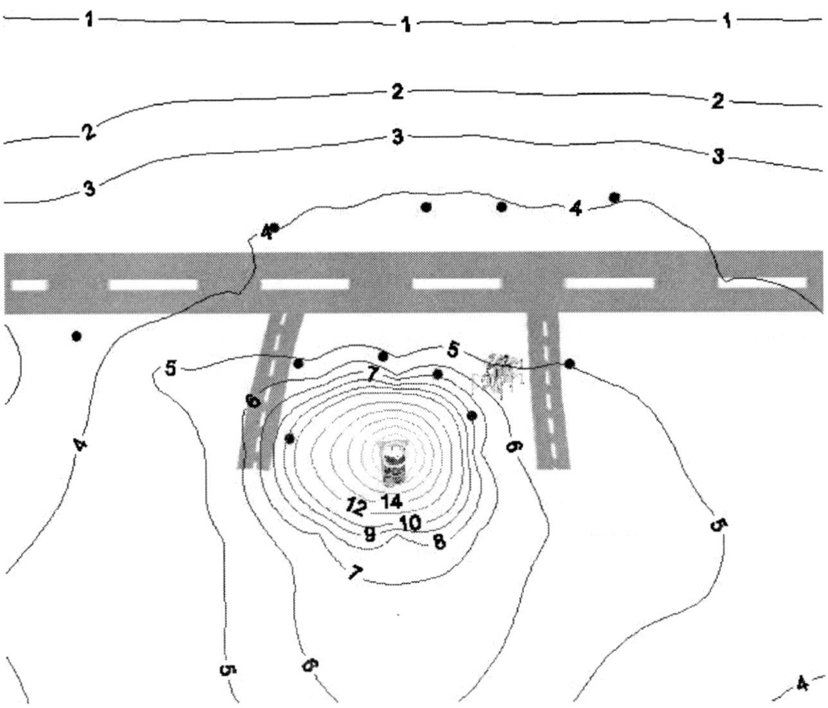

Figure 15. Rural filling station benzene isoconcentration lines ($\mu g/m^3$).

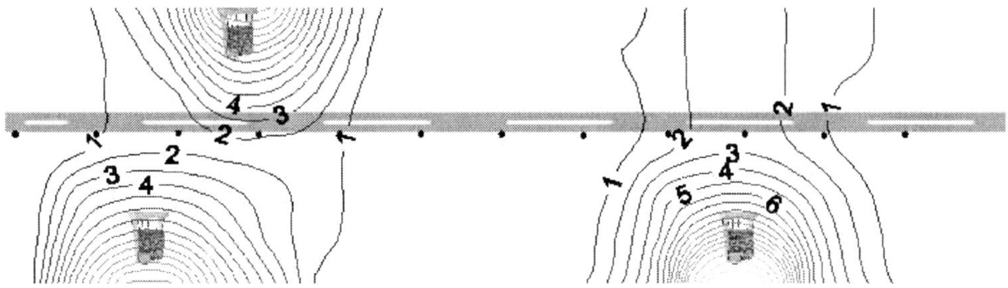

Figure 16. Suburban filling station benzene isoconcentration lines (μg/m^3).

Finally, in the suburban road (Figure 16), the effect of the gas stations is smaller, but this is relative, because in this case the spatial distribution was examined only in one dimension (along the length of the road) and in a larger scale. Another fact that is affecting the concentrations is the mixing conditions. Even though that the total amount of fuels being distributed by the urban gas station is smaller than this in the suburban stations and consequently the benzene emissions are lower, the effect in the concentrations is higher, a fact that can be attributed only in the parameters that impede the pollutants dispersion in an urban environment.

From the above results, excess environmental interest has the contribution of the urban gasoline station to the wider vicinity, because it is affecting the air quality of several people living or working in the area.

4. Personal Exposure Results

4.1. General

Personal exposure is equal to the average concentration of a pollutant that a person is exposed to over a given period of time, e.g. 1 day, 1 month or 1 year. If over the given period of time, *T*, the person passes through *n* locations, spending a fraction *fn* of the period *T* in location *n* where the concentration of the pollutant under consideration is *Cn*, then the personal exposure for this period *T*, represented by the concentration C_T, is given by (Hinwood et al., 2006) :

$$C_T = \sum_n f_n \cdot C_n \tag{18}$$

Thus, to evaluate the contribution of the sources to personal exposure, the amount of time a person spends in each location, along with the concentration of the compound in each location needs to be measured (active sampling).

Due to the fact that in most cases it is practically impossible to measure the pollutants concentrations of every microenvironment in context to the residence time, results of average personal exposure during a work shift indicate in a more precise way than simple outdoor or indoor measurements the running risk for the several population categories (passive sampling).

In this study, in order to determine with precise the total exposure during the work shifts in combination to the special conditions that comprise that exposure value, both sampling methodologies were used.

4.2. Control Group

Exposure to benzene (Figure 17) is anticipated to be related to a person's life style and the types of environments that are encountered. The parameters that strongly influence the exposure levels are often using of the car, walking nearly heavily traffic streets and the residence in indoor places with smokers. Home location of the exposed subject was also of great importance, as in general the highest recorded values were determined by people living near the centre. The highest observed value (18,2 μg/m^3) was noticed by a person that was living in an area with moderate benzene levels (7,4 μg/m^3), who used move on foot and by car for at least 1 hour per day in the morning for several occupations in the city. The lowest value (4,7 μg/m^3) was observed by a non smoking subject living in the periphery of the city (5,2 μg/m^3). The rest of the subjects had a great variety of activities, and no clear conclusion beside them, referred above, could be attributed by the questionnaires.

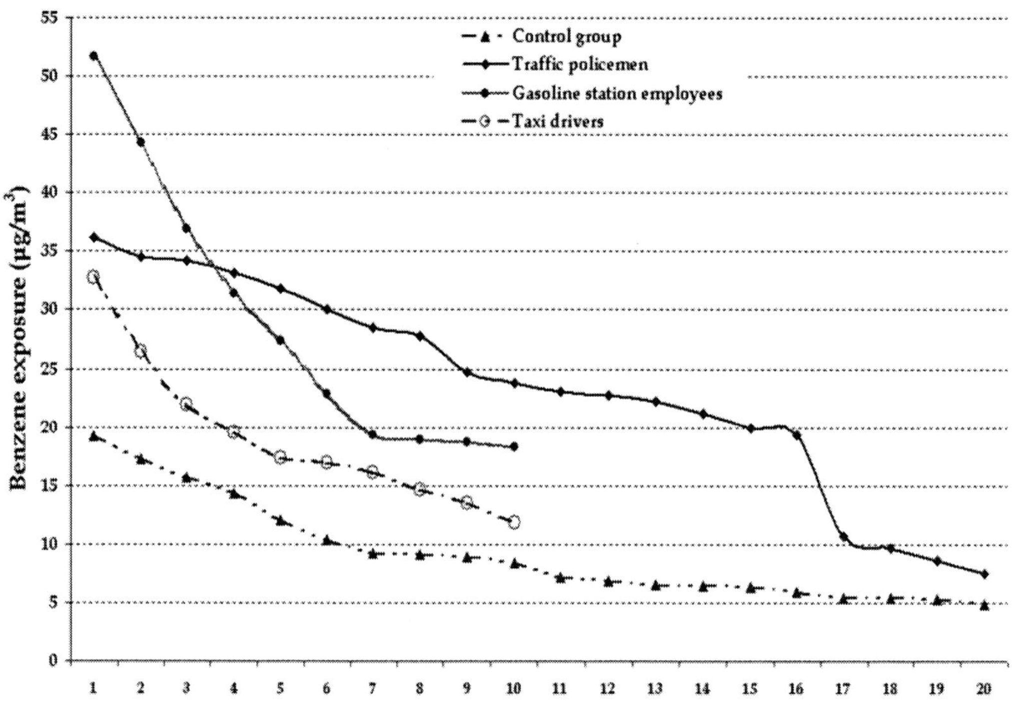

Figure 17. Average weekly exposure values for the studied population groups (μg/m^3).

4.3. Policemen

A close examination at the results (Figure 17) reveals that policemen are exposed to benzene at 3-5 times higher levels in comparison to the non occupationally exposed subjects and the results are similar to a previously campaign contacted in Bolognia (Maffei et al., 2005). This observation was expected having in mind that policemen working outdoor are highly exposed to traffic pollutants. It is necessary to clarify that almost all of the policemen were performing a combination of activities (e.g patrol, traffic regulation, working in offices), the duration of each of them contributed to the total amount of exposure; for example policemen that belong to patrol (in vehicle or on motorcycle) also executed traffic regulation or spend some time in the offices.

The contribution of each activity of the policemen to quantified pollutants was estimated by the information gained from the questionnaires and the measurements of personal exposure. Policemen doing patrol in vehicles, are highly exposed to benzene (29 $\mu g/m^3$) because they spend a lot of time in the car circulating in the centre of the city, where the pollution is increased and pollutants are more likely to be accumulated in the vehicles cabin (Jo and Park, 1998; Manini et al., 2006). Traffic regulation is the activity with the higher exposure potency, because the policemen stay in the middle of the intersections and they are exposed to the direct emissions of vehicles, where the concentrations are elevated.

Similar is the situation for policemen doing patrol on motorcycles. They are exposed all the working time to direct emissions of the vehicles, either circulating or regulating the traffic. This explains the high average benzene exposure values (28 $\mu g/m^3$). This value could be also higher, if they do not spend significant time in the police building offices (2 hours, in comparison to the 1 hour spent by policemen patrolling in vehicles).

Policemen doing patrol on foot or guarding the police station are exposed to the urban concentrations and their average exposure is similar to the ambient air concentrations on the pavements of the central roads. It is important also to mention that policemen doing patrol on foot probably inhale an increased dose of benzene than this imposed by the exposure measurements, due to the increased inhalation rate caused of walking (O'Donoghue et al, 2007).

Finally, benzene exposure of the people working in the police station building was minimized compared to the other policemen. The exposure value (9,1 $\mu g/m^3$) is almost equal to the test group exposure, a relatively high value, having in mind that is an indoor exposure; this is attributed to the location of the police building that is next to a road with elevated benzene values (14 $\mu g/m^3$). The interchange between both environments is regulated by the ventilation in the building, which depends either on microclimatic factors or on the intrinsic condition of the building.

4.4. Filling Station Employees

4.4.1. Introduction

Employees in fuel stations can easily be exposed to high levels of gasoline vapours during refuelling as well as high concentrations of emissions from tail pipe exhausts. Therefore, service station employees are a vulnerable sector of the population because of their exposure during the 8h shifts (Jo and Song, 2001). Preventive measures adopted during the last few years, such as a reduction in the quantity of benzene in gasoline (CONCAWE, 2000), the installation of systems to extract solvent vapours on gasoline pumps (CONCAWE, 2002) and the introduction of automatic quantities measured directly into vehicle tanks, all mean that the exposure to benzene by filling station employees is lower than before (Franco and Bisio, 1995; Giacopetti, 2000). Despite this fact, exposure to benzene for filling station employees remains relatively high compared to the general population, also due to the unexceptionable reduction of ambient benzene concentrations occurring in the last years in all kind of microenvironments (Aleksic et al, 2005).

Until now, several remarkable studies were performed determining filling station employees' personal exposure. These studies usually included exposure monitoring by using personal samplers (Cruz-Nunez et al, 2003) and blood biomarkers analyses (Hotz et al, 1997; Brugnone et al, 1999). Knowing that measured levels of biomarkers in the blood are affected by other sources inducing benzene in the body (like smoking) and also they don't have a direct time response in the exposure levers, correlation with the exposure values is usually low. Thus, measurements of direct personal exposure still constitute the most reliable methodology accessing health impacts to hazardous air pollutants (Carrieri et al, 2006).

In the present study, a campaign of benzene personal exposure measurements performed in two filling stations employees by passive sampling. In order to reveal the contribution of the main activities related to the operation of the filling station to the employees exposure patterns, active sampling was performed.

The active sampling results in combination with other environmental parameters were used to develop a set of ANNs. ANNs have been largely used to predict atmospheric pollutants concentrations, especially the reactive ones like NO_X and O_3 (Gardner and Dorling, 1998). The main reason for this is their remarkable ability to handle non linear systems. The non reactive pollutants are also non linear systems (in a much smaller degree than photochemical pollutants) since they are highly depend on atmospheric dispersion conditions and vehicles emissions. Recently, non reactive pollutants forecasting via ANNs (Gardner and Dorling, 1999; Dorling et al, 2003; Karakitsios et al, 2006), has gained an increased interest, because ANNs can handle the large amounts of data, now routinely acquired through modern monitoring techniques. If these data (e.g. air quality, meteorological and traffic) are used for training an ANN, this can lead to reliable prediction of atmospheric concentrations of emitted pollutants in a way easier and faster than a deterministic model (Nagendra et al, 2004), but according to our knowledge, ANNs have not been used so far for the prediction of human exposure. In the present study, the developed set of ANNs is aiming to produce an exposure predicting model, quantifying also the contribution of the parameters that constitute the exposure pattern.

4.4.2. Methodology

The measurements campaign was performed during two sampling periods (August and December, 2006) to investigate the seasonal variation of benzene exposure values.

The study was performed at two filling stations (the same mentioned before in a previous study), an urban one and a rural one, at a distance of 15 km of the city limits. Both of them were equipped with Stage I fuel evaporation recovery systems.

Personal exposure to benzene by the method of passive sampling was performed to 15 volunteers each season for two weeks during the working hours (8:00-16:00). One sampler per day for each volunteer was used. The volunteers group included 10 employees in the urban filling station and 5 in the rural one.

People working in filling stations were doing three major activities· cars refuelling, miscellaneous activities (like car washing or oil checking) and cash machine occupying in the filling station office. Most of the employees were doing a combination of these activities, especially in the rural filling station.

In order to examine the daily exposure variation and the significance of the activities in the configuration of the total amount of exposure for the filling station employees, active sampling was also performed. In this case, sampling pumps were attached on the employees, and the samplers were placed in the height of the chest, just like the commonly known passive samplers. This was attainable due to the relatively small size and weight of the sampling pumps. The employees carrying the pumps were executing strictly only one kind of activity and consequently were employees refueling cars, employees of miscellaneous activities or cashiers. The pump was set to work for 30 minutes in a flow rate of 100 ml/min. Each sample was attached for half hour, and one sample was used per hour, so until the end of the 8h shift, 8 samples per employee were taken. In each filling station, one employee of each category was carrying the active sampler. Active sampling was performed day per day. Totally, 224 samples were taken.

Apart from active sampling, the following parameters were measured, and all the values were recorded in a 30' basis:

- The amount of the gasoline that was traded.
- Wind speed (m/sec) and direction (degrees).
- Ambient air temperature.
- Traffic flow from two independent observers, where the vehicles were registered in seven main categories (catalytic passenger vehicles, non catalytic passenger vehicles, diesel passenger vehicles, light duty passenger vehicles, heavy duty passenger vehicles, buses and motorcycles) every half an hour. In the urban site the measurements were continuous from 8:00 to 16:00. Vehicles speed was calculated by the quotient of the distance (a part of the road) and the time, which was needed to cover that distance.
- Urban background concentration.

4.4.3. Artificial Neural Network (ANN) Modeling Development

Intending to obtain a way to predict the exposure pattern of the filling station employees, the development of a set of ANNs was introduced. The main concept was to use as input parameters (i) the amount of gasoline sold, (ii) the traffic flow of the road in front of the filling station, (iii) the background concentration (only in the urban filling station because in the rural it was almost zero), (iv) the wind speed, and (v) the ambient temperature, in order to predict as output the exposure pattern of the three main categories of employees mentioned above. Consequently, three ANNs were employed, one for each group of workers. Furthermore, the ANNs can provide significant information for environmental policy purposes, after quantifying the parameters that constitute the exposure pattern, in terms of the Relative Importance (% contribution values).

In our approach a feedforward network architecture was developed, which is widely used in ANN applications. This is a multilayer architecture, meaning that the artificial neurons are organized in layers. Typically, a standard feedforward ANN has an input layer, where data are introduced to the ANN, one or more hidden layers, where processing is realised, and an output layer, which generates the final results. Another important characteristic of feedforward ANNs is that neuron activation (or signalling) propagates towards one direction only and specifically from the input layer to the hidden layers and finally to the output layer.

In this study, 3 sets of ANNs were developed: one for the urban station, one for the rural and one for both. The number of the neurons corresponds to each input parameter. The input parameters were the amount of gasoline traded, wind speed, ambient temperature and traffic flow of the road in front of the filling station and benzene urban background concentration. The difference between the first and the other two sets of ANNs are that in this case, the first (input) layer consisted of 5 neurons while for the rural one consisted of 4, due to the lack of background concentration. In the third set, data gained from both stations were used for training and testing. In this case, the lack of the background concentration as an input parameter was not so determinant for the urban stations' predictions, because it is indirectly included in the parameters of traffic flow and wind speed. We should note that in all cases the ANNs' architectures were kept the same. Specifically, we met in the input layer four or five nodes, in the hidden layer 10 and in the output one (Figure 18).

In particular, the second (hidden) layer consists of 10 neurons that implement the hyperbolic tangent sigmoid transfer function. This function generates outputs between -1 and 1 as the neuron's net input goes from negative to positive infinity and is more appropriate for ANN applications on function approximation

As for the number of the hidden neurons, various architectures were tested (different numbers of neurons and hidden layers) but the choice of one hidden layer with 10 neurons provided consistently the best results. Finally, the third (output) layer consists of three linear neurons that correspond to the three predicted exposure concentrations for each one for each exposed group of workers.

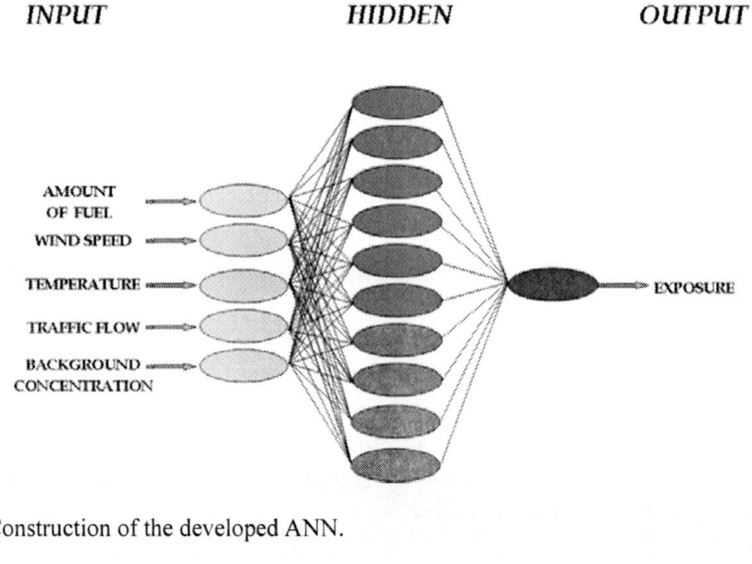

Figure 18. Construction of the developed ANN.

Network training was performed using the Bayesian regularisation algorithm (MacKay, 1992, Foresee & Hagan, 1997), which is already excessively described in an above sector.

For training and testing the proposed ANN model we followed the procedure of Leave-One-Out Cross-Validation (LOOCV). According to this schema, iteratively one sample is kept for testing while the rest are used for training the model, until all samples are finally tested (Witten and Frank, 2005). There are many ways to implement cross-validation, however LOOCV is a very commonly used validation schema and is considered to better resemble real practice where the proposed model is trained with all the samples at hand and then is applied to the specific application.

4.4.4. Measurements Results

4.4.4.1. Environmental Data-Amount of Gasoline Traded

Given the fact that the measurements were performed at the same filling stations of the study to the effect on the vicinity benzene concentrations, information about traffic and meteorological data is already provided.

The amount of the gasoline traded during the working hours was recorded in both filling stations. In the urban filling station, the amount of gasoline ranged from 550 to 950 l/h, having an average value of 700 l/h, while for the rural station ranged from 150 to 300 l/h, having an average value of 200 l/h. Daily and seasonal variation of the gasoline amount was proportional to the adjacent road traffic flow variation.

4.4.4.2. Personal Exposure Results

Exposure pattern for the filling station employees is complex, presenting significant variation of the recorded values. In Figures 19-20 the weekly averaged results of passive sampling are presented. The values ranged from 51 to 18,3 $\mu g/m^3$ in summer (Figure 19) and from 41 to 17,5 $\mu g/m^3$ in the winter (Figure 20) for the urban filling station. In the rural filling station, the values range from 29,5 to 10,3 $\mu g/m^3$ and 26,8 to 9,1 $\mu g/m^3$ for summer (Figure 20) and winter (Figure 21) respectively. The higher exposure levels in the urban filling station may be explained by the total amount of gasoline daily distributed (6000-12000 liters of gasoline in comparison to 2000-3000 liters) and to the absence of any other significant source of benzene in the rural filling station. On the contrary, the emissions from the central road in front and the contribution of the urban background concentration in the urban filling station seem to be significant additive parameters.

A seasonal variation was observed, especially in the urban filling station. The increased benzene concentrations in personal exposure values, while also in the vicinity of the filling station are explained by the higher temperatures occurring in summer (average temperature 23 ^oC, while in winter is 7 ^oC), due to stronger fuel evaporation. In the urban filling station this observation is more intense, due to the bigger amount of gasoline distributed (and consequently evaporated). Furthermore, the corresponding increase in background concentration and direct traffic emissions from the adjacent road, concurrent to the recorded variation among summer and winter measurements.

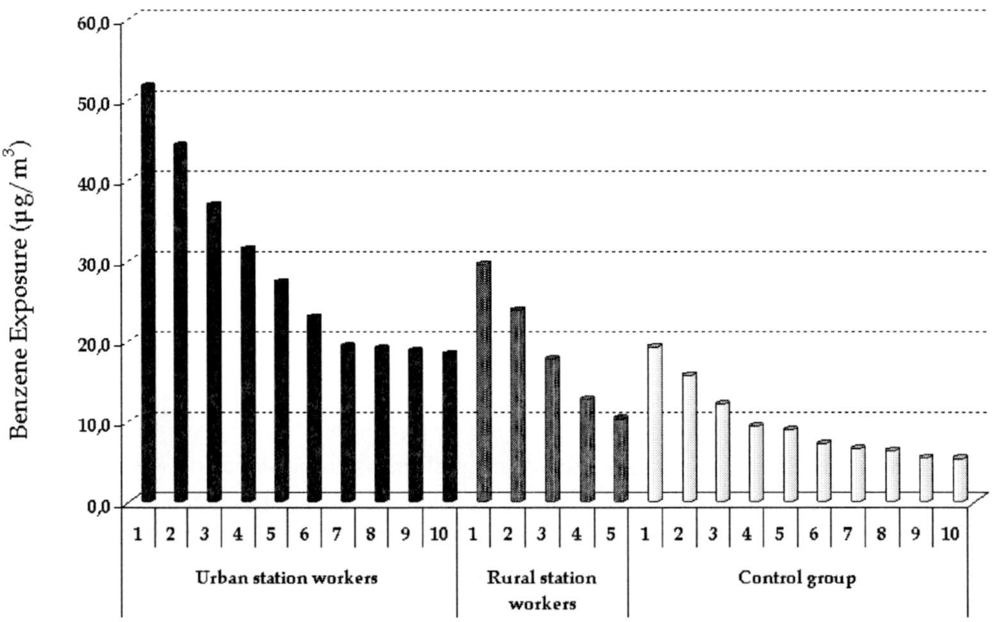

Figure 19. Average exposure of filling station employees in summer (μg/m^3).

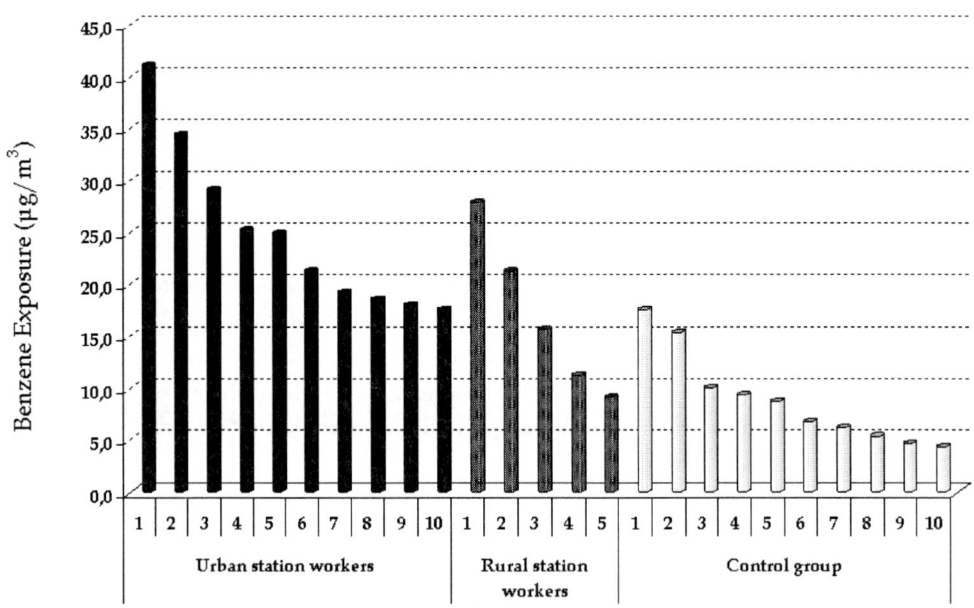

Figure 20. Average exposure of filling station employees in winter (μg/m^3).

Results of active sampling measurements were necessary to clarify the importance among the activities occurring in a filling station. In Figures 21-22, the hourly averaged weekly active sampling measurements results for each category of employees are presented.

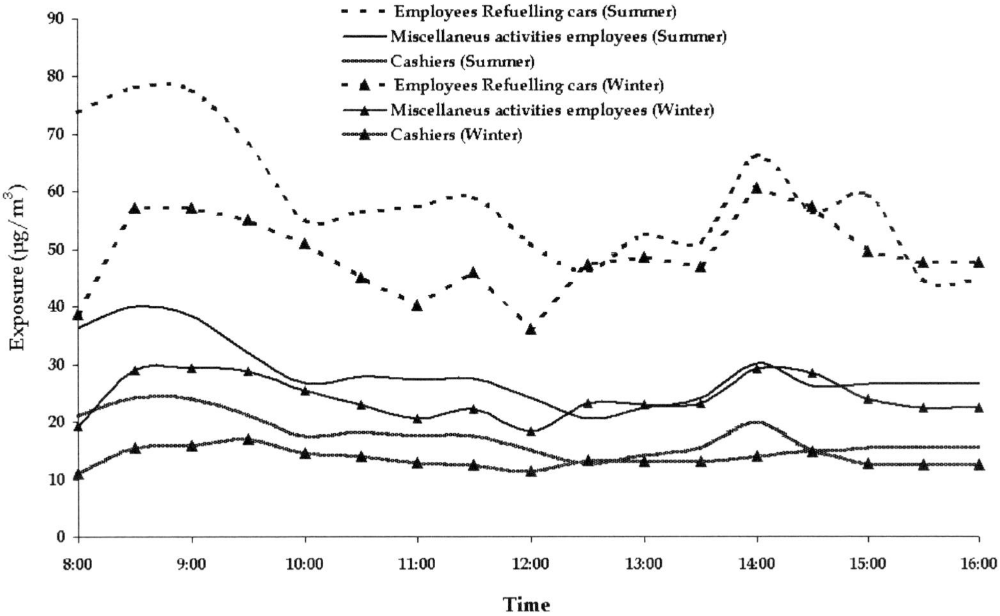

Figure 21. Hourly averaged weekly active sampling results in summer (μg/m^3).

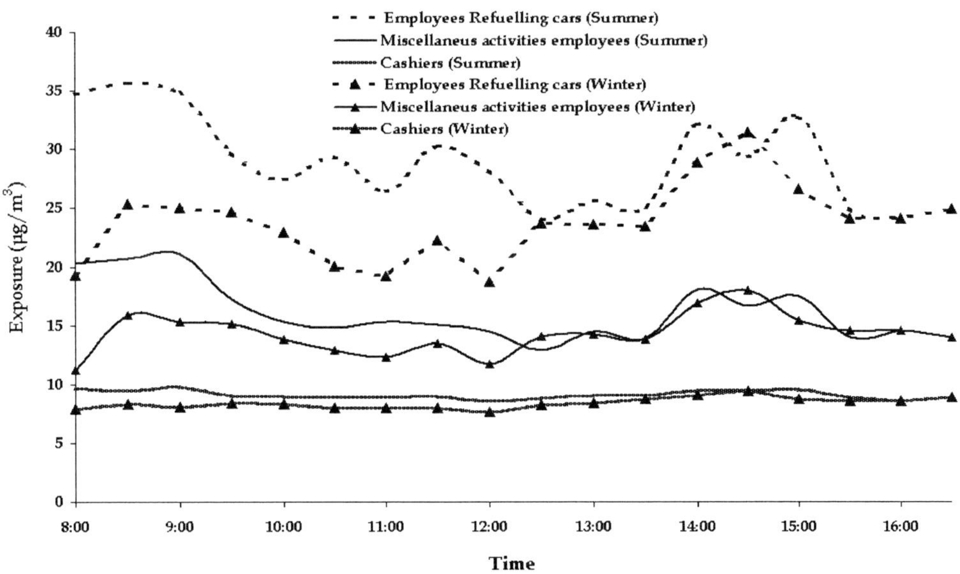

Figure 22. Hourly averaged weekly active sampling results in winter (μg/m^3).

From the results it is understood that employees refueling cars present the higher exposure levels in benzene. This was an expected result because the dominant source of benzene is car refueling. Although that newer fuel pumps with automatic fuel counters do not demand the presence of the employee above the fuel hose during the refueling, the fact they resident or move closely to the pumps wider area where the concentrations are high, leads to the recorded elevated exposure values.

Daily variation of exposure values of employees working outdoor, seemed to be highly correlated to the amount of gasoline distributed, an observation confirming the results from older studies (Periago, 1997). Gasoline trading rate is higher in the morning (8-9:00) and evening hours (14-15:00), related to the transportation of citizens to working places. This incident is not clearly observed in the rural filling station, were only in the morning an increase of the distributed gasoline is noticed. Because the variation of fuel traded is in linked up to traffic flow, maybe by mistake someone overestimate or underestimate the importance of traffic flow in the configuration of the exposure pattern. It is important to clarify in this moment of the study, that the increased traffic flow of the adjacent road leads more customers to visit the filling station and consequently the increase in fuel traded.

Ambient air temperature also affected the exposure levels. When temperatures are near 0 ^{o}C (a fact occurring at the early morning hours in the winter time), the exposure levels are lower than usual for the same amount of fuel traded (Figure 21-22).

The presence of wind seems to reduce exposure levels, especially to miscellaneous activities employees. No strong correlation among wind speed and exposure levels of employees refueling cars and cashiers was observed.

Finally, in the urban station, background concentration seems well correlated to the exposure levels of all the categories of employees, but the reason of that correlation is the same with the traffic flow, because when the traffic flow is raised, this occurs in the whole urban area and consequently the urban air pollution is elevated. The quantification of the importance of the above parameters is further investigated next through the ANNs modeling.

Compared to employees refueling cars (Figure 21-22), employees of miscellaneous activities are exposed to significantly lower benzene concentrations, because they rarely go near to the pumps, staying mainly in the area of car washing machines. This means that they work in an average distance of 15-25 meters from the pumps, where the concentrations of benzene are significantly reduced.

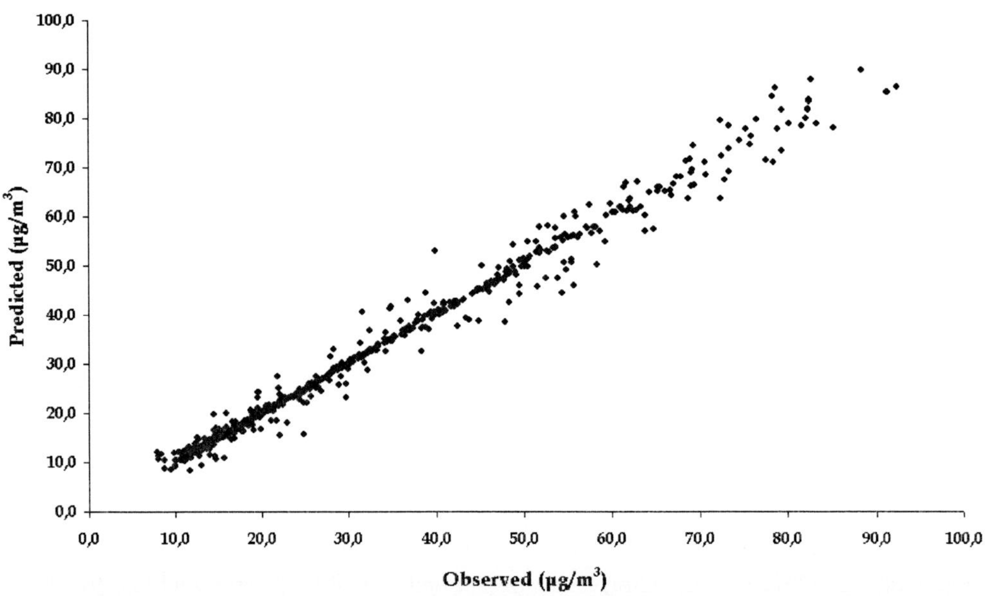

Figure 23. ANN predicted vs observed values (urban filling station).

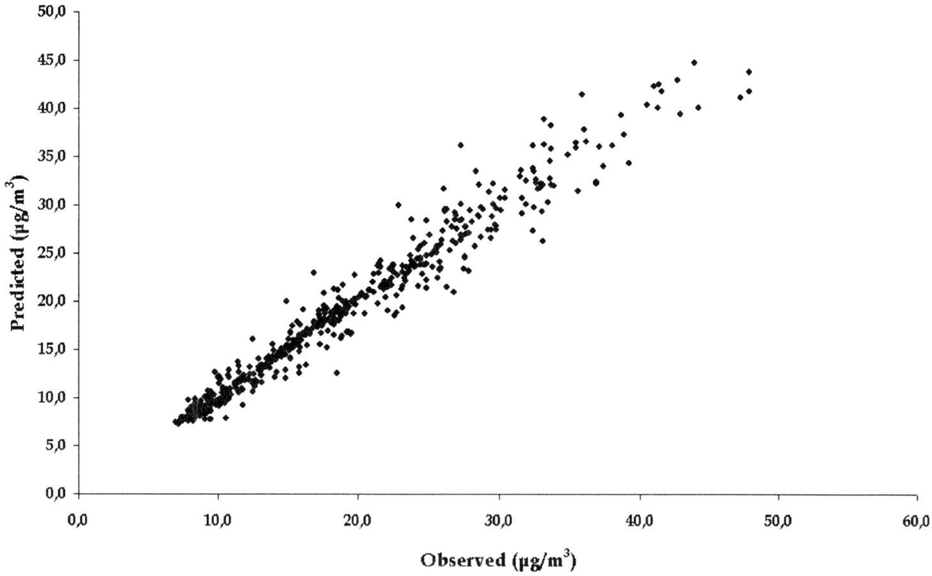

Figure 24. ANN predicted vs observed values (rural filling station).

The cashiers are working in an indoor environment, which is also affected by the same parameters that configure concentrations in the wider vicinity. Thus, the recorded values are lower to the employees of miscellaneous activities. The only remarkable difference at the rural filling station (in comparison to the urban one) is the absence of any significant variation among summer and winter exposure measurements of the cashiers. This maybe occurs due to an additive indoor source of benzene affecting the concentrations, possibly smoking, which is permitted inside the filling station office.

4.4.5. Artificial Neural Network Modelling Results

The results among the observed and the predicted values are shown in Figures 23-24. The evaluation parameters RRMSE (Relative Root Mean Square Error) and R^2 (Correlation) are presented in Table 9.

From the above, it is easily observed that the performance of the ANN applied to the urban filling station data is slightly better than the ANN applied to the rural one for the first two categories of employees. On the contrary, the performance referring to cashiers is worst in the urban station. This occurs because the indoor concentration of benzene presents a time lag following the outdoor concentrations that determine the exposure of the other two categories of working people and is strongly affected by the general benzene concentrations that affect the wider area.

This is also the reason, that among the three categories of employees, the best performance of the model is for the miscellaneous activities employees. Exposure pattern of these employees is directly affected by the referred input parameters, located in the area of the station that does not appear to have acute changes in the observed benzene concentrations. On the contrary, this occurs to the employees refueling cars during the refueling process, having as a consequence sometimes the quickly elevated levels of exposure to be hardly predicted with the same degree of accuracy.

Table 9. ANNs evaluation parameters

		Employees refuelling cars	Miscellaneous activities employees	Cashiers
RRMSE	**Urban**	0,0678	0,0352	0,1081
	Rural	0,0880	0,0822	0,0501
	Urban + Rural	0,0900	0,0814	0,0556
R^2	**Urban**	0,9729	0,9921	0,9379
	Rural	0,9437	0,9447	0,8136
	Urban + Rural	0,9714	0,9919	0,9441

The results of the third ANN are also very good, presenting similar evaluation values. This reveals that the model in its general form is very well fitted in filling stations with different characteristics of the emission sources. Having in mind the simplicity of the model (one input parameter fewer) and the fact that for both the filling stations one model is required, for applications aiming to predict the exposure levels of the employees, this is the intended model.

4.4.6. Relative Importance of the Parameters Constituting the Exposure Pattern

The main reason to apply two sets of ANNs (for each filling station) comes from the exigency to quantify the relative importance of the parameters that determine the exposure levels. Network modelling can assess the importance of each of the input variables by using the network weights. The method for partitioning the connection weights proposed by Garson (1991) was used. The technique involves partitioning the hidden–output connection weights of each neuron into components associated with each input neuron (Goh, 1995). The results of the calculations are shown in Figures 25-26, displayed as columns representing the relative importance of the various input variables.

These results in combination with the active sampling ones clarify the relations of the emission sources (car refuelling and the corresponding amount of gasoline traded, direct traffic emissions and background concentration) with the meteorological parameters (wind speed and temperature).

In both filling stations, the amount of the fuel traded is the dominant parameter of exposure for all the categories of employees. The importance of that parameter is decreasing, in a way proportional to the distance of the fuel pumps an employee is occupying, thus this parameter is more important for employees refuelling cars, especially them working in the urban station, were the fuel amounts are higher.

Wind speed is also an affecting parameter for the employees working outdoor. This is explained by the fact that the employees working outdoor are directly exposed to wind which affect influential the ambient benzene concentrations. The importance of the wind is stronger in the rural filling station, because the dominant source emitting benzene (car refuelling) is not so strong like in the urban station and the ambient concentrations are more affected to dispersion parameters.

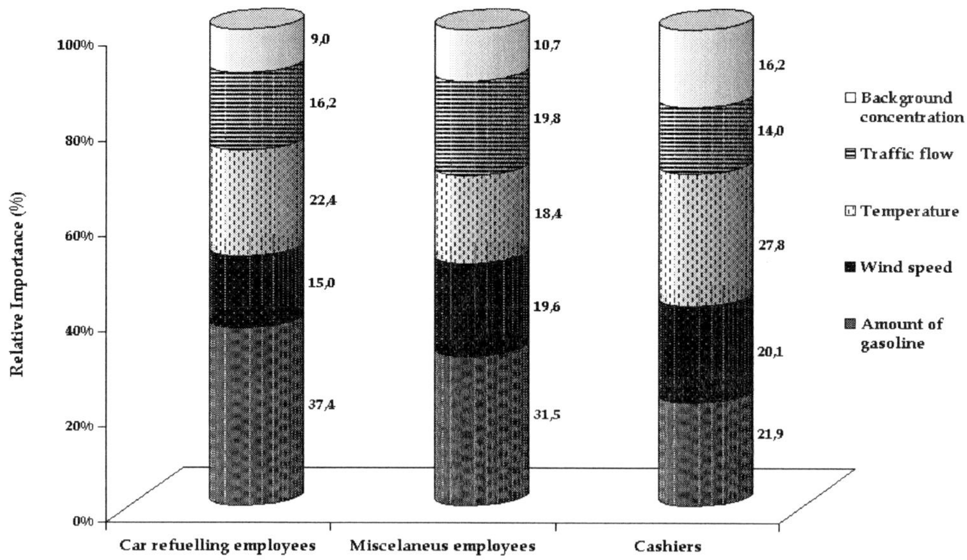

Figure 25. Relative importance of the input parameters (urban filling station).

Temperature is a determinant parameter, affecting fuel evaporation and consequently the presence of benzene in air. The effect of temperature in both filling stations is stronger than winds' effect, not only based on the daily variation, but also based in the average exposure values, that in summer are in general elevated. An exception of this acceptance occurs for the employees refuelling cars in the rural filling station.

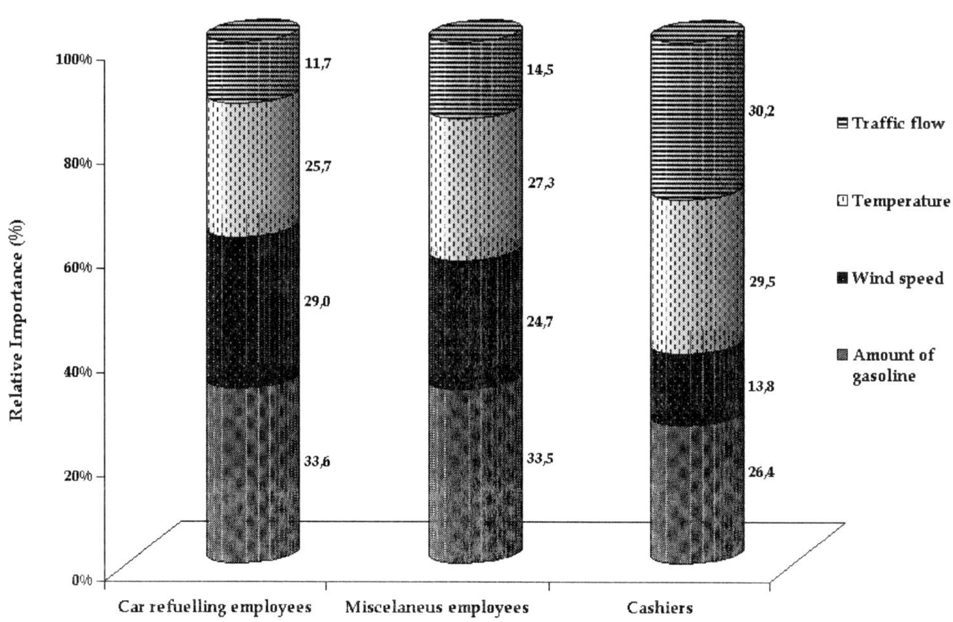

Figure 26. Relative importance of the input parameters (rural filling station).

Traffic flow importance of the road adjacent to the filling stations, has a smaller effect than the other parameters, because the dominant source of benzene in the filling stations has local origin. Some differences are observed related to the distance of the road that the employees are located. Because urban station is larger, employees occupying miscellaneous activities are located in greater distance from the adjacent road than these from the rural one. On the contrary, the emission rate from the urban station adjacent road is significantly higher, resulting in the increased importance compared to the rural one. In the rural filling station, this parameter seems to have a significant effect for the cashiers, which is practically impossible. In this parameter the network probably include all the unknown parameters that determine the exposure pattern of that category of employees.

Finally, urban background concentration appears to have a small effect in the configuration of the exposure values, affecting mainly the cashiers. Background concentration express the ambient benzene levels in the wider area, and this value present a slow response to the parameters affecting ambient benzene levels in a way similar to the exposure of cashiers to the exposure parameters. This parameter, as mentioned above, could be excluded from the network for reasons of simplicity, without decisive loss in the accuracy of the predictions.

4.5. Taxi Drivers

Taxi drivers were also significantly exposed to benzene (Figure 17), and the exposure levels ranged from 32 to 12 $\mu g/m^3$. The variance among the drivers was explained by the questionnaires. Drivers that spent most of the shift time circulating around the centre of the city presented to be more exposed in comparison to their colleagues that circulate away from the 'hot spots' or around the suburbs of the city.

The seasonal variation was almost undetectable compared to the other exposed subjects (slightly higher exposure values in summer). This is explained by the fact, that during the wintertime, the windows of the cars are all the time closed, and although the ambient air concentrations are lower, the continuous benzene accumulation in the cabinet leads to elevated in car concentrations. In summer also air conditioning is used, mostly working with recycling indoor air, a fact that prevents from excessive accumulation.

4.6. Active Sampling Personal Exposure

By the active sampling measurements, benzene concentrations in several microenvironments and under different conditions were determined. Four main activities were performed, which include driving, walking on urban streets, indoor activities (home, coffee-bars) and filling station employee's activities. The first three kinds of activities were chosen because they are common for all the population groups, while the fourth set of activities is related only to that category of employees and the detailed description of the parameters affecting the exposure pattern of that category of employees was presented above.

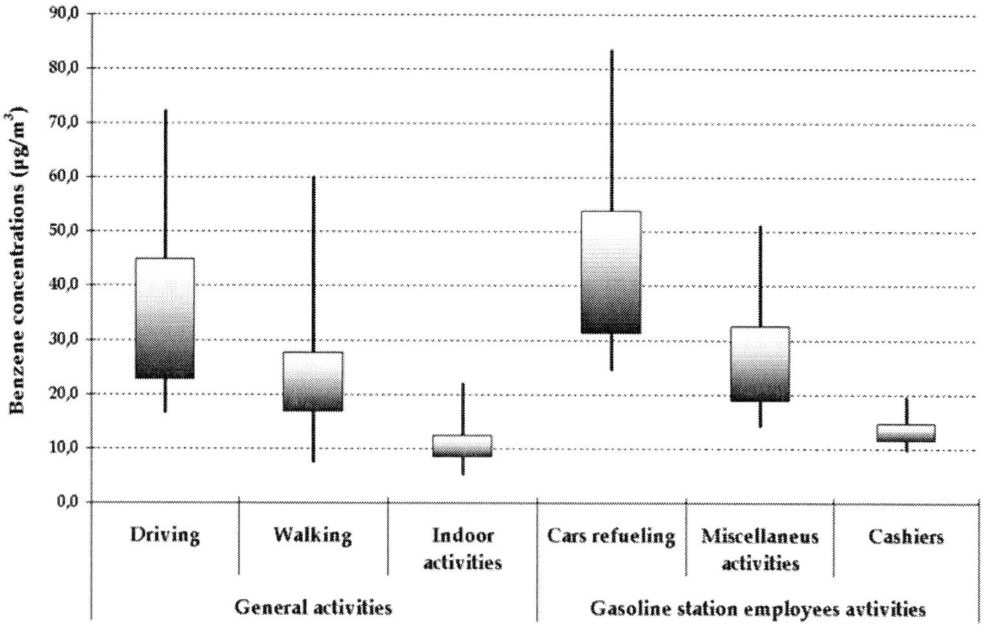

Figure 27. Active sampling results of personal exposure.

For all kind of activities 30 samples were taken (totally 120 per season). The measurements results are presented in Figure 27.

Driving was performed in a EURO 4 passenger car equipped with a catalytic converter (in order to eliminate the own car in cabinet emissions) under several driving conditions and in several roads. The highest values (up to 80 μg/m^3) were observed when the car was driven for a respectable amount of time circulating under congestion traffic conditions (especially in canyons) and there were calm wind conditions. Under these circumstances, ambient benzene concentrations are already increased and the pollutants are accumulated in the cabinet, especially when the windows are closed, resulting to the recorded high values. Driving to less polluted streets leads to significantly lower values (30-15 μg/m^3), especially when the travel time is sort and the pollutants accumulation is minimized (10 μg/m^3). The dependence on the wind speed is not so strong in comparison to the measured levels of ambient air benzene, because the pollutants inserting in the cabinet come mainly from direct (and consequently non dispersed) traffic emissions.

Almost the same things occur for walking, despite the fact that in this case no pollutants accumulation may occur, resulting in lower levels of exposure. The highest observed value (58 μg/m^3) was recorded when the exposed subject was walking in the city's most polluted street canyon under calm wind conditions and congested traffic (2:00-2:30 pm).

Walking in less pollutant streets, with smaller traffic volume and not in street canyons, results to lower recorded exposure values, especially when the wind is not in calm. Dependence on the wind is stronger than driving, and slightly lower to ambient concentrations.

Indoor exposure values were in general lower (25-4 μg/m^3), indicating that traffic is the main source of benzene in the ambient air of the city. The variability among the recorded values in the same houses was narrow, while also the response to outdoor concentrations is slow. In general, house proximity to busy road leaded to elevated benzene exposure values. More specific, in the street mentioned above, the measured values taken into an apartment of first floor ranged from 19 to 12 μg/m^3, having in mind that the outdoor average weekly concentration at the level of the street is 22 μg/m^3. Exposure values recorded in houses located away from the city centre are significantly lower (11-4 μg/m^3), and in all the occasions the dependence on the wind speed was small. Smoking is a well known significant indoor source of benzene, and the presence of smokers in the room, where the expose subject was staying during the sampling time, seemed to raise the measured values up to 5 μg/m^3, depending on the number of smokers and the smoking rate. A few samples were also taken in a coffee-bar, but the well function of the ventilation system did not allowed any abnormal value of benzene to appear and the measured values ranged from 12 to 6 μg/m^3, almost equal to the proximity street average concentration (10 μg/m^3).

5. Cancer Risk Assessment

Exposure studies are of major importance when they are linked to health effects implications (Nieuwenhuijsen et al, 2006). The estimated cancer risk ***R*** related to the exposure at a toxically substance is calculated using the equation (EPA, 1986):

$$R = C \bullet IUR \qquad (19)$$

where C is the concentration in μg/m^3 of the substance and IUR is the inhalation unit risk estimate (probability of cancer for a 70-year exposure to 1 μg/m^3). The WHO, in its update of the Air Quality Guidelines for Europe (WHO, 2000), used data from the updated Pliofilm cohort and models based on relative risk and cumulative exposure to calculate IUR for benzene in the range of $4.4 - 7.5 \cdot 10^6$ per μg/m^3 with a geometric mean of $6 \cdot 10^6$ per μg/m^3.

For the control group, risk is calculated easily by equation 4. For the working groups, risk calculation is slightly more complicated, because the parameter of occupational exposure C_O has to insert in the equation. Having in mind that the people work for 30 years, for 5 days a week and 8 hours a day, and for the rest of their lifespam they are exposed to the same concentrations C_T of the control group, the equation for an occupant is configured:

$$R = C_O \cdot \frac{30y \cdot 5d \cdot 8h}{70y \cdot 7d \cdot 24h} \cdot IUR + C_T \cdot \left(\frac{40y}{70y} + \frac{30y \cdot 2d \cdot 16h}{70y \cdot 7d \cdot 24h} \right) \cdot IUR \qquad (20)$$

The results from the above calculations are presented in Figure 28. From the results it is easily understood that total risk due to benzene exposure for the studied occupational groups is not directly proportional to the measured concentrations and this is due to the complexity of equation 20. Nevertheless, the increased risk of the above occupational groups is 16%, 22% and 10% for the policemen, the filling station employees and the taxi drivers respectively.

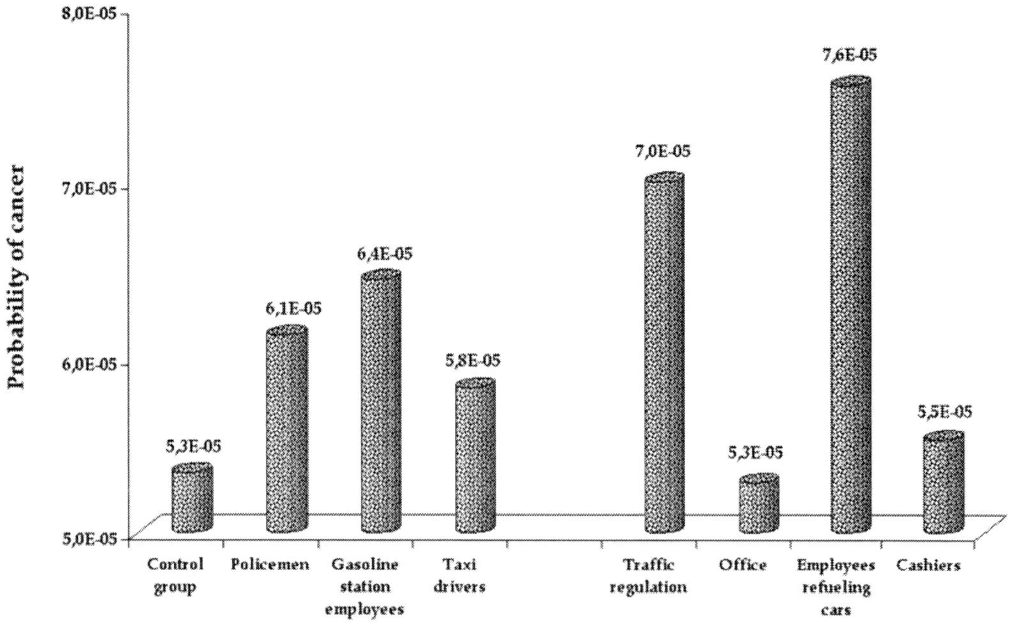

Figure 28. Risk assessment evaluation among the examined population groups.

In a more detailed analysis, some subcategories of each group of occupants appear to be exposed in higher risk due to the nature of the performed activities. Reliant to this acceptance, policemen executing traffic regulation and filling station employees refuelling cars are exposed to a significant higher risk equal to 32% and 42% respectively, compared to their colleagues working indoor (office policemen and cashiers) or to the general population. These results lead us to the conclusion that the interchange of roles among an occupational group is required, in order to avoid accumulated exposure of some subcategories of employees.

6. The Effect of Changes in Traffic Flow Patterns on Emissions by the Means of Environmental Policy

With the rapid increase of the motor vehicles' population in the recent years, the vehicle exhaust emissions have become one of the most important sources that cause the continued worsening of urban areas air quality. This occurs due to the very simple reason that the increase rate of vehicle fleet is significantly extended compared to development rate of the related traffic networks. Although, traffic pollution issues remain of major concern, the primary focus of studies in the field of transportation has been on the traffic congestion mitigation in stead of on the traffic emission reduction, usually without the expected results. Traffic control strategies are widely applied to mitigate traffic congestions, smooth traffic flows, and reduce travel times, which are rarely considered for being used to reduce emissions. Furthermore, traffic emissions are largely influenced by traffic conditions such as speed and acceleration of the motor vehicles. On the congested roads, the frequent start-and-stop, accelerate, and decelerate will result in higher emission concentrations than other roads. It has been widely known that an alternative traffic control and management strategy will likely change a vehicle's instantaneous modal activities, and thus influence a vehicle's instantaneous speed, which further result in different vehicle exhaust emissions. The latest studies have indicated that the traffic control strategies with an objective to mitigate traffic congestions and with an objective to reduce traffic emissions are usually different, and sometimes conflictive. Therefore, the effective evaluation of traffic control strategies should consider their impacts on both improving the traffic performance and reducing the traffic emissions (Kun and Lei, 2007). As a conclusion, vehicle technology improvement and fleet renewal are seen as the key strategy for emissions reduction.

In the urban area under study, a significant part (almost 25%) of passenger vehicle cars fleet is not equipped with a catalytic converter. Beside this, most of the passenger cars equipped with catalytic converters they are not last generation technology of Euro III and EURO IV emission standards. Furthermore in Greece, diesel passenger vehicles (except taxis) which have in general low benzene emissions are not allowed in Athens and Thessaloniki that are the largest cities, a fact that is discouraging for the buyers of new cars in the whole country to choose a diesel vehicle. The problem of the old technology vehicles concern all vehicle categories (also including buses, trucks, motorcycles), but the special need for changes in vehicle patterns mainly is needed for passenger vehicles (and secondary to motorcycles and mopeds) as they constitute the greatest amount of circulating vehicles.

In order to quantify the effect on traffic emissions by suggested changes in fleet composition, COPERT III methodology (Ntziachristos and Samaras, 2000) was enabled.

The suggested changes include five different "what if" scenarios, representing different speeds of technological renewal or changes in diesel passenger vehicles legislation. A summary of the different what if scenarios is presented in Table 10.

In the first scenario, all non catalytic passenger vehicles are replaced by catalytic passenger cars, in a proportion of subcategories (EURO emission legislation) similar to the existent of that vehicles category.

In the second scenario non catalytic passenger vehicles and first generation catalytic passenger vehicles (EURO I) are replaced by EURO IV catalytic passenger cars.

In the third scenario, half of the amount of non catalytic passenger vehicles are replaced by catalytic passenger cars and the rest are replaced by diesel passenger vehicles.

In the fourth scenario, non catalytic passenger vehicles and first generation catalytic passenger vehicles (EURO I) are replaced by gasoline (EURO IV) and diesel (EURO IV) new generation cars.

Finally, in the fifth scenario the amount of passenger vehicles includes only gasoline (EURO IV) and diesel (EURO IV), thus motorcycles and mopeds are replaced by newer generation, assorting to the earliest emissions legislations. The last three scenarios have an increased interest, because the retraction of the restriction mentioned above about the diesel passenger vehicles is under study by the apposite Greek authorities.

Table 10. Fleet composition under the 'what if' scenarios

	Percentage %	Catalytic (61.28	Non Catalytic	Diesel vehicles	Motorcycles (8.84 %)
Present fleet composition	Pre ECE till Conventional		100	10	100
	EuroI	59		24	
	EuroII	26		35	
	EuroIII	12		29	
	EuroIV	3		2	
	Percentage %	Catalytic (73.69	Non Catalytic (0	Diesel vehicles	Motorcycles (8.84 %)
1st Scenario	Pre ECE till Conventional			10	100
	EuroI	0		24	
	EuroII	21		35	
	EuroIII	10		29	
	EuroIV	69		2	
	Percentage %	Catalytic (73.69	Non Catalytic (0	Diesel vehicles	Motorcycles (8.84 %)
2nd Scenario	Pre ECE till Conventional			10	100
	EuroI	59		24	
	EuroII	26		35	
	EuroIII	12		29	
	EuroIV			2	
	Percentage %	Catalytic (67.48	Non Catalytic (0	Diesel vehicles	Motorcycles (8.84 %)
3rd Scenario	Pre ECE till Conventional			10	100
	EuroI	59		24	
	EuroII	26		35	
	EuroIII	12		29	
	EuroIV	3		2	
	Percentage %	Catalytic (43.20	Non Catalytic (0	Diesel vehicles	Motorcycles (8.84 %)
4th Scenario	Pre ECE till Conventional			6	100
	EuroI	0		13	
	EuroII	21		19	
	EuroIII	10		16	
	EuroIV	27		46	

Table 10. Continued.

	Percentage %	Catalytic (43.20	Non Catalytic (0 %)	Diesel vehicles (13.98 %)	Motorcycles (8.84 %)
5th Scenario	Pre ECE till Conventional				
	EuroI				100
	EuroII				
	EuroIII				
	EuroIV	100		100	

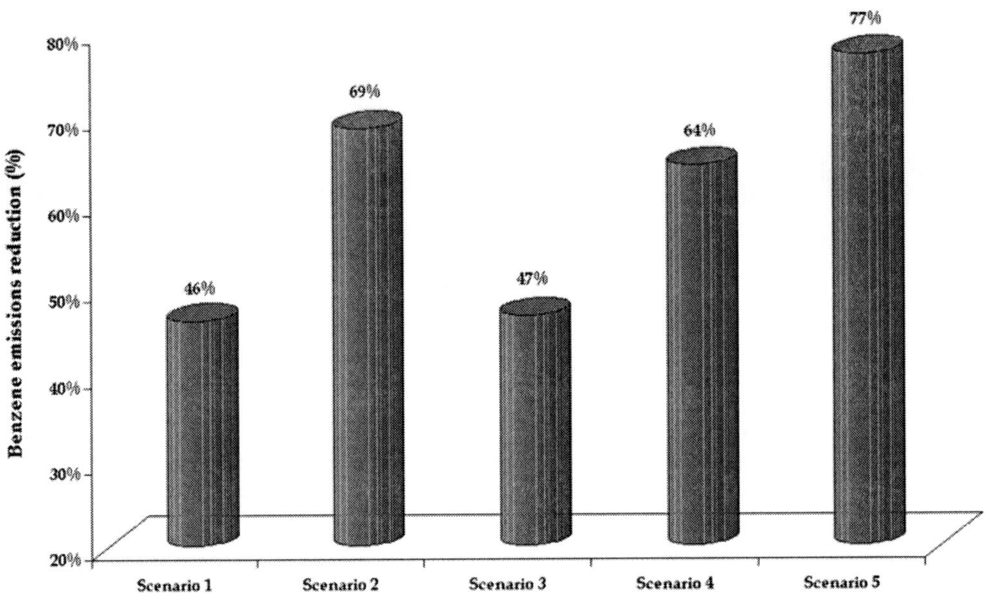

Figure 29. Percentage reduction of emissions according to the 'what if' scenarios.

All results are presented in Figure 29. It is remarkable that even in the first scenario which is the simplest (and consequently the one with the lowest socioeconomic cost) a significant reduction of the emissions (almost 50%) is noticed.

That means that is strongly recommended non catalytic passenger vehicles to be retired from the circulation and to be replaced with vehicles equipped with catalytic converters, even with used ones, in order to remain the cost of the replacement low and affordable for all the parts of the population.

The improvement in the vehicles emissions technology is noticed in the second scenario, where the replacement of the non catalytic and first generation catalytic vehicles (EURO I) by vehicles minded to the emissions directive EURO IV, leads to a 69% reduction and this is due to the fact that the first generation catalytic vehicles have a significantly higher benzene emission rate. This is further extended considering the fade of efficiency of the emissions control compartments of the engine through mileage of old vehicles.

In the third scenario, the importance of the diesel vehicles to the reduction of benzene emissions is found. Although non catalytic vehicles are not replaced by completely newer emissions technology vehicles, the reduction is equal to first scenario. Depending on the results of this scenario and based on the importance of benzene on public health, the legislation related to diesel vehicles in Greece should be reconsidered.

In the fourth scenario the results are similar to the second one, verifying for once more the improvement of newer technology cars using gasoline, as far as them using diesel.

As it was expected the greatest benzene emission reduction (77%) is noticed in the fifth scenario, where a hypothetical renewal of the traffic fleet is done. Although this scenario is with the highest cost, still remains a clarifying guidepost for the effective directives in order to setup a sustainable environment by the perspective of urban air quality.

From all the above, the importance of older vehicles renewal, is more than clear. If such a renewal occurs, then the target of the 2000/69/EC directive can easily be achieved. Furthermore, the presence of diesel vehicles can lead to a further reduction of ambient air benzene, which is a priority pollutant. A good combination of cost – effectiveness measure in order to reduce benzene emissions is the provision of motivations to be replaced all the non catalytic passenger vehicles with used catalytic cars.

7. Conclusions

In this work the results of an integrated study for ambient benzene concentrations in a medium size south European city were presented and discussed.

The detailed campaign was consisted of field measurements (in various urban environments), modeling efforts (classical statistical, neural networks, models to compute emissions of pollutants in the roads, Gaussian models to estimate the dispersion of air pollutants etc), human exposure measurements, test of various "what if scenarios" to examine the composition of air pollutants emitted under various existing/future political regulations in terms of environmental management.

The study presented in detailed above, has concluded to the following results:

The measuring campaign found that in the congested traffic roads significant hot spot of ambient benzene concentrations were conducted. Specifically during the morning rush hour (e.g. 08.00-10.00 LT) the mean benzene ambient concentrations were 2 times higher than the threshold imposed by the European Union (~11 $\mu g/m^3$), but in less congested roads or under urban background conditions the measured concentrations were significantly lower not exceeding 3 $\mu g/m^3$ for urban background conditions and 6 $\mu g/m^3$ for less congested roads. In total ambient benzene concentrations were recorded in 9 sites in the city.

It is strange, comparing to similar campaigns, that the highest values were found to be during summer months, instead of winter, but this finding was due to the fact that the climatic conditions in the area during winter months (November-April) does not allow the use of motorcycles, which are permitted and preferred during summer months. The extended measurements of the distribution of vehicles' fleet found that 17% of the vehicles are motorcycles in summer instead of 8% in winter.

The composition of the vehicles fleet in the area shows that the non-catalytic vehicles are still 12-14% and higher from the national mean value. A remarkable percentage of diesel vehicles were also detected (9%) while in the center of the city the more heavy vehicles were

8% (light and heavy duty vehicles and buses). Finally the catalytic cars were about 60% (most of them EURO I and EURO II).

A detailed trial was then made to find the dispersion of benzene in the air and specifically in the city's street canyons. For this purpose we firstly computed the traffic density through a traffic model (STREET), secondly we estimated the emissions in the area and finally we computed the dispersion of benzene over the canyon. The later was made with the aid of both a dispersion model and an ANN model. The evaluation of the results of the two models, as well as, the comparison of the outcome of the models with the real measurements revealed that ANN model succeeded to simulated better the measurements comparing to the traditional statistical model.

Extended measurements through passive/active sampling were also made near gas stations located inside and outside the city (in the periphery). The results showed significant levels that may affect the health of the people living around the filling stations. To understand the ambient benzene concentrations around filling stations and to isolate the contribution of them to the overall benzene levels in their neighborhood we used the CALINE 4 Gaussian model for roads. Significant levels of ambient benzene concentrations were revealed from the CALINE 4 model especially for the urban filling station.

In order to evaluate the personal exposure to ambient benzene levels in the city we organized a control group and some occupation oriented groups of population as policemen, taxi drivers and filling station employees. The results showed that for filling station employees the detected values were ranged from 51-18.3 $\mu g/m^3$ in summer and 41-17.5 $\mu g/m^3$ in winter when the urban filling station was inside the city and for 29.5-10.3 $\mu g/m^3$ for summer and 26.8-9.1 $\mu g/m^3$ for winter when it is located in the periphery of the city. The relevant exposure values for taxi drivers were 32-12 $\mu g/m^3$ while for policemen were 40-9.1 $\mu g/m^3$. The people living near the urban filling station were generally exposed to high benzene concentrations.

The cancer risk for these exposures was higher for employees refueling cars ($7.4*10^{-5}$) and the lower for the control group ($5.3*10^{-5}$), activities increasing the cancer risk are also the traffic regulation and taxi driving. It must be noted that these numbers shows probability of cancer for 70-y exposure to 1 $\mu g/m^3$ of benzene concentration.

The effect of changes in traffic flow patterns on emissions by means of environmental policy was finally studied. Five different scenarios were examined for this purpose. All the examined scenarios showed a reduction on benzene emission, ranging from 46 to 77%, strongly depended on the extent of newer technologies applied and the grade of diesel vehicles inserted in the fleet.

Although that the measurements of ambient benzene concentrations are lower from the EU imposed threshold (5 $\mu g/m^3$ mean daily value in urban background conditions), the hot spots found and the rapid increase of the city will create in the near future conditions of possible exceedances if abatement measurements are not imposed.

8. References

ACGIH, 2001. Documentation of the Threshold Limit Values and Chemical Substances, 7th ed., *American Conference of Governmental Industrial Hygienists,* Cincinnati OH, USA.

Aleksic, N., Boynton, G., Sistla, G., Perry, J., 2005. Concentrations and trends of benzene in ambient air over New York State during 1990–2003. *Atm. Env.* 39 7894–7905.

ATSDR. 1997. *Toxicological profile for benzene.* Atlanta, GA: U.S. Department of Health and Human Services. Public Health Service, Agency for Toxic Substances and Disease Registry.

Balaguer Ballester, E., Camps i Valls, G., Carrasco-Rodriguez, J.L., Soria Olivas, E., Valle-Tascon, A. (2002). Effective 1-day ahead prediction of hourly surface ozone concentrations in eastern Spain using linear models and neural networks. *Ecological Modelling,* 156, 27-41.

Ballesta, P.P., Field, A.R., Connolly, R., Cao, N., Caracena, B.A., De Saeger, E. (2006). Population exposure to benzene: One day cross-sections in six European cities. *Atmospheric Environment,* 40, 3355–3366.

Battiti, R. (1992). First and second order methods for learning: Between steepest descent and Newton's method. *Neural Computation,* 4 (2), 141-166.

Benson, P.E. (1979). CALINE 3 — a versatile dispersion model for predicting air pollutant levels near highways and arterial streets. *FHWA/CA/TL-79/23.* State of California Department of Transportation Division of New Technology and Research, Sacramento, CA.

Benson, P.E. (1984). CALINE 4 - *A dispersion model for predicting air pollutant concentrations near roadways.* Office of Transportation Laboratory, California Department of Transportation, Sacramento, USA.

Benson, P.E. (1992). A review of the development and application of the CALINE 3 and 4 models. *Atmospheric Environment,* 26B (3), 379–390.

Berkowicz, R., Hertel, O., Larsen, S.E., Sørensen, N.N., Nielsen, M. (1997). *Modelling traffic pollution in streets.* Ministry of Environment and Energy, National Environmental Research Institute.

Bono, R.; Scursatone, E.; Schiliro, T.; Gilli, G. (2003). Ambient air levels and occupational exposure to benzene, toluene and xylenes in Northwestern Italy. *J. Toxicol. Environ. Health* 66:519–531.

Brocco, D., Fratarcangeli, R., Lepore, L., Petricca, M., Ventrone, I. (1997). Determination of aromatic hydrocarbons in urban air of Rome. *Atmospheric Environment,* 31, 557-566.

Brugnone, F., Perbellini, L., Romeo, L., Cerpelloni, M., Bianchin, M., Tonello, A. (1999). Benzene in blood as a biomarker of low level occupational exposure. *Science of the Total Environment,* 235, 247–252.

Capleton, A.C., Levy, L.S. (2005). An overview of occupational benzene exposures and occupational exposure limits in Europe and North America. *Chemico-Biological Interactions,* 153–154, 43–53.

Carr, D., Ehrenstein, O., Stephan, W., Claudia, W., Wellie, O., Thomas, N.A., Mutius, E. (2002). Modeling Annual Benzene, Toluene, NO2, and Soot Concentrations on the Basis of Road Traffic Characteristics. *Environmental Research Section* A, 90, 111-118.

Carrieri, M., Bonfiglio, E., Scapellato, M.L., Macca, I., Tranfo, G., Faranda, P., Paci, E., Bartolucci, G.B. (2006). Comparison of exposure assessment methods in occupational exposure to benzene in gasoline filling-station attendants. *Toxicology Letters,* 162, 146–152.

Chatzis, C., Alexopoulos, E., Linos, A. (2005). Indoor and outdoor personal exposure to benzene in Athens, Greece. *Science of the Total Environment,* 349, 72–80.

Cocheo, V.; Sacco, P.; Boaaretto, C.; De Saeger, E.; Perez Ballesta, P.; Skov, H.; Goelen, E.; Gonzalez, N.; Baeza Caracena, A. (2000). Urban Benzene and Population Exposure. *Nature,* 404, 141-142.

Colvile, R.N., Hutchinson, E.J., Mindell, J.S., Warren, R.F. (2001). The transport sector as a source of air pollution. *Atmospheric Environment,* 35, 1537–1565.

CONCAWE, *A Review of European Gasoline Exposure Data for the Period 1993–1998,* CONCAWE, Brussels, Belgium, 2000.

CONCAWE, *A Survey of European Gasoline Exposures for the Period 1999–2001,* CONCAWE, Brussels, Belgium, 2002.

Corani, G. (2005). Air quality prediction in Milan: feed-forward neural networks, pruned neural networks and lazy learning. *Ecological Modelling,* 185, 513-529.

Cruz-Nunez, X., Hernandez-Solis, J.M., Ruiz-Suareza, L.G. (2003). Evaluation of vapour recovery systems efficiency and personal exposure in service stations in Mexico City. *Science of the Total Environment,* 309, 59–68.

Dennis, J.E. and Schnabel, R.B. (1983). *Numerical Methods for Unconstrained Optimization and Nonlinear Equations.* Englewood Cliffs, NJ: Prentice-Hall.

Dorling, S.R., Foxall, R.J., Mandi, D.P., Cawley, G.C. (2003). Maximum likelihood cost functions for neural network models of air quality data. *Atmospheric Environment,* 37, 3435–3443.

DePaul, F.T., Sheih, C.M. (1985). A tracer study of dispersion in an urban street canyon. *Atmospheric Environment,* 19, 555–559.

Dirks, K.N., Johns, M.D., Hay, J.E., Sturman, P.A. (2003). A semi-empirical model for predicting the effect of changes in traffic flow patterns on carbon monoxide concentrations, *Atmospheric Environment,* 37, 2719–2724.

Duarte-Davidson, R., Courage, C., Rushton, L., Levy, L. (2001). Benzene in the environment: an assessment of the potential risks to the health of the population. *Occup. Environ. Med,* 58, 2–13.

Foresee, F.D., Hagan, M.T. (1997). Gauss-Newton approximation to Bayesian regularization. *International Joint Conference on Neural Networks,* 1930-1935.

Franco, G., Bisio, S. (1995). Valutazione e previsione della esposizione occupazionale a benzene. In: Minoia, C., Apostoli, P., Bartolucci, G.B. (Eds.), *Il Benzene: Tossicologia, Ambienti di Vita e di Lavoro.* Morgan, Milano, pp. 89–97.

E.C. (2000). Council Directive 2000/69/EC relating to limit values for benzene and carbon monoxide in ambient air. Off. *J. Eur. Communities.* (OJ L. 163, 29.06.1999), pp. 0041–0021.

Edwards, R.D. Jantunen, M.J. (2001). Benzene Exposure in Helsinki, Finland. *Atmospheric Environment,* 35, 1411–1420.

Edwards, R.D., Jurvelin, J., Saarela, K., Jantunen, M.J. (2001a). VOC concentrations measured in personal samples and residential indoor, outdoor and workplace microenvironments in EXPOLIS-Helsinki, Finland. *Atmospheric Environment,* 35, 4531–4543.

Edwards, R.D. Jurvelin, J. Koistinen, K. Saarela, K. Jantunen, M.J. (2001b). VOC source identification from personal and residential indoor, outdoor and workplace microenvironment samples in EXPOLIS-Helsinki, Finland. *Atmospheric Environment,* 35, 4829–4841.

Environmental Protection Agency (EPA), 1997. National Ambient Air Quality Standards for Particulate Matter, *Federal Register,* 62, no. 138.
European Environment Agency (EPA), 2001. *Air Quality in Larger Cities in the European Union. Copenhagen 7 EEA;* 2001 Topic report 3.
Ferrari, C.P., Kaluzny, P., Roche, A. (1998). Aromatic hydrocarbons and aldehydes in the atmosphere of Grenoble, France. *Chemosphere*, 37, 1587–1601.
Fenger, J. (1999). Urban air quality. *Atmospheric Environment,* 33, 4877–4900.
Foresee, F.D., Hagan, M.T. (1997). Gauss–Newton approximation to Bayesian learning. In: *Proceedings of the 1997 International Joint Conference on Neural Networks* 1930–1935.
Gardner, M.W., Dorling, S.R. (1998). Artificial neural networks (the multi-layer perceptron)- a review of applications in the atmospheric sciences. *Atmospheric Environment,* 32, 2627-2636.
Gardner, M.W., Dorling, S.R. (1999). Neural network modelling and prediction of hourly NO_X and NO_2 concentrations in urban air in London. *Atmospheric Environment,* 33, 709-719.
Garson, G.D. (1991). Interpreting neural-network connection weights. *AI Expert,* 6, 47–51.
Giacopetti, D. (2000). Il rischio sanitario dell'addetto all'erogazione di carburante: la posizione dell'unione petrolifera. *G. Ital. Med. Lav.,* 22 (3), 262–264.
Goh, A.T.C. (1995). Back-propagation neural networks for modelling complex systems. *Artificial Intelligence in Engineering,* 9, 143–151.
Gonzalez-Flesca, N., Bates, M.S., Delmas, V., Cocheo, V. (2000). Benzene exposure assessment at indoor outdoor and personal levels. The French contribution to the Life Macbeth programme. *Environmental Monitoring and Assessment,* 65, 59–67.
Gonzalez-Flesca, N., Cicolella, A., Bates, M., Bastin, E. (2007). Pilot Study of Personal, Indoor and Outdoor Exposure to Benzene, Formaldehyde and acetaldehyde. Env. Sci. Poll. Res., 6 (2), 95-102.
Greenshields, B.D. (1943). *The photographic method of studying traffic behaviour.* Proceedings of the 13th Meeting of Highway Research Board.
Guerra, A.I., Lerda, D., Martines, C. (1995). Benzene emissions from motor vehicle traffic in the urban area of Milan: hypothesis of health impact assessment. *Atmospheric Environment,* 29, 3559-3569.
Hagan, M.T., Menhaj, M. (1994). Training feedforward networks with the Marquardt algorithm. *IEEE Trans Neural Networks,* 5, 989–993.
Hagerman, L.M., Aneja, V.P., Lonneman, W.A. (1997). Characterization of non-methane hydrocarbons in the rural Southeast United States. *Atmospheric Environment,* 31, 4017–4038.
Held, T., Chang, D.P.Y., Niemeier, D.A. (2003). UCD 2001: an improved model to simulate pollutant dispersion from roadways. *Atmospheric Environment,* 37, 5325–5336.
Hertel, O., de Leeuw, F.A., Raaschou-Nielsen, O., Jensen, S.S., Gee, D., Herbarth, O., Pryor, S., Palmgren, F., Olsen, E. (2001). Human exposure to outdoor air pollution. *Pur. and App. Chem.,* 73, 933–958.
Hinwood, A.L., Berko, H.N., Farrar, D., Galbally, I.E., Weeks, I.A. (2006). Volatile organic compounds in selected micro-environments. *Chemosphere,* 63, 421–429.
Hotz, P., Carbonnelle, P., Haufroid, V., Tschopp, A., Buchet, J.P., Lauwerys, R. (1997). Biological monitoring of vehicle mechanics and other employees exposed to low concentrations of benzene. *Occup. Environ. Health,* 70, 29–40.

Hunter, L.J.; Johnson, G.T.; Watson, I.D. (1992). An investigation of three-dimensional characteristics of flow regimes within the urban canyon. *Atmospheric Environment,* 26, 425–432.

IARC, 1987. IARC monographs on the evaluation of carcinogenic risks to humans. Overall evaluations of carcinogenicity: An updating of IARC Monographs volumes 1 to 42. Suppl 7. Lyons, France: World Health Organization. *International Agency for Research on Cancer,* 38-74, 1987.

IARC, 2004. *IARC Monographs on the evaluation of carcinogenic risk to humans.* International Agency for Research on Cancer; 2004.

Infante, P.F. (2001). Benzene: an historical perspective on the American and European occupational setting, in: *Late Lessons from Early Warnings: The Precautionary Principle 1896-2000 (European Environment Agency; Environmental Issue Report No.22),* Office for Official Publications of the European Communities, Luxembourg, 38–51.

Jo, W.K, Park, K.H. (1998). Concentrations of volatile organic compounds in automobiles' cabins while commuting along a Korean urban area. *Environment International,* 3, 259-265.

Jo, W.K., Song, K.B. (2001). Exposure to volatile organic compounds for individuals with occupations associated with potential exposure to motor vehicle exhaust and gasoline vapour emissions. *Science of the Total Environment,* 269(1–3), 25 –37.

Johnson, W.B., Ludwig, F.L., Dabberdt, W.F., Allen, R.J. (1973). An urban diffusion simulation model for carbon monoxide. *Journal of the Air Pollution Control Association,* 23, 490–498.

Karakitsios, S., Papaloukas, C., Kassomenos, P., Pilidis, G. (2006). Assessment and prediction of benzene concentrations in a street canyon using artificial neural networks and deterministic models. Their response to "what if" scenarios. *Ecological Modelling,* 193, 253–270.

Karakitsios, S., Delis, V., Kassomenos, P., Pilidis, G. (2007). Contribution to ambient benzene concentrations in the vicinity of petrol stations: Estimation of the associated health risk. *Atmospheric Environment,* 41, 1889–1902.

Karakitsios, S. Papaloukas, C. Kassomenos, P. Pilidis, G. (2007). Assessment and prediction of exposure to benzene at filling station employees. *Atmospheric Environment,* 41, 9555-9569.

Kassomenos, P.A., Karakitsios, S.P., Pilidis G.A. (2004). A simple semi-empirical approach to modeling benzene concentrations in a street canyon. *Atmospheric Environment,* 38, 6073-6078.

Kotzias, D. (2005). Indoor air and human exposure assessment – needs and approaches. *Exper. and Toxic. Path.*, 57, 5–7.

Kourtidis, K., Ziomas, I., Rappenglueck, B., Proyou, A., Balis, D. (1999). Evaporative traffic hydrocarbon emissions, traffic CO speciates HC traffic emissions from the city of Athens. *Atmospheric environment,* 33, 3831-3842.

Kourtidis, K.; Ziomas, I.; Zerefos, C.; Kosmidis, E.; Symeonidis, P.; Christophilopoulos, E.; Karathanassis, S.; Mploutsos, A. Benzene, toluene, ozone, NO_2 and SO_2 measurements in an urban street canyon in Thessaloniki, Greece. *Atmospheric Environment,* 36, 5355–5364.

Kun, C. and Lei, Y., 2007. Microscopic Traffic-Emission Simulation and Case Study for Evaluation of Traffic Control Strategies. *J Transpn Sys Eng & IT,* 7(1), 93−100.

Lek, S., Guegan, J.F. (1999). Artificial neural networks as a tool in ecological modelling, an introduction. *Ecological Modelling,* 120, 65–73.

Leung, P., Harrison, R.M. (1998). Evaluation of personal exposure to monoaromatic hydrocarbons. *Occup. Environ. Med.* 55, 249– 57.

Levitin, J., Harkonen, J., Kukkonen, J., Nikmo, J. (2005). Evaluation of the CALINE 4 and CAR-FMI models against measurements near a major road. *Atmospheric Environment,* 39, 4439–4452.

Lippmann, R.P. (1987). An introduction to computing with neural nets. *IEEE Acoust. Speech Signal Process,* 4 (2) 4-22.

MacKay, D.J.C. (1992). Bayesian interpolation. *Neural Computation,* 4, 415–447.

Mantis, J., Chaloulakou, A., Samara, C. (2005). PM_{10}-bound polycyclic aromatic hydrocarbons (PAHs) in the greater area of Athens, Greece. *Chemosphere,* 59, 593-604.

Maffei, F., Hrelia, P., Angelini, S., Carbone, F., Forti, G.V., Barbieri, A., Sanguinetti, G., Mattioli, S., Violante, F.C. (2005). Effects of environmental benzene: Micronucleus frequencies and haematological values in traffic police working in an urban area. *Mut. Res.,* 583, 1–11.

Maitre, A., Soulat, J.M., Masclet, P., Stoklov, M., Marques, M., de Gaudemaris R. (2002). Exposure to carcinogenic air pollutants among policemen working close to traffic in an urban area. *Scand. J. Work. Environ.,* 28, 402– 410.

Manini, P., De Palma, G., Andreoli, R., Poli, D., Mozzoni, P., Folesani, G., Mutti, A., Apostoli, P. (2006). Environmental and biological monitoring of benzene exposure in a cohort of Italian taxi drivers. *Toxicology Letters,* 167, 141-152.

McMichael, A.J., Spirtas, R., Kupper, L.L., Gamble, J.F. (1975). Solvent exposure and leukemia among rubber workers: an epidemiologic study. *J. Occup. Med.,* 17, 234–239.

Medinsky, M.A., Kenyon, E.M., Schlosser, P.M. (1995). Benzene: a case study in parent chemical and metabolite interactions. *Toxicology,* 105 (2–3), 225–233.

Moller, M.F. (1993). A scaled conjugate gradient algorithm for fast supervised learning. *Neural Networks,* 6, 525-533.

Moschonas, N., Glavas, S. (1996). C_3–C_{10} hydrocarbons in the atmosphere of Athens, Greece. *Atmospheric Environment*, 30, 2769-2772.

Na, K., Kim, Y.P., Moon, K.C., Moon, I., Fung, K. (2001). Concentrations of volatile organic compounds in an industrial area of Korea. *Atmospheric Environment,* 35, 2747–2756.

Nagendra, S.M.S., Mukesh Khare, M. (2004). Artificial neural network based line source models for vehicular exhaust emission predictions of an urban roadway. *Transportation Research Part D,* 9, 199–208.

Nicholson, S.E. (1975). A pollution model for street-level air. *Atmospheric Environment,* 9, 19–31.

Nguyen, D., Widrow, B., Improving the learning speed of 2-layer neural networks by choosing initial values of the adaptive weights. In: *Proceedings of the International Joint Conference on Neural Networks* 3 21–26.

Nunnari, G., Nucifora, A.F.M., Randieri, C. (1998). The application of neural techniques to the modelling of time-series of atmospheric pollution data. *Ecological Modelling* 111 (2–3), 187–205.

Ntzaichristos, L. Samaras, Z. COPERT 3 – *Computer Programme to calculate Emissions from Road Transport, Methodology and emission factor (version 2.1).* European Environment Agency, Copenhagen; 2000.

O'Donoghue, R.T., Gill L.W., McKevitt R.J., Broderick B. (2007). Exposure to hydrocarbon concentrations while commuting or exercising in Dublin. *Environment International,* 33, 1–8.

OSHA, 1977. *Occupational exposure to benzene: emergency temporary standards, hearing May 3, Part IV, Fed. Reg.* 42 22516–22529.

Papageorgopoulou, A., Manoli, E., Touloumi, E., Samara, C. (1999). Polycyclic aromatic hydrocarbons in the ambient air of Greek towns in relation to other atmospheric pollutants. *Chemosphere,* 39, 2183-2199.

Periago, J.F., Zambudio, A., Prado, C. (1997). Evaluation of environmental levels of aromatic hydrocarbons in gasoline service stations by gas chromatography. *J. of Chrom.* A, 778, 263-268.

Petrakis, M., Psiloglou, B., Kassomenos, P., Cartalis, C. (2003). Summertime measurements of benzene and toluene in Athens using a differential optical absorption spectroscopy system. *J. of Air and Waste Manag. Assoc.,* 53, 1052– 1064.

Pilidis, G., Karakitsios, S., Kassomenos, P. (2005). BTX measurments in a medium sized European city. *Atmospheric Environment,* 39, 6051-6065.

Reich, S.L., Gomez, D.R., Dawidowski, L.E. (1999). Artificial neural network for the identification of unknown air pollution sources. *Atmospheric Environment,* 33, 3045-3052.

Riain, C.M.N., Fisher, B., Martin, C.J., Littler, J. (1998). Flow field and pollution dispersion in a central London street. *Environmental Monitoring and Assessment,* 52, 299–314.

Riedmiller, M. and Braun, H. (1993). A direct adaptive method for faster backpropagation learning: The RPROP algorithm. Proc. IEEE Intern. Conf. on Neural Networks.

Robbins, H., Monro, S. (1951). A stochastic approximation method. *Annals of Mathematical Statistics,* 22, 400–407.

Savitz, D.A., Andrews, K.W. (1997). Review of epidemiogical evidence on benzene and lymphatic and haematopoietic cancer. *Am. J. Ind. Med.,* 31, 287-295.

Schnatter, R. (2000). Petroleum worker studies and benzene risk assessment. *J. Toxicol. Environ. Health Part A,* 61, 433–437.

Skov, H., Hansen, A.B., Lorenzen, G., Andersen, H.V., Lofstrom, P., Christensen, C.S. (2001). Benzene exposure and the effect of traffic pollution in Copenhagen, Denmark. *Atmospheric Environment.* 35, 2463–2471.

Son, B., Breysse, P., Yang, W. (2003). Volatile organic compounds concentrations in residential indoor and outdoor and its personal exposure in Korea. *Environment International,* 29, 79– 85.

Qin, Y., Kot, S.C. (1993). Dispersion of vehicular emission in street canyons, Guangzhou city, South China (P.R.C.). *Atmospheric Environment,* 27, 283–291.

TOXNET, 2000. *Hazardous Substances Data Bank. National Library of Medicine*, Internet Edition, http://toxnet.nlm.nih.gov.

U.S. Environmental Protection Agency (USEPA), 1986. Guidelines for carcinogen risk assessment. *Fed. Reg. (September 24)* 51, 33992-35003; 1986.

U.S. Environmental Protection Agency (USEPA), 2000. 'Fuel controls': draft technical support document: control of emissions of hazardous air pollutants from motor vehicles and motor vehicle fuels, Washington DC: Office of Air and Radiation, *EPA* 420-D-00-003, 2000.

Vardoulakis, S., Gonzalez-Flesca, N., Fisher, B.E.A. (2002). Assessment of traffic-related air pollution in two street canyons in Paris: implications for exposure studies. *Atmospheric Environment,* 36, 1025–1039.

Vardoulakis S., Fisher, E.A.B., Pericleous, K., Gonzalez, N.F. (2003). Modelling air quality in street canyons: a review. *Atmospheric Environment,* 37, 155–182.

Verma, D.K., Tombe, K. (2002). Benzene in gasoline and crude oil: occupational and environmental implications. *Am. Ind. Hyg. Assoc. J.,* 63, 225–230.

World Health Organization (WHO), 1987. *Air Quality Guidelines for Europe, European Series 23*. World Health Authority Regional Publications, Copenhagen, Denmark.

World Health Organization (WHO), 2000. *Air Quality Guidelines for Europe, 2*nd ed., WHO Regional Office for Europe, Copenhagen, WHO Regional Publication, European Series no. 91; 2000.

Xie, S., Zhang, Y., Qi, L., Tang, X. (2003). Spatial distribution of traffic-related pollutant concentrations in street canyons. *Atmospheric Environment,* 37, 3213–3224.

In: Air Quality in the 21st Century
Editors: G. C. Romano and A. G. Conti, pp. 67-114
ISBN 978-1-60456-793-9

Chapter 2

METEOROLOGICAL ASPECTS OF AIR QUALITY

G. J. Steeneveld and A. A. M. Holtslag
Wageningen University, Meteorology and Air Quality Group, P.O. Box 47,
6700 AA Wageningen, The Netherlands

ABSTRACT

This paper provides a survey on the role of the atmospheric boundary layer and its role on air quality. Based on the evaluation of a weather forecast model it is found that the nighttime or stable boundary layer is difficult to forecast. We briefly summarize the relevant processes in this boundary layer prototype, and we discuss their interactions. Next, a climatology of flow meandering and intermittency of the atmospheric turbulence, and its spatial extent under these conditions, is presented based on field observations. Finally, a new formula for the stable boundary layer height is derived.

1. INTRODUCTION

The atmosphere in which we live, consists of a mixture of natural gases (nitrogen, oxygen, water vapor, carbon dioxide and some other spore gases). Human activity leads to several emissions of non-natural pollutants, e.g. by road and air traffic (Charron and Harrison, 2005), industry, agriculture and power plants. Well-known anthropogenic emissions are sulfur dioxide, carbon dioxide, nitrogen oxide, ammonia, methane, volatile organic compounds, and particulate matter. After a primary gas release, several chemical reactions may result in (harmful) reaction products (i.e. secondary pollutants, e.g. Reitebuch et al, 2000), e.g. ozone. Accidents with industrial processes or in traffic might result in a sudden strong emission of pollutants. For example, in the Bhopal (India) disaster a pesticide plant released ~40 tons of methyl isocyanate gas (MIC), killing approximately 3,800 people instantly, and 8,000 others died within two weeks. The incident took place in the *early hours* of the morning of 3 December 1984 in the heart of the city of Bhopal. Studies showed that the local meteorology (together with the local topography) played an important role in the dispersion of the MIC over the city of Bhopal (Boybeyi et al, 1995; Sharan et al., 1996). First, the MIC release occurred under calm, clear sky conditions, in a shallow layer close to the ground. Second, a circulation induced by the urban heat island effect between the city and the surroundings, resulted in a flow from the pesticide plant towards the city. As such, it is clear that the local weather can impact on the local air quality.

In addition to anthropogenic sources, a variety of natural sources as forests and wetlands can release several gases and constituents, especially volatile hydrocarbons, ozone and NO_x, but pollen and spores as well. Finally, dust emissions due to wild fires, dust storms and volcanic eruptions can have substantial impact on the regional and global air quality and climate (e.g. Mt Pinatubo, 1992).

Some of these primary and secondary emissions can be harmful for human health, natural vegetation and agricultural crops. As an example: long-term exposure of sensitive nature to ammonia and other nutrient rich pollutants has lead to a substantial reduction of the heather vegetation in the Netherlands (Roelofs, 1994; Van Eerden et al., 1998). Also, short term exposure to smog has shown to increase mortality rates in sensitive groups (i.e. the elderly, young children and people with cardiovascular diseases). Stedman (2006) found that for the heat wave in Europe in the summer of 2006, it was due to the poor air quality in the London area that ~21-38 % of the victims were probably not caused by heat stress, but by smog and particulate matter.

Regarding the role of meteorology on air quality, the air mass and the large-scale meteorology determine background concentrations. In general, air masses from an Arctic or oceanic origin have low concentrations, while air masses with a continental origin are more polluted. Persistent high pressure systems over land collect near surface emissions for a long time period since efficient removal processes (washout and rainout) are absent, and because the air is slowly subsiding. Subsidence causes the air to stabilize and 'lock' the near polluted surface air under a so called inversion. On the contrary, in low pressure systems washout and rainout are efficiently active, and the mean upward motions help to remove near surface emissions into the free troposphere aloft.

Given the air mass, the atmospheric flow near the surface consists of irregular and semi-chaotic swirls and vortices of a variety of dimensions (ranging from 1 mm-1 km). These motions are called eddies, and are responsible for the turbulent transport and mixing of the pollutants within a reservoir layer or mixing height. The turbulent atmospheric mixing also provides lateral dispersion of the released emissions (e.g. from a stack) over a certain width. Thus, one may state that to first order the concentration (C) is of an air pollutant due to the release of a *point* source is:

$$C \sim Q/\left(Uh\sigma_y\right) \tag{1.1}$$

with U the wind speed (ms^{-1}), h the mixing height (m), σ_y the dispersion parameter (m), and Q the source strength (kg s^{-1}) (Pasquill and Smith, 1983). Thus, high concentrations occur for large Q, and for small wind speed, small mixing height and small dispersion conditions. These conditions occur typically during clear calm nights. Therefore, air pollution concentrations are strongly determined by the basic atmospheric conditions. As such, it is essential to understand the atmospheric processes that govern pollutant concentrations to enable skillful air quality forecasting. A reliable air quality forecast allows for on time and effective counter measures to reduce the harmful effects on pollutant emissions, and an effective warning of potential victims. As an illustration, Figure 1.1 shows a thin layer of brownish pollution (probably NO_2) on a calm morning in January, with a typical mixing depth of 50 m. The ingredients that govern the air pollutant concentrations are wind speed, mixing height and dispersion parameters. This paper provides a survey on the meteorological issues for air quality. In the next chapter, we introduce the diurnal cycle of the atmospheric boundary layer and its characteristic for wind speed, mixing layer depth and dispersion. Section 3 illustrates the common practice of high resolution weather forecasting for air quality forecasting, and provides an evaluation and identification of typical problematic features in modeling. Section 4 provides an overview of difficulties in understanding the stable boundary layer. Section 5 explains atmospheric dispersion and intermittency of the turbulence in calm nighttime conditions. Section 6 presents some methods for estimation of the nighttime boundary-layer height. A summary of our results will be given in section 7.

Figure 1.1. Illustration of a highly polluted thin stable boundary layer in the early morning of 19 January 2008 in Wageningen, The Netherlands. Photo: G.J. Steeneveld.

2. THE ATMOSPHERIC BOUNDARY LAYER

The Atmospheric Boundary Layer (ABL) is defined as the atmospheric layer adjacent to the Earth's surface that directly 'feels' the effect of the diurnal cycle of wind and temperature at the surface (Stull, 1988). The main characteristic of the ABL is turbulence, which normally mixes air pollutants effectively. Therefore, in air quality literature this layer is often referred to as the mixing height. During daytime, solar radiation heats the surface, which consequently triggers convection by thermals. Figure 2.1 shows a typical diurnal cycle of this layer by observations of wind speed and potential temperature (the temperature of an air parcel when it is adiabatically moved to the surface, $\theta \approx T + 0.0981z$) from the Cabauw meteorological tower (in the Netherlands, Beljaars and Bosveld, 1997). First, we find that during nighttime strong vertical gradients exist in the potential temperature, and in wind speed. Cold air is built up close to the surface due to radiative heat loss to space, and potentially warmer air remains aloft. This is a *stable* boundary layer or the *inversion* layer. Inversion layers inhibit vertical transport and are thus harmful for air quality, so therefore this chapter will largely focus on stable conditions. Stable boundary layers prevail at night, but also during daytime in winter in mid-latitudes as well as in polar regions (Yagüe and Redondo, 1995), and during daytime over irrigated regions with advection. The stable stratification in the morning of the Bhopal accident was one of the causes for the high MIC concentration. On the contrary, during the day all tower levels above 20 m show a nearly uniform wind speed and potential temperature, although it is warmer very close to the surface (*unstable stratification*).

a)

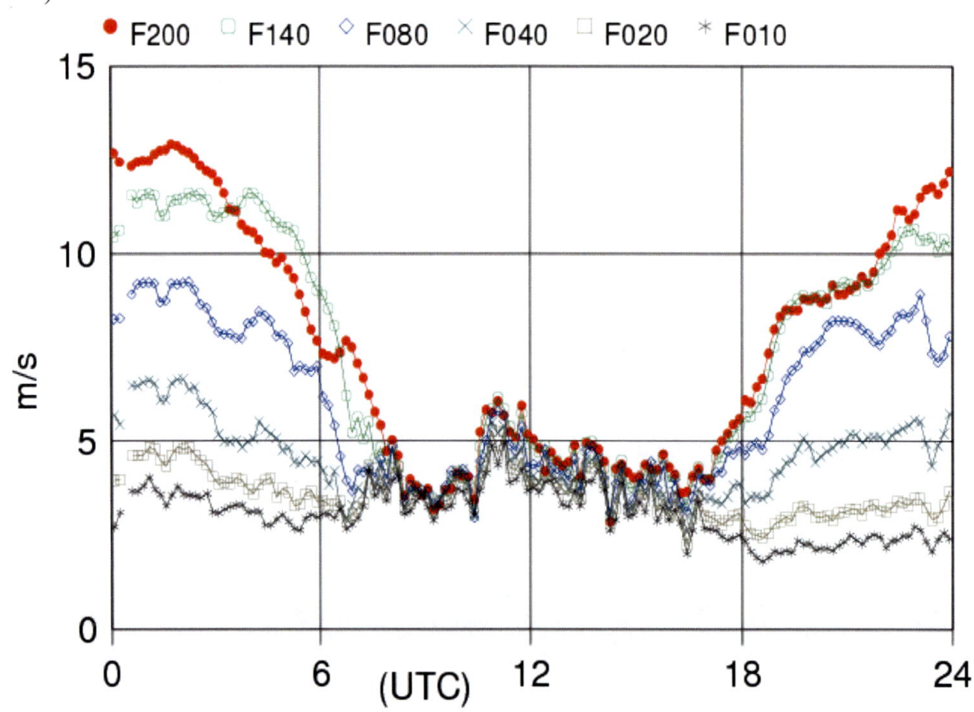

b)

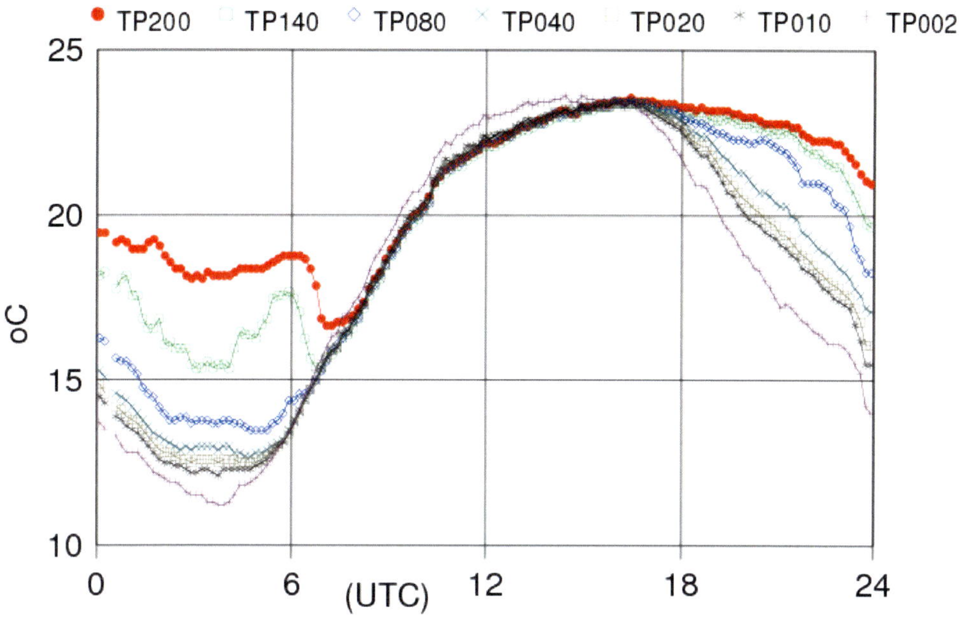

Figure 2.1. A typical diurnal cycle on a clear day for wind speed (a) and potential temperature (b) as recorded by a 200 m tower in Cabauw, The Netherlands, for 7-05-2008. (Courtesy: Fred Bosveld, KNMI).

To understand the ABL diurnal cycle, we need to examine the turbulent nature of the flow in the ABL. Turbulence is the key process for transport and dilution of constituents. Turbulence consists of irregular swirls of motions of different scales superimposed on each other. Turbulence is generated by two mechanisms: *forced convection* (mechanical turbulence) and *buoyancy*. Forced convection occurs when a flow travels over a rough area (with obstacles as grass, trees, cities, hills), and is forced to pass these objects. Wind shear is the source for turbulence in this case. Buoyancy is the effect that warm air parcels have a smaller density than surrounding colder parcels, and therefore they start to ascend. A characteristic property of turbulence is its effectiveness in transportation of heat, moisture and pollutants. Therefore temperature and wind speed are well-mixed in the ABL, because the turbulence very easily transports heat from the surface upward, and momentum from the free atmosphere downward. Strong boundary layer convection often results in eddies that are sufficiently energetic to mix free tropospheric air (above the ABL) into the ABL. This is called *entrainment*, and is an important contributor to the ABL ventilation, because the imported air is typically cleaner than the ABL air. One can also imagine that the ABL plays a vital role in the exchange of natural (e.g. CO_2 and other greenhouse gases) and anthropogenic (e.g. pollutant emission; Neu, 1995) contaminants from the Earth's surface to the free atmosphere above the ABL (e.g. Pino et al., 2006b; Górska et al., 2008).

In the daytime boundary-layer, the sun heats the surface and the turbulence is dominantly driven by buoyancy. The boundary layer rapidly grows in the morning, and large convective eddies provide vigorous vertical mixing over typically 1-2 km depth. Then surface emissions are diluted in a deep layer. Godish (2003) shows an evident difference in ABL height climatology between locations in the USA. Northern and coastal station have an average summertime mixing height of 1000-1200 m, while in desert areas as in Phoenix and Denver the mean mixing height reaches 2600 m. Thus, land use substantially influences the ABL height.

In contrast, at night buoyancy suppresses the turbulence intensity, so only forced convection is a source of turbulence. Therefore, the nighttime boundary layer is much shallower, typically 100-200 m deep, but even 10-50 m is possible for weak winds. Then very high concentrations of (possibly harmful) species may build up in the SBL. During weak winds the suppression of turbulence by buoyancy is larger than the production by wind shear, so as a result the turbulence vanishes. Then, other transport processes than turbulence appear to govern the boundary-layer structure. These are radiation divergence, gravity waves, drainage flows, and heat conduction from the soil, although their role and interactions are not a priori clear for each process until now, and some of them has been only qualitatively understood. Questions such as "What determines the nocturnal cooling timescale at 2 m?" are still not completely answered (Pattantyús-Ábrahám and Jánosi, 2004).

In the next section we will evaluate the mesoscale weather forecast model WRF, which is a state-of-the-art model to provide meteorological information for air quality models.

3. Mesoscale Models for Air Quality Forecasts

Transport and dispersion of pollutants in the atmosphere occurs at first instance at horizontal scales that are much smaller than covered by typical global circulation models/weather forecast models (grid size typically 25 km). Therefore, high resolution weather forecast models, so called meso-scale models or limited area models, are used to forecast the local wind and temperature at much finer resolution (up to 1 km). In that case, the mesoscale model is used to zoom in a specific area of interest. As an example, a sea breeze circulation is a phenomenon on the so called meso-β scale (~100 km), and is usually forecasted with mesoscale model instead of operational global forecast models, because of the restricted resolution of the latter.

Many mesoscale models have been developed during the recent years (e.g. RAMS, MM5, WRF, HIRLAM, AROME, MESO-NH). These models provide the user the possibility to use different calculation methods for the physical processes: the boundary layer scheme, land surface scheme, the radiation scheme, cloud processes, gravity wave drag, and atmospheric radiation. The availability of a myriad parameterizations for each physical process already indicates that no unanimity has been reached for the optimal schemes. We will illustrate some characteristic differences of the forecasted mean and turbulent ABL structure.

3.1. Case Study and Model Set up

In this section, we analyze the ability of the limited area model WRF (version 2.2) to forecast the structure of the ABL for a case study in the Netherlands. We will validate our model forecasts against Cabauw tower observations (e.g. Beljaars and Bosveld, 1997, 51.97°N, 4.93°E, -0.7 m AGL, Figure 3.1), but also with wind profiler data, soundings and ceilometer observations in De Bilt (20 km north of Cabauw). A ceilometer is an instrument that uses lidar techniques to determine the ABL height.

We selected the period of 30-6-2006, 12.00 UTC to 4 July 2006, 0.00 UTC because this is a clear sky summer period, and many observations are available for model validation. Note that a part of this period is under further study within GABLS3 (Bosveld et al., 2008). Herein column models for the ABL are given the same forecasting task. Their results are evaluated and compared in order to understand model deficiencies and to enable further model development. In this period, The Netherlands are located under a high-pressure system with winds from the southeast. The area consists mainly of grassland and is absolutely flat and relatively homogeneous. Also, the area is characterized by a large water supply and thus a high soil moisture availability. For these simulations the initial and boundary conditions (every 6 hours) were provided by NCEP Final Analysis. The model was run in an area of 1600 x 1600 km with a grid size of 54 km. In this domain, we nested 3 domains with a grid spacing of 18, 9 and 2 km respectively to minimize model errors due to lack of horizontal resolution (Figure 3.1). Moreover, the U.S. Geological Survey provided the land surface properties for the model simulations, such as soil moisture availability, surface roughness, and land use.

The model was run for 3 different boundary layer schemes. The first scheme is the so-called MRF scheme (Troen and Mahrt, 1986; Hong and Pan, 1996). This is a first order closure scheme, and the main characteristic of the scheme is that the height dependence of the eddy diffusivity profile is prescribed with a cubic function, and that its magnitude depends on the characteristic velocity scale at the surface layer. Another characteristic is that the scheme allows for non-local counter gradient transport for heat during the day. This extension is needed to represent transport by large eddies on the scale of the boundary layer itself, instead of local transport. A well known disadvantage of this widely used scheme is excessive daytime entrainment at the boundary-layer top, and overestimation of the turbulent transport at night (e.g. Vila et al., 2002; Steeneveld et al., 2008).

The second scheme has recently been introduced in WRF and is an extension of the MRF scheme. The extensions consist of a) inclusion of prescribed entrainment rate at the top of the boundary layer, b) counter gradient transport of momentum (wind), and c) Prandtl number (i.e. the ratio of diffusivity for momentum and heat) depending on height (see also Noh et al., 2006). As such, we will evaluate whether these modifications circumvent the deficiencies in the MRF scheme. We will refer to this scheme as the YSU scheme.

Finally the third scheme is a 1.5 order closure scheme (MYJ) and uses a prognostic equation for the turbulent kinetic energy (see Stull, 1988; Steeneveld et al., 2008). Then the eddy diffusivity is determined by multiplication of the turbulent kinetic energy and a length scale. The NOAH land surface scheme has been used (Ek et al., 2003). For completeness we mention that we use the Kain-Fritsch cumulus convection scheme, the RRTM scheme for longwave radiation, the Dudhia scheme for shortwave radiation, and the WSM 3-class simple ice microphysics scheme. In the surface layer we use Monin-Obukhov theory as in Janjic (2000), and the 5 layer simple soil scheme is utilized for the soil.

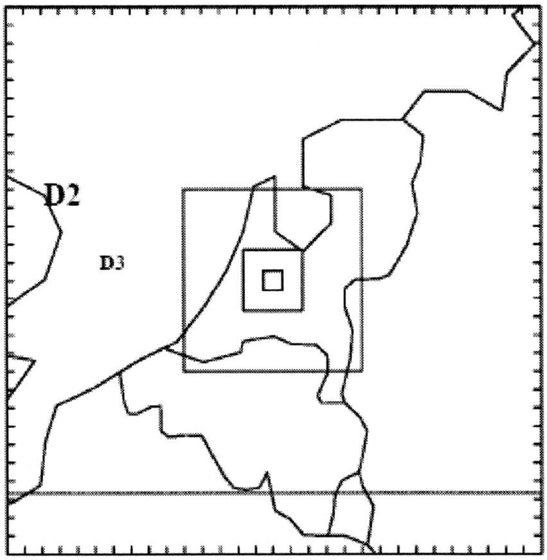

Figure 3.1. Model configuration for the evaluation of the WRF model. The center of the domain corresponds to the location of the Cabauw meteorological observatorium.

3.2. Results

We will evaluate model forecasts with eddy correlation observations of surface fluxes of sensible heat (H) and evapotranspiration (L_vE), and next also the profiles of wind speed (U), potential temperature (θ), and specific humidity (q). Also, we assess the model's ability to reproduce the 10 m wind speed and ABL height. Since the model probably needs sufficient time to adapt the initial field to its own model physics, we analyze the forecasts after 24 h for the convective ABL and after 36 h for the stable ABL.

Figure 3.2 shows a clear difference in the forecasted profiles during daytime between MYJ on one hand and the other schemes on the other. A characteristic difference is that MYJ forecasts a boundary layer that is typically too shallow and too cold compared to observations. The non-local schemes MRF and YSU simulate an ABL of 1400 m depth, while it is 700 m with MYJ. Furthermore, in the bulk of the ABL MYJ is typically 2 K colder, and 3 g kg^{-1} more humid than YSU and MRF, and it is evident that MRF and YSU are in closer agreement with the observations. These findings support previous results (e.g. Berg and Zhong, 2005; Steeneveld et al., 2008). At the same time it is realized that also the wind speed profiles differ substantially. MYJ and MRF forecast similar wind speed profiles close to the surface, while the YSU model provides stronger wind speed close to the surface, and seems to be in closer agreement with the tower observations. This also holds for the individual components, but the wind direction is correctly forecasted by all schemes.

At night the situation is different (Figure 3.3). Although the YSU scheme was intended to improve the daytime representation of the ABL, it also has a clear impact on the full diurnal cycle, and especially on the representation of the nocturnal low-level jet. Normally this phenomenon is a difficult feature to forecast correctly (e.g. Cheinet, 2005, Steeneveld et al., 2008). The observed LLJ of 12.5 ms^{-1} at 200 m altitude is best reproduced by YSU, which forecasts a LLJ of 10 ms^{-1} at 200 m height. This is substantially better than MYJ and MRF that forecast a weaker LLJ. The improved representation of the LLJ with YSU originates from its daytime modifications, and it can be explained from the Ekman equations. Stull (1988) explains that the sudden decrease of turbulence friction at sunset starts a rotation of the ageostrophic component around the geostrophic wind. As a consequence a LLJ develops. The amplitude of this LLJ is proportional to surface friction at the end of the day. Since this value is larger with YSU, due to the nonlocal momentum mixing, also the friction drop at the end of the afternoon is larger, resulting in a faster LLJ. Consequently the representation of the nocturnal wind maximum improves.

Contrary to the representation of wind speed, the temperature structure and nighttime cooling is better forecasted by MYJ than by the non-local schemes MRF and YSU. This is due to the fact that the MYJ scheme has a smaller turbulent mixing in the boundary layer, which is preferable for SBL modeling (Steeneveld et al., 2008). At the same time the surface specific humidity is larger with MYJ and in closer agreement with the observations.

We conclude that in the current study the YSU scheme should be preferred for modeling the diurnal cycle. However, especially the representation of the nighttime boundary layer requires improvement.

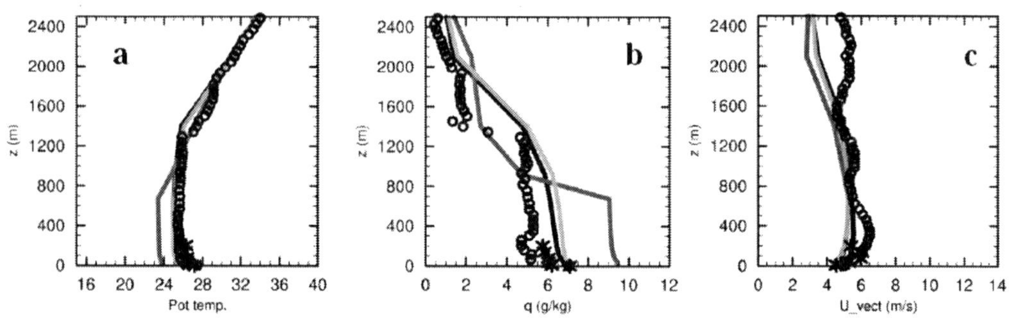

Figure 3.2. Modeled and observed potential temperature (a) and specific humidity (b) for 1 July 2006, 12 UTC. Black line YSU, dark grey line: MYJ, light grey line: MRF. Asterisk: Cabauw tower observations, open circles: De Bilt tower observations.

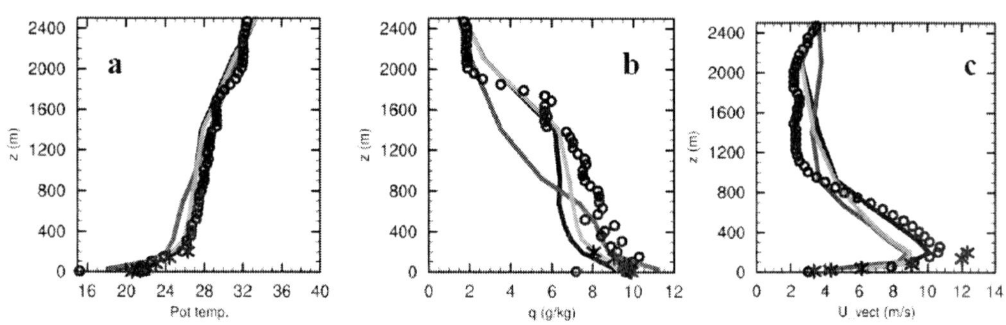

Figure 3.3: Modeled and observed potential temperature (a), specific humidity (b), and wind speed (c) for 2 July 2006, 00 UTC. Black line YSU, dark grey line: MYJ, light grey line: MRF. Asterisk: Cabauw tower observations, open circles: De Bilt radiosonde observations.

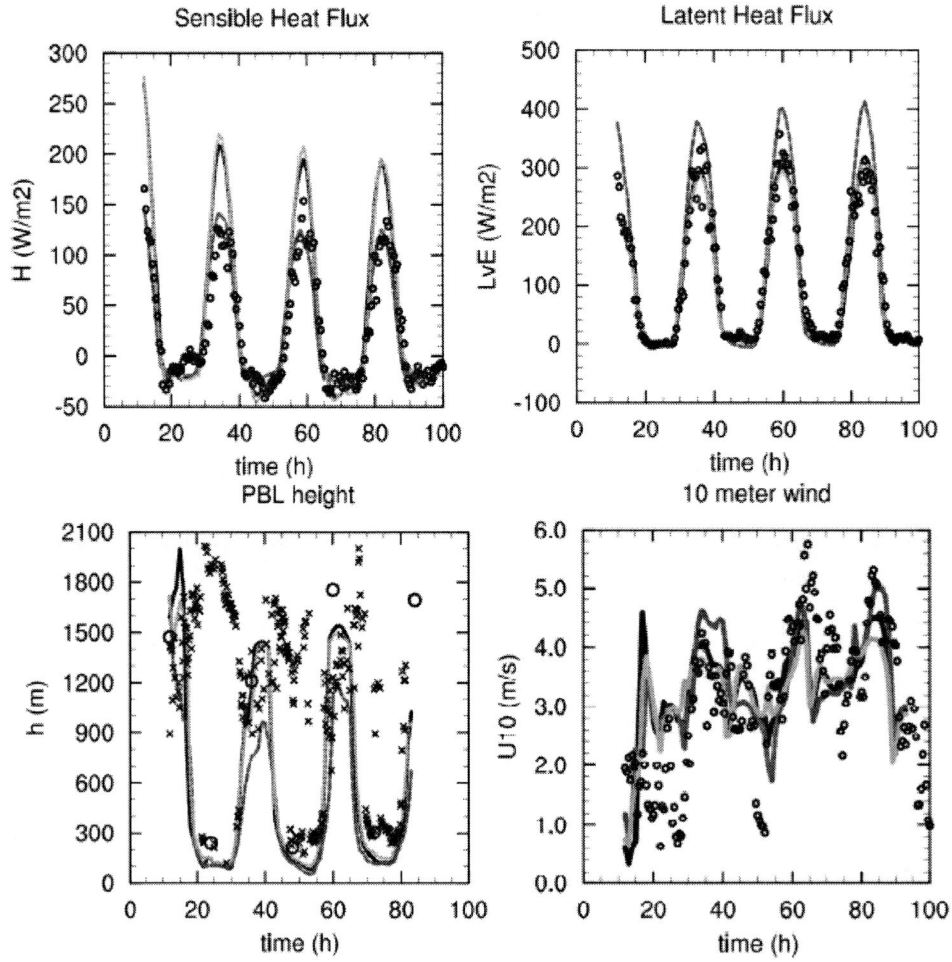

Figure 3.4. Modeled and observed surface sensible heat flux, latent heat flux, and boundary layer height for 30-6-2006_12 UTC to 3-7-2006_12 UTC. Black line: YSU, dark grey line MYJ, light grey line: MRF. Dots: observations at the Wageningen field, + : observations at Cabauw. X = ceilometer De Bilt.

Figure 3.4 shows the modeled and observed near surface variables. The diurnal cycle of the sensible heat flux is well modeled by WRF, although it is overestimated by ~50 Wm^{-2} by YSU and MRF at noon. The nighttime flux is correctly modeled, although this is usually problematic for models. The relatively high performance may be attributed to the fact that this case has relative high geostrophic wind speed (~7 ms^{-1}), and forecast skill breaks down for weak winds.

WRF forecasts the latent heat flux well, although MYJ evapotranspirates more than observed, which is a well known model deficiency of MYJ (e.g. Steeneveld et al., 2008).

The diversity in modeled sensible heat flux is reflected in the forecasted ABL height: MRF and YSU forecast a deeper daytime ABL that is in close agreement with ceilometer observations. At night the model underestimates the ABL height, both compared to the radiosonde and to the ceilometer observations. However, one should realize that no consensus exists on the determination of the stable boundary-layer height from direct model output

(Vogelezang and Holtslag, 1996 VH96 hereafter; see also section 5). Finally we note that the 10 meter wind speed is well forecasted by all schemes in WRF.

To summarize we find that mesoscale models are well able to reproduce (at least for the current case study) the main characteristics of the boundary layer and its diurnal cycle. However, the model results for stable conditions diverge due to difference in model assumptions. Further research is needed, especially in nighttime conditions (Holtslag, 2006).

4. Difficulties of the Stable Boundary Layer

The most relevant processes in the SBL are turbulence, radiative transfer, an elevated nighttime wind maximum ("low-level jet") near the top of the stable boundary layer (Garratt, 1985), heat conduction in the soil and vegetation, and gravity waves (e.g. Steeneveld et al., 2008). We will now briefly point out these processes and explain their role.

-Turbulence

Except for the aspect of dispersion and dilution (see section 1), turbulence also plays a role in the determination of the surface sensible heat flux. This is the amount of heat that is transported by the turbulent eddies between the land surface and the atmosphere. The sign of this variable is extremely relevant for air pollution dispersion, since it determines the atmospheric stability. In unstable (daytime) conditions the sensible heat flux is positive (towards the atmosphere), and pollutants are easily vertically mixed and diluted. In stable (nighttime) conditions the sensible heat flux is negative and the sensible heat flux limit the turbulent intensity. Therefore, in stable conditions, pollutants are only hardly mixed or diluted. Thus the sensible heat flux directly governs the mixing regime.

-Longwave Radiative Transfer

Each object that is warmer than the absolute temperature minimum of 0 K emits radiation. This also holds for the air in the atmosphere and in the ABL. Different air layers contain different amounts of absorbing material, such as water vapor, carbon dioxide, ozone etc, and this results in emissivity differences between air layers. The emitted and absorbed long wave radiation differs therefore between the different layers, which consequently results in a net radiative flux divergence, and consequently in net cooling. Especially in the stable boundary layer, the temperature gradients near the surface can become extremely large, and consequently the emitted radiation differs strongly by the different layers (e.g. Hoch et al., 2007; Drüe et al., 2007). The potential temperature at a certain level in the atmosphere is amongst others governed by the divergence of the net long wave radiative flux (Rodgers, 1967; Anfossi et al., 1976). Usually a three layer structure is seen in the nocturnal boundary layer: close to the surface a sharp inversion develops due to radiative cooling. In the middle of the SBL, turbulence is dominant and the stratification is limited. Finally, at the SBL top radiative cooling causes a sharp inversion at the SBL top (Garratt and Brost, 1981).

-Soil And Vegetation

At night, conduction enables upward heat transfer from the soil to the surface, and counteracts the surface cooling. On clear nights with weak winds, turbulence cannot be maintained and therefore both the turbulent sensible and latent heat fluxes vanish. Accordingly, the boundary-layer energy budget is governed by other processes than turbulence. At the surface, the energy budget is in that case solely governed by the net radiation and the soil heat flux. The soil heat flux through the soil depends on the soil material (sand, silt, clay, peat) and the water moisture content of the soil. In the real world, the soil moisture, and thus the soil thermal properties can differ on very small horizontal scales. However, for model applications (see e.g. section 3) the soil is assumed to be homogeneous.

-Elevated Nighttime Wind Maximum

Near the top of the SBL, the wind speed profile is in the idealized case characterized by a small air vertical layer with wind speeds larger than the geostrophic wind speed in the free atmosphere above the ABL. This may occur over large horizontal distances. During daytime, the mean flow is determined by the pressure gradient force, the Coriolis force and the friction that is generated by the ABL turbulence. However, during sunset the turbulent drag suddenly decreases rapidly and the equilibrium is disturbed. As a consequence, the air accelerates in response to the lack of friction in the ABL. The resulting low-level jet (LLJ) or nocturnal wind maximum can act as a second source (beyond the production by near surface wind shear) of turbulence in the stable boundary layer due to the shear above and below the wind speed maximum. On the other hand, the LLJ can also act as a lid on the SBL, because transport across the LLJ is inhibited (Mathieu et al., 2005; Cuxart and Jiménez, 2007), and the LLJ height is thus a relevant height for atmospheric dilution.

-Gravity Waves

A common feature in stably stratified geophysical flows is the ability to support gravity wave propagation (Nappo, 2002). Einaudi et al. (1978) define gravity waves as essentially coherent structures (mainly horizontally propagating) with a speed mostly much less than of the mean wind with horizontal scales less than 500 km and times scales less than a few hours. Gravity waves can be generated by a variety of features: sudden surface roughness changes, convection, and undulating topography, on which we will focus here. Indeed these wave (type) motions are widely observed during special observational campaigns in CASES-99 (Newsom and Banta, 2003). Since gravity waves are able to redistribute energy and momentum, they are important in determining the vertical structure of the atmosphere and the coupling of mesoscale motions to the microscale phenomena.

- Interactions

Figure 4.1 illustrates the above mentioned processes in the SBL, and their interactions (as far as understood at the moment). The main SBL forcings are the pressure gradient force, the Coriolis force, cloud cover, and free flow stability. For example, an increased geostrophic wind speed will enhance the turbulent mixing, and thus give reduced stratification. A reduced stratification will reduce the magnitude of the surface sensible heat flux in the weakly stable regime, and also limits the radiation divergence and thus the clear air radiative cooling. However, in the very stable regime, a reduction of the stratification might result in increased surface sensible heat flux. In both cases the surface energy budget is also altered, resulting in a modified soil heat flux. In the case of ceasing turbulence, the magnitude of the soil heat flux increases and vice versa. Moreover, this will alter the surface temperature and therefore the outgoing long wave radiation, and so the stratification.

In addition, increased geostrophic wind will under certain conditions increase the impact of wave drag over due to the orography, which at first increase the cyclone filling and thus reduce the geostrophic wind. On the other hand, it will also enhance the low-level jet wind speed. This consequently might result in additional downward turbulent mixing from the jet, which impacts on the stratification again.

One can imagine that the myriad of complex interactions between processes in the SBL have contributed the fact that our current understanding is limited. Therefore, also the representation of the stable boundary layer in atmospheric models in use for weather forecasting, climate and air quality is limited. This regularly results in biased forecasts of weather, air quality and climate (see also section 3).

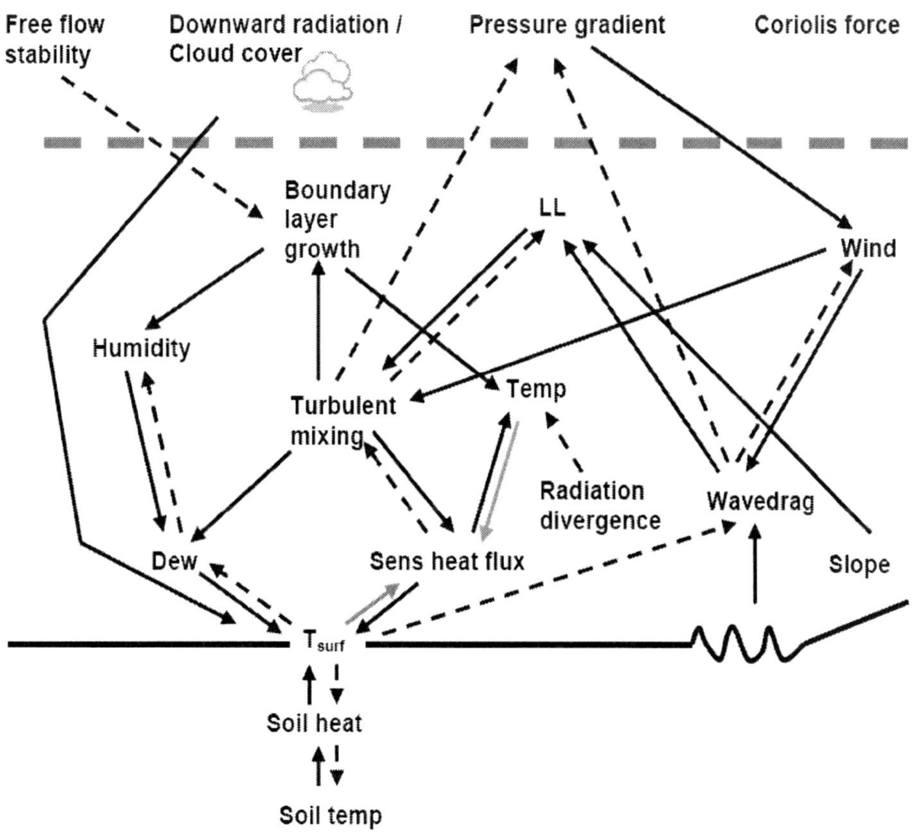

Figure 4.1. Overview of processes in the stable atmospheric boundary layer and their interactions. Positive interactions are full lines, negative feedbacks are dashed lines (Steeneveld, 2007).

5. Dispersion and Meandering in Stable Boundary Layers

Air quality forecasts are made based on the meteorological variables forecasted by mesoscale models as WRF in Section 3. Apart from the translation of pollutants by the mean wind, pollutants are also mixed away from the plume centerline due to the turbulent nature of the flow. Therefore, the weather dependent dispersion parameters in the *x*, *y* and *z* direction σ_x, σ_y, and σ_z have to be determined to forecast the dispersion of pollutants, such as with Eq. (1.1).

The numerical values of the dispersion parameters depend on the isotropic behavior of the turbulence. In neutral conditions, the dispersion occurs at the same rate in all directions and the turbulence is isotropic. Contrary, for unstable conditions with large thermals in the ABL, the turbulence is evidently anisotropic and σ_x σ_y, << σ_z. However, in stably stratified conditions, vertical motions are suppressed and the turbulence is anisotropic (σ_x σ_y, >> σ_z). Since vertical displacement in stratified conditions is limited, one can expect that the dispersion will take place dominantly in the lateral direction. Taylor (1921) derived that for lateral dispersion parameter:

$$\sigma_y^2 = 2\sigma_v^2 \int_0^t (t-\tau)R_1(\tau)d\tau \qquad (5.1)$$

In which σ_v^2 is the variance of the lateral wind speed, t is the time after particle release and τ the time lag over which the Lagrangian auto-correlation function $R_1(\tau)$ has to be determined. For well behaved continuous turbulence, this function is a decaying exponential function. This means that the longer the time difference between particle release is, the lower is their correlation in lateral spread. Note that $T_L = \int_0^\infty R_1(\tau)d\tau$ is defined as the Lagrangian time scale. This is a measure how quickly flow direction of emitted particles become uncorrelated with itself (Stull, 2000). For an exponential decay the integral converges to a finite value, and is therefore T_L is a useful parameter in dispersion models.

At the moment the behavior of $R_1(\tau)$ and σ_v^2 under stable condition is under investigation. In this section we determine $R_1(\tau)$ and σ_v^2 from field observations, as function of classes of atmospheric stability. We utilize long-term observations of a sonic anemometer at 3.44 m above a short grass field in Wageningen, The Netherlands (Jacobs et al., 2003, 2006). The eddy covariance processing software of Van Dijk et al. (2004) is utilized to obtain turbulent fluxes. For this particular study we use the observations for 2006.

The σ_v^2 can be parameterized from mesoscale model output (since the mesoscale models provide friction velocity u_*, sensible heat flux H, and thus the Obukhov length L) as follows:

$$\sigma_v / u_* = f(z/L) \qquad (5.2)$$

The functional form of f is uncertain for stable conditions due to large measurement uncertainties for weak wind conditions (e.g. Mahrt et al., 1998; Mahrt, 1999), and due to self correlation between the dependent and independent variables (Hicks, 1978; Baas et al., 2006). To overcome measurement uncertainties, we selected atmospheric conditions with $U_{10} > 3.0$ ms^{-1} in the following analysis.

Figure 5.1 shows $\sigma_v/u_* \sim 2.2$ for neutral conditions (z/L=0), which is consistent with findings of Stull (2000), slightly above the range in a review by Banta et al. (2006) who found that $1.6< \sigma_v/u_*<2.1$. More interesting is that for larger stability (i.e. larger z/L), σ_v/u_* increases, and thus that the variability of the lateral wind speed increases. Finally, we also mention the large uncertainties of these type of observations as shown by the error bars. This uncertainty also increases with stability, due to the fact that turbulent fluctuations are more difficult to measure accurately for stronger stability. It is also therefore that several authors report different empirical formula for Eq. (5.2). The current data analysis suggests $\sigma_v/u_* = 2.2 + 0.86\dfrac{z}{L}$. However, earlier work by Pahlow et al. (2001) propose $\sigma_v/u_* = 2 + 4\left(\dfrac{z}{L}\right)^{0.6}$. Note that from a theoretical point of view the dimensionless ratio σ_v/u_* should become constant for strong stratification (Nieuwstadt, 1984), although none of the parameterizations show this behavior. This is probably caused by the instrumental problems for large z/L. Future developments in improving scaling relations for stable conditions will depend on instrument development and refinement of data treatment procedures. For example Hiscox et al. (2006) measured the plume spread under stable conditions with the innovative lidar technique. This technique has the advantage that dispersion parameter can be obtained with high temporal frequency and directly for a volume.

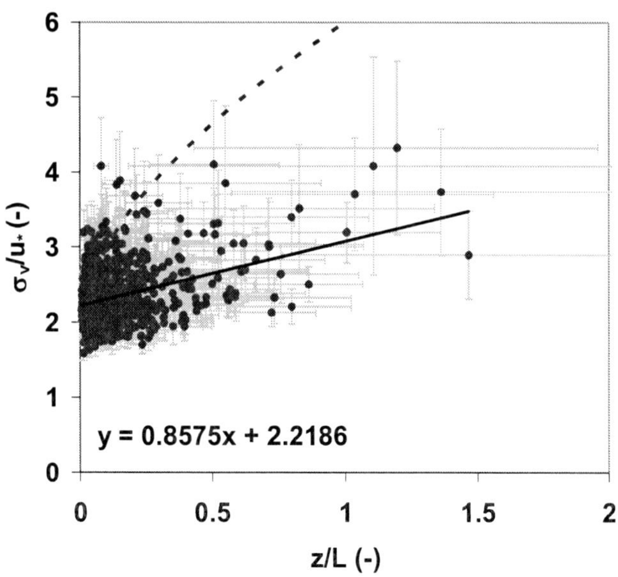

Figure 5.1. Observed relation between dimensionless lateral velocity variance and atmospheric stability as expressed by z/L. Observations for 2006 at the Wageningen field, The Netherlands, are 30 min. averages, and after data screening (see text). Dashed line: Pahlow et al. (2001).

Although the dispersion is controlled by the Lagrangian autocorrelation function, $R_l(\tau)$ is often derived from the Eulerian autocorrelation function by a conversion in time (e.g. Tennekes, 1982). For well behaved turbulent conditions, as for Fig 5.1, the autocorrelation function $R_l(\tau)$ shows a quick decay with delay time. This is illustrated in Fig 5.2a that shows $R_l(\tau)$ for 0-6 UTC at DOY 61 at the Wageningen station. The mean wind amounts about 3-4 ms^{-1} and is from western directions. After approximately 300 sec the $R_l(\tau)$ amounts about 0, as in agreement with the findings of Anfossi et al. (2005).

This picture changes when one considers a more stably stratified night, e.g. 2 Jan 2006 (Fig 5.2b), with mean 10 meter winds of 0.5-1 ms^{-1} from eastern directions. $R_l(\tau)$ now shows a clearly different: $R_l(\tau)$ decreases with τ, but much more slowly than in Fig 5.2a, which means that the wind speed and direction of released particles persistent over a longer time period than for the continuous turbulence. At about 600 sec $R_l(\tau)$ shows a zero crossing. The physical meaning of $R_l(\tau) < 0$ is that particles become anti-correlated, i.e. a particle that has recently been released travels eastwards relative to the coordinate system, while the parcel that was released 800-1000 sec before travels westwards. This is an indication that the flow in the boundary layer is meandering. Finally, only after 1400 sec $R_l(\tau)$ tends to zero. Note that our findings correspond to findings in Anfossi et al. (2005) (their Figure 3).

Meandering flows complicate our ability to model dispersion of pollutants in the atmosphere, because the common methods (i.e. Gaussian Plume Model) are not equipped for this flow type because they assume a linear plume center line. The previous analysis stresses that accurate air quality forecasting needs information on the presence of meandering and if possible also on the time scale of the meandering, i.e. the period.

Figure 5.3 relates the observed minimum value of $R_l(\tau)$ to the wind speed, using half year of observations. It is clear that low wind speeds support substantial negative $R_l(\tau)$, and thus meandering, while for wind larger than 4 ms^{-1} R_{min} is large. Between wind speeds of 1.5 and 4 ms^{-1} large uncertainty exists on R_{min}, because meandering can apparently not only be described by wind speed, but likely also atmospheric stability is relevant. The current findings for R_{min} correspond to earlier findings of Anfossi et al. (2005)

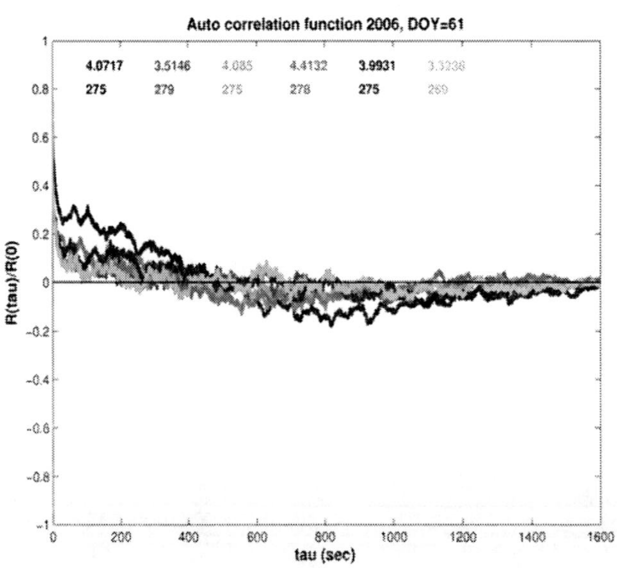

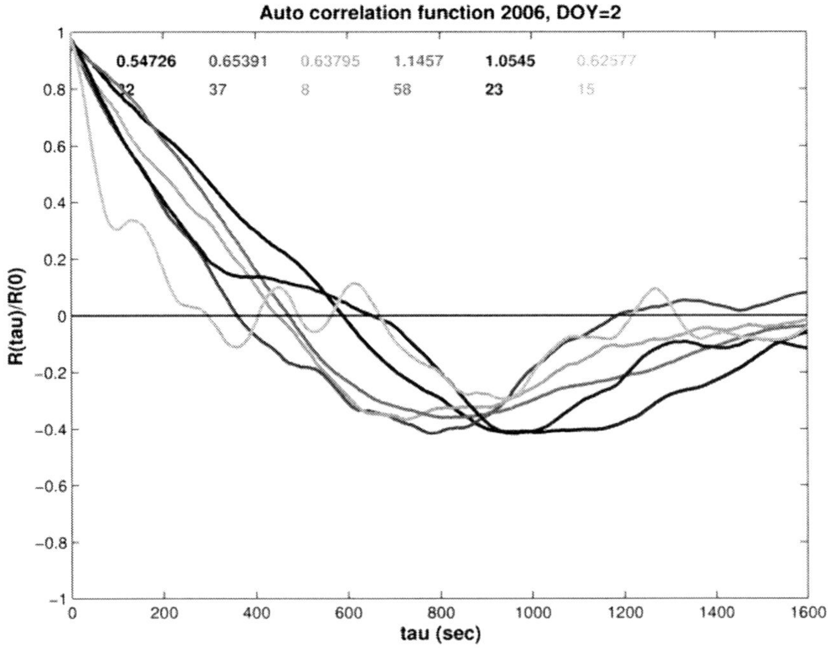

Figure 5.2. Observed autocorrelation functions of lateral wind speed component for 1 March 2006 (a) and 2 Jan 2006 (b), 0-6 UTC in Wageningen, The Netherlands. Numbers on top of the graph indicate mean wind speed (first row) and direction (second row) for the corresponding autocorrelation function for the consecutive hours.

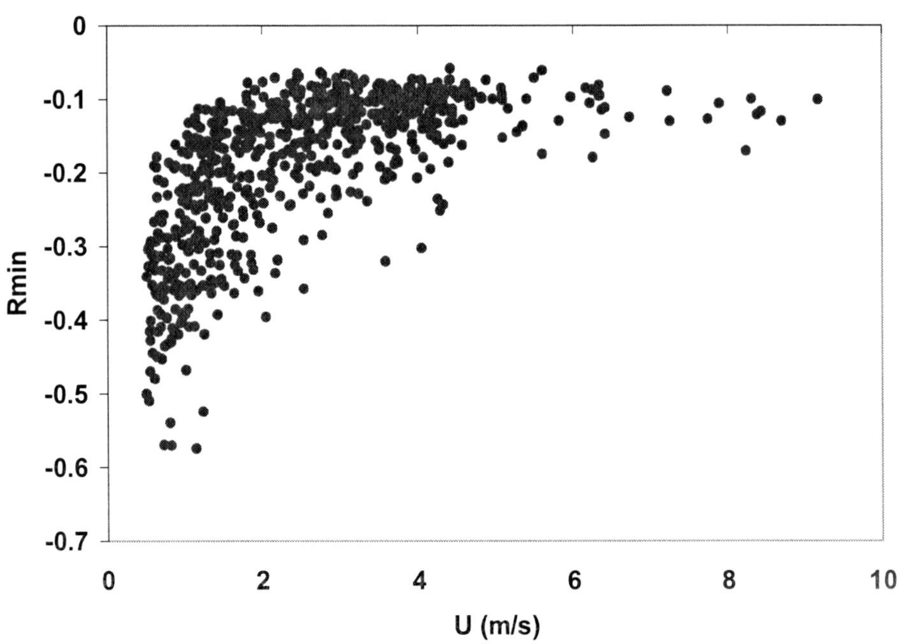

Figure 5.3. Observed Rmin for the year 2006 (0-6 UTC) at the Wageningen station as function of 10 meter wind speed.

5.2. Intermittency in Stable Boundary Layers

Until now we found a clear difference in understanding of stable boundary layers: fully turbulent stable boundary layers are reasonably understood, but our understanding is poor for calm and very stratified conditions. However, for weak wind conditions ($U < 3$ ms^{-1}), turbulence is often discontinuous, and of intermittent nature (Holtslag and Nieuwstadt, 1986), and the exchange of heat and momentum may occur in so-called "bursts", i.e. short episodes of increased turbulence (e.g. Kondo 1978; Weber and Kurzeja, 1991; Van de Wiel, 2002), also known as *global intermittency.* This coincides with rapid changes in wind speed and temperature. Different physical mechanisms have been suggested to explain the global intermittency. First large shear stress at the low-level jet (LLJ) causes locally increased turbulence that can be transported downwards (Mahrt, 1999). Secondly, passing gravity waves alter the wind speed during passage, and might generate turbulence (Duynkerke, 1991; Nappo, 1991). Thirdly, Businger (1973) explains the global intermittency by decoupling of air in strong stratification so that momentum fluxes are suppressed. Subsequently the pressure force accelerates the flow, and recouples the flow with the surface for several times.

Beyond the exact mechanism, the horizontal extent of this phenomenon is unknown, although this is very relevant for turbulent flux calculations in atmospheric models and dispersion calculations. Intermittently turbulent patches might only occur in a part of a grid cell (Mahrt, 1987). Then, the grid averaged wind speed and temperature gradients are not uniquely related to the mean surface fluxes, and atmospheric models break down (Mahrt, 1998)

Let us illustrate the phenomenon and spatial extent of the intermittent night of 15/16 November 2002 in The Netherlands. The Netherlands were situated in between four weak low-pressure areas and experienced weak winds (typically 2-3 ms^{-1}) from the southeast.

A) Single Station Observations: The Wageningen Field

At 1830 UTC the screen level temperature drops sharply (from 7 to 2.5°C), and ~1 hour later increases again to 7°C, although the forcing by the long wave incoming radiation (e.g. clouds) does not change (Figure 5.4). Longwave outgoing radiation follows these temperature oscillations. At 2100 UTC another temperature decrease was observed of the same magnitude. Striking is that the temperature decrease occurs while the clouds are coming in, as can be seen in the LWD. After 2200 UTC the temperatures at 10 cm and screen level are the same. The recorded temperature oscillations coincide with oscillations in wind speed, soil heat flux and wind direction (not shown): high temperatures coincide with a temporary wind speed increase. The surface ground heat flux is less negative during a burst (period of increased turbulence) than during calm conditions. Wind direction shows a small veering when a temperature is increased. Both fluxes of momentum, sensible heat and latent heat are strongly reduced when low temperatures are observed. This overall picture seems to be typical for the Businger (1973) mechanism.

Because little is known whether this phenomenon is locally driven or is present on a larger scale, we use the routine weather observations in The Netherlands (a 300 x 400 km area) to examine the horizontal scale of these events. Observations from coastal station were discarded because they might be influenced by the sea-land transition.

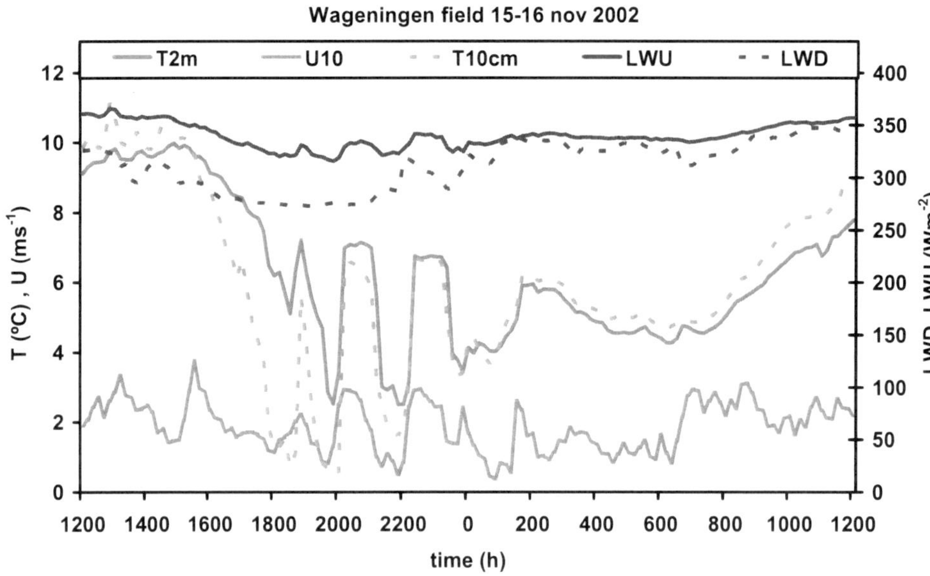

Figure 5.4. Observations of screen level wind speed (U), temperature (T at 1.5 m and 10 cm.), longwave upwelling (LWU) and downwelling (LWD) radiation components at the Wageningen field.

B) Spatial Development

To investigate the horizontal extent of the observed oscillations, the 10 min. averaged temperature signal from the 32 automatic weather stations were moving average (MA) filtered with a 2 hour window to suppress the fluctuations of higher frequency. Next, we applied a 30 min MA filter. The resulting signal is used to determine the amplitude, period and number of oscillations at various stations. Figure 5.5 shows the 2m temperature as observed in Wageningen, Soesterberg and De Bilt (see Figure 5.5d for the locations). Although the different stations are located far apart, and also located over different landscapes and land use, we find a surprising correspondence of the frequency and amplitude of the temperature oscillation.

a)

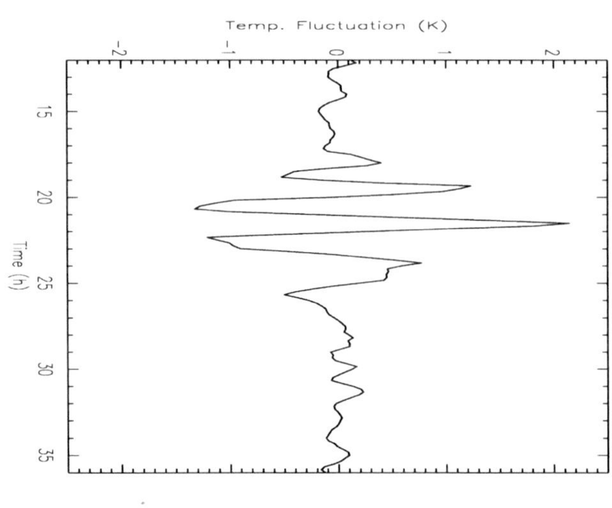

b)

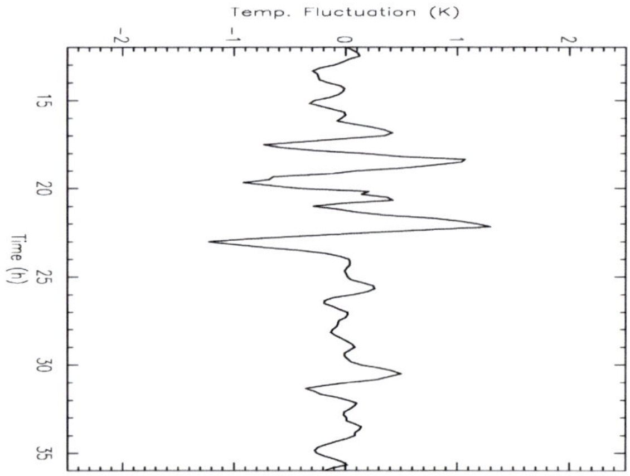

c)

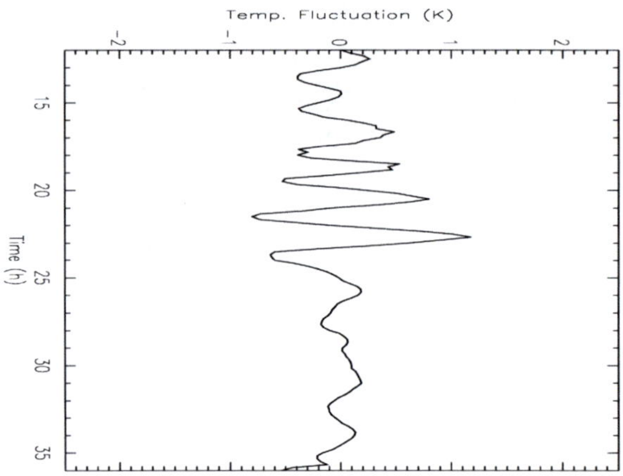

Figure 5.5: MA filtered temperature signal in Wageningen (a), Soesterberg (b) and De Bilt (c) for the night of 15-16 November 2002. d) Map of The Netherlands.

a)

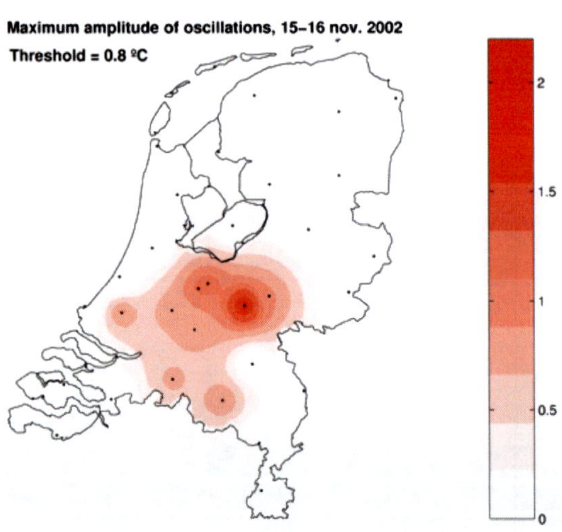

b)

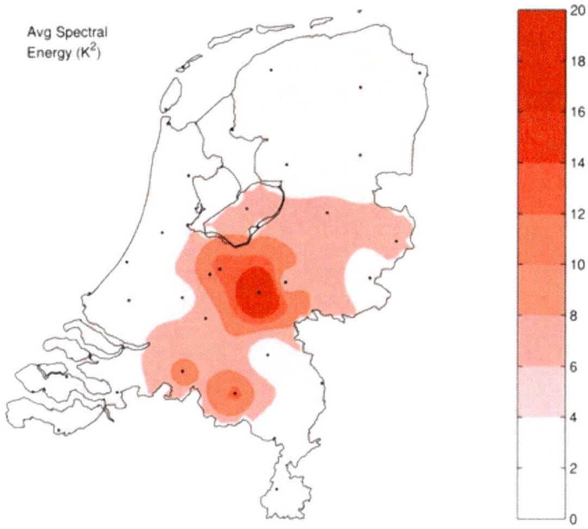

c)

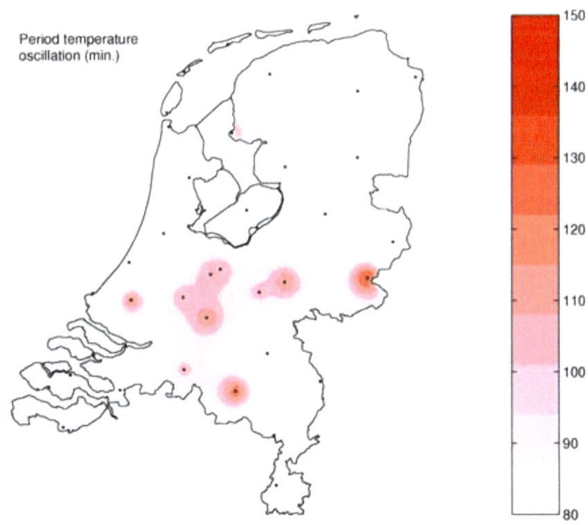

Figure 5.6. Contour plots of number of observed oscillations per station with 0.8 K as threshold anomaly (a), spectral energy (b) and wave period (c).

C) Wavelet Analysis

To analyze the results in section b) with a more objective measure for its intensity, we applied a Morlet wavelet analysis on the temperature observations (Torrence and Compo, 1998). To summarize the information stored in the observed variance, Figure 5.6c shows the integrated variance between 1800 UTC and 2200 UTC over the time scales between 0 and 160 minutes (about 3 hours) for each stations. In this manner we neglect the high frequent transition form day to night and the high frequent temperature change due to incoming clouds. The spatial development of the time integrated temperature variance clearly shows that this method confirms the results from section b). In this case intermittency is seen in a larger area than by Kurzeja and Weber (1991) who found intermittency in an at least 30 x 30 km area.

To summarize, global intermittency can occur over large spatial scales (here at least 100 x 100 km). Unfortunately, we are unable identify the physical mechanism of the oscillations. Local observations at the Wageningen field indicate the Businger mechanism as a possible mechanism. However, because we can translate the oscillation to De Bilt and Soesterberg, and because we also found that the phenomenon travels (phase speed about 10 ms^{-1}, not shown) faster than the typical wind speed (2 ms^{-1}), we also have strong indication that a wave motion may initiates the oscillations.

D) A Climatology of Intermittency of Nighttime Turbulence.

For parameterization purposes in atmospheric models, it is important to know the relative occurrence of the intermittency in the stable boundary layer, since established methods fail under these conditions. In very stable conditions, the total turbulent heat transfer can occur in short bursts: only a few time slots are responsible for the main transport. Then, universal relations between vertical gradients of wind speed and temperature on one hand and turbulent fluxes on the other hands are absent, and a more statistical approach could be fruitful (Poulos and Burns, 2003). Here we will briefly summarize the available tools in the literature, and provide a climatology of these intermittency indices for a grass field in the Netherlands. Statistical tools for describing intermittency intensity are:

a) Coulter and Doran (2002) define the *Intermittency fraction* (IF) as the time in which the individual 1-min fluxes make up the first 50% of the total integrated flux when accumulated over a certain averaging time. The index ranges from 0 to 0.5, and the lower the fraction, the stronger the degree of intermittency. In the current analysis we use 5 min averaged fluxes to examine the degree of intermittency in hourly records.
b) Kurzeja et al. (1988) define a *Gravity wave index* (GWI) for indicating the intermittency of the turbulence. It is the standard deviation of the vertical velocity spectrum with a time scale larger than 90 seconds, divided by the total vertical standard deviation. This index must be derived from power spectra, and if gravity waves are important the GWI will be much smaller than 1, while GWI will approach 1 for very turbulent cases.
c) Bercowiz et al. (1988) also defines a *Meandering index* (MI) the standard deviation of the lateral velocity spectrum with a time scale larger than 90 seconds, divided by the total standard deviation (Nieuwstadt, 1984; Kurzeja et al., 1988). This index must also be determined from power spectra. If meandering occurs, then MI $\ll 1$, and the MI will approach 1 for very turbulent cases without meandering.
d) Van de Wiel et al. (2002) define a Π parameter that is based budget equations for a bulk model of the stable boundary layer. A bifurcation analysis showed for this system dynamics approach that $\Pi < 1$ indicates intermittent oscillations. Despite the solid theoretical background of this index, it needs many parameters as input that are usually not or only known with limited accuracy. These are e.g. pressure gradient, boundary layer height, surface resistance of the vegetation layer, heat capacity of the vegetation.

The availability of different indices already indicates that a single index has not proven to be decisive until now.

The observed histograms for the listed indices are depicted in Figure 5.7 and 5.8. We find that the IF has its median around 0.40, and the distribution is heavily skewed. Only 20% of the data has an IF < 0.32. Thus, only a small percentage of the studied nights shows a strong intermittency. The same holds for the Π parameter, which is skewed towards positive values. The criterion for intermittency has been fulfilled for 6.3% of the nights. For MI, only 4.7% of the observations have an MI < 0.7, indicating that meandering is rather scarce according to the criterion. From the analysis of Figure 5.3 we found that 36.6% of the observations shows $R_{min} < -0.2$, and 7.9% showed $R_{min} < -0.4$. As such, MI and R_{min} analysis seem to provide consistent results. The GWI is smaller than 0.9 for 6.6% of the time, and the occurrence of intermittency is thus scarce according to the GWI definition.

For all these indices, one must be aware that under very stable conditions, i.e. for which these phenomena are likely to occur, it is more difficult to perform accurate measurements. Therefore one should be careful with the interpretation of the observations.

It is important to remark here that for direct applications in atmospheric models, the degree of intermittency should be expressed in model variables. We attempted to do so by linking the above indices to for example the Richardson number. Although a clear correlation between IF and Ri was found (smaller IF for larger Ri), the substantial data scatter impeded us to find a mathematical expression with sufficient degree of confidence.

The current data analysis was only possible for a single observational field. It is of substantial scientific interest to further study the spatial and temporal development of the intermittency phenomenon. The climatological study Pielke et al. (2008) shows that the nighttime stratification in the Oklahoma mesonet observations has substantially decreased during the last decade. Therefore, also climatological trends of intermittency could be expected, and should be analyzed in the near future.

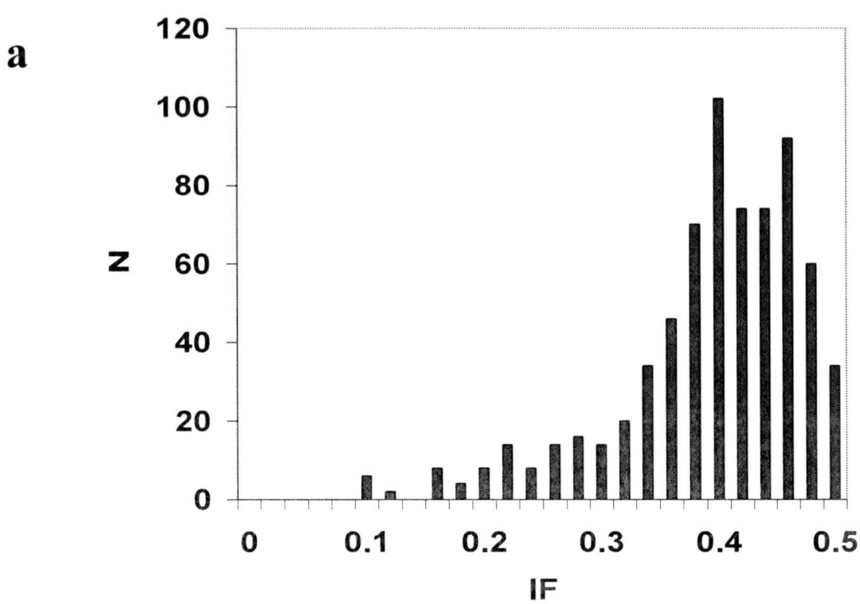

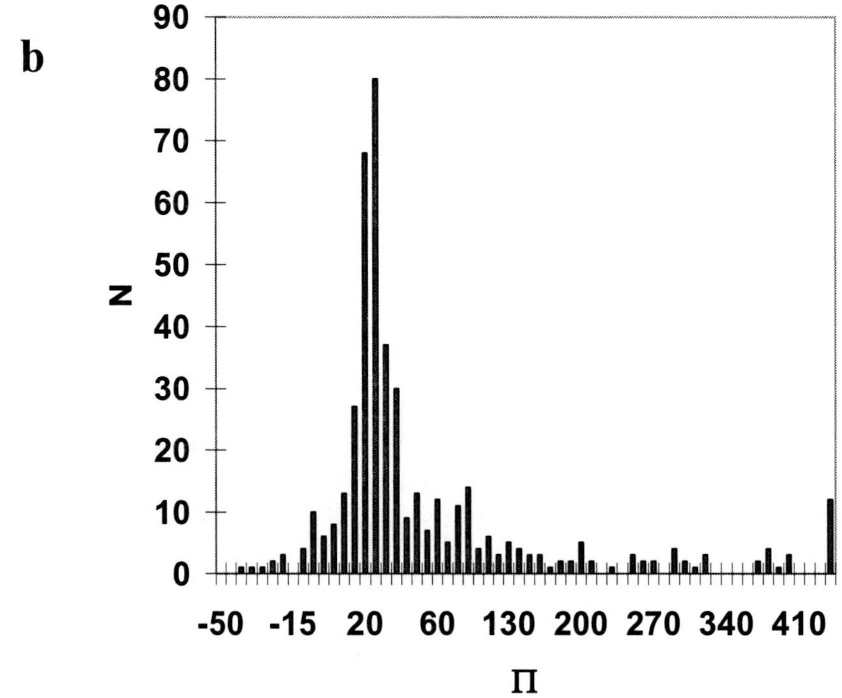

Figure 5.7. Histogram of the intermittency fraction IF and PI as calculated from field observations in Wageningen, The Netherlands over the years 2001-2003.

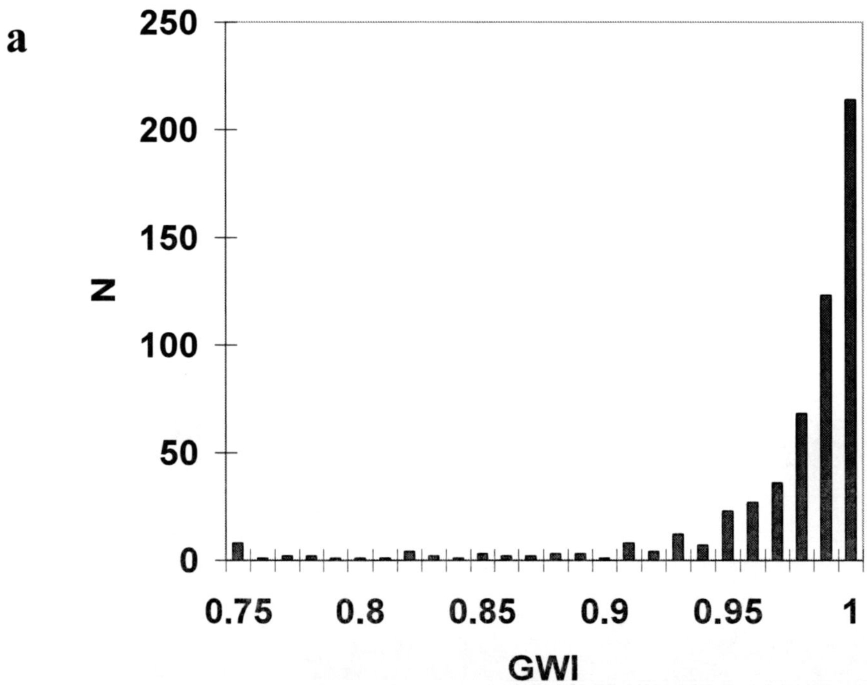

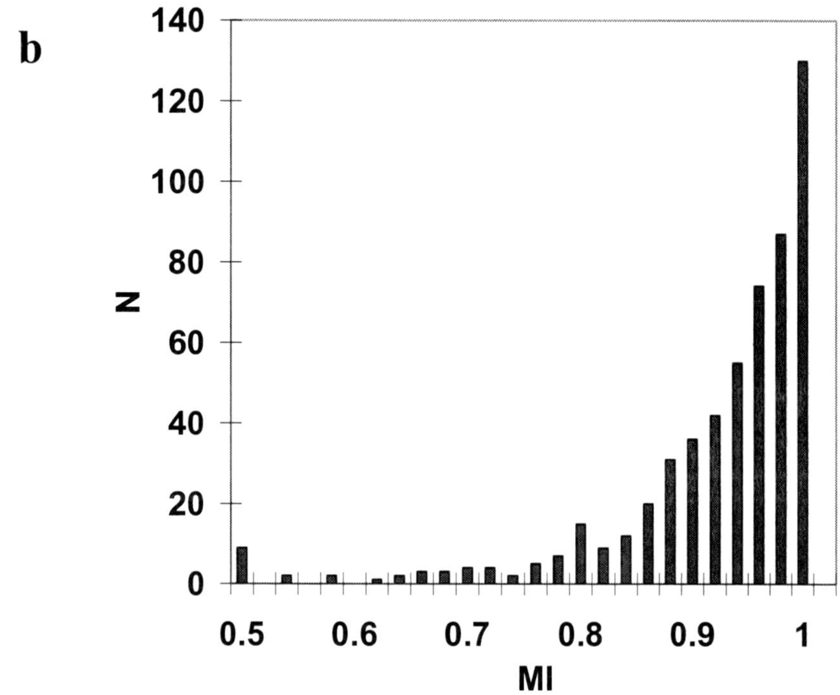

Figure 5.8. Histogram of the intermittency fraction IF and PI as calculated from field observations in Wageningen, The Netherlands over the years 2001-2003.

6. THE STABLE BOUNDARY-LAYER HEIGHT

The stable boundary-layer (SBL) height (h) is an important quantity to describe the relevant processes that govern the SBL development and its vertical structure (Holtslag and Nieuwstadt, 1986). Clearly, the stable boundary-layer height has an impact on the mixing properties of the SBL.

The dispersion of pollutants during stable stratification is strongly affected by h (e.g. Salmond and McKendry, 2005). Release of pollutants below h during periods of weak winds and consequently weak vertical mixing, may result in very high concentrations of primary and secondary pollutants, causing serious consequences for life. Therefore, h is a critical quantity to estimate for meteorological preprocessors in air quality models (Venkatram, 1980; Gryning et al, 1987; Lena and Desiato, 1999; Seibert et al., 2000; Karppinen et al., 2001).

Observing the SBL height (h_{obs}) is not straightforward because turbulence is gradually suppressed during stable conditions and it may also show intermittent behavior (e.g. Holtslag and Nieuwstadt, 1986). Vickers and Mahrt (2004, VM04 from now on) found a classical SBL with well-defined surface based turbulence for only 22% of the time for the CASES-99 field campaign. For non-well defined cases, problems occur in measuring h_{obs} because a universal relationship between the profiles of temperature, wind speed and turbulence is lacking. The interpretation of wind speed and temperature profiles is not straightforward in that case, and it is therefore not surprising that several definitions for h are in use nowadays (VM04, Beyrich, 1994).

Despite the uncertainties mentioned above, much work has been done in modeling (scaling) the SBL height (h_{mod}) from profile information or from turbulent variables at the surface. However, many of these models have been validated for a limited range of boundary-layer heights or for a single data set. The aim is to present a robust and practical formulation for h based on the governing variables, and *based on formal dimensional analysis instead of inverse interpolation*. Finally, we discuss and assess the relevance of the Coriolis parameter in practical estimates for h.

6.1. Background

The subject of defining and modeling the SBL height has a long history. Well-known definitions for h are (e.g. VM04):

a) a fraction of the layer through which turbulence exists (Lenschow et al., 1988),
b) the top of the downward turbulent sensible heat flux (Caughey et al, 1979),
c) the lowest maximum of the wind speed (often referred to as the Low-Level Jet, LLJ, Melgarejo and Deardorff, 1974), and
d) the top of the temperature inversion or the first discontinuity in the potential temperature profile (Yamada, 1979).

Seibert et al. (2000) state that no final answer is received to the question what is the most suitable height scale to characterize the vertical mixing in the SBL..

In principle also subsidence and baroclinicity determine the value of h (Zilitinkevich and Baklanov, 2002; Zilitinkevich and Esau, 2003). However, routine observations of subsidence and baroclinicity are usually unavailable and are therefore not taken into account in this study. Note that the technique of dimensional analysis we use in the second part of the paper is also applicable to other definitions of h.

Inspired by the fact that consensus on the SBL height modeling is currently lacking, we use dimensional analysis by using the Buckingham Π-theorem, without prescribing some particular shape. As governing variables for h we choose (inspired by ZM96) the surface turbulent buoyancy flux B_s, the surface turbulent friction velocity (u_*), the Coriolis parameter f, the Brunt-Väisälä frequency above the ABL. Note that the Π-theorem is semi-empirical and only applicable in the range of available observations.

6.2. Observations

Many models for h have been proposed based on a single dataset or on datasets biased towards shallow boundary layers (VM04), and thus universality is not a priori guaranteed for these models. Contrary, the analysis in this paper is based on six observational data sets over different terrain types according to their surface roughness and land use:

a) CASES-99

This measurement campaign was held October 1-31, 1999 near Leon, east of Wichita, Kansas, U.S.A. (37.6486° N, -96.7351° E, and 430 m ASL, Poulos et al, 2002). The area is relatively flat prairie grassland, free of obstacles. We determined h for the 101 profiles using the method in Joffre et al. (2001). They identified h subjectively by inspecting together the wind speed, potential temperature and Richardson number profiles for clear changes below the inversion height that would indicate a change in the structure of the lower atmosphere. This is illustrated in Figure 6.1 for 0700 UTC 23 Oct 1999. The SBL height was based on (1) the LLJ height and (h_{LLJ}) and (2) the height of the first discontinuity in the potential temperature profile (h_θ), from curved to linear in Figure 6.1a. Figure 6.1b illustrates the first peak in the gradient Richardson number at that level. If h_{LLJ} and h_θ differ more than 60 m, the data points were rejected. In addition, we selected data for $N > 0.015\ s^{-1}$, since weaker N could not be determined accurately. The reliability of eddy covariance measurements (10 min. averages) for weak turbulence is questionable, and therefore we disregarded data with friction velocity $u_* < 0.04\ m\ s^{-1}$ and sensible heat flux $H > -2\ Wm^{-2}$.

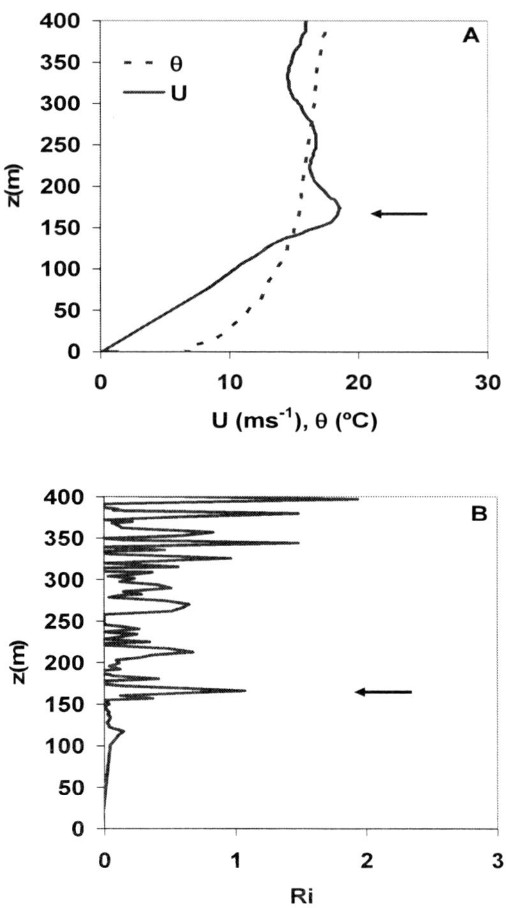

Figure 6.1. Profiles of potential temperature, wind speed (a) and gradient Richardson number (b) for 23 October 0700 UTC during CASES-99. The arrow indicates the stable boundary-layer height.

b) Sodankylä

This radiosonde dataset is based on the intensive campaign of the international NOPEX/WINTEX program at the observatory of Sodankylä, Finland (67.4° N, 26.7° E, 180 m ASL) between 10-21 March 1997 (Halldin, 1999). The observations were carefully selected by Joffre et al (2001). The terrain around the site is mostly flat, characterized by isolated gently rolling hills (altitude differences 50-150 m), and is covered by sparse forest (mean tree height of ~8 m around the site). The immediate vicinity of the mast was characterized by semi-open pine forest with a mean height $H_r \approx 8$ m.

c)Cabauw

The first Cabauw dataset has been gathered in the period 1977-1979, The Netherlands (51.971 °N, 4.927 °E; -0.7 m ASL), by Nieuwstadt (1980b), and used in VH96. The data were carefully selected (e.g. filtered for gravity waves) before in VH96.

The Cabauw area is flat and covered with grass with an overall roughness length of 0.20 m (De Rooy and Holtslag, 1999; Van Ulden and Wieringa, 1996). Contrary to the other datsets, the SBL height for this dataset has been observed with a sodar with an uncertainty of about 40% (VH96). However, Arya (1981) and Hicks et al. (1977) found that h_{LLJ} is probably a suitable alternative to represent the observed h with a sodar.

A second Cabauw dataset ('Cabauw2') was gathered between August 2003 - April 2004. In this case the observed SBL height was derived from the method described in VH96 instead of the use of a sodar.

d)SHEBA

The SHEBA dataset was obtained over the Arctic ice pack north of Alaska between October 1997 and October 1998. Ice Station SHEBA drifted from approximately 75°N, 144°W to 80°N, 166°W. For this site, h_{LLJ} as observed from radiosondes was taken as h_{obs} provided that this height was also supported by a 'discontinuity' in the θ profile. Similar selection criteria as for CASES-99 have been applied, but with the additional restriction that RH < 0.985 to eliminate the frequently occurring saturated conditions.

e) SABLES98

The final dataset was gathered during the SABLES98 measurement campaign at the CIBA site (41.49°N, 5.47°W, see Cuxart et al., 2000). The observations consist of wind and temperature profiles from regularly launched radiosondes, and from sonic anemometers at 5.8, 13, and 32 m mounted on a tower.

6.3. Using Dimensional Analysis for Derivation an Equation for the Stable Boundary Layer Height

A) Three Dimensionless Groups

On the basis of earlier works, we identify that the relevant quantities to describe h are u_*, f, B_s and N. Using the Buckingham Π theory (e.g. Langhaar, 1951) we find three dimen -

sionless groups:

$$\Pi_1 = \frac{|B_s|}{hfu_* N}, \; \Pi_2 = \frac{kh|B_s|}{u_*^3} = \frac{h}{L}, \text{ and } \qquad \Pi_3 = \frac{N}{f}.$$

Consequently we may determine the functional form of the surface that describes the relationship between Π_1, Π_2 and Π_3 from observations. This should be a universal relationship if all relevant quantities are included.

Figure 6.2a shows Π_1 versus Π_2 on a linear scale for different classes of Π_3 using Sodankylä observations. Despite the small number of data per class, Π_2 clearly increases with Π_1, but levels off at different values for different classes of *N/f*. This relevance of *N/f* was already mentioned by Kitaigorodskii and Joffre (1988). Note that no data are available for $N/f < 50$ and $N/f > 300$. Furthermore we remark that the plotted dimensionless groups in Figure 6.2 have common terms, and the risk of spurious correlation exists (e.g. Baas et al., 2006). However, it turns out that by randomizing the current datasets the scatter increases. Moreover, below we also use dimensional plots to confirm the performance.

On a log-log scale (Figure 6.2b), we can determine the different slopes for different classes of *N/f*. We propose to fit the data according to $\Pi_1 = \alpha \Pi_2^{\beta_1 - \beta_2 \Pi_3}$. Applying this result and after some re-arrangement we find for *h*:

$$h = L \left(\frac{\left| \frac{g}{\theta} \overline{w\theta_s} \right|}{\alpha u_* f N L} \right)^{\lambda} \tag{5.1}$$

with $\alpha = 3$, $\lambda = (C_1 - 0.001 N/f)^{-1}$, and $C_1 = 1.8$. The calibration for α and C_l is discussed in detail below, L is the classical Obukhov length (thus including the Von Kármán constant). Note that the innovative aspect of this equation is that the exponent is not constant, but it depends on one of the dimensionless groups. The numerical value 0.001 in λ was found by plotting the slopes in Figure 6.5b for different classes of *N/f* (not shown). In addition, the obtained coefficients (based on Sodankylä observations) were confirmed by using each half of every dataset for calibration and the other halves for validation.

Considering the applicability range of Eq. (6.1), we have to realize that the denominator in the exponent should be positive, hence $N/f < 1800$. With the N=0.076 s^{-1} the maximum free flow stability in the dataset, this corresponds to a latitude $|\phi| > 16°$. In addition both N and L need to be larger than zero.

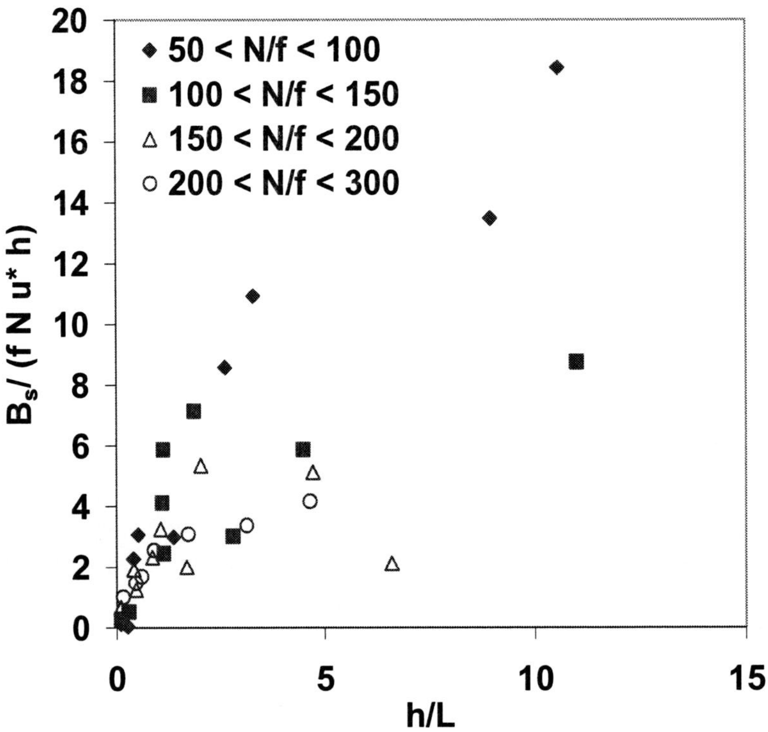

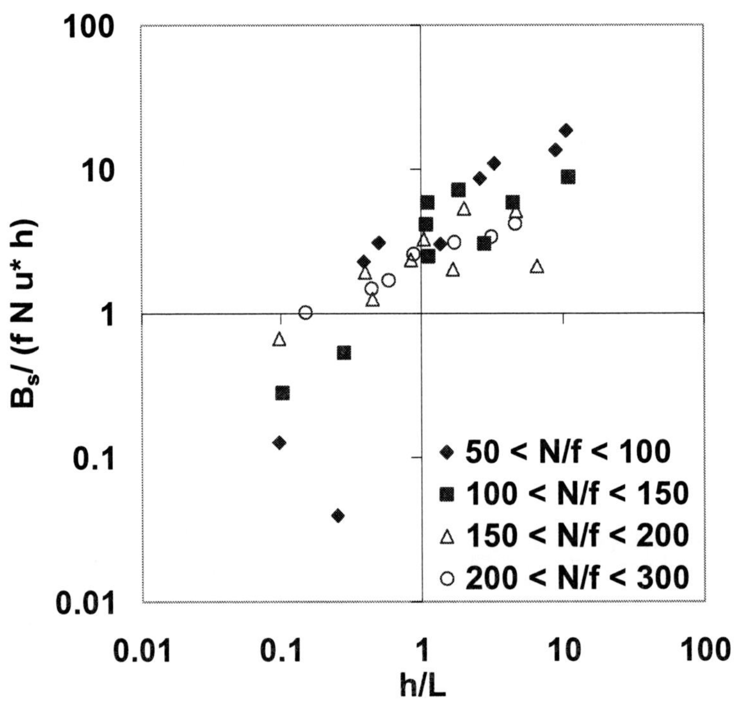

Figure 6.2. Dependence of observed dimensionless groups $B_s/(u_* fNh)$ versus h/L for different classes of N/f on a linear (a) and logarithmic scale (b).

A Monte-Carlo strategy was followed to estimate α and C_1 (on Sodankylä (Figure 6.3a) and on the whole dataset (Figure 6.3b)). A clear minimum in the meae is found in the contour plots with $C_1 = 1.8$ and $\alpha = 3$ as optimal estimates and is confirmed using the other statistical quality measures (not shown), and the parameters in Eq. (6.1) can thus be determined with good confidence.

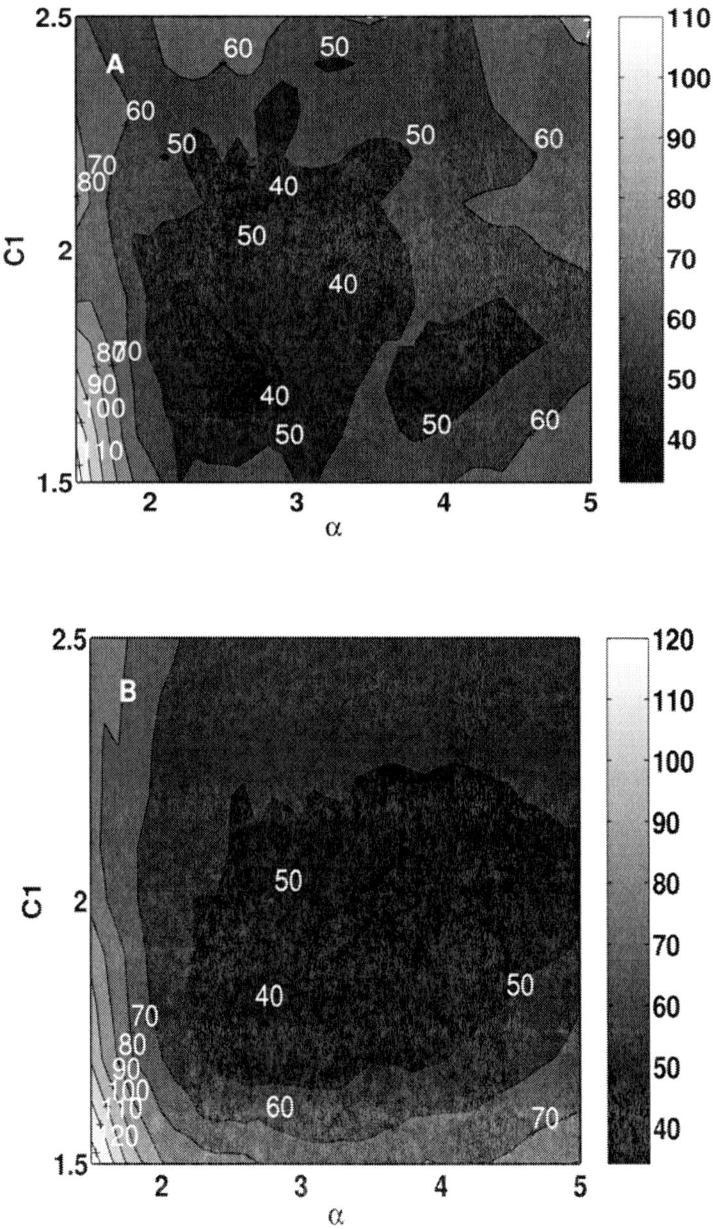

Figure 6.3. Contour plot of the median of absolute error (MEAE in m) for a range of C1 and □ for Sodankylä (a) and the total dataset (b).

b) Verification

In this Section we will *verify* the performance of Eq. (6.1) against the independent data from Cabauw, CASES-99, SHEBA (Figure 6.4b-d, Table 6.1) and a cross validation for Sodankylä (Figure 6.4a). The good performance for Sodankylä is obvious, since these are the same data as used for the calibration. Nevertheless, it seems that the data collapse onto a *single* curve. This is not trivial and it gives confidence in the method and the variables that we selected. The model agrees well with the CASES-99 (rmse = 53.6 m, but is largely unsystematic) and Cabauw observations (rmse = 80.3 m with rmse-u = 62.6 m), although the scatter is larger for the Cabauw dataset than for the other datasets. This relatively large scatter is probably inherent to the sodar based observations for Cabauw instead of radiosondes profiles for the other datasets (Section 6.3). For SHEBA the model performance is good (rmse = 40.3 m of which 37.6 m is unsystematic), although the model seems to underestimate the observations slightly (Figure 6.4d). Overall, the magnitude of FB < 0.15, which is not achieved for any of the other proposals, and IoA is larger than for Eqs. (6.1) and (6.2). The main improvement by Eq. (6.1) is achieved for shallow boundary layers. In addition to the evaluation of Eq (6.1) in Steeneveld et al (2007) we performed additional testing on a dataset of Cabauw from In addition we tested the Eq. (6.1) on the SABLES98 dataset (Cuxart et al., 2000). The last dataset is gathered in September 1998 over a flat plateau in Spain.

C) On the Relevance of the Coriolis Parameter F: Two Dimensionless Groups Only

The recent literature discusses the variables that govern h (Kosovic and Curry, 2000; Kosovic and Lundquist, 2004). Although f should theoretically play a role in governing h based on it presence in the Ekman equations (at least for neutral boundary layers), we can discuss the relevance of f as compared with N in practical application for the SBL as mentioned in VH96 and Zilitinkevich and Baklanov (2002).

Since free flow stability is always present in the atmosphere and is of order $O(10^{-2})$ while f is typically $O(10^{-4})$, VH96 suggest that the impact of f can be neglected in practice. Mahrt and Heald (1979) and VM04 also argue that the Coriolis parameter is of minor importance since the SBL development is governed by an inertial oscillation. In that case, a pure Ekman boundary layer does not exist and thus the use of f is doubtful.

In addition, if we analyze sodar observations during so-called intermittent nights (e.g. during CASES-99, Van de Wiel et al., 2003), we find that the boundary-layer turbulence and also the boundary-layer height responds quickly (< 10 min) to a change of the near surface turbulence intensity if decoupling from the surface occurs. This suggests that the SBL can react on a timescale much shorter than f^{-1} which is believed to be the governing timescale for the SBL (Nieuwstadt and Duynkerke, 1996). This implies that f^{-1} is not a priori the most dominant timescale for the SBL growth, but that timescales that originate from the interaction with the surface may be more important. Estournel and Guedalia (1990) and VM04 suggest that the roughness length for momentum (z_0) is a relevant quantity. Consequently, the relevance of f and z_0 will be examined below.

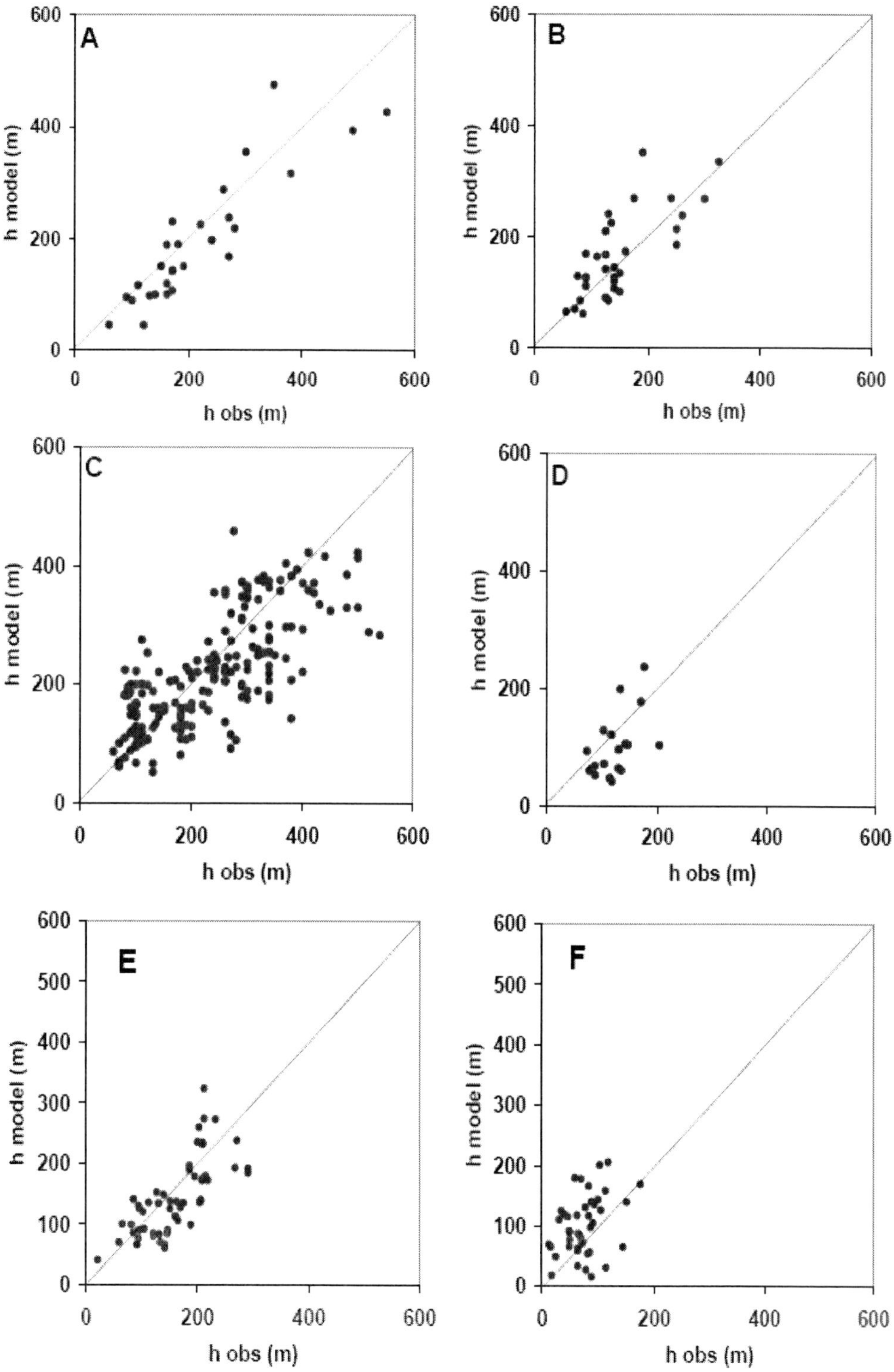

Figure 6.4. Observed and modeled stable boundary-layer height using Eq. (6.1) for Sodankylä (a), CASES-99 (b), Cabauw (c), SHEBA (d), Cabauw2 (e), and SABLES98 (f).

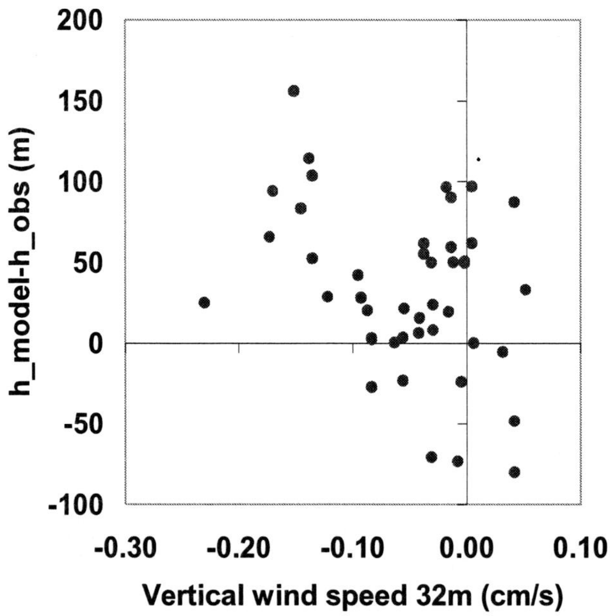

Figure 6.5. Difference between modeled and observed stable boundary layer height during SABLES98 versus vertical wind speed at 32 m.

Due to the variable nature of h, it is useful to adopt statistical techniques to gain insight into the relevant quantities that govern h. In this section, we perform a principal component analysis (PCA) on h_{obs} from all datasets to obtain information about the relative impact of the different variables on h_{obs}. Recall that PCA is a statistical technique in which the total variance of a dataset is decomposed along orthogonal vectors by determining the eigenvalues and eigenvectors of the covariance matrix between the variables. These eigenvectors are sorted in descending order according to the eigenvalues. The eigenvector associated with the largest eigenvalue is called the first principal component (FPC), the second large is called second principal component (SPC), etc. Finally, the data are transformed back into real space, and correlated to the original variables.

Table 6.1. Overview of statistical measure for the new proposals for the SBL height

Equation (6.1)	mae	rmse	rmse-s	rmse-u	meae	FB	IoA
Cabauw	61.5	80.3	50.3	62.6	49.9	-0.108	0.843
Sodankylä	71.1	109.7	77.1	78.1	45.5	-0.146	0.831
CASES-99	41.1	53.6	19.0	50.1	33.5	0.114	0.855
SHEBA	35.2	40.3	16.0	37.6	39.8	-0.137	0.746
Cabauw2	40.6	48.2	22.0	42.9	36.0	0.108	0.823
SABLES98	43.4	54.3	30.6	44.8	43.0	-0.297	0.562
Equation (6.2)							
Cabauw	65.1	86.7	57.4	65.0	46.4	0.042	0.795

Equation (6.1)	mae	rmse	rmse-s	rmse-u	meae	FB	IoA
Sodankylä	166.4	271.9	123.3	242.4	61.8	0.390	0.606
CASES-99	55.7	74.3	20.7	71.4	40.4	-0.125	0.753
SHEBA	225.0	349.2	193.2	290.9	86.5	0.857	0.135
Cabauw2	90.9	123.2	73.5	98.8	62.4	-0.368	0.584
SABLES98	62.3	91.7	30.8	86.4	33.2	-0.333	0.477
KB88: 700 u_*							
Cabauw	89.9	111.3	29.9	107.2	71.2	0.117	0.808
Sodankylä	79.0	105.3	44.0	95.6	60.0	0.092	0.880
CASES-99	60.7	72.4	35.3	63.2	51.8	-0.274	0.767
SHEBA	22.9	27.3	16.9	21.4	23.5	-0.147	0.853
Cabauw2	49.1	58.9	37.6	45.4	46.4	0.234	0.755
SABLES98	32.5	39.1	21.3	32.7	31.3	-0.058	0.618

Table 6.2 shows the absolute values of the correlation coefficients between the observed u_*, N, $\overline{w\theta_s}$, f and z_0 and the FPC. The FPC explained 99.9% of the variance, so the higher principal components can be neglected safely. It appears that the correlation coefficient between u_* and the FPC is large compared to the other coefficients. The quantities $\overline{w\theta_s}$ and N show a considerable smaller correlation but are still relevant. The Coriolis parameter and z_0 have a correlation coefficient of only 0.15 and 0.16 with the FPC respectively. Therefore, the latter quantities are *in practice* relatively unimportant for h estimation (at least for these datasets). Note that the dominance of u_* gives support to the simple estimate by Koracin and Berkowicz (1988).

Given the discussion above, we may exclude f and z_0 from the list of relevant variables, at least for estimation of h in practical applications. Then only two dimensionless groups remain, namely hN/u_* and h/L^* (Figure 6.6a, all data are used). Two regimes can be clearly distinguished. For $h/L^* < 1$ (towards the near neutral limit) $h \propto u_*/N$, in accordance with Kitaigorodskii and Joffre (1988), Van Pul et al., (1994), and VH96.

For $h/L^* > 1$ (towards the very stable limit) the two groups are linearly related on the log-log scale. This means that $h \propto \sqrt{|B_s|/N^3}$. Although, it seems that h is independent of u_* in this regime, this is not really the case.

Table 6.2. Correlation coefficient *r* between the relevant quantities and the First Principal Component

Parameter	u_*	N	$\overline{w\theta_s}$	f	z_0
r	0.75	0.47	0.37	0.15	0.16

The dependence of u_* comes in via B_s, because in the very stable regime the turbulent temperature scale θ_* is linearly dependent on u_* (Holtslag and De Bruin, 1988; Van de Wiel, 2002). Thus making B_s quadratically dependent on u_* means that h is again proportional to u_*, but now with a different factor depending on N. Thus our diagnostic equation for h based on the current analysis reads as:

$$h = \begin{cases} 10u_*/N & \text{for} \quad u_*^2 N/|B_s| > 10 \\ 32\sqrt{|B_s|/N^3} & \text{for} \quad u_*^2 N/|B_s| < 10 \end{cases} \tag{6.2}$$

Obviously the application of Eqs. (6.1) and (6.2) is limited to cases where N>0 s^{-1}.

An alternative formulation is developed with formal dimensional analysis with the same quantities as in the multi-limit equations. The proposed formulation is robust and appears to reduce a significant model bias for shallow boundary-layer heights in comparison with the existing formulations. As such, the proposed formulation appears applicable also for high stability conditions in contrast to existing formulations that primary are derived for weakly stable cases.

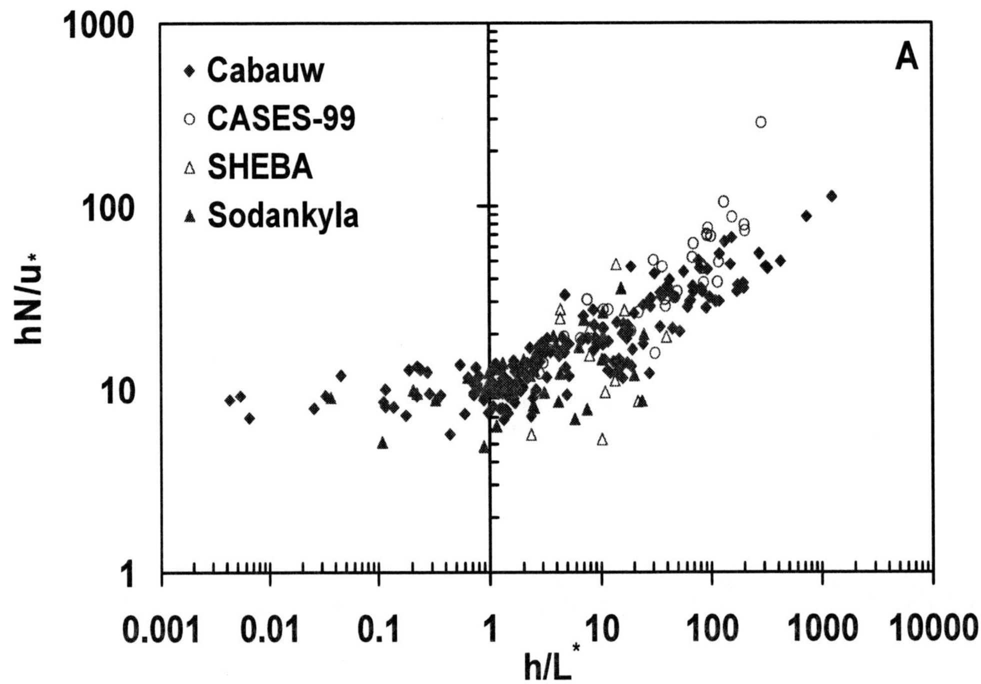

Figure 6.6a: Relationship between dimensionless groups hN/u_* and h/L^*.

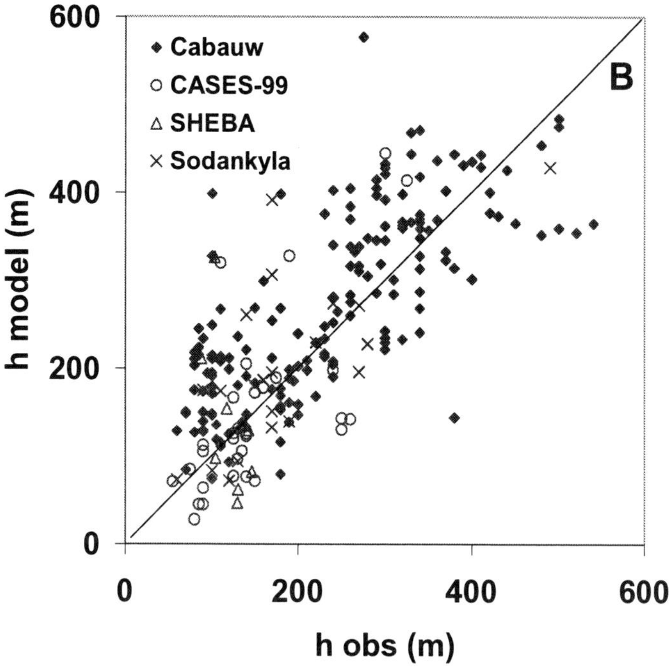

Figure 6.6b. Modeled and observed boundary-layer heights for Equation (6.2).

6.4. An Alternative Theoretically Based Formulation for the Stable Boundary Layer Height

Besides the observational point of view, the SBL height can also be studied more theoretically. The equilibrium SBL height h_E, and its relevance for predicting the SBL structure has been discussed intensively (e.g. Zilitinkevich and Esau, 2003, henceforth ZE03; Steeneveld et al., 2007). Recently, several papers (Zilitinkevich and Mironov, 1996; Zilitinkevich and Calanca, 2002; Zilitinkevich and Baklanov, 2002; ZE03) discuss the relevant processes that govern the SBL height in equilibrium conditions. In these studies, the basic variables governing h_E are the surface friction velocity u_* the surface buoyancy flux $B_s = \frac{g}{\theta}\overline{w\theta_s}$, the Coriolis parameter f and the free flow stability N. Based on these variables, ZE03 identified three boundary-layer prototypes: the truly neutral ($B_s = 0$ and $N = 0$), the conventionally neutral ($N \neq 0$ and $B_s = 0$) and the nocturnal boundary-layer (N=0 and $B_s \neq 0$).

6.4.1. Background

Following the reasoning by ZE03, the SBL depth is defined as the Ekman layer depth (h_*), which is given by the eddy diffusivity K_M and the Coriolis parameter f (Stull, 1988):

$$h_* = \sqrt{K_M / f} \qquad (6.3)$$

For the eddy viscosity K_M ZE03 distinguish between three different boundary layer types, and for each type a characteristic velocity scale u_T and length scale l_T:

Truly neutral $$K_M = u_T l_T = u_* h_* \tag{6.4}$$

Conventionally neutral $$K_M = u_T l_T = u_*^2 / N \tag{6.5}$$

Nocturnal $$K_M = u_T l_T = u_* L \tag{6.6}$$

with $L = -u_*^3 / B_s$ the Obukhov length (note the Von Karman constant is not included here). To obtain an equilibrium height that accounts for all three combined prototypes, the bulk diffusivity K_M can directly be written as:

$$\frac{1}{K_M} = \frac{1}{u_* h_*} + \frac{1}{u_*^2 / N} + \frac{1}{u_* L}, \tag{6.7}$$

Here the proportionality constants are taken 1 for convenience. Consequently,

$$K_M = \frac{u_*^2 h_* L / N}{u_* h_* / N + h_* L + u_* L / N} \tag{6.8}$$

Combining Eq. (6.8) in Eq. (6.3), solving for $h_* = h_E$ and choosing the physical solution in the quadratic equation, we obtain:

$$h_* = \alpha \frac{u_*}{N}, \tag{6.9}$$

where

$$\alpha = \frac{-1 + \sqrt{1 + 4\left(\frac{u_*}{fL} + \frac{N}{f}\right)}}{2\left(\frac{u_*}{NL} + 1\right)} \tag{6.10}$$

The format of Eq. (6.9) was already found earlier in many studies. VH96 found α to be a function of the shear and Richardson number across the SBL, while Steeneveld et al. (2007) derived Eq. (6.10) with α solely depending on the free-flow stability. In any case, Eqs. (6.9) and (6.10) show that α is related to the traditional parameter groups $u_* / (fL)$ (the Monin-Kazanski parameter) and N/f (Kitaigordskii and Joffre, 1988). The numerical value of α is typically 7-13 (e.g. VH96).

Here we compare our result with the proposal from ZE03 that reads as:

$$h_E = C_R \frac{u_*}{f}\left(1 + \frac{C_R^2 C_{uN}}{C_S^2}\frac{N}{f} + \frac{C_R^2}{C_S^2}\frac{u_*}{fL}\right)^{-1/2} \tag{6.11}$$

with C_R= 0.5; C_{uN}/C_S^2= 0.56 and C_S =1.0. Both proposals will be evaluated against the observations described in the next section.

6.4.3. Results

Results obtained with Eq. (6.11) and Eq. (6.9) with Eq. (6.10) are shown in Figs. 6.7 and 6.8, respectively. Table 6.3 summarizes some statistical quantities for model performance, i.e. mean absolute error (MAE), Systematic RMSE (RMSE-S), median of the mean absolute error (MEAE) and the index of agreement (IoA, Willmott, 1982. The IoA equals 1 for a perfect model performance). Equation (6.9) gives a substantial reduction of the RMSE-S, and an increased IoA compared to Eq. (6.11). Note that for shallow SBLs, mesoscale effects may become important and these may contribute to the bias, since mesoscale effects are not incorporated in the current model. Unfortunately, the proposed interpolation method cannot avoid the negative bias for shallow SBLs.

We realize that the evaluation of the above equations for the equilibrium depth with field data may be troublesome, due complexity of making observations in stable conditions and due to the fact that in reality conditions cannot be controlled. Alternatively, we may consider to explore Large Eddy Simulation results for more controlled testing (as in Esau, 2004). In that case however, we must be aware of the fact that especially in very stable conditions, LES results (profiles of mean and turbulent quantities) are strongly dependent on the model resolution (Beare and MacVean, 2004). Also longwave radiation divergence plays an important role, which is usually not taken into account by LES. Note that the field data used in this study cover a wide range of conditions, including nonturbulent effects such as radiation divergence (e.g. André and Mahrt, 1982).

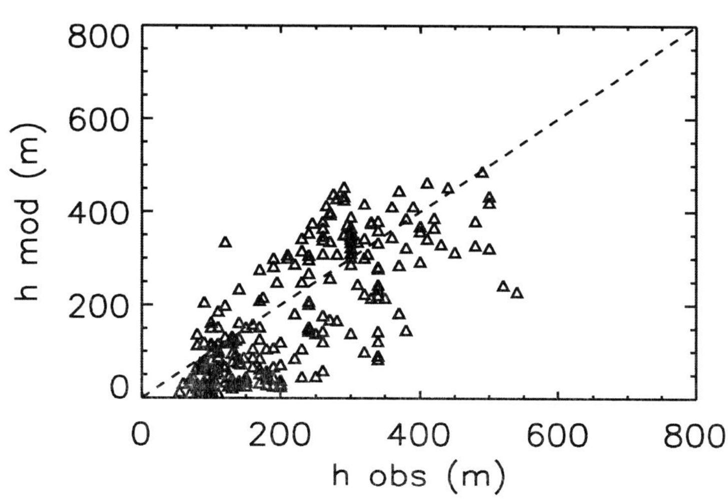

Figure 6.7. Modeled (Eq. 6.9) vs. observed stable boundary-layer height.

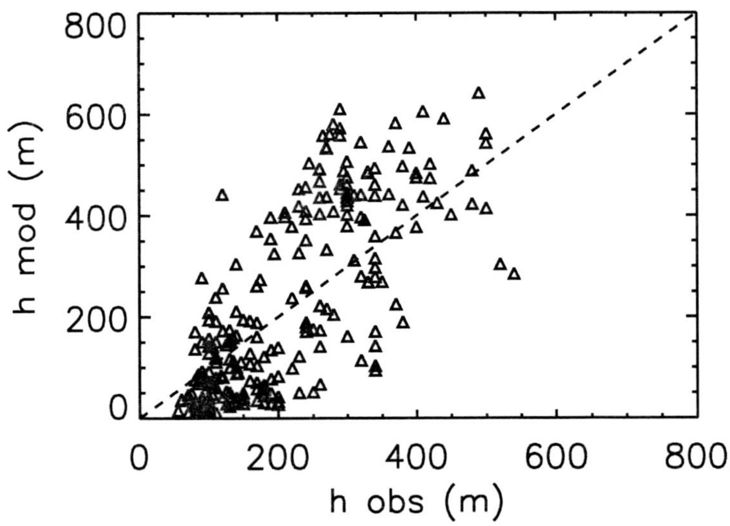

Figure 6.8. Modeled (Eq. 6.11) vs. observed stable boundary-layer height.

Table 6.3. Statistical Evaluation of SBL height proposals

Model	*MAE (m)*	*RMSE-S (m)*	*MEAE (m)*	*IoA (-)*
Eq. (6.10)	78.7	62.9	67.9	0.84
Eq. (6.11)	100.9	99.3	83.0	0.80

6.5. CONCLUSIONS

We propose an alternative method to derive a formula for the stable boundary-layer height when more than one stable boundary-layer prototype contributes to the final boundary-layer height. We directly interpolate the eddy diffusivities of each prototype. The alternative formulation performs well, and reduces the bias of the predicted stable boundary-layer height compared to the original formulation.

7. CLOSURE

This chapter discussed the role of the atmosphere, and especially the atmospheric boundary layer on local air quality. We have seen that the background air quality is controlled by the air mass, but that on the local scale the emissions are dispersed under the influence of the boundary layer wind speed, turbulence intensity and the mixing height. These are to a large extent determined by the atmospheric stratification, and these ingredients are required for accurate air quality modeling.

Mesoscale numerical models are used to forecast the local weather conditions that are needed for air quality forecasts. Different model strategies provide different outcome for turbulence intensity and mixing height. Especially for nighttime conditions (i.e. during stratification) the forecasts diverge between schemes. As such, a research effort into nighttime atmospheric conditions and model improvement is recommended. For example, the intermittent behavior of the nighttime turbulence was shown to be of considerable interest.

Atmospheric dispersion is normally expressed in terms of a weather dependent dispersion parameter. For continuous turbulent conditions, the correlation of lateral movement between released particles soon drops, and becomes approximately zero. However, field observations for calm conditions show meandering flows and intermittent behavior of the turbulence for substantial fraction of nights.

Finally, we discussed the currently available expressions for the nighttime boundary layer height. Recent updates have been validated against field observations, and are ready for use in dispersion models.

Air quality affects the human health and natural and agricultural vegetation, and accurate modeling is requested for correct warnings, and for testing the fulfillment of air quality legislation. Therefore, the current outcome may contribute to further understanding and improved modeling.

ACKNOWLEDGEMENTS

We would like to acknowledge the technical staff of our department for operating the Haarweg field in Wageningen. Furthermore, we thank our MSc students Bram van Kesteren and Miranda Braam for their contribution in section 6. Also, we would like to thank the Royal Netherlands Meteorological Institute for providing the necessary observations from the Cabauw tower. Prof. Joan Cuxart and Daniel Martinez are thanked for providing the SABLES98 observations. The authors wish to thank Sylvain Joffre and Markku Kangas for providing the Sodankylä dataset and their careful selection of the dataset. We thank the SHEBA Atmospheric Surface Flux Group, Ed Andreas, Chris Fairall, Peter Guest, and Ola Persson for help collecting and processing the data. The National Science Foundation supported this research with grants to the U.S. Army Cold Regions Research and Engineering Laboratory, NOAA's Environmental Technology Laboratory, and the Naval Postgraduate School. Furthermore, we thank our colleague Oscar Hartogensis for gathering the surface observations in CASES-99. We thank the Royal Netherlands Meteorological Institute for gathering and providing the Cabauw observations. We also acknowledge the GABLS community, whose LES model results have been used in this study. Finally we thank our colleague Dr. Leo Kroon his valuable comments on our work. Finally G.J. Steeneveld acknowledges the project Climate Changes Spatial Planning (Klimaat voor Ruimte).

REFERENCES

André, J.C., 1983: On the variability of the nocturnal boundary-layer height. *J. Atmos. Sci.*, 40, 2309-2311.

André J.C., and L. Mahrt, 1982: The nocturnal surface inversion and influence of clear-air radiative cooling. *J. Atmos. Sci.*, 39, 864-878.

Anfossi, D., P. Bacci and A. Longhetto, 1976: Forecasting of vertical temperature profiles in the atmosphere during nocturnal radiation inversions form air temperature trend at screen height, *Quart. J. R. Met. Soc*, 102, 173-180.

Anfossi, D., D. Oettl, G. Degrazia, and A. Goulart, 2005: An Analysis of sonic anemometer observations in low wind conditions, *Bound.-Layer Meteor.*, 114, 179-203.

Arya, S.P.S., 1981: Parameterizing the height of the stable atmospheric boundary layer. *J. Appl. Meteor.*, 20, 1192-1202.

Baas, P., G.J. Steeneveld, B.J.H. van de Wiel, and A.A.M. Holtslag, 2006: Exploring Self-correlation in Flux-Gradient Relationships for Stably Stratified Conditions. *J. Atmos. Sci.*, 63, 3045-3054.

Banta, R.M., Y.L. Pichugina, W.A. Brewer, 2006: Turbulent velocity profiles in the stable boundary layer generated by a nocturnal low-level jet, *J. Atmos. Sci.*, 63, 2700-2719.

Beare, R.J. and M.K. MacVean, 2004: Resolution sensitivity and scaling of large-eddy simulations of the stable boundary –layer, *Boundary-Layer Meteorol.*, 112, 257-281.

Beljaars, A.C.M., and F.C. Bosveld, 1997: Cabauw Data for Validation of Land Surface Parametrization Schemes. *J. Climate*, 10, 1172-1193.

Berg, L.K., and S. Zhong, 2005: Sensitivity of MM5 Simulated Boundary-Layer Characteristics to Turbulence Parameterization. *J. Appl. Meteor.*, 44, 1467-1483.

Beyrich, F., 1994: Sodar observations of the stable boundary-layer height in relation to the nocturnal low-level jet. *Meteor. Z.*, 3, 29-34.

Beyrich, F., and V. Kotroni, 1993: Estimation of Surface Stress over a Forest from Sodar measurements and its use to parameterize the stable boundary layer height. *Bound.-Layer Meteor.*, 66, 93-103.

Bosveld, F.C., E.I.F. de Bruijn, and A.A.M. Holtslag: Intercomparison of single-column models for GABLS3 – preliminary results. *18th Symp. on Bound. Layers and Turb.*, 9-13 June 2008, Stockholm, Sweden, P. A8.5.

Boybeyi Z., S. Raman, P Zanetti, 1995: Numerical investigation of possible role of local meteorology in Bhopal gas accident, *Atmos. Environ.*, 29, 479-496.

Braun, S.A., and W.K. Tao, 2000: Sensitivity of high-resolution simulations of hurricane Bob (1991) to planetary boundary layer parameterizations. *Mon.Wea.Rev.*, 128, 3941-3961.

Bright, D.R., and S.L. Mullen, 2002: The Sensitivity of the Numerical of the Southwest Monsoon Boundary Layer to the Choice of PBL Turbulence Parameterization in MM5. *Wea. Forecasting*, 17, 99-114.

Charron, A. and R.M. Harrison, 2005: Fine ($PM_{2.5}$) and Coarse ($PM_{2.5-10}$) Particulate Matter on A Heavily Trafficked London Highway: Sources and Processes. *Environ. Sci. Technol.*, 39, 7768 -7776.

Cheng, W.Y.Y., and W.J. Steenburgh, 2005: Evaluation of surface sensible weather forecasts by the WRF and Eta models over the western United States. *Wea. For.*, 20, 812-821.

Coulter, R.L., and J.C. Doran, 2002: Spatial and temporal occurrences of intermittent turbulence during CASES-99. *Bound.-Layer Meteor.*, 105, 329-349.

Couvreux, F., F. Guichard, J.-L. Redelsperger, C. Kiemle, V. Masson, J.-P. Lafore, and C. Flamant, 2005: Water-vapour variability within a convective boundary-layer assessed by large-eddy simulations and IHOP-2002 observations. *Quart. J. Roy. Meteor. Soc.*, 131, 2665-2693.

Cuxart, J., M.A. Jiménez, 2007: Mixing proceses in a nocturnal low-level jet, an LES study. *J. Atmos. Sci.*, 64, 1666-1679.

Cuxart, J., C. Yagüe, G. Morales, E. Terradellas, J. Orbe, J. Calvo, A. Fernández, M.R. Soler, C. Infante, P. Buenestado, A. Espinalt, H.E. Joergensen, J.M. Rees, J. Vilà, J.M. Redondo, I.R. Cantalapiedra, and L. Conangla, 2000: Stable Atmospheric Boundary-Layer Experiment in Spain (SABLES 98): A Report. *Bound.-Layer Meteor.,* 96, 337-370.

Cuxart, J., A.A.M. Holtslag, R.J. Beare, E. Bazile, A.C.M. Beljaars, A. Cheng, L. Conangla, M. Ek, F. Freedman, R. Hamdi, A. Kerstein, H. Kitagawa, G. Lenderink, D. Lewellen, J. Mailhot, T. Mauritsen, V. Perov, G. Schayes, G.J. Steeneveld, G. Svensson, P.A. Taylor, W. Weng, S. Wunsch, and K-M. Xu, 2006: A single-column model intercomparison for a stably stratified atmospheric boundary layer. *Bound.-Layer Meteor.*, 118, 273-303.

Dijk, A. van, A.F. Moene, and H.A.R. de Bruin, 2004: The principles of surface flux physics: theory, practice and description of the ECPACK library. Wageningen Univ. Internal report 2004/1, Wageningen, The Netherlands, 99 pp.

Drüe, C., and G. Heinemann, 2007: Characteristics of intermittent turbulence in the upper stable boundary layer over Greenland, *Bound.-Layer Meteor.*, 124, 361-381.

Eerden, L,. van, W. de Vries, H. van Dobben: Effects of Ammonia Deposition on Forests in The Netherlands. *Atmos. Environ.*, 32,525-532.

Einaudi, F., and J.J. Finnigan, 1981: The interaction between an internal gravity waves and the planetary boundary layer. Part I: The linear analysis, *Quart. J. Roy. Meteor. Soc.*, 107, 793-806.

Ek, M.B., K.E. Mitchell, Y. Lin, E. Rogers, P. Grunmann, V. Koren, G. Gayno, and J.D. Tarpley, 2003, Implementation of the Noah land surface model advances in the National Centers for Environmental Prediction operational mesoscale Eta model. *J. Geophys. Res*, 108 (D22), 8851, doi:10.1029/2002JD003296.

Esau, I., 2004: Simulation of Ekman boundary layers by large eddy model with dynamic mixed sub-filter closure. *Env. Fluid Mech.*, 4, 273-303.

Estournel, C., and D. Guedalia, 1990: Improving the diagnostic relation for the nocturnal boundary layer height. *Bound.-Layer Meteor.*, 53, 191-198.

Garratt, J.R., 1982: Surface fluxes and the nocturnal boundary layer height. *J. Appl. Meteor.*, 21, 725-729.

Garratt, J.R., 1985: The inland boundary layer at low latitudes, I. The nocturnal jet, *Bound.-Layer Meteor.*, 32, 307-327.

Garratt, J.R., 1992: *The Atmospheric Boundary Layer*, Cambridge Univ. Press, 316 pp.

Garratt, J.R., and R.A. Brost, 1981: Radiative Cooling effects within and above the nocturnal boundary-layer. *J. Atmos. Sci.* 38, 2730-2746.

Gassmann, M.I., and N.A. Mazzeo, 2001: Nocturnal stable boundary layer height model and its application. *Atmos. Res.*, 57, 247-259.

Godish, T., 2003: *Air Quality*, 4th edition, Lewis Publishers, New York, USA, 460 p.

Gryning, S.E., A. A. M. Holtslag, J. S. Irwin, and B. Siversten, 1987: Applied dispersion modelling based on meteorological scaling parameters. *Atmos. Environ.*, 21, 79-89.

Górska, M., J. Vilà-Guerau de Arellano, M.A, LeMone, C.C van Heerwaarden, 2008: Mean and flux horizontal variability of virtual potential temperature, moisture and carbon dioxide: Aircraft observations and LES study. *Mon.Wea.Rev.*, in press.

Halldin (ed.), 1999: *Final Report for WINTEX*, NOPEX Technical Report No. 29,70 pp. (Available from: NOPEX Central Office, Institute of Earth Sciences, Uppsala University, Norbyvägen 18B, SE-75236 Uppsala, Sweden).

Hanna, S.R., and R. Yang, 2001: Evaluation of meso scale models' simulations of near-surface winds, temperature gradients, and mixing depths. *J. Appl. Meteor.*, 40, 1095-1104.

Hartogensis, O. K., H. A. R. de Bruin, 2005: Monin-Obukhov similarity functions of the structure parameter of temperature and turbulent kinetic energy dissipation rate in the stable boundary layer. *Bound.-Layer Meteor.*, 116, 253-276.

Hicks, R.B., D. Smith, P.J. Irwin, and T. Mathews, 1977: Preliminary studies of atmospheric acoustic sounding at Calgary. *Bound.-Layer Meteor.*, 12, 201-212.

Hiscox, A.L., C.J. Nappo, and D.R. Miller, 2006: On the Use of Lidar Images of Smoke Plumes to Measure Dispersion Parameters in the Stable Boundary Layer. *J. Atmos. Oceanic Technol.*, 23, 1150–1154.

Hoch, S.W., P. Calanca, R. Philipona, and A. Ohmura, 2007: Year-Round Observation of Longwave Radiative Flux Divergence in Greenland. *J. Appl. Meteor. Clim.*, 45, 1469-1479.

Holtslag, A.A.M., 2006: Special issue for boundary-layer meteorology: GEWEX Atmospheric Boundary-Layer Study (GABLS) on stable boundary layers. *Bound.-Layer Meteor.*, 118, 243-246.

Holtslag A.A.M., F.T.M. Nieuwstadt, 1986: Scaling the atmospheric boundary layer. *Bound.-Layer Meteor.*, 36, 201-209.

Holtslag, A.A.M., and Bruin, H.A.R. de, 1988: Applied modelling of the nighttime surface energy balance over land. *J. Appl. Meteor.* 27, 689-704.

Hong, S.Y., and H.L. Pan, 1996: Nonlocal boundary layer vertical diffusion in a Medium-Range Forecast model. *Mon. Wea. Rev.*, 124, 2322-2339.

Howell, J.F., and J. Sun, 1999: Surface-Layer Fluxes in Stable Conditions. *Bound.-Layer Meteor.*, 90, 495-520.

Hunt, J.C.R., J.C. Kaimal, and J.E. Gaynor, 1985: Some observations of turbulence structure in stable layers. *Quart. J. Roy. Meteor. Soc.*, 111, 793-815.

Jacobs, A.F.G., B.G. Heusinkveld, and A.A.M. Holtslag, 2003: Carbon dioxide and water vapour flux densities over a grassland area in the Netherlands, *Int. J. Clim.*, 23, 1663-1675.

Jacobs, A.F.G., B.G. Heusinkveld, R. J. Wichink Kruit, and S. M. Berkowicz, 2006, Contribution of dew to the water budget of a grassland area in the Netherlands, *Water Resour. Res.*, 42, W03415, doi:10.1029/2005WR004055.

Janjić, Z.I., 1990: The step-mountain Coordinate: Physical Package. *Mon. Wea. Rev.*, 118, 1429-1443.

Joffre, S.M., M. Kangas, M. Heikinheimo, and S.A. Kitaigorodskii, 2001: Variability of the stable and unstable boundary layer height and its scales over a boreal forest. *Bound.-Layer Meteor.*, 99, 429-450.

Joffre S., and M. Kangas, 2002: Determination and scaling of the atmospheric boundary layer height under various stability conditions over a rough surface. In: Rotach M., Fisher B., Piringer M. (Eds.), COST *Action 715 Workshop on Urban Boundary Layer Parameterisations*, Zurich, 24-25 May 2001. Office for Official Publications of the European Communities, EUR 20355, pp. 111-118.

Kain J. S., and J. M. Fritsch, 1993: Convective parameterization for mesoscale models: The Kain–Fritsch scheme. The Representation of Cumulus Convection in Numerical Models. *Meteor. Monogr.*, 24, Amer. Meteor. Soc., 165-170.

Karppinen, A., S. Joffre, J. Kukkonen, and P. Bremer, 2001: Evaluation of inversion strengths and mixing heights during extremely stable atmospheric stratification. *Int. J. Environment and Pollution*, 16, No 1-6, 1-11.

Kitaigorodskii, S.A., and S. Joffre, 1988: In search of a simple scaling for the height of the stratified atmospheric boundary layer. *Tellus,* 40A, 419-433.

Koracin, D., and R. Berkowicz, 1988: Nocturnal boundary-layer height: observations by acoustic sounders and prediction in terms of surface-layer parameters. *Bound.-Layer Meteor.*, 43, 65-83.

Kosovic, B., and J.A. Curry, 2000: A Large Eddy Simulation study of a quasi-steady, stably stratified atmospheric boundary-layer. *J. Atmos. Sci.*, 57, 1052-1068.

Kosovic, B., and J.K. Lundquist, 2004: Influences on the height of the stable boundary layer, *16th Symposium on Boundary-layer and Turbulence*, Portland, Maine, U.S.A., 9-13 Aug. 2004, Amer. Meteorol. Soc., Boston, p. 4.20.

Kurzeja, R.J., S. Berman, and A.H. Weber, 1991: A climatological study of the nocturnal planetary boundary layer. *Bound.-Layer Meteor.*, 105, 105-128.

Langhaar, H.L., 1951: *Dimensional Analysis and Theory of Models*, Wiley, New York, 166p.

Lena, F., and F. Desiato, 1999: Intercomparison of nocturnal mixing height estimate methods for urban air pollution modeling. *Atmos. Environ.* 33, 2385-2393.

Mahrt, L., 1981: Modelling the height of the stable boundary-layer. *Bound.-Layer Meteor.*, 21, 3-19.

Mahrt, L., 1987: Grid-averaged surface fluxes. *Mon. Wea. Rev.* 115, 1550-1560.

Mahrt, L., J. Sun, W. Blumen, T. Delany, S. Oncley, 1998: Nocturnal boundary-layer regimes, *Bound. Meteor.*, 88, 255-278.

Mahrt, L., 1998: Stratified atmospheric boundary layers and breakdown of models, *Theo. Comp. Fluid. Phys.*, 11, 263-279.

Mahrt, L., 1999: Stratified atmospheric boundary layers. *Bound.-Layer Meteor.*, 90, 375-396.

Mathieu, N., I.B., Strachan, M.Y. Leclerc, A. Karipot and E.Pattey, 2005: Role of low-level jets and the boundary-layer properties on the NBL budget technique, *Agr. For. Meteorol.*, 135, 35-43.

Melgarejo, J.W., and J.W. Deardorff, 1974: Stability Functions for the Boundary-Layer Resistance Laws Based upon Observed Boundary-Layer Heights. *J. Atmos. Sci.*, 31, 1324-1333.

Mellor, G.L., and T. Yamada, 1974: A Hierarchy of Turbulence Closure Models for Planetary Boundary Layers. *J. Atmos. Sci.*, 31, 1791-1806.

Nappo, C.J., 1991: Sporadic breakdowns of stability in the PBL over simple and complex terrain. *Bound.-Layer Meteor.*, 54, 69-87.

Nappo, C.J., 2002: *An Introduction to Atmospheric Gravity Waves*, Academic Press, London, 276 pp.

Nappo, C.J., and G. Chimonas, 1992: Wave Exchange between the Ground Surface and a Boundary-Layer Critical Level. *J. Atmos. Sci.*, 49, 1075-1091.

Nappo, C.J., and P.-E. Johansson, 1999: Summary of the Lövånger international workshop on turbulence and diffusion in the stable planetary boundary layer. *Bound.-Layer Meteor.*, 90, 345-374.

Nappo, C.J., H-Y Chun, and H.J. Lee, 2004: A parameterization of wave stress in the planetary boundary layer for use in mesoscale models. *Atmos. Environ.*, 38, 2665-2675.

Neu, U., 1995: A parameterisation of the nocturnal ozone reduction in the residual layer by vertical downward mixing during summer smog situations using sodar data. *Bound.-Layer Meteor.*, 73, 189-193.

Newsom, R.K., and R.M. Banta, 2003: Shear-flow instability in the stable nocturnal boundary layer as observed by Doppler lidar during CASES-99. *J. Atmos. Sci.*, 60, 16-33.

Nieuwstadt, F.T.M, 1980a: A rate equation for the inversion height in a nocturnal boundary layer. *J. Appl. Meteor.*, 19, 1445-1447.

Nieuwstadt, F.T.M., 1980b: Some aspects of the turbulent stable boundary layer. *Bound.-Layer Meteor.*, 30, 31-55.

Nieuwstadt, F.T.M., 1984: The turbulent structure of the stable, nocturnal boundary-layer. *J. Atmos. Sci.* 41, 2202-2216.

Nieuwstadt, F. T. M., and H. Tennekes, 1981: A rate equation for the nocturnal boundary-layer height. *J. Atmos. Sci.*, 38, 1418-1429.

Nieuwstadt, F.T.M., and P.G. Duynkerke, 1996: Turbulence in the atmospheric boundary layer. *Atmos. Res.*, 40, 111-142.

Noh, Y., W.G. Cheon, S.Y. Hong, S. Raasch, 2003: Improvement of the K-profile model for the planetary boundary layer based on large eddy simulation data, *Bound.-Layer Meteorol.*, 107, 401-427.

Pahlow, M., M.B. Parlange and F. Porte-Agel, 2001: On Monin-Obukhov similarity in the stable atmospheric boundary layer, *Bound.-Layer Meteor*, 99, 225-248.

Pattantyús-Ábrahám, M., and I.M. Jánosi, 2004: What determines the nocturnal cooling timescale at 2m? *Geophys. Res. Lett.*, 31, L05109, doi:10.1029/2003GL019137.

Lin, X. Pielke, R.A. Hubbard, K.G. Crawford, K.C. Shafer, M.A. Matsui, T., 2007: An examination of 1997-2007 surface layer temperature trends at two heights in Oklahoma. *Geophys. Res. Lett.*, 34, L24705-L24708.

Pasquill, F., and F.B. Smith, 1983: *Atmospheric Diffusion*, Ellis Horwood, Chichester, England, 437 pp.

Pino, D., H.J.J. Jonker, and J. Vila-Guerau de Arellano, 2006a: Role of shear and the inversion strength during sunset turbulence over land: characteristic length scales, *Bound.-Layer. Meteor.*, 121, 537-556.

Pino, D., J. Vilà-Guerau de Arellano, and S.W. Kim, 2006b: Representing sheared convective boundary layer by zeroth- and first-order-jump mixed-layer models: Large-Eddy Simulation verification. *J. Appl. Meteor. Clim.*, 45, 1224–1243.

Pollard, R. T., P. B. Rhines, and R.O.R.Y. Thompson, 1973: The deepening of the wind-mixed layer. *Geophys. Fluid Dyn.* 3, 381-404.

Poulos, G.S., and co-authors, 2002: CASES-99: A comprehensive Investigation of the stable nocturnal boundary layer. *Bull. Am. Met. Soc.*, 83, 555-581.

Poulos, G.S., and S.P. Burns, 2003: An evaluation of Bulk Ri-Based Surface Layer Flux Formulas for Stable and Very Stable Conditions with Intermittent Turbulence. *J. Atmos. Sci*, 60, 2523-2537.

Pul, W.A.J. van, A.A.M. Holtslag, and D.P.J. Swart, 1994: A Comparison of ABL heights inferred routinely from LIDAR and radiosondes at noon. *Bound.-Layer Meteor.*, 68, 173-191.

ReVelle, D.O., 1993: Chaos and "bursting" in the planetary boundary-layer. *J. Appl. Meteor.* 32, 1169-1180.

Reitebuch, O., A Strassburger, S. Emeis, W. Kuttler, 2000: Nocturnal secondary ozone concentration maxima analysed by sodar observations and surface measurements, *Atmos. Environ.*, 34, 4315-4329.

Roelofs, J.G.M., 1986: The effects of airborne sulphur and nitrogen deposition on aquatic and terrestrial heathland vegetation, *Experentia*, 42, 372-377.

Rooy, W.C. de, and A.A.M. Holtslag, 1999: Estimation of surface radiation and energy flux densities from single-level weather data. *J. Appl. Meteor.*, 38, 526-540.

Salmond, J.A., and I.G. McKendry, 2005: A review of turbulence in the very stable boundary layer and its implications for air quality. *Progress in Phys. Geography*, 29, 171-188.

Seibert, P., F. Beyrich, S.E. Gryning, S. Joffre, A. Rasmussen, and P. Tercier, 2000: Review and intercomparison of operational methods for the determination of the mixing height. *Atmos. Environ.*, 34, 1001-1027.

Sharan, M., S. Gopalakrishnan, R. McNider, and M. Singh, 1996: Bhopal Gas Leak: A Numerical Investigation of the Prevailing Meteorological Conditions. *J. Appl. Meteor.*, 35, 1637–1657.

Stedman, J.R., 2004. The predicted number of air pollution related deaths in the UK during the August 2003 heat wave. *Atmos. Environ.* 38, 1087-1090.

Steeneveld, G.J., B.J.H. van de Wiel, and A.A.M. Holtslag, 2006a: Modelling the Arctic nocturnal stable boundary layer and its coupling to the surface. *Bound.-Layer Meteor.*, 118, 357-378.

Steeneveld, G.J., B.J.H. van de Wiel, and A.A.M. Holtslag, 2006b: Modeling the evolution of the atmospheric boundary layer for three contrasting nights in CASES-99. *J. Atmos. Sci.*, 63, 920-935.

Steeneveld, G.J., B.J.H. van de Wiel, and A.A.M. Holtslag, 2007: Diagnostic Equations for the stable boundary layer height: evaluation and dimensional analysis, *J. Appl. Meteor. Clim.,* 46, 212-225.

Steeneveld, G.J., A.A.M. Holtslag, C.J. Nappo, B.J.H. van de Wiel, and L. Mahrt, 2008: Exploring the role of small-scale terrain drag on stable boundary layers over land, *J. Appl. Meteor. Clim.,* 47, 2518-2530.

Stull, R.B, 1988: *An Introduction to Boundary-layer Meteorology*, Kluwer Academic Publishers, Dordrecht, 666 pp.

Taylor, G.I., 1921: Diffusion by continuous movements. *Proc. London Math. Soc.*, 20, 196-202.

Tennekes, H. 1982: Similarity Relations, Scaling laws and spectral dynamics, in: *Atmospheric Turbulence and Air Pollution Modelling*, Eds: F.T.M. Nieuwstadt and H. van Dop, Reidel Publ, Dordrecht, Netherlands, 358 pp.

Troen, I.B., and L. Mahrt, 1986: A simple model of the atmospheric boundary layer; sensitivity to surface evaporation. *Bound.-Layer Meteor.*, 37, 129-148.

Torrence, C., and G. P. Compo, 1998: A practical guide to wavelet analysis, *Bull. Amer. Meteor. Soc.*, 79, 61-78.

Ulden, A.P. van, and A.A.M Holtslag, 1985: Estimation of atmospheric boundary layer parameters for diffusion applications, *J. Clim. Appl. Meteor.*, 24, 1196-1207.

Ulden, A. P. van, J. Wieringa, 1996: Atmospheric boundary layer research at Cabauw. *Bound.-Layer Meteor.*, 78, 39-69.

Vickers, D., and L. Mahrt, 2004: Evaluating formulations of the stable boundary layer height. *J. Appl. Meteor.*, 43, 1736-1749.

Vilà-Guerau de Arellano, J., Vellinga, O.S., Holtslag, A.A.M., Bosveld, F.C., and H. Klein Baltink, 2001: Observational evaluation of PBL parameterization modeled by MM5, *11th PSU/NCAR Mesoscale Model Workshop Users's workshop, 25-27 June Boulder, Colorado,* 102-104.

Vogelezang, D.H.P., and A.A.M. Holtslag, 1996: Evaluation and model impacts of alternative boundary-layer height formulations. *Bound.-Layer Meteor.*, 81, 245-269.

Wiel, B.J.H van de, 2002: Intermittent turbulence and oscillations in the stable boundary layer over land, PhD thesis, Wageningen University, Wageningen, 129p.

Willmott, C.J., 1982: Some Comments on the Evaluation of Model Performance, *Bull. Amer. Meteor. Soc*, 63, 1309-1313.

Yagüe, C., and J.M. Redondo, 1995: A case study of turbulent parameters during the Antarctic winter. *Antarctic Sci.*, 7, 421-433.

Yamada, T., 1979: Prediction of the nocturnal surface inversion height. *J. Appl. Meteor.*, 18, 526-531.

Yu, T.W., 1978: Determining height of the Nocturnal Boundary Layer. *J. Appl. Meteor.*, 17, 28-33.

Zilitinkevich, S.S., 1972: On the determination of the height of the Ekman boundary layer. *Bound.-Layer Meteor.*, 3, 141-145.

Zilitinkevich, 2002: Third-order transport due to internal waves and non-local turbulence in the stably stratified surface layer. *Quart. J. Roy. Meteor. Soc.*, 128, 913-925.

Zilitinkevich S.S., and D.V. Mironov, 1996: A multi-limit formulation for the equilibrium height of a stably stratified boundary layer. *Bound.-Layer Meteor.*, 81, 325-351.

Zilitinkevitch, S.S., and P. Calanca, 2000: An extended similarity theory for the stably stratified atmospheric surface layer. *Quart. J. R. Met. Soc.*, 126, 1913-1923.

Zilitinkevich, S.S., and A. Baklanov, 2002: Calculation of the height of the stable boundary layer in practical applications. *Bound.-Layer Meteor.*, 105, 389-409.

Zilitinkevich, S.S., and I.N. Esau, 2003: The effect of baroclinicity on the equilibrium depth of the neutral and stable planetary boundary layers. *Quart. J. Roy. Meteor. Soc.*, 129, 3339-3356.

Zilitinkevich, S.S., V.L. Perov, and J.C. King, 2002: Near-surface turbulent fluxes in stable stratification: Calculation techniques for use in general circulation models. *Quart. J. Roy. Meteor. Soc.*, 128, 1571-1587.

Zilitinkevich, S.S., A. Baklanov, J. Rost, A.S. Smedman, V. Lykosov, and P. Calanca, 2002: Diagnostic and prognostic equations for the height of the stably stratified Ekman boundary layer. *Quart. J. Roy. Meteor. Soc.*, 128, 25-46.

In: Air Quality in the 21st Century
Editors: G. C. Romano and A. G. Conti, pp. 115-138
ISBN 978-1-60456-793-9

Chapter 3

A QUANTITATIVE COMPARISON OF ANGSTROM'S TURBIDITY PARAMETERS (α,β) RETRIEVED IN DIFFERENT SPECTRAL RANGES BASED ON SPECTRAL SOLAR EXTINCTION MEASUREMENTS

C. P. Jacovides and D. N. Asimakopoulos
Department of Environmental Physics-Meteorology, Athens University Campus, Build PHYS-V, Athens 157 84, Greece.
E–mail address: kiakovid@phys.uoa.gr, Tel: ++30/210-7276931, FAX: ++30/ 2107295281

ABSTRACT

Using ground-based spectral solar extinction data taken in the greater urban Athens atmosphere during a field survey, experimental and modeled aerosol optical depths have been retrieved. The Angstrom's turbidity parameters (**α, β**) were derived via three widely used techniques: the Volz, the logarithmic and the direct fitting methods to the experimental aerosol optical depths. This chapter investigates the ability of the different methods to determine similar Angstrom's turbidity parameters further addressing their dependence on the spectral ranges considered for their derivation. The results obtained reveal that the various techniques result in different pairs of the Angstrom's parameters mainly at the shorter wavelengths. It also was found that the Angstrom's turbidity parameters obtained by any of the three methods at shorter wavebands are not representative of the whole spectrum exhibiting large uncertainties especially under low turbidity levels. From the overall analysis it is established that the logarithmic method predict almost identical results as those obtained via the direct fitting method, also possessing the least wavelength dependence.

Keywords*:* Aerosol optical depth; Angstrom parameters; Wavelength-turbidity dependence; Different techniques.

INTRODUCTION

Atmospheric aerosol particles affect the Earth's radiation budget directly by scattering and absorbing solar and terrestrial radiation and indirectly by modifying the physical and radiative properties of clouds (Charlson et al. 1991; Penner et al. 2002). However, quantification of the change in radiation is difficult because aerosol distribution varies greatly in type, size, space, and time (Power, 2001). Tropospheric aerosols, through the combined direct and indirect effects, are believed to have the potential to cause a climate forcing (IPCC, 2001) comparable in magnitude, but opposite in sign, to that of greenhouse gases (Charlson et al. 1991; Jacobson, 2001). Moreover, aerosol particles and air gaseous pollutants may incur significant detrimental effects on human health (Colls, 1997).

From the existing literature, it is apparent there is a growing awareness of the need to consider aerosol particles in climate modeling and global climate change. One of the most important parameters to consider, as originally portrayed by Angstrom (1929), is the aerosol optical depth (AOD)-$\tau_\alpha(\lambda)$. This is defined as the extinction of solar irradiance caused by the atmospheric aerosols and it is an index representing the aerosol column burden. It is empirically given through the Angstrom (1929) formulation:

$$\tau_a(\lambda) = \beta\lambda^{-\alpha} \tag{1}$$

where the exponent **α** relates to the aerosol particle size, whilst **β** is known as the turbidity. AOD and turbidity are essentially synonymous quantities both being logarithmic indices of atmospheric optical attenuation to a vertical beam. Recently, AOD/turbidity has been continuously monitored on a global scale via satellite platforms (Penner et al., 2002). However, satellite-based techniques as well as their associated retrieval algorithms still have limitations and, in many instances, in-situ observations are crucial for reliable interpretation. As a result of recent concern over possible climate changes caused by the increased aerosol loading and the lack of available aerosol extinction data to quantitatively address the issue, there is a need for quantifying turbidity parameters especially over regions where such data is scarce. In this chapter, results obtained from a field survey in the urban polluted Athens atmosphere provide insights on the Angstrom parameters and aerosol optical depths.

Over the years, different indices have been used to monitor the atmospheric turbidity levels as the first approach to quantifying aerosol loading and properties (Linke, 1922; Angstrom, 1929; Unsworth-Monteirh, 1972). Nevertheless, the Angstrom parameters (**α**, **β**) have long been used as turbidity indices that in turn can be derived via different techniques: i) Pyrheliometric measurements of broadband direct beam solar irradiance (Angstrom, 1961; Louche et al., 1987; Jacovides et al., 1994; Cucumo et al., 1999; Power, 2001; Tadros et al., 2002; Ogunjobi et al., 2003; Zakey et al., 2004); ii) Dual wavelengths sun-photometric measurements at wavelengths where molecular absorption is either absent or is negligible (King et al., 1978; Harrison et al., 1994; Holbern et al., 1998); and iii) Multi wavelengths spectroradiometric measurements over several spectral segments of solar spectrum (Riordan et al., 1989; Martinez-Lozano et al., 1998; Esposito et al., 1998; Cachorro et al., 2000, 2001; Jacovides et al., 1999, 2000). For further details see the review article of Power (2003). All these measurements use the Bouguer-Beer-Lambert law as a monochromatic approach for determination of the aerosol optical depths. In the present chapter further insights into derivation of Angstrom's turbidity parameters (**α**, **β**) and aerosol optical depths are investigated, through spectral aerosol extinction data obtained from a field survey in the greater urban Athens atmosphere.

Experimental Measurements

During May 1995, a joint research experiment between the Athens University and the Nottingham University was set up to study the effects of aerosols-air gaseous pollutants on the solar spectral energy distribution. The experiment was deployed in the greater Athens metropolitan area (37^0N, 24^0E), a city known for its high pollution level, from 16 to 27 May. Spectral solar global and diffuse irradiance measurements at the ground were obtained using a portable Li-Cor LI-1800 model (Lincoln, NE, USA) scanning spectroradiometer (wavelength range 300-1100 nm; resolution 6 nm and spectral accuracy ±2 nm). Instrumentation and data acquisition details may be found elsewhere (Jacovides et al., 2000). Calibration results for the LI-1800 revealed total uncertainty over several spectral segments: ±23% for 300-320 nm, ±5% for 320-380 nm, ±3% for 380-900 nm, and ±4% for 900-1100 nm, which are comparable to those reported by Riordan et al. (1989) for a Li-Cor LI-1800 model.

During the period of observations, stagnating conditions allowed polluted urban air over the Athens center. Morning and afternoon spectral extinction data were used here (air mass range 1.22-2.28, i.e. zenith angle was $35\text{-}64^0$), coinciding with atmospheric stratification due to weak variation of the meteorological parameters: air temperature varied from 19 to 22^0C, relative humidity remained almost constant at 58-52%, and light winds of 1.5 ms^{-1} were blowing, resulting in relatively low optical depth uncertainties (Carlund et al., 2003). The air mass range employed here further reduces the Angstrom parameters (**α**, **β**) uncertainties (Wagner and Silva, 2008). All spectra data and related meteorological quantities were recorded on a 20-min basis. Moreover, due to the complex topography of the Athens basin the aerosol optical depth at 500 nm (AOD_{500}) was found to range between 0.084 and 0.628. For this reason, the use of spectral irradiance measurements to study the effects of aerosol-gaseous pollutants on the spectral radiant energy distribution is extremely important in the Athens area.

Further, the direct solar beam spectra on a plane normal to the solar rays are obtained as the difference between global and diffuse spectra that is valid under atmospheric stratification. As shown by various authors (de La Casinier et al., 1995; Cachorro et al., 2000) the procedure for measuring spectral diffuse component by the shade disc technique needs to be improved over the range 320-900 nm. For this reason, a spectral correction factor has been derived and applied on the diffuse spectra following the approach of de La Casinier et al. (1995). To further ensure the validity of the assumption used in obtaining the direct-beam component this was further calibrated by comparing Li-Cor's integrated direct beam with that of Linke-Feussner's pyrhiliometric readings in the 380-710 nm band. The comparison showed that the two methods reached similar results, displaying an average difference of less than 2.1% that in turn confirms reliability of the measured-corrected diffuse spectra. In order to minimize further possible measurements errors, some additional restrictions have been imposed on the spectral data. Thus, by applying statistical processing criteria, e.g. χ^2-test at the 95% confidence level, data with large errors in spectral optical depth determination, were rejected. After this filtering, 78 solar irradiance spectra were available for analysis. It is worth noting here that no temperature correction on the data was necessary as the instruments were carried to the field only when measurements needed to be performed. A broadband correction factor was also applied to pyrheliometric readings in order to eliminate any parasitic circumsolar contribution (Geuymard, 1994).

THEORETICAL BACKGROUND AND METHODOLOGY

Definitions

The atmospheric total optical depth $\tau(\lambda)$, at a central wavelength λ, can be derived by inverting the Bouguer-Beer-Lambert law:

$$E_{b\lambda} = E_{0\lambda}\exp[-\tau(\lambda).m] \tag{2}$$

Here, $E_{b\lambda}$ is the measured direct-beam spectral irradiance corrected for the sun-earth distance. It is common to subtract the molecular-Rayleigh scattering component from $\tau(\lambda)$, so that the remainder the so-called "reduced" spectral optical depth characterizes both aerosol attenuation and molecular absorption. The relative optical air mass, m, computed via Kasten and Young's (1989) formula was corrected for pressure variations. It is assumed, as in Utrillas et al. (2000) that the optical air mass leads to error values lower than 0.1%. $E_{0\lambda}$ is the extraterrestrial spectral irradiance that is determined through the Langley plot method (Jacovides et al. 2000).

The spectral aerosol optical depth $\tau_\alpha(\lambda)$-AOD already defined as the extinction of solar irradiance caused by atmospheric aerosols, can be obtained experimentally by subtracting the estimated components of Rayleigh scattering and trace gases absorptions from the measured $\tau(\lambda)$. Thus, $\tau_\alpha(\lambda)$ is given by:

$$\tau_a(\lambda) = \tau(\lambda) - \tau_r(\lambda) - \tau_{O_3}(\lambda) - \tau_{NO_2}(\lambda) - \tau_{H_2O}(\lambda) - \tau_{O_2}(\lambda) \tag{3}$$

where, $\tau_r(\lambda), \tau_{O_3}(\lambda), \tau_{NO_2}(\lambda), \tau_{H_2O}(\lambda), \tau_{O_2}(\lambda)$ are the respective components due to molecular-Rayleigh scattering, ozone, nitrogen dioxide, water vapor, and oxygen absorptions. The molecular-Rayleigh optical depth is well modeled as (Bird and Riordan, 1986): $\tau_r(\lambda)=(P_S/P_0)x0.008735\lambda^{-4.08}$; here P_S is being the atmospheric site pressure in hPa and P_0 its sea level value. Eck et al. (1999) reported a maximum error in estimated $\tau_\alpha(\lambda)$ at 340 nm of up to 0.021, at 380 nm of about 0.013 and at 440 nm as low as 0.007 assuming a 3% maximum departure from the mean surface pressure. To this respect, as pressure measurements are quite accurate, further errors in the $\tau_\alpha(\lambda)$ determination caused by errors in $\tau_r(\lambda)$ values are negligible.

Further, ozone and nitrogen dioxide optical depths follow a similar law (Gueymard, 2001): $\tau_{O_3NO_2\lambda} = a_{O_3NO_2}(\lambda)u_{O_3NO_2}$,with $a_{O_3NO_2}(\lambda)$, being the respective spectral absorption coefficients for O_3 and NO_2, whereas $u_{O_3NO_2}$ are the respective O_3 and NO_2 column densities in atm-cm. Ozone spectral absorption coefficients were taken from Molina and Molina (1986) over UV spectral band (300-345 nm) and from Anderson and Mauersberger (1992) in the Chappuis band (440-770 nm). Ozone column density u_{O3} was obtained from a Brewer instrument (in DU unit) located at the Athens University building in downtown Athens area. NO_2 spectral absorption coefficients were extracted from Schneider et al. (1987) in the 350-500 nm range and smoothed to 2 nm intervals; whereas NO_2 reduced path-length was derived from its surface concentrations ($\mu g\ m^{-3}$) via simple thermodynamic calculations (Jacovides et al., 2000) by assuming that the NO_2 scale height is equal to that of aerosol.

Eck et al. (1999) reported that departures from the climatological ozone values as high as 50% resulted in additional uncertainty in estimating $\tau_\alpha(\lambda)$ of about 0.0036 at 340 nm, 0.0045 at 500 nm and up to 0.0063 at 675 nm. Therefore, these errors in obtaining $\tau_\alpha(\lambda)$ are almost negligible. The errors in the determination of $\tau_\alpha(\lambda)$ arise mainly from errors in the measured direct-beam irradiances. The largest sources of error in $\tau_\alpha(\lambda)$ from any instrument are the direct-beam irradiance measurements and the determination of the extraterrestrial irradiance using Langley calibration since the errors by subtracting the other components are an order of magnitude lower. Interestingly, as in Jacovides et al. (2000), the uncertainty in the Langley calibration coefficient was estimated to be less than 2.5% over the measurement period in May 1995, with a resultant uncertainty in optical depths of approximately 0.025 at the lower air mass 1.22. It was also found, as in Jacovides et al. (2005), that the errors in estimating $\tau_\alpha(\lambda)$ are lowest for highly as opposed to less polluted air. As reported by several workers (Martinez-Lozano et al. 1998; Cachorro et al., 2001) since $\tau_\alpha(\lambda)$ is highly wavelength-dependent, spectral solar extinction data are confined to absorption free narrow bands. That is, for more accurate analysis, the spectral intervals 680-746, 754-774, 786-844, and 872-1014 nm, corresponding to H_2O and O_2 absorption bands, are excluded. In addition, spectral measurements in the range 300-320 nm are omitted because of the low signal-to-noise ratio in this region. Moreover, an overall uncertainty in $\tau_\alpha(\lambda)$ was found to range between 0.01 and 0.032 possessing both wavelength and turbidity dependence.

To further visualize the definitions of the retrieved aerosol optical depths, Figure 1 shows: i) spectrally reduced optical depths (total minus Rayleigh component) of the atmosphere retrieved from ground-based spectral extinction observations over the Athens atmosphere; ii) the corresponding spectral aerosol curve, $\tau_\alpha(\lambda)$, determined via Equation (3); and iii) the Angstrom-modeled aerosol $\tau_\alpha(\lambda)$ smooth curve via Equation (1). The differences between the experimental total Rayleigh-corrected and the spectral aerosol optical depths might stem from absorption effects over UV and NIR bands. Strong absorption due to O_3 and NO_2 is revealed by UV spectral signature. For λ>700 nm significant errors due to imprecise O_2 and H_2O absorption bands are noticed. In this chapter, retrieved spectral aerosol optical depth values are obtained via Equation (3) having omitted the O_2 and H_2O components due to difficulties involve (Cachorro et al., 2000; Jacovides et al., 2000). Equation (1) provides the modeled $\tau_\alpha(\lambda)$ value by inserting the **α-β** pair derived in specified spectral range.

The Angstrom Parameters

As already mentioned, the spectral aerosol optical depth is expressed through Angstrom's power law (Equation 1). The Angstrom exponent **α** is commonly used to characterize the wavelength dependence of $\tau_\alpha(\lambda)$ providing some information on the aerosol size distribution.

Both $\tau_\alpha(\lambda)$ and $\boldsymbol{\alpha}$ exhibit spectral variation commensurate to the aerosol physical and chemical properties (Eck et al., 1999). The $\boldsymbol{\alpha}$ exponent has been the subject of numerous studies during the last two decades in relation to its application in remote sensing (Pedros et al., 2003). Further, both $\boldsymbol{\alpha}$ and $\boldsymbol{\beta}$ in Equation (1), are assumed independent to each other and constant with wavelength, revealing that any spectral interval may be used for their derivation. In practice, however, both Angstrom's parameters are not independent attaining, also, wavelength dependence. Following definitions given by Shifrin (1995), Equation (1) can be re-written as,

$$\tau_a(\lambda) = \beta(\lambda)(\lambda/\lambda_0)^{-\alpha(\lambda)} \tag{4}$$

In this expression, $\boldsymbol{\alpha}$ and $\boldsymbol{\beta}$ possess a dependence on λ. It is also obvious that,

$$\alpha(\lambda) = \frac{-\partial \ln \tau(\lambda)}{\partial \ln(\lambda/\lambda_0)} + \frac{\partial \ln \beta(\lambda)}{\partial \ln(\lambda/\lambda_0)} \tag{5}$$

Under various assumptions discussed by Cachorro et al. (2001), Equation (5) can be converted to the original formula of Equation (1). As the Angstrom's formula does not fit $\tau_\alpha(\lambda)$ accurately, several authors have investigated this fitting (Martinez-Lozano et al., 1998; Cachorro et al., 2001; Pedros et al., 2003; Jacovides et al., 2005; Kaskaoutis et al., 2006). All these studies suggested that the fit is more accurate for highly as opposed to less turbid air. Also, as shown earlier (Jacovides et al., 2000) in both borders of the solar spectrum, (UV) and (NIR), the fit significantly deviated due to the singular values of Angstrom parameters used. Therefore, the Angstrom formula (Equation 1) is better suited for relatively narrow spectral range; by contrast, when it is expanded in wider spectral regions, departure from linearity may be expected in real aerosols (Eck et al., 1999; Jacovides et al., 2005). This reveals a more complex representation of the aerosol particle volume spectra, far from Junge (1955) or mono-modal log normal distributions (Cachorro et al., 2001; Kaskaoutis et al., 2006).

For the purpose of determining Angstrom's parameters ($\boldsymbol{\alpha}$, $\boldsymbol{\beta}$) different procedures may be applied (Cachorro et al., 2000). The Volz (1959) method applies Equation (1) to sun photometric data measured at two different wavelengths of a spectral band. To this respect, $\boldsymbol{\alpha}$ is obtained by,

$$\alpha = -\frac{d \ln \tau_\alpha}{d \ln \lambda} = -\frac{\ln\left(\dfrac{\tau_{\alpha 2}}{\tau_{\alpha 1}}\right)}{\ln(\dfrac{\lambda_2}{\lambda_1})} \tag{6}$$

λ_1 and λ_2 are the effective wavelengths in the spectral interval considered. In Equation (6), $\boldsymbol{\alpha}$ is the negative slope or the negative of the first derivative of aerosol optical depth, $\tau_\alpha(\lambda)$, in a log-log scale. Since $\boldsymbol{\alpha}$ is derived, the $\boldsymbol{\beta}$ is evaluated for the same wavelength interval using $\boldsymbol{\alpha}$ and $\tau_\alpha(\lambda)$ values at a certain λ.; that is, $\boldsymbol{\beta}=\tau_\alpha(\lambda_1)/\lambda_1^{-\boldsymbol{\alpha}}$ or $\boldsymbol{\beta}=\tau_\alpha(\lambda_2)/\lambda_2^{-\boldsymbol{\alpha}}$. It is important that there are several choices of wavelength pairs to use in these equations given the multiple channels employed by the spectroradiometer. It also is reasonable to choose two wavelengths that are as far apart as possible so as to minimize errors. Through this method the errors in **α** and **β**, result from the errors in $\tau_\alpha(\lambda)$ determination as well as the selected pairs $\tau_{\alpha 2}/\tau_{\alpha 1}$ and λ_1/λ_2. Cachorro et al. (1987), commenting on the Volz method, suggested that large errors occurred because of the sensitivity of the method on the selected wavelengths. Nevertheless, several workers tried to establish an optimum pair of wavelengths with contradictory results (Cachorro et al., 1987; Bokoye et al., 1997; Pedros et al., 2003); these studies further suggested that the optimum pair depends on the instrument, the spectral range and the resolution used as well as the atmospheric conditions.

Further, as mentioned above, $\tau_\alpha(\lambda)$ is determined through Equation (3) by removing from $\tau(\lambda)$ the contributions of Rayleigh scattering and molecular absorptions. As Figure 1 reveals, large uncertainties in the strong water vapor absorption bands in NIR band, are expected. Once $\tau_\alpha(\lambda)$ is determined via Equation (3), a direct fitting of Equation (1) to the experimental $\tau_\alpha(\lambda)$ curve can predict single values of **α** and **β** valid for the spectral band considered. Henceforth, this method is quoted as the direct fitting method (DF). The errors in **α** and **β** though this method, are likely caused by fitting divergences that in turn depend on the personal experience of the researcher; relative large uncertainties occurred under low turbidity conditions (Jacovides et al., 2005; Kaskaoutis et al. 2006). It is worth noticing here that the $\tau_\alpha(\lambda)$ uncertainties in the narrow strong absorption bands do not significantly affect the results.

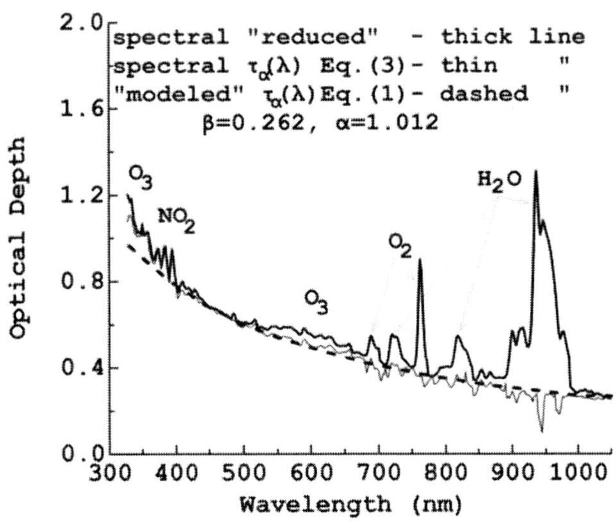

Figure 1. Spectral reduced (total minus Rayleigh) optical depth (thick solid line) and the retrieved experimental spectral aerosol optical depth-AOD (Equation 3) accounting for ozone, nitrogen dioxide and water vapor absorptions (thin solid line). The modeled-AOD smooth curve via Equation (1) is also shown (dashed line). The most important O_3, NO_2, O_2 and H_2O absorption bands are indicated on the upper curve.

There is, however, a possibility of omitting $\tau_\alpha(\lambda)$ values in the strong absorption bands and deriving the Angstrom parameters at the remaining wavelengths. This method is referred to as "spectral window method" and has been applied earlier by several workers (Cachorro et al., 2000, 2001; Jacovides et al., 1999, 2000, 2005). For that purpose, the spectral intervals 680-746, 754-774, 786-844, and 872-1000 nm have been excluded as they correspond to strong absorption bands of H_2O and O_2. This fact constitutes the main difference from the above direct fitting method (DF). Further, linearly fitting the log-log transform of Equation (1) with spectral $\tau_\alpha(\lambda)$ data to the remaining wavelengths, is the usual method to determine the Angstrom (**α, β**) parameters from spectroradiometric measurements (Martinez-Lozano et al. 1998); that is:

$$\ln \tau_a(\lambda) = -\alpha \ln \lambda + \ln \beta \qquad (7)$$

where the gradient of the line yields **α**, while the intercept provides **β**. In this chapter, this method is applied after performing the "spectral window method" and constitutes the third method, namely logarithmic method (LM), for deriving the Angstrom's parameters. This method has often been used to determine aerosol characteristics because of its simplicity (Cachorro et al., 1987, 2001; Adeyewa and Balogun, 2003; Jacovides et al., 2000, 2005). The logarithmic linear fit is the least imprecise method although results obtained may also depend on the spectral interval considered. Using the least-squares fit for the spectral $\tau_\alpha(\lambda)$ data and excluding the aforementioned spectral intervals the Angstrom parameters were obtained in exactly the same wavelength intervals used in the Volz and direct fitting methods. Errors in **α** and **β**, result from the inaccuracy of the fitting mainly in the UV band.

In this chapter, for the purpose of determining **α-β** pair, the three methods described above, were applied over five discrete spectral intervals, UV (340-400 nm), UV-VIS (340-676 nm), UV-NIR (340-862 nm), VIS (400-676 nm), VIS-NIR (400-862 nm), as well as over the whole spectrum 320-1020 nm. Applying the three different techniques the Angstrom parameters were determined and compared in order to establish whether the spectral aerosol optical depths were well represented by the power law, thus rendering the methods' accuracy to deriving similar values.

RESULTS

Before proceeding with the analysis on the reliability of the three methods employed to predicting Angstrom's turbidity parameters (**α, β**), the contrasting sensitivities of the goodness of Angstrom law fits to changes in atmospheric conditions are further illustrated. Figure 2 shows Equation (1) on a log-log plot with the aerosol optical depths ($\tau_\alpha(\lambda)$-AOD) at selected wavelengths versus λ are displayed, for two different turbidity levels occurred over Athens center. Figure 2a presents spectral data for 0900 LST "rush hour" with high aerosol loading, $\tau_\alpha(500)$-AOD_{500}=0.496, whereas Figure 2b provides similar spectral results obtained at 1500 LST under low turbidity level ($\tau_\alpha(500)$-AOD_{500}=0.179). In these figures the aerosol optical depth as well as the Angstrom exponent **α** values, are also shown.

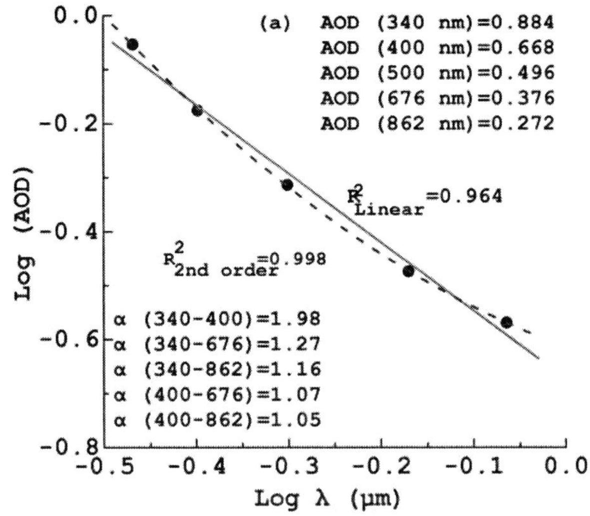

Figure 2a.

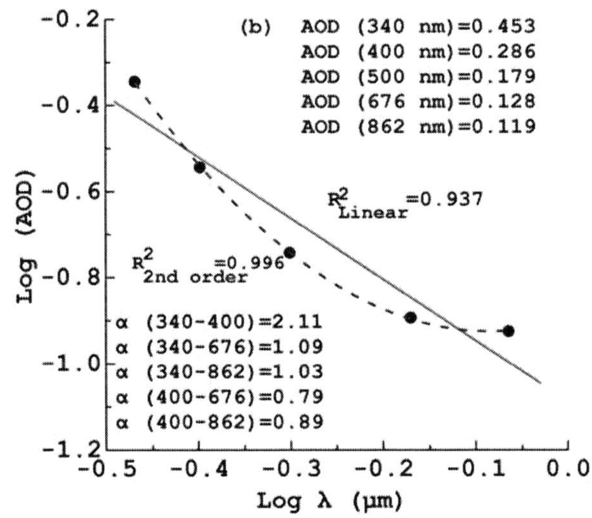

Figure 2b.

Figure 2. Log-log plot of the aerosol optical depth $\tau_\alpha(\lambda)$-AOD vs. λ at selected wavelengths over the spectral range 340-862 nm, at Athens center: a) for high turbidity; and b) for low turbidity. The linear and the second order polynomial fits to the spectral data as well as the Angstrom exponent **α** values obtained in various spectral segments are shown.

Both graphs clearly indicate the occurrence of significant differences in the aerosol parameters. During morning hours, when turbidity is high, lower **α** values, in the shorter-wavelength band occur compared to those determined during early afternoon. In contrast, **α** values determined at the longer wavelengths were higher for polluted than for clear air conditions. These features support the premise that under highly polluted air the wavelength dependence of the exponent **α** is less significant, whereas the log $\tau_\alpha(\lambda)$ versus log λ curve tends to be linear (Cachorro et al., 2001; Jacovides et al., 2005), that in turn, further suggests that the Angstrom law fit is better suited for highly as opposed to less turbid air. Furthermore, the second order polynomial fitting represents the measured aerosol optical depth $\tau_\alpha(\lambda)$ with an excellent agreement, whilst the linear fit results in higher differences than the polynomial approximation. The straight line underestimates the measured $\tau_\alpha(\lambda)$ values at shorter and longer wavelengths, while it overestimates them in the visible spectrum due to the shape of the curvature of log $\tau_\alpha(\lambda)$ versus log λ. The linear fit strongly depends on the atmospheric conditions, exhibiting significant uncertainties under low turbidities.

Comparisons of Angstrom α Values

In view of the above, the three methods described earlier, are employed for the purpose of determining **α** values: the Volz method (VM) deriving **α** values through Equation (6), the direct method (DF) by applying a direct fitting of Equation (1) to the experimentally determined $\tau_\alpha(\lambda)$, and the logarithmic method (LM), fitting $\tau_\alpha(\lambda)$ via Equation (7) in non-absorbing spectral bands. In Figure 3, **α** values derived over the whole spectrum (320-1020 nm) via the three methods are plotted as VM and LM versus DF. It is clear that both LM and DF techniques match almost similar **α** values, while VM overestimates them. That is, LM gives a mean **α** value of 1.059, with a standard deviation of the mean ±0.24, while DF predicts **α**=1.057± 0.22. The correlation between the two is well described by a linear fit with a coefficient of determination up to 98.9%, at the 97.5% confidence level, whereas the relative differences between the two are low, reaching only at 2.8% at extreme cases. The mean **α** value derived in this spectral range relate well with the 1.070 one reported in Jacovides et al. (2000) for the 320-862 nm band.

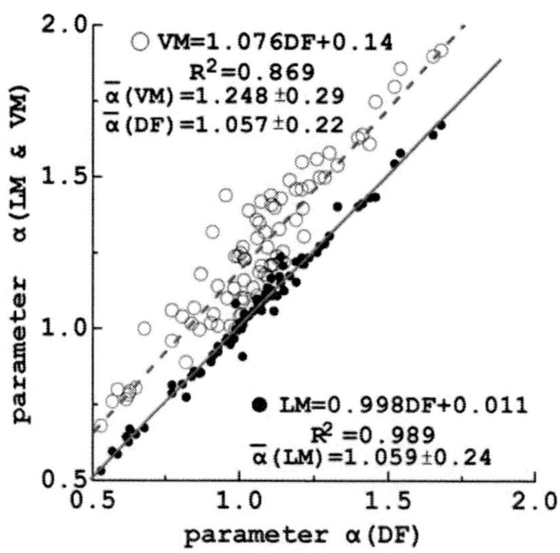

Figure 3. Correlation results between the Angstrom **α** values determined though LM, VM and DF methods for the whole spectrum, 320-1020 nm.

In contrast, VM predictions result in higher **α** value (1.248±0.29) with relatively large scatter of data points. The linear regression between VM and DF is associated with 86.9% of the variance, while the relative differences are substantially larger than those found between

LM vs. DF. However, VM predictions are in line with those reported earlier (Jacovides et al. 2000) for moderate turbidity levels and for the spectral band 320-862 nm. The α values from VM are larger in comparison to those derived through the other two methods under low turbidity levels; the overall differences range between 9.8 and 18.2% under high and low turbidity conditions, respectively. These findings are in line with Martinez-Lozano et al.'s (1998) results, showing that lower differences in α values occurred under high turbidity levels, when comparing the LM and DF techniques. They reported also that the α values derived in the 400-670 nm spectral interval through DF were larger than those obtained via LM, while their differences ranged between 3% for lower α (hence higher β) to 10% for higher α values (hence lower β).

A more detailed analysis of the derived α in several spectral bands could reveal significant features. Therefore, values of α determined through the three methods employed over the five spectral bands considered are compared. First, the mean α values together with their standard deviations, obtained via VM, LM and DF methods in the spectral segments considered, are given in Figure 4. All methods seem to provide similar α values in the spectral ranges VIS, VIS-NIR and UV-NIR; by contrast, VM predicts significantly higher α values at shorter wavelengths (UV) as well as standard deviations. Interestingly, the standard error at this waveband is very large for all the three methods. This is likely associated to the experimental $\tau_\alpha(\lambda)$ exhibiting a wavy pattern at this specific band. Also, Figure 4 reveals that the use of VM leads to a stronger wavelength dependence of α, as reported earlier by Cachorro et al. (2001) and Pedros et al. (2003).

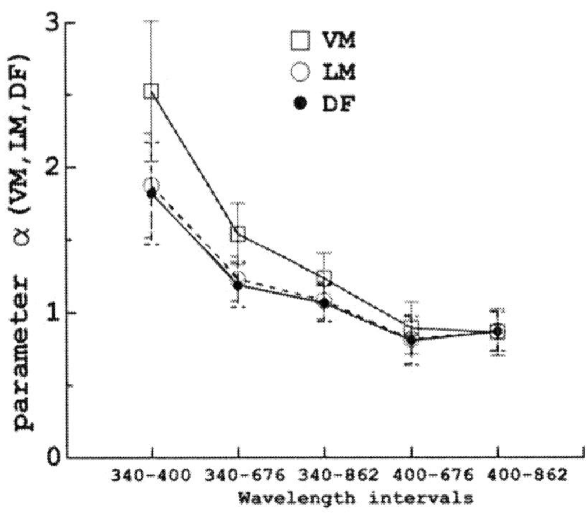

Figure 4. Mean α value obtained through LM, VM and DF methods in the five spectral ranges considered. The error bars are one standard deviation.

Correlation results between α derived via VM and LM over the spectral bands considered, are displayed in Figure 5. This figure clearly shows different values of α, depending on the wavelength and the method employed. There is also no systematic over/under estimation of α value using any of the two methods. The regression functions $\alpha_{LM}=f(\alpha_{VM})$ for the five spectral intervals considered are:

$$\alpha_{LM}=0.71\alpha_{VM}+0.16 \quad (8a)$$
$R^2=0.62$ (340-400 nm)

$$\alpha_{LM}=0.68\alpha_{VM}+0.39 \quad (8b)$$
$$R^2=0.89 \text{ (340-676 nm)}$$

$$\alpha_{LM}=0.74\alpha_{VM}+0.27 \quad (8c)$$
$$R^2=0.87 \text{ (340-862 nm)}$$

$$\alpha_{LM}=0.96\alpha_{VM}+0.032 \quad (8d)$$
$$R^2=0.97 \text{ (400-676 nm)}$$

$$\alpha_{LM}=0.93\alpha_{VM}+0.12 \quad (8e)$$
$$R^2=0.94 \text{ (400-862 nm)}$$

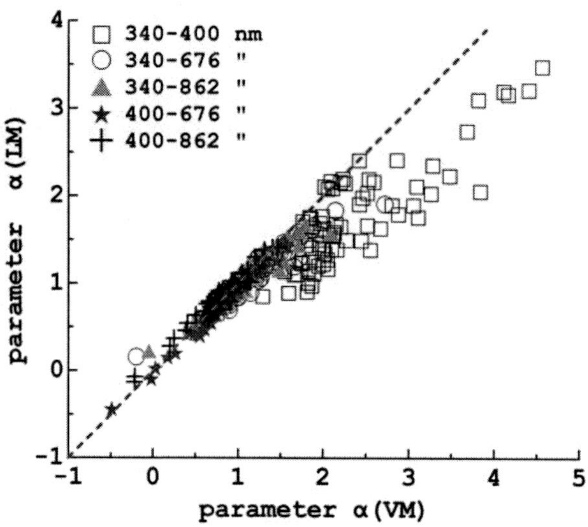

Figure 5. Correlation results between the Angstrom **α** values, determined via LM and VM in the five spectral ranges considered. The diagonal line (dashed) is given.

These equations clearly illustrate that the two methods strongly disagree in the UV spectral band (340-400 nm), where our data exhibit relatively low correlation; on the other hand, the scatter of the data points is significantly far from the 1:1 line (Figure 5). Our data also indicate that the differences in **α**, predicted via both VM and LM, exhibit a clear wavelength and turbidity dependence, as illustrated in Figure 6. This figure reveals that in all spectral bands considered, the values of **α** tend to be roughly similar only under high turbidity conditions, as expressed through aerosol optical depth (AOD) at 500 nm. On the contrary, for lower turbidity levels the differences are significant large, mainly in the UV band, exceeding 1.5. It also is clear that almost similar differences in **α** values, may assume over the whole turbidity range for both VIS and the expanded spectral bands VIS-NIR and UV-NIR.

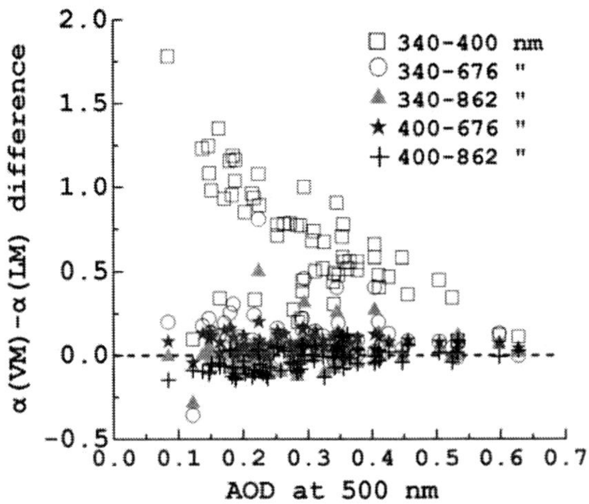

Figure 6. Differences in **α** obtained through LM and VM methods in the five spectral ranges considered versus aerosol optical depth (AOD) at 500 nm.

To further illustrate the results obtained via the two methods, Figure 7 gives the differences between VM and LM as a function of **α** values predicted via VM. This figure further clarifies that the differences in **α** values obtained through both methods increase with increasing **α** mainly in the wavelength intervals 340-400 nm and 340-676 nm, since the scatter of the data points increases with increasing **α**. It also is clear from Figure 7 that the narrow spectral interval (340-400 nm) provides the most pronounced differences between the two methods.

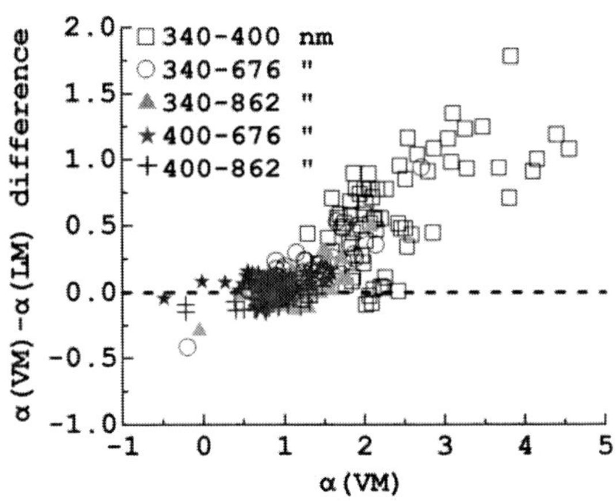

Figure 7. Differences in **α** derived through LM and VM methods in the five spectral intervals considered versus **α** (VM) values.

Another interesting feature results from Figure 8, where the differences in **α** values determined through LM and DF methods as a function of **α**, are plotted. In this case, the differences significantly reduced in comparison to that provided by Figure 7. Figure 8 further indicates that as **α** reaches higher values (hence lower turbidity) the two methods diverge mainly at shortest wavelengths, where the LM provides lower **α** values.

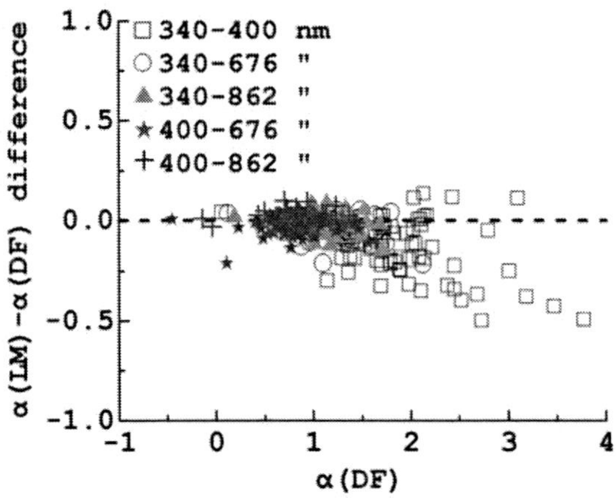

Figure 8. Differences in **α** determined through LM and DF methods in the five spectral intervals considered versus **α** (DF) values.

To further isolate the effects of atmospheric conditions on LM and DF predictions, Figure 9 illustrates the differences between the two as a function of aerosol optical (AOD) at 500 nm. This graph clearly suggests that the lower the turbidity the greater the differences between them, especially in the shortest wavelengths domain. This is likely associated with the inaccuracies of LM under low turbidity conditions as discussed earlier (see Figure 2). In the present chapter, a mean difference between LM and DF of up to 8.6% was found for lower turbidities, whereas for highly polluted conditions this difference was as low as 2.05%. The present findings agree well with those reported in Martinez-Lozano et al. (1998). In general, the diversity of the **α** values over the five discrete spectral ranges, gives support to Martinez-Lozano et al.'s (1998) and Cachorro et al.'s (2001) assertions, according to which the results can not be extrapolated from one spectral range to another because different aerosol types and loadings can significantly affect the results. Consequently, the five different values of **α** obtained here may serve as valuable information for the real aerosol characterization of the Athens atmosphere, although the data sample is limited.

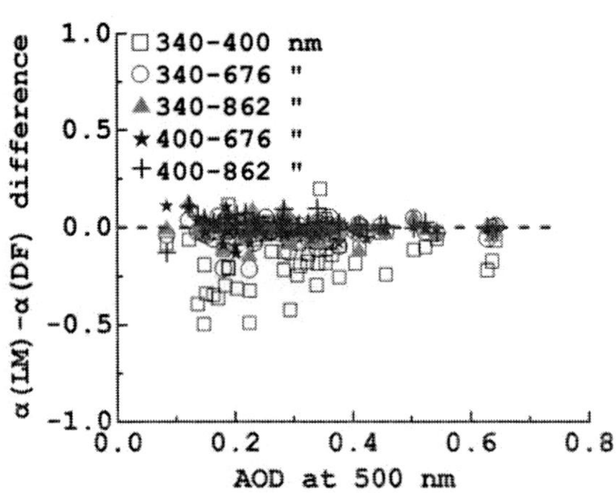

Figure 9. Differences in **α** obtained through LM and DF methods in the five spectral ranges considered versus aerosol optical depth (AOD) at 500 nm.

In view of the above and in conjunction with Figure 2, the contrasting sensitivities of standard errors in **α** values, determined via LM, to changes in air conditions are further examined. These errors, plotted in Figure 10 as a function of the aerosol optical depth at 500 nm (AOD_{500}), further address their dependence on atmospheric turbidity, with low values for high turbidity levels and high and more variable values under low turbid conditions. More specifically, in the UV band, the error's variation ranges between 0.721 (st.dev=0.122) under low turbidity ($AOD_{500}<0.2$) and 0.565±0.088 under high turbidity ($AOD_{500}>0.4$), with a mean value of 0.612±0.098 for the whole turbidity range. In the expanded spectral UV-VIS band, the errors vary from 0.389±0.086 to 0.33±0.062 when comparing lower and higher AOD's values, with a mean value of 0.356±0.075. In the spectral UV-NIR band, the errors further reduce to 0.205±0.038 and 0.152±0.022 at the respective low and high AOD's values, resulting in a mean error value of 0.183±0.029.

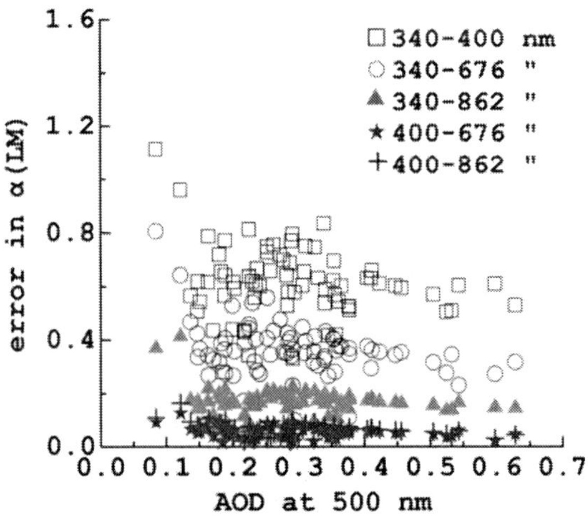

Figure 10. The standard errors in **α** obtained via LM versus aerosol optical depth (AOD) at 500 nm.

Further, the VIS spectral range exhibits the lowest errors in **α** values; the respective errors were 0.067±0.011 for $AOD_{500}<0.2$ and 0.046± 0.0012 for $AOD_{500}>0.4$, with a mean value of 0.056±0.0016 for the whole turbidity range; whereas the expanded visible VIS-NIR (400-862 nm) follows with the errors ranging from 0.076±0.018 for low turbid air to 0.054 ±0.002 for high turbid air, giving a mean error value of 0.065±0.002. As reported earlier (Cachorro et al., 2001; Jacovides et al., 2005), in the visible spectral range the Angstrom power law posses a quite accurate fit on the spectral aerosol optical depth $\tau_{\alpha}(\lambda)$ curve. It is clear that these findings further suggest that the LM formula of the Angstrom power law fit is better suited for highly as opposed to less turbid conditions discussed in Figure 2.

Comparisons of Angstrom β Values

Following exactly the same methodology applied above, the Angstrom turbidity parameter **β** was derived for the whole spectrum and the five individual spectral ranges. It also is important to note here that as reported by several workers (Cachorro et al., 1987; Martinez-Lozano et al., 1998), the **β** values derived though the VM may be different depending on the $\tau_\alpha(\lambda)$ and λ values employed in a specific spectral band; for example, the **β** value in the 340-400 nm spectral interval may be different if $\tau_\alpha(340)$ at λ=0.34 μm or $\tau_\alpha(400)$ at λ=0.4 μm used. Nevertheless, the entries in Table 1 reveal that our spectral data for highly polluted atmosphere provide **β** values that are identically similar. The entries in Table 1 also indicate that, for slightly turbid air some differences do have occurred, mainly at the shorter spectral (340-400 nm) band and are vanished towards longer wavelengths. One can also note that the difference in **β** values obtained is mostly on the fourth decimal point. Overall, the present findings disagree with those reported earlier by several other investigators.

Table 1. Derived □ values through the Volz method from equation $\beta = \tau_{\alpha i}(\lambda_i)\lambda_i^{\alpha i}$

Spectral interval	τ_{320}	τ_{340}	τ_{400}	τ_{676}	τ_{862}	τ_{1020}
$\beta_{340-400}$		0.0987* (0.0397)	0.0987* (0.0408)			
$\beta_{340-676}$		0.1934* (0.0992)		0.1934* (0.0998)		
$\beta_{340-862}$		0.157* (0.0712)			0.157* (0.0716)	
$\beta_{400-676}$			0.208* (0.112)	0.208* (0.113)		
$\beta_{400-862}$			0.159* (0.0736)		0.159* (0.0733)	
$\beta_{320-1020}$	0.2032* (0.0883)					0.2032* (0.0884)

*β values derived under highly polluted conditions.
β values in parentheses correspond to slightly polluted conditions.

Further, the **β** values determined over the whole spectrum (320-1020 nm) through the three methods are plotted in Figure 11, as VM and LM versus DF. It is noted that using VM, the parameter **α** is obtained via Equation (6) and then through Equation (1) **β** is estimated; the **α** values used are those found above, for the 320-1020 nm spectral band. Furthermore, it is clear that both LM and DF give almost similar **β** values, while VM slightly underestimates them. The linear regression between LM and DF explains 98.9% of the variance at the 97.5% confidence level, whereas the differences are very low, reaching only 2.7% in their maximum value. No clear trend of under/over estimation of one method over the other was found. Similarly, the linear regression between VM versus DF explains 91.9% of the variance, whereas the differences are larger than those found for LM vs. DF regression. The LM method provides a mean **β** value of 0.169±0.022, DF gives **β**=0.168±0.019, while VM predicts a mean **β** value of 0.162±0.036 with relatively large scatter of the data points. It is noted here that all these mean **β** values relate well with climatological values of this urban area (Jacovides and Karalis, 1996).

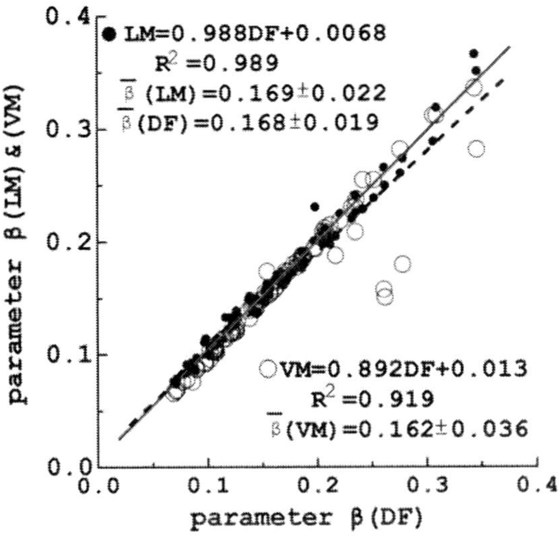

Figure 11. Correlation results between the Angstrom **β** values determined through LM, VM and DF methods in the whole spectrum, 320-1020 nm.

In general, the differences between **β** values increase with decreasing turbidity, but with no clear statistically significant trend. Martinez-Lozano et al. (1998) reported differences between **β** ranging from 0.004 to 0.008. In contrast to the **α** values, where the differences are lower under turbid air conditions, no such a trend was obvious from Martinez-Lozano et al.'s results, when comparing the **β** values determined through LM and VM. This agrees with our findings, although the aforementioned authors reported systematically higher **β** values through LM.

The mean **β** values derived through VM, LM and DF techniques together with their standard deviations are shown in Figure 12. Substantial differences between VM versus LM and DF predictions occur at the shorter wavelength band, 340-400 nm. Another interesting feature that is revealed from Figure 12 is that the **β** values derived from VM exhibit a stronger wavelength dependence compared to the LM. These findings give support to the Cachorro et al.,'s (2001) and Pedros et al.'s (2003) assertions, according to which VM exhibits a strong dependence on the spectral range considered. In the followings, comparisons between **β** values derived through LM and VM are further considered because both LM and DF predict almost identically the **β** values; that is the same conclusions can be drawn when comparing VM and DF.

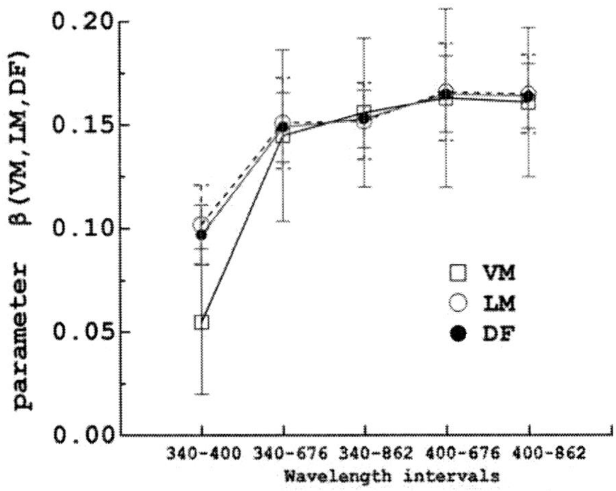

Figure 12. Mean **β** value obtained through LM, VM and DF methods in the five spectral ranges considered. The error bars are one standard deviation.

The **β** values, obtained through both VM and LM at the different spectral ranges considered are displayed in Figure 13. It is clear that significant differences appear in the shorter wavelength band (340-400 nm) due to the high uncertainties in **α** and **β** determination via both methods. As the wavelength increases the two methods tend to predict similar **β** values, even if in several cases the inaccuracies may be large. The best fit between the **β** values, via both methods, is provided in both VIS and the expanded VIS-NIR and UV-NIR spectral bands. In most cases, LM overestimates **β,** whereas this tendency is more pronounced in the shortest wavelengths domain (340-400 nm). At this spectral band, the VM provides very low **β** values due to the fact that the values of **α** for the same spectral range are significant large (see Figure 5). Correlations results for the **β** values obtained through both VM and LM in the different spectral bands are given by:

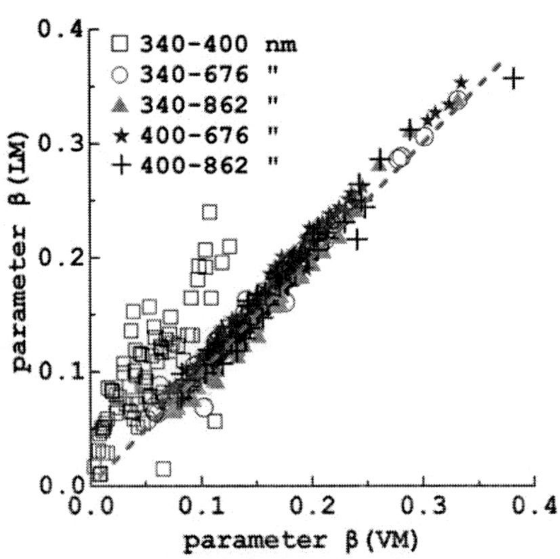

Figure 13. Correlation results between the Angstrom **β** values determined via LM and VM in the five spectral ranges considered. The diagonal line (dashed) is shown.

$$\beta_{LM}=1.079\ \beta_{VM}+0.043 \quad (9a)$$
R^2 =0.63 (340-400 nm)

$$\beta_{LM}=1.009\ \beta_{VM}+0.0044 \quad (9b)$$
R^2 =0.952 (340-676 nm)

$$\beta_{LM}=1.04\ \beta_{VM}-0.017 \quad (9c)$$
R^2 =0.982 (340-862 nm)

$$\beta_{LM}=1.029\ \beta_{VM}+0.0073 \quad (9d)$$
R^2 =0.985 (400-676 nm)

$$\beta_{LM}=1.023\ \beta_{VM}-0.0004 \quad (9e)$$
R^2 =0.989 (400-862 nm)

From these equations, it is evident that there exists a strong correlation between **β** values in most of the five spectral segments, with an exception at the shorter UV band. It is obvious, therefore, that the use of narrow spectral bands, mainly at the shorter wavelengths, for the derivation of **β** results in unrealistic values for the whole spectrum.

To further illustrate results predicted via VM and LM fittings, the differences in **β** values derived through both methods as a function of the turbidity, expressed through aerosol optical depth at 500 nm, are plotted in Figure 14. This figure is in line with Figure 6 predictions, since the differences in **β** increase under low turbidity conditions in all the spectral ranges considered. Also, Figure 14 clearly reveals that in all spectral intervals, the **β** values determined via VM and LM tend to be roughly similar only under highly turbidity conditions. On the contrary, for low turbidity levels the differences are significantly larger in the spectral UV band. It is also clear that roughly similar differences in **β** values, but substantially lower than that over the UV, occur over the whole turbidity range, for both VIS and the expanded spectral VIS-NIR and UV-NIR ranges that in turn further indicate that VM and LM predict almost identical results at the same spectral intervals.

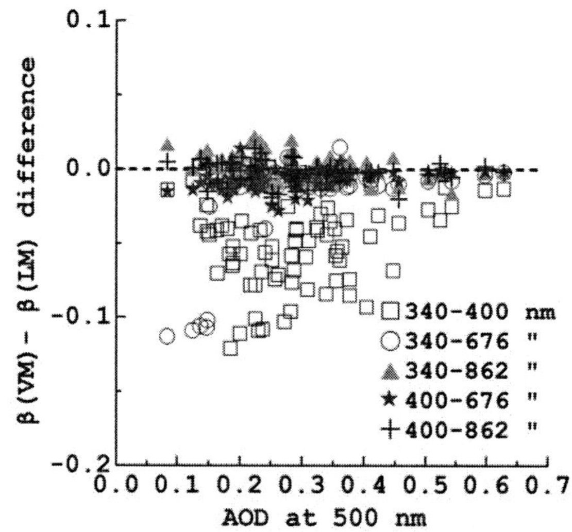

Figure 14. Differences in **β** values obtained through LM and VM methods in the five spectral ranges versus aerosol optical depth (AOD) at 500 nm.

Further, the contrasting sensitivities of standard errors in **β** values, determined via LM, to changes in air conditions are further investigated. The errors, as a function of turbidity, expressed through aerosol optical depth at 500 nm, are plotted in Figure 15. It is clear that the errors are lower for highly as opposed to less turbid air conditions. As the turbidity increases the fit of the LM on the $\tau_a(\lambda)$ curve is more accurate (see Figure 2) and the standard error of the fit is lower (Jacovides et al., 2005). The decrease in standard error of **β** with increasing atmospheric turbidity is more pronounced in the spectral UV band, whereas in the VIS band the error is low and the derivation is very accurate.

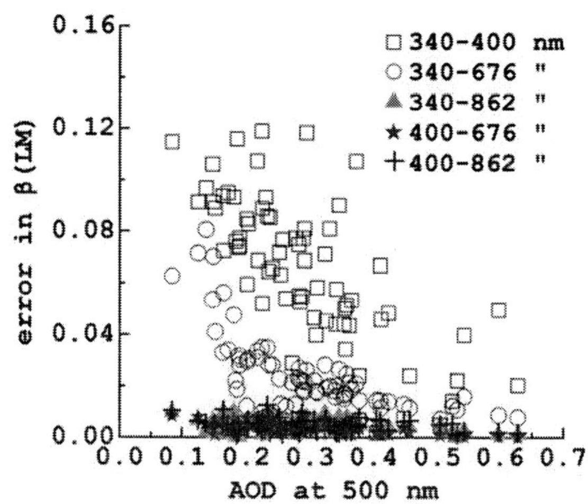

Figure 15. The standard errors in **β** values obtained via LM versus aerosol optical depth (AOD) at 500 nm.

More specifically, in the UV band the error's variation ranges between 0.082 (st.dev=0.032) under low turbidity (AOD_{500}<0.2) and 0.028±0.022 under high turbidity

(AOD_{500}>0.4), with a mean value of 0.064±0.028 for the whole turbidity range. In the expanded spectral range UV-VIS, the error values vary from 0.0396±0.0056 to 0.017±0.0031 when comparing lower and higher AOD's values, with a mean value of 0.0238±0.004; whereas in the wider spectral UV-NIR band, the errors further reduce to 0.0156±0.0028 and 0.0061±0.0015 for the respective low and high AOD values, resulting in a mean error value of 0.013±0.0021. The spectral VIS interval exhibits the lowest errors in **β** values; the respective errors were 0.0053±0.0018 for AOD_{500}<0.2 and 0.0011±0.0006 for AOD_{500}>0.4, with a mean value of 0.0042±0.001 for the whole turbidity range. The wider visible VIS-NIR band follows with the errors ranging from 0.0059±0.0021 for low turbid air to 0.0018± 0.0008 for high turbid air, giving a mean error value of 0.0048±0.0012. These findings are similar to the errors in exponent **α** (Figure 10), and agree well with those reported in Martinez-Lozano et al.(1998).

The overall analysis reveals sensitivities of the Angstrom's turbidity parameters, **α** and **β**, to both wavelength intervals used and methods employed for their determination. Despite differences of methods' predictions, viewing further Figures 4 and 5 in conjunction with Figures 12 and 13, additional notable conclusions can be drawn. To this respect, results obtained clearly indicate different aerosol volume spectra and retrieved mean **α** and **β** values. The spectral UV band (Figures 4 and 12) attains a higher **α** value (hence lower **β**) as it gives rise to higher spectral slopes of $\tau_\alpha(\lambda)$ versus λ, while the spectral VIS range exhibits flatter spectral behavior (Cachorro et al., 2001), resulting in lower **α** value (hence higher **β**). Consequently, the expanded wavebands UV-VIS, UV-NIR and VIS-NIR give lower values of **α** (hence higher **β**) in comparison to that obtained in the shorter wavebands UV and VIS. Further, the results obtained clearly reveal that different aerosol types, e.g. sub-micron aerosol particles from automotive exhaust, may have a greater effect on the shorter wavelengths, while larger aerosol particles from industrial activity strongly affect the longer wavelengths. Detailed analysis of our solar spectral extinction data showed a positive correlation between **α** and **β** in several cases; this correlation may be explained partly by aerosol processes that are associated largely with sub-micron hygroscopic particles. Nevertheless, the tendency for larger values of **α** in a highly polluted atmosphere than in a moderately one may be attributed to the influence of radiatively active gases that are present in highly polluted urban atmospheres. The increase of **α** with **β** may constitute an additional characteristic of the photochemically active Athens atmosphere.

CONCLUSION

From the existing literature there is a lack of both extensive and local aerosol data as well as analytical studies quantifying their influence on the Earth's radiative budget and climate system. This is largely because aerosol properties/turbidity measurements are difficult to make. The determination of the turbidity parameters over different polluted atmospheres and spectral ranges are valuable in aerosol studies for assessing the merits of applying the Angstrom law. This research clearly demonstrates the necessity of defining a standard spectral interval where the Angstrom turbidity parameters can be determined as well as the ability of various well-known techniques to predict similar values of these parameters.

More specifically, for the purpose of determining the Angstrom parameters (**α, β**) three different techniques, namely Volz (VM), logarithmic (LM) and direct fitting (DF), and five spectral intervals as well as the whole spectrum, were used. From this analysis it is clear that a direct comparison between different pairs of the Angstrom turbidity parameters is not an easy task, because of the different techniques and spectral ranges employed in their determination. Furthermore, regarding the whole spectrum, both LM and DF techniques predict similarly the **α** and **β** values. On the other hand, the VM slightly overestimates these parameters. On the contrary, when individual spectral intervals employed, the different techniques used lead to substantial differences mainly at the shorter wavelengths domain, whereas the associated errors in both **α** and **β** significantly increase. The results obtained reveal that in all five spectral bands considered, the Angstrom parameters (**α, β**) determined via the three techniques tend to be roughly similar under high turbidities, whereas under low turbidities, the differences are significantly larger mainly at the shorter wavelengths domain. Finally, weather conditions play an important role on the turbidity values; therefore, because of the low levels of relative humidity, our results may be representative of dry aerosol conditions.

Reviewed By

Reviewed by Associate Professor John Boland, Schools of Mathematics & Statistics, University of South Australia, Mawson Lakes, Australia; E-mail address: John.Boland@unisa.edu.au.

References

Anderson, S.M. ; Mauersberger, K. *Geophys. Res. Lett.* 1992, 19, 933-936.
Angstrom, A. *Geogr. Ann. A.* 1929, 2**,** 156 – 166.
Angstrom, A. *Tellus* 1961, 13, 214-223.
Adeyewa, Z.D. ; Balogun, E.E. *Theor. Appl. Climatol.* 2003, 74, 105-122.
Bird, R.E. ; Riordan, C.J. *J. Clim. Appl. Meteorol.* 1986, 25, 87-97.
Bokoye, A.I.; de La Casiniere, A.; Cabot, T. *J. Geophys. Res.* 1997, 102, 21905-21914.
Cachorro, V.E.; de Frutos, A.M.; Casanova, J.L. *Appl. Opt.* 1987, 26, 3069-3076.
Cachorro, V.E.; Duran, P.; Vergaz, R.; de Frutos, A.M. *J. Aerosol Sci.* 2000, 31, 687-702.
Cachorro, V.E.; Vergaz, R.;de Frutos, A.M. *Atmos. Environ.* 2001, 35, 5117-5124.
Carlund, T.;,Landelius, T.; Josefsson, W. *J. Appl. Meteorol.* 2003, 42, 1598-1610.
Charslon, R.J.; Langner, J.; Rodhe, H.; Leovy, C.B.; Warren, S.G. *Tellus* 1991, 43AB, 152-163.
Colls, J.J. *Air Pollution – An Introduction;* E&FN SPON: London, UK, 1997; 225-250.
Cucumo, M.; Morineli, V.; Oliveta, G. *Renew. Energ.* 1999, 17, 397-410.
de La Casinier, A.; Cabot, T.; Benmansour, S. *Sol. Energy* 1995, 54,173-182.
Eck, T.F.; Holbern, B.N.; Reid, J.S.; Dubovic, O.; Smirnov, A.; O'Neill, N.T.; Slutsker, I.; Kinne, S. *J. Geophys. Res.* 1999, 104 (D24), 31333-31349.

Esposito, F.; Serio, C.; Pavese, G.; Auriemma, G.; Satriano, C. *J. Aerosol Sci.* 1998, 29, 1212-1218.

Gueymard, C. (1994). Updated transmittance functions for use in fast spectral direct beam irradiance models. *Proc. Solar '94, 23rd ASES Annual Conf.*, San Jose, CA, American Solar Energy Society, 355-360.

Gueymard, C. *Sol. Energy* 2001, 71, 325-346.

Harrison, L.; Michalsky, J.; Berndt, J. *Appl. Opt.* 1994, 22, 5118-5125.

Intergovernmental Panel on Climate Change (IPCC), 2001: Climate Change 2001: the scientific basis. Houghton, J. T.; Ding, Y.; Griggs, D.J.; Noguer, M.; van der Linden. P.J.; Dai, X.; Maskell, K.; Johnson, C.A., New York: Cambridge Univ. Press, 881 pp.

Jacobson, M.Z. *Nature* 2001, 409, 695-697.

Jacovides, C.P.; Varotsos, C.; Kaltsounides, N.A.; Petrakis, M.; Lalas, D.P. *Renew. Energ.* 1994, 4(5), 465-470.

Jacovides, C.P.; Karalis, J.D. *Int. J. Climatol.*1996, 16, 107-119.

Jacovides, C.P.; Asimakopoulos, D.N.; Steven, M.D. *Atmos. Environ.* 1999,33, 3427-3431.

Jacovides, C.P.; Steven, M.D.; Asimakopoulos, D.N. *J. App. Meteorol.* 2000, 39, 917- 930.

Jacovides, C.P.; Kaltsounides, N.A.; Asimakopoulos, D.N.; Kaskaoutis, D.G. *Theor. Appl. Climatol.* 2005, 81, 161-167.

Junge, C.E. *J. Meteorol.* 1955, 12, 13-25.

Kaskaoutis, D.G.; Kambezidis, H.D.; Jacovides, C.P.; Steven, M.D. *Atmos. Res.* 2006, 80, (2-3), 237.

Kasten, F.; Young, A.T. *Appl. Opt.* 1989, 28, 4735-4738.

King, M.V.; Byrne, D.M.; Herman, B.M.; Reagan, J.A. *J. Atmos. Sci.* 1978,35, 2153-2167.

Louche, A.; Maurel, M.; Simonnoy, G.; Peri, G.; Iqbal, M. *Sol. Energy* 1987, 38, 89-96.

Martinez-Lozano, J.A.; Utrillas, M.P.; Tena, F.; Cachorro, V.E. *Sol. Energy* 1998, 63, 303-311.

Molina, L.T.; Molina, M.J. *J. Geophys. Res.*1986, 95, 14501-14508.

Ogunjobi, K.O., Kim, Y.J.; He, Z. *Theor. Appl. Climatol.* 2003, 76, 65 - 75.

Pedros, R.; Martinez-Lozano, J.A.; Utrillas, M.P.; Gomez-Amo, J.L.; Tena, F. *J. Geophys. Res.* 2003, 108 (D18), 4571-4587.

Penner, J.E.; Zhang, S.Y.; Chin, M.; Chuang C.C.; Feichter, J.; Feng, Y.; Geogdzhayev, I.V.; Ginoux P.; Hersog, M.; Higurashi, A.; Koch, D.; Land, C.; Lohmann, U.; Mishchenko, M.; Nakajima, T.; Pitari, G.; Soden, B.; Tegen, I.; Stowe, L. *J. Atmos. Sci.* 2002, 59, 441-460,

Power, H. C. *Atmos. Environ.* 2001, 35, 125 – 134.

Power, H. C. *Prog. Phys. Geog.* 2003, 27(4), 502-547.

Reid, J.S.; Eck, T.F.; Christopher, S.A.; Hobbs, P.V.; Holben, B.N. *J. Geophys. Res.* 1999, 104, 27473 – 27489.

Riordan, C.J.; Myers, D.; Rymes, M.; Hulstrom, R.; Marion, W.; Jennings, C.; Whitaker, C. *Sol. Energy* 1989, 42, 67-79.

Schneider, W.; Moortgat, G.K.; Rymes, M.; Hulstrom, R.; Marion, W. *J. Photoch. Photobio.-A.* 1987, 40, 195-217.

Shifrin, K.S. *Appl. Opt.* 1995, 34, 4480-4485.

Tadros, M.T.Y.; El-Metwally, M.; Hamed, A.B. *Renew Energ.* 2002, 621-645.

Unsworth, M.H.; Monteith, J.L. *Q. J. Roy. Meteor. Soc.* 1972, 98, 778-797.

Utrillas, M.P.; Martinez-Lozano, J.A.; Cachorro, V.E.; Tena, F.; Hernadez, S. *Sol. Energy* 2000, 68, 197-205.

Volz, F. *Arch. Meteor. Geophy. A* 1959, 10, 100-131.

Wagner, F.; Silva, A.M. *Atmos. Chem. Phys.* 2008, 8, 481-489.

Zakey, A.S.; Abdelwahab, M.M.; Makar, P.A. *Atmos. Environ.* 2004, 38, 1579-1591.

In: Air Quality in the 21st Century
Editors: G. C. Romano and A. G. Conti, pp. 139-209
ISBN 978-1-60456-793-9

Chapter 4

SIMULTANEOUS EVALUATION OF ODOR EPISODES AND AIR QUALITY. METHODOLOGY TO IDENTIFY AIR POLLUTANTS AND THEIR ORIGIN COMBINING CHEMICAL ANALYSIS (TD-GC/MS), SOCIAL PARTICIPATION, AND MATHEMATICAL SIMULATIONS TECHNIQUES

***E. Gallego*[1,*], *F. J. Roca*[1], *J. F. Perales*[1] *and X. Guardino*[2,†]**

[1]Laboratori del Centre de Medi Ambient. Universitat Politècnica de Catalunya, Avda. Diagonal, 647. E-08028 Barcelona, Spain

[2]Centro Nacional de Condiciones de Trabajo. INSHT. Dulcet 2-10. E 08034 Barcelona, Spain

ABSTRACT

Odor episodes and environmental air quality are topics of worldwide concern, mainly due to the fact that industrial facilities are often located very close to inhabited areas. A newly developed methodology based on the simultaneous application of meteorological analysis techniques, social control, chemical control and numerical modeling makes it possible to identify and evaluate air quality at different points in urban and industrial areas, as well as to identify the emission source of the odor episodes. Social control makes it possible to build databases of odor episodes, obtain sensory measurements (by determining the annoyance index) and obtain samples during odor episodes. This is done by switching on air samplers at the beginning and end of odor episodes. When databases of odor episodes are treated statistically, a valuable tool is obtained: annoyance index maps, which can be used to calibrate the dispersion modeling

* E-mail: Lcma.Info@upc.edu

† E-mail: cnctinsht@mtas.es

results. The chemical compounds are identified by means of chemical control. A validated analytical method, based on thermal desorption (TD) coupled with gas chromatography (GC) and mass spectrometry (MS), is used to determine a wide range of volatile organic compounds (VOCs) that cause odor nuisances and affect air quality in indoor and outdoor air, including alkanes, aromatic hydrocarbons, aldehydes, alcohols, chlorides, esters, ketones, terpenes, amides, carbon disulfide and isocyanates. If the meteorological conditions and main VOC emission sources of a given location are known, numerical modeling can be used to create impact prediction maps that closely resemble the impact maps created after analysis of the VOCs. These impact maps are obtained with a Gaussian-type model developed at the Environmental Center Laboratory (LCMA) of the Department of Chemical Engineering at the Polytechnical University of Catalonia (UPC). They make it possible to study the dispersion of odor compounds in the atmosphere surrounding emission sources. These maps are a three-dimensional representation of time-averaged accumulative concentrations for a particular period of time around the emission sources. In addition, meteorological models can be used to track, backwards in time, the source of the air mass responsible for the discomfort, mainly to find possible VOC sources outside the urban area. The procedure combines an analytical approach based on the acquisition of samples, which requires the participation of the affected population (i.e. social participation used as a scientific tool), with a modeling approach.

1. INTRODUCTION

Odor episodes pose a frequent environmental problem in urban areas today. Several atmospheric pollutants, mainly volatile organic compounds (VOCs), are responsible for odor episodes with different degrees of annoyance. Detectable odors have a significant impact on people's daily lives, as they can affect moods and have psychological and physiological impacts. The number of emission sources, the magnitude of emission, the continuous/discontinuous nature of the emissions, the distance from the source, the meteorological conditions and the geography of the particular area determine the effects of outdoor VOC concentrations on the affected population. The generalized absence of regional planning in countries around the world allows urban and residential areas to be built near industrial zones, thereby generating conflictive hot spots. The abovementioned factors cause a wide range of complex situations in urban areas, such as problems derived from multiple-source emissions during short and long episodes, simultaneous emissions from different activities, and uncontrolled activities. These problems require control techniques capable of identifying the compounds responsible for the odor episodes as well as the emission sources.

Odor episodes are generally evaluated by their annoyance index, which measures odor units using olfactometric techniques. This simplified evaluation does not determine the specific compounds that generate the odor episode, nor does it determine any other compounds that do not have an odorous component, some of which may be toxic and potentially carcinogenic. The annoyance index must therefore be complemented by an exhaustive evaluation of the possible presence of toxic compounds in significant concentrations [Roca, 2006].

Using a methodology based on the simultaneous application of meteorological analysis techniques, social participation, chemical control and numerical modeling, it is possible to identify and evaluate air quality at different points in urban and industrial areas and identify the emission source of odor episodes. Social participation makes it possible to create databases of odor episodes, obtain sensory measurements (by determining the annoyance index) and obtain samples during odor episodes. This is done by switching on air samplers at the beginning and end of odor episodes. The chemical compounds are identified by means of chemical control. A validated analytical method, based on thermal desorption (TD) coupled to gas chromatography (GC) and mass spectrometry (MS), is used to determine a wide range of VOCs that cause odor nuisances and affect air quality in indoor and outdoor air. If the meteorological conditions and main VOC emission sources of a given location are known, numerical modeling can be used to create impact prediction maps that closely resemble the impact maps created after analysis of the VOCs. These impact maps make it possible to study the dispersion of odorous compounds in the atmosphere surrounding emission sources. In addition, meteorological models can be used to track, backwards in time, the source of the air mass responsible for the discomfort, mainly to find possible VOC sources outside the urban area. The procedure combines an analytical approach based on the acquisition of samples, which requires the participation of the affected population (i.e. social participation used as a scientific tool), with a modeling approach.

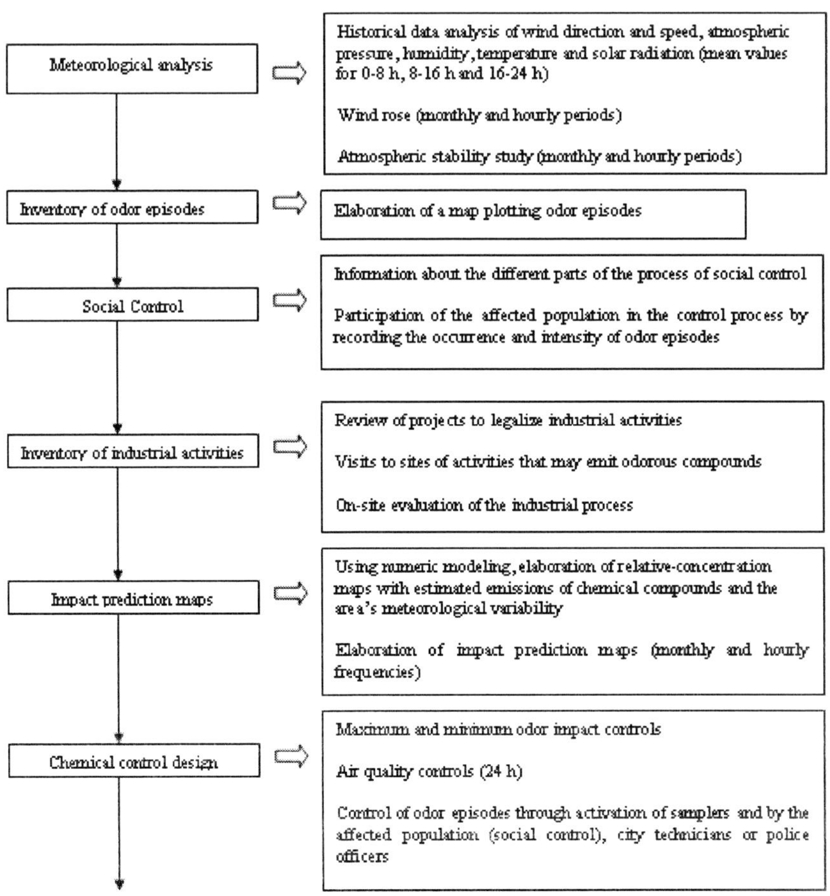

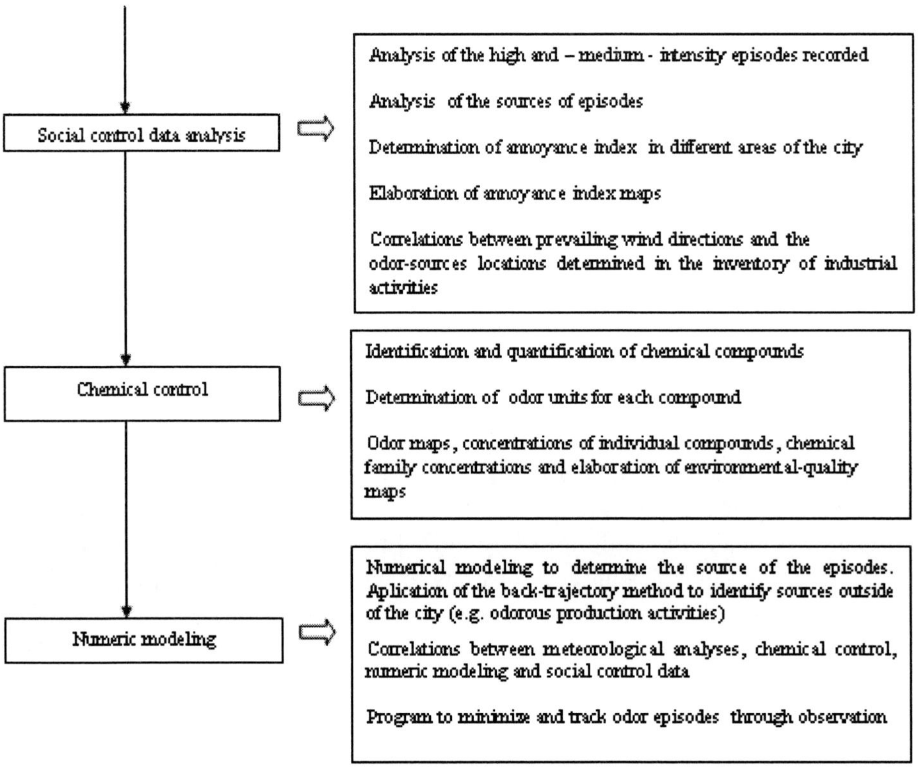

Figure 1. Methodological stages for the simultaneous control of odor and air quality.

2. Odor Impact and Air Quality Assessment and Management for Urban Areas

Figure 1 shows the various stages and expected results of the methodology described herein. The social participation stage can be carried out simultaneously with the inventory of emitting activities and the inventory of episodes, if these data are not already available. The methodology includes a final validation process, in which the results of the various techniques are converged.

3. Meteorological Analysis

Meteorological analysis is one of the first tasks in air quality evaluation. Several atmospheric factors that characterize the study area and could influence the transport and distribution of air pollutants must be determined. At the local scale, with dimensions in the range of several square kilometers, wind can be considered uniform in areas where the land is flat, since orography plays a major role in wind direction and velocity in the lower layers of the atmosphere, which are the most important layers at the local scale.

Urban areas usually have one or more surface meteorological stations. The data that these stations measure and record are used as inputs for the meteorological analysis described in this section. The measured variables are usually the following:

- Wind direction at a height of 10 m in degrees (1-360°) or in the 16 cardinal directions (N, NNE, etc.)
- Wind velocity at a height of 10 m ($m\ s^{-1}$)
- Solar radiation ($w\ m^{-2}$)
- Relative humidity (%)
- Precipitation (liters m^{-2})
- Atmospheric pressure (mbar)

Other variables that are sometimes measured, especially in the United States, are the opaque sky cover (portion of sky covered by clouds) and the ceiling height (height of the lowest clouds), but these measurements are rarely taken in Europe.

The variables tracked with modern meteorological stations are measured frequently and saved in electronic files. All data acquired in each half-hour or hour period is averaged.

Validation is the first step in processing the meteorological data received from these stations. When there is more than one station, this can be done by comparing data from different stations. One problem that has arisen is the presence of obstacles near anemometers, for example deciduous trees that cause inaccurate measurements in the spring and summer due to the presence of leaves.

Due to orography, the parameters that suffer the greatest variation at these stations are wind direction and velocity. In flat areas, these measurements tend to be accurate. Other variables—such as cloud cover, relative humidity, solar radiation, atmospheric pressure and precipitation—are more homogeneous, because these meteorological phenomena usually vary at scales larger than the local scale.

In the total absence of cloud cover or smog, solar radiation values can be estimated with heliographic models as a function of the date and time of the day. These models determine the sun position, azimuth and elevation angles, and then the theoretical radiant energy received by the ground.

Wind roses translate the local wind in the area. In coastal areas, the wind goes from land to sea at night and from sea to land during the day. In a valley, the wind flows along its axis. In mountainous areas, the wind is generally katabatic at night and anabatic during the day, due to changes in solar radiation. Figure 2 shows an example of a wind rose from Benicarló, a Spanish city on the Mediterranean coast. The prevailing winds flow from land to sea and vice versa, as expected.

At different hourly intervals (diurnal/nocturnal or night/morning/afternoon), wind fields show different statistical behaviors, with periodic outlines that cause differentiated concentration ranges (air quality) due to the impact of air pollution sources in urban areas.

The smaller the variability in wind direction, the more pronounced the wind rose is. The ground impact of air pollution sources can initially be estimated qualitatively using only the information contained in wind roses by following wind directions from the air pollution sources. The wind frequencies and velocities affect the probability, continuousness and strength of the impact of pollutants in the areas indicated by the wind directions. Wind roses are a graphic nexus between cause (air pollutant sources) and effect (ground impact zones that can be quantified by air dispersion models on impact maps), as shown in Figures 21 and 32.

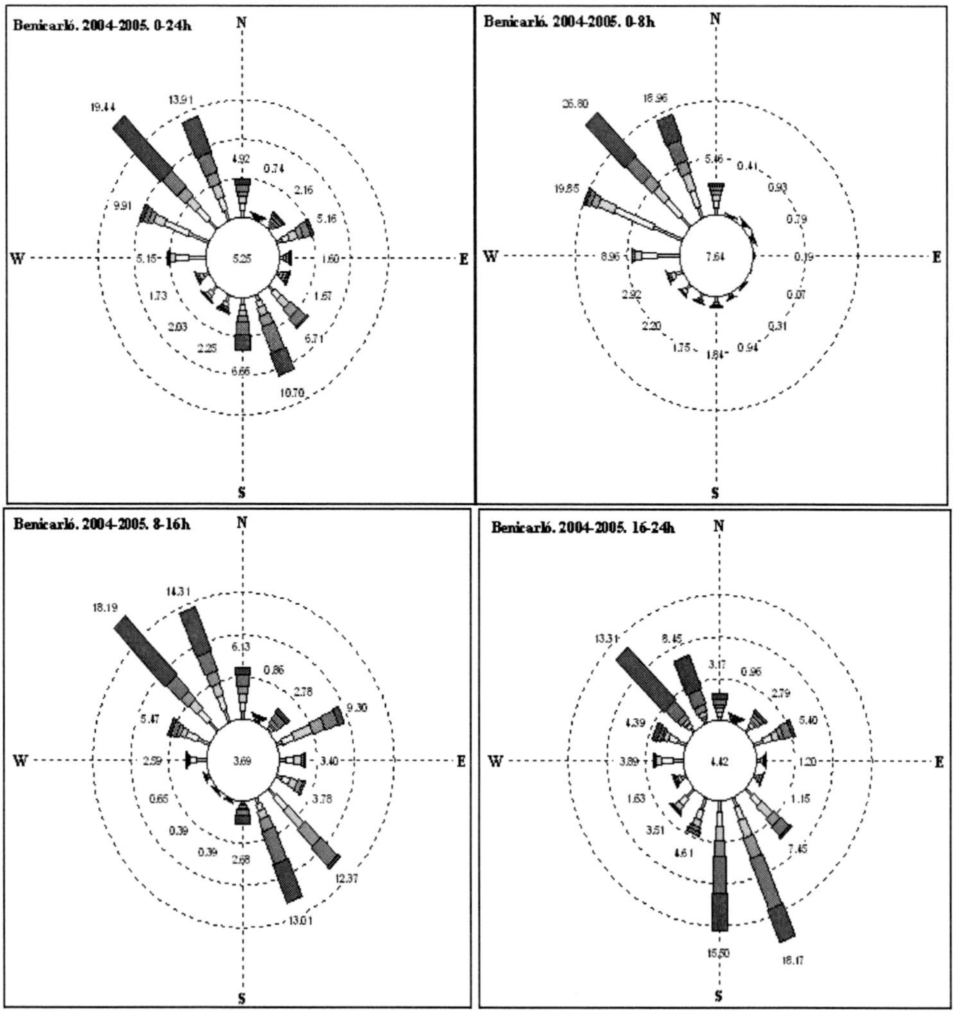

Figure 2. Wind roses with daily and hourly intervals for the city of Benicarló.

In Figure 3, the average hourly values and standard deviation of wind direction and velocity are plotted. The figure shows certain regular trends over the course of the day that cannot be appreciated using just wind roses.

Some general rules regarding wind velocity can be observed in more complete wind roses. Usually, wind velocities tend to be greater in certain directions. Wind velocity also tends to increase with solar radiation due to the increase in turbulence created by convection, as shown in Figure 4.

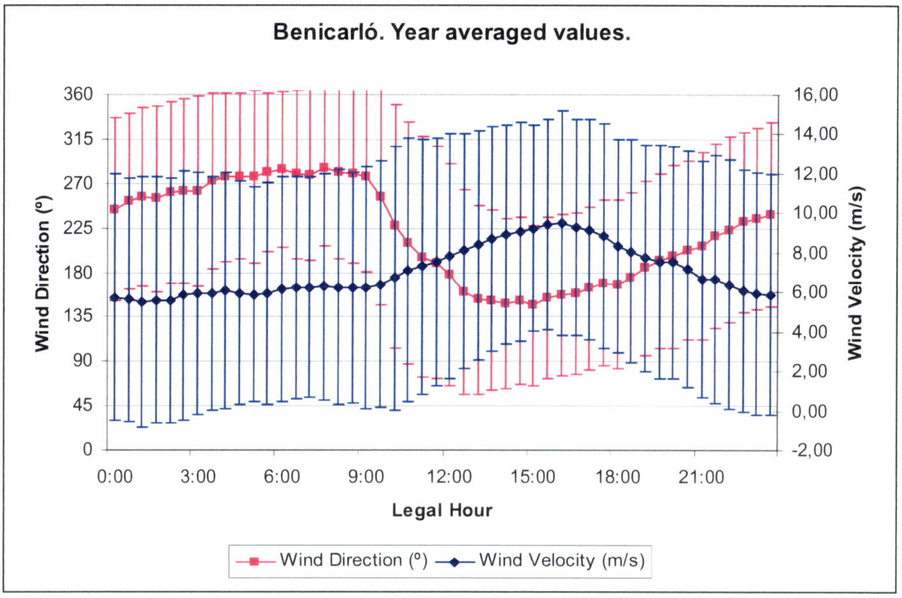

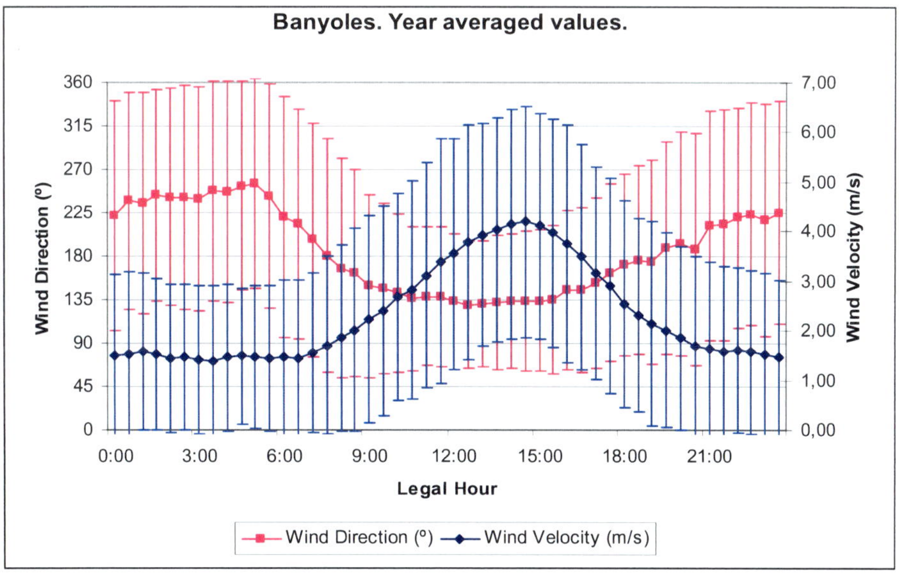

Figure 3. (Continued)

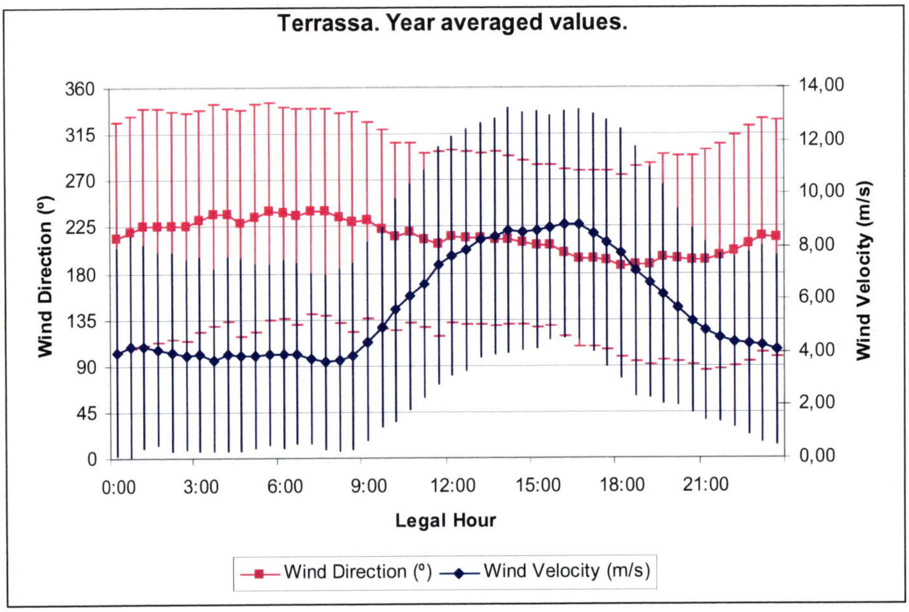

Figure 3. Hourly wind data for three cities, showing trends over the course of the day (data comprise several years).

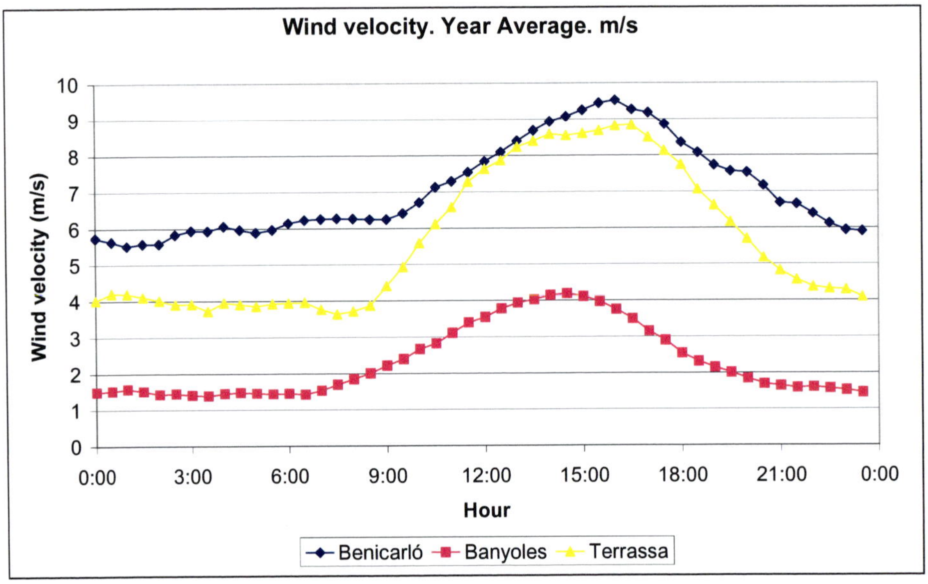

Figure 4. Comparison of daily averaged velocities for three cities. The data were obtained from a database with data for a full year. All times are local.

Wind directions and velocities at the earth's surface maintain general trends from one year to the next, as shown by the example in Figure 5 for the city of Banyoles, where four full years are plotted separately.

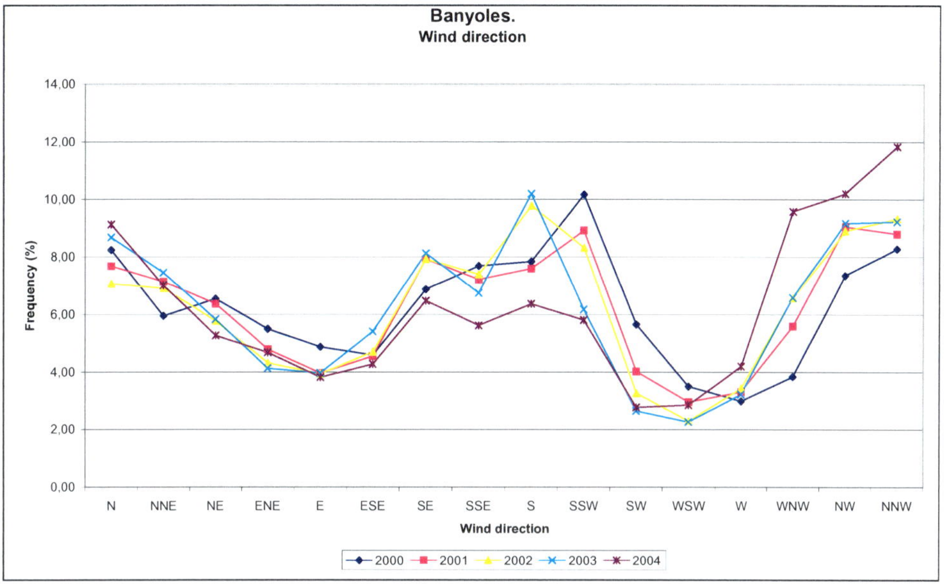

Figure 5. Interannual wind direction comparison for the city of Banyoles.

These general trends support the validity of the results obtained using historical meteorological data in dispersion models at the local scale.

Atmospheric stability is another parameter that strongly affects the dispersion of pollutants in the air. Atmospheric stability is not measured directly by surface meteorological stations, but instead is calculated based on the wind velocity and solar radiation variables recorded on the ground.

As explained in the section on impact map calculation, there are two ways to represent meteorological data: sequential and average (also called short-term or long-term, respectively). These methods are named for the type of calculation being done and roughly correspond with dynamic and steady-state calculations. In the sequential representation, each time period (hour or half-hour) is defined by the meteorological variables described, and the calculations are made with the values of the results of the previous time period taken as initial and contour values. Once the calculation has been finished for the entire period, the results can be processed in order to obtain the average and maximum concentrations, etc., for every geographical location in the area.

In long-term studies, each situation is defined by the frequency with which certain value ranges of wind velocity, wind direction and atmospheric stability occur. Steady state is assumed in the dispersion study. The concentration value obtained for each point or surface node is the average concentration value attained according to the time frequency. In this case, the meteorological data must be prepared as presented in Table 1. Several years of meteorological data can be processed in this way in order to increase the validity of the long-term results.

Both methods can be used to calculate average or maximum concentrations, as they yield similar results, but dynamic studies can only be carried out with the short-term method.

Table 1. Frequency array.

FREQUENCY DATA							
		WIND VELOCITY (m/s)					
		0-1.6	1.6-3.2	3.2-4.8	....	22.5-24.1	>24.1
PASQUILL CATEGORY	WIND DIRECTION						
A	N	0.00170	0.00170	0.00850	....	0.00000	0.00000
A	NNE	0.00170	0.00000	0.00510	....	0.00000	0.00000
A	NE	0.00170	0.00000	0.00340	....	0.00000	0.00000
A	E	0.00170	0.00510	0.01361	....	0.00000	0.00000
....	....	....	....	....	....	....	....
A	NW	0.00000	0.00170	0.01361	....	0.00000	0.00000
A	NNW	0.00170	0.00170	0.00340	....	0.00000	0.00000
B	N	0.00510	0.00000	0.01020	....	0.00000	0.00000
B	NNE	0.00000	0.00000	0.00170	....	0.00000	0.00000
B	NE	0.00000	0.00000	0.00510	....	0.00000	0.00000
....	....	....	....	....	....	....	....
B	NW	0.00340	0.00340	0.01361	....	0.00000	0.00000
B	NNW	0.00000	0.00850	0.01871	....	0.00000	0.00000
....	....	....	....	....	....	....	....
F	N	0.00510	0.00510	0.01020	....	0.00000	0.00000
F	NNE	0.00340	0.00510	0.03912	....	0.00000	0.00000
F	...	....	....	....	....	....	....
F	NNW	0.00000	0.00340	0.00000	....	0.00000	0.00000

The frequency array used in the calculation has a number of rows (96) equal to the number of Pasquill-Gifford stability classes (6) multiplied by the number of wind directions (16). The number of columns corresponds to the number of wind velocity intervals considered (17), which cover the range from 0 m s^{-1} to > 24 m s^{-1}. Each array element is the time frequency of the periods defined by wind velocity, wind direction and atmospheric stability. A total of 1,632 array elements were used in the LCMA studies.

Hence, an area's atmospheric situation is defined by dispersion studies at the local scale with 1,632 numerical values, which resume the meteorological data from the surface station(s) located in the study area.

The atmospheric situation usually makes most of the frequency array null values. During calculation, the dispersion from the different sources is computed for all possibilities marked with a statistical non-zero value in the frequency array. The concentration results obtained for each case are then averaged according to these frequencies and the corresponding impact maps are obtained as explained in Part 7 for long-term studies.

4. Social Participation

Odor control methods have incorporated several different types of participation on the part of the affected individuals (henceforth referred to as "social participation" or "social control"). This participation has taken the form of telephone or email questionnaires [Ministry for the Environment, 2003, McGinley, M.A., 1995]; complaint protocols [Department of Environment and Conservation NSW, 2006]; the creation of sensory databases based on annoyance scales, detected effects and potential generating sources; and questionnaires in which odor emissions can be correlated with hourly meteorological data [Nicolas et al., 2007]. The assignment of specific values for odor intensity [Aitken et al., 1992, McKenzie et al., 2004] is one way to quantify to what extent odors annoy people over time. Because factors such as differences in sensitivity to odors can influence participants' responses, these control processes are not considered very objective. Nevertheless, citizen participation can provide a lot of very important information about episodes of low-quality air and aid in the identification of the chemical compounds responsible for odor episodes.

4.1. Model of Social Participation for Controlling Odor Episodes and Air Quality: Identifying Chemical Compounds and their Sources

A social participation model obtains data on the occurrence and duration of episodes by using sensory data that meet several criteria related to participation level, discernment and representativity and aid in the process of generating chemical data (chemical compounds identification) in a later phase. This model is a fundamental tool in evaluating odor episodes in multi-source areas.

4.1.1. Participation Levels

As explained below, social participation is regulated in order to obtain significant data on episode sources and identify chemical compounds for episodic and daily (24-hour) periods:

Questionnaire model to register odor episodes

Personal data
Name:
Address:
Telephone:
Annoyance data
Date of the odor episode
Hour of beginning of the odor episode
Hour of finishing of odor episode
Location of odor
Odor description
Odor intensity each half-hour according to the next scale:

1 No odor
2 Lightly perceptible odor
3 Clearly perceptible odor
4 Strong odor
5 Very strong odor

Figure 6. Simplified model of a questionnaire for recording odor episodes.

Level 1: Sensory and occurrence data
Level 2: Detection of odor episodes and determination of background level

4.1.1.1. Sensory and Occurrence Data

The database for this level can be created by using questionnaires as simple as the one shown in Figure 6, or by using more complete ones that include data related to detected effects such as headaches, nausea, irritation of the respiratory system, etc., and personal data (time of presence in the area, type of work, etc.)

In order to create a more precise record and obtain simultaneous sensory and occurrence data, participants in the impact zones must fill out forms of types A and B, shown in Figures 7 and 8. Before they fill out these forms, participants should take part in preparatory sessions that explain the rules for recording odors cover and common problems. Integration in a continuous information process helps to maintain a high ongoing participation level throughout a control process that can last between three and six months, depending on the complexity of the study area.

Type A forms are used mainly in areas with few pollutant emission sources. Although the recorded data is discontinuous, its analysis allows the correlation of meteorological variables with production data from industrial facilities and annoyance data. This analysis indicates the source of the air contaminants and changes over time, so it shows the effect of measures taken by emitters to reduce impact [Roca, 2006]. Due to the half-hourly control, type B forms produce large databases (60,000 pieces of data in 6 months) [Roca et al., 2005], which can be used to make episode roses (showing the frequency of episodes along the 16 cardinal directions, like wind roses) to determine the contribution of known emitting sources and to identify sources that were initially unknown or were not included in the group of emitters [Roca et al., 2006]. The sources are ultimately identified by the complementary use of data from chemical analysis of air samples and by mathematical modeling.

4.1.2. Annoyance Level Evaluation

Through social participation, data are provided on two types of forms: type A, where affected people can record the maximum annoyance produced by a smell in three hourly intervals, according to the cited scale; and type B, where perceptions are recorded in half-hour intervals in order to obtain the annoyance index). This is done according to the following validation criteria [Aitken et al., 1992, Roca, 2006]:

a. Participation level: The participant is eliminated if his/her response is lower than 70% (fewer than 456 annotations in a six-month period).

b. The participant is eliminated if the discernment coefficient (CD) is smaller than 0.3 and F(1) = 0 or F(2)+F(3) = 0.

$$\mathbf{CD} = \sigma_{2-5} / \sigma_{max}$$

Where:

σ_{2-5}: Standard deviation of the number of responses in the different ranges comprised between 2 (light annoyance) and 5 (very high annoyance).

σ_{max}: Maximum possible standard deviation for the total number of responses with an annoyance level between 2 and 5.

Odor Formulary

1: NO ODOR CODE RECEPTOR:
2: LITGHLY PERCEPTIBLE ODOR FIRST NAME: SURNAME:
3: CLEARLY PERCEPTIBLE ODOR ADDRESS
4: STRONG ODOR
5: VERY STRONG ODOR MONTH/YEAR:

Date	**day**	**01:00-09:00**					**09:00-17:00**					**17:00-01:00**				
1	Saturday	1	2	3	4	5	1	2	3	4	5	1	2	3	4	5
2	Sunday	1	2	3	4	5	1	2	3	4	5	1	2	3	4	5
3	Monday	1	2	3	4	5	1	2	3	4	5	1	2	3	4	5
4	Tuesday	1	2	3	4	5	1	2	3	4	5	1	2	3	4	5
5	Wednesday	1	2	3	4	5	1	2	3	4	5	1	2	3	4	5
6	Thursday	1	2	3	4	5	1	2	3	4	5	1	2	3	4	5
7	Friday	1	2	3	4	5	1	2	3	4	5	1	2	3	4	5
8	Saturday	1	2	3	4	5	1	2	3	4	5	1	2	3	4	5
9	Sunday	1	2	3	4	5	1	2	3	4	5	1	2	3	4	5
10	Monday	1	2	3	4	5	1	2	3	4	5	1	2	3	4	5
11	Tuesday	1	2	3	4	5	1	2	3	4	5	1	2	3	4	5
12	Wednesday	1	2	3	4	5	1	2	3	4	5	1	2	3	4	5
13	Thursday	1	2	3	4	5	1	2	3	4	5	1	2	3	4	5
14	Friday	1	2	3	4	5	1	2	3	4	5	1	2	3	4	5
15	Saturday	1	2	3	4	5	1	2	3	4	5	1	2	3	4	5
16	Sunday	1	2	3	4	5	1	2	3	4	5	1	2	3	4	5
17	Monday	1	2	3	4	5	1	2	3	4	5	1	2	3	4	5
18	Tuesday	1	2	3	4	5	1	2	3	4	5	1	2	3	4	5
19	Wednesday	1	2	3	4	5	1	2	3	4	5	1	2	3	4	5
20	Thursday	1	2	3	4	5	1	2	3	4	5	1	2	3	4	5
21	Friday	1	2	3	4	5	1	2	3	4	5	1	2	3	4	5
22	Saturday	1	2	3	4	5	1	2	3	4	5	1	2	3	4	5
23	Sunday	1	2	3	4	5	1	2	3	4	5	1	2	3	4	5
24	Monday	1	2	3	4	5	1	2	3	4	5	1	2	3	4	5
25	Tuesday	1	2	3	4	5	1	2	3	4	5	1	2	3	4	5
26	Wednesday	1	2	3	4	5	1	2	3	4	5	1	2	3	4	5
27	Thursday	1	2	3	4	5	1	2	3	4	5	1	2	3	4	5
28	Friday	1	2	3	4	5	1	2	3	4	5	1	2	3	4	5
29	Saturday	1	2	3	4	5	1	2	3	4	5	1	2	3	4	5
30	Sunday	1	2	3	4	5	1	2	3	4	5	1	2	3	4	5

Figure 7. Type A formulary to registry odor intensity by hourly intervals [Aitken, 1992].

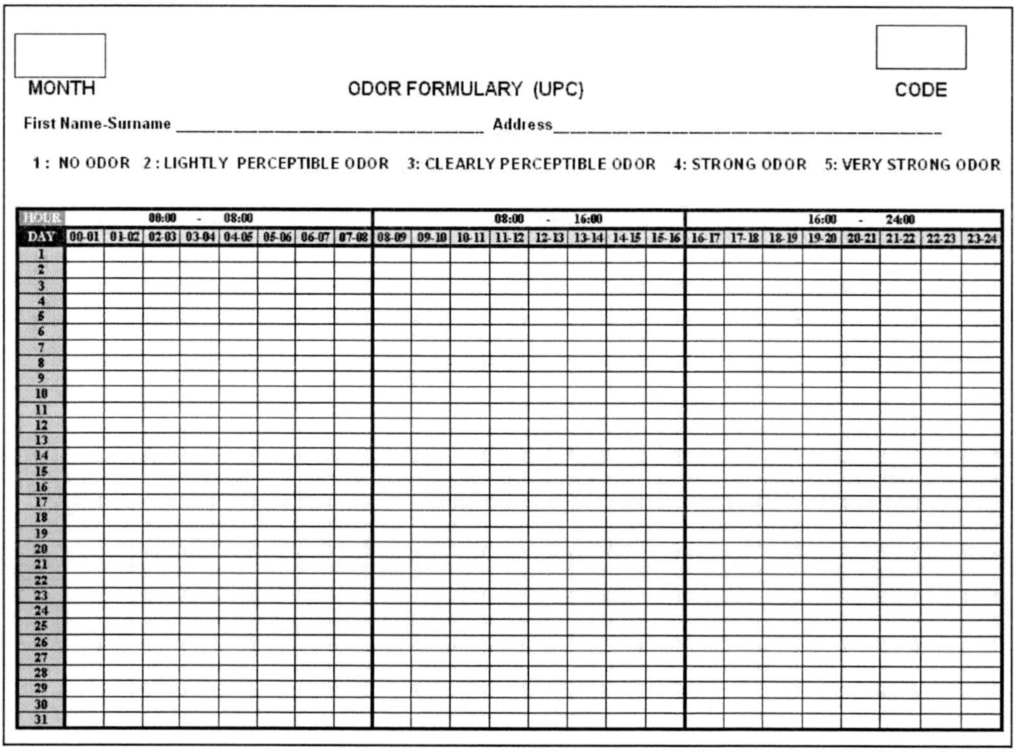

MONTH

ODOR FORMULARY (UPC)

CODE

First Name-Surname ______________________ Address ______________________

1: NO ODOR 2: LIGHTLY PERCEPTIBLE ODOR 3: CLEARLY PERCEPTIBLE ODOR 4: STRONG ODOR 5: VERY STRONG ODOR

HOUR	00:00 - 08:00								08:00 - 16:00								16:00 - 24:00							
DAY	00-01	01-02	02-03	03-04	04-05	05-06	06-07	07-08	08-09	09-10	10-11	11-12	12-13	13-14	14-15	15-16	16-17	17-18	18-19	19-20	20-21	21-22	22-23	23-24
1																								
2																								
3																								
4																								
5																								
6																								
7																								
8																								
9																								
10																								
11																								
12																								
13																								
14																								
15																								
16																								
17																								
18																								
19																								
20																								
21																								
22																								
23																								
24																								
25																								
26																								
27																								
28																								
29																								
30																								
31																								

Figure 8. Type B form used to record odor episodes [Roca et al., 2005].

F(1): Number of annotations for annoyance index 1 (no annoyance).
F(2): Number of annotations for annoyance index 2 (light annoyance).
F(3): Number of annotations for annoyance index 3 (moderate annoyance).

c. The participant is eliminated for the month if:

$$X_i \notin [X - S, X + S] \text{ or } S_i > 2S$$

Where:

X_i: Average value of the participant's sensory annotations for the range 1 to 5.
X: Average value for all participants in the area for the range 1 to 5.
S: Standard deviation of the annotations of all participants in the area.

Based on the resulting data after the aforementioned criteria have been applied, the annoyance index (AI) is calculated by applying a weighted value for each annoyance value: 0 for value 1, 25 for value 2, 50 for value 3, 75 for value 4 and, finally, 100 for value 5 [VDI 3883, 1993]. The final result, expressed as an annoyance percentage (scale: 0 to 100) is calculated using the following expression:

$$AI(\%) = 1/N_k \times SUM\ \Sigma_i(W_I N_{ik})$$

Where:

AI: Annoyance index for the period under study
N_k: Total number of annotations
I: Annoyance value (1-5)

W_I: Weighting factor
N_{iK}: Number of annotations for each annoyance value

The AI values can be calculated for daily, hourly or monthly intervals for each of the affected areas.

Let us apply the aforementioned criteria to a real case using type A forms for an area affected by industrial activities (possible odor emitters include manufacturers of cleaning products and perfumes and a water treatment plant) and by a municipal solid waste disposal site. Thus, we are able to obtain the possible emitters' contribution to the odor episodes perceived in the nearby residential area.

The initial data are as follows:

- Meteorological data: Half-hourly data for wind direction and velocity, atmospheric pressure, temperature, humidity and precipitation (Table 2).
- Impact sectors: This division is made according to the relative location of the residential areas, the emitting sources and prevailing wind directions (Table 3).
- Zones where odor episodes are detected (Figure 9):
  - Zone 1: Located 1400 m from the urban solid waste disposal site (Source 1) and 2100 m from the industrial site (Source 2).
  - Zone 2: Located 1200 m from Source 1 and 1550 m from Source 2.
  - Zone 3: Located 1750 m from Source 1 and 1200 m from Source 2.
  - Zone 4: Located 2500 m from Source 1 and 300 m from Source 2.

Potential source data: Volume of urban solid waste entering the disposal site (Table 4). There are no data for the emitters at the industrial site; most are surface and discontinuous emitters (e.g. industrial water treatment plant).

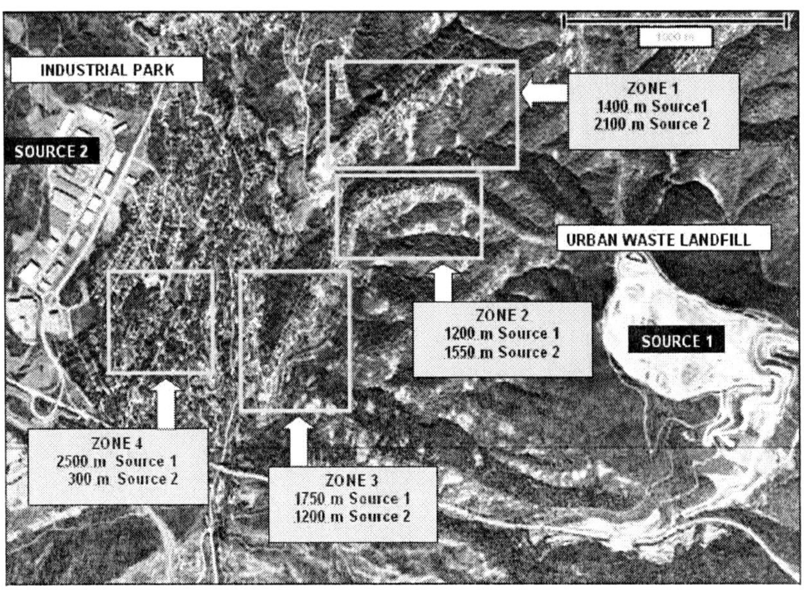

Figure 9. Different zones of a residential area that experiences odor episodes.

Table 2. Historical meteorological data for the evaluation zone.

WDIR	Jan	Feb	Mar	Apr	Mai	Jun	Jul	Aug	Sep	Oct	Nov	Dec
N	3.8	8.2	3.9	2.5	2.8	2.4	4.0	3.5	4.9	11.1	14.7	16.2
NNE	0.4	0.2	1.1	0.5	0.6	0.5	0.2	0.6	0.9	0.5	0.1	0.3
NE	0.5	0.7	1.6	0.3	1.1	0.2	0.0	0.2	1.0	1.8	3.6	4.6
ENE	1.2	0.3	2.7	1.0	1.9	0.9	0.5	0.4	1.3	0.5	0.3	0.6
E	1.1	1.3	3.4	3.3	3.6	1.0	0.5	0.8	3.1	1.1	0.8	0.9
ESE	0.7	1.4	5.3	3.6	5.2	2.1	2.5	1.8	4.5	2.4	1.0	0.6
SE	1.3	2.7	6.8	6.6	9.1	4.0	8.5	3.7	6.5	3.1	0.8	1.1
SSE	2.3	6.4	10.9	16.2	17.8	16.1	20.7	10.0	10.6	4.9	2.1	2.0
S	1.3	2.5	7.5	13.4	9.5	12.1	14.5	14.2	14.9	6.9	3.2	2.5
SSW	1.0	1.2	1.8	2.9	1.6	1.9	3.4	2.4	3.6	2.8	1.8	0.6
SW	0.5	0.6	0.4	0.1	0.3	0.8	0.4	0.4	0.9	0.3	0.3	0.2
WSW	2.9	2.5	1.1	1.5	0.9	1.3	0.4	0.8	1.6	1.8	0.7	0.4
W	4.4	3.9	1.3	1.5	1.4	1.6	0.7	0.4	2.0	1.7	1.0	0.4
WNW	8.6	8.0	4.9	4.1	3.9	3.8	2.4	1.8	3.4	3.8	3.1	2.3
NW	16.3	13.3	8.2	8.7	8.2	5.5	6.3	6.8	9.1	12.8	11.0	9.3
NNW	12.7	13.8	7.3	6.2	7.8	4.4	3.8	3.3	8.4	9.6	6.9	7.0
CALM	41.0	33.0	31.8	27.6	24.3	41.5	30.9	49.0	23.1	35.0	48.7	51.2
TEMP	3.7	8.4	10.7	12.7	14.5	20.6	22.3	20.8	18.6	15.3	9.7	6.8
VEL	1.0	1.4	1.4	1.8	1.7	1.2	1.8	1.0	1.3	1.1	0.9	0.9
HUM	n.d.	76.8	78.7	77.6	75.3	64.5	67.7	77.0	82.9	82.4	85.5	86.9
RAIN	n.d.	7.2	32.2	93.8	61.2	34.0	11.8	136.4	51.8	167.2	91.8	72.8
ATMP	986	980	978	971	971	975	9	975	980	977	972	975

WDIR: Wind direction (%)
TEMP: Temperature (ºC)
HUM: Relative Humidity (%)
RAIN: Rainfall (mm H_2O)
ATMP: Atmospheric Pressure (mmHg)

As seen from the evolution of monthly wind direction frequencies, the period with the greatest impact from the urban solid waste disposal site is March to September, and the greatest impact from the industrial site is October to February. In order to evaluate the impact of the most significant surface source (the urban solid waste disposal site), the affected population was invited to informative meetings in January and February and the control process was explained. As a result of these informative sessions, 24 neighbors (11 in Zone 1, 6 in Zone 2, 7 in Zone 3 and 3 in Zone 4) decided to participate in the task of recording sensory data on the forms.

According to the validation criteria (examples in Tables 5 and 6), during the study period (March to September) effective social participation was obtained from 15 people (4 in Zone 1, 5 in Zone 2, 5 in Zone 3 and 1 in Zone 4) (Figure 10). Most eliminations were due to the fact that the participants did not meet the participation criteria or due to the discernment coefficient. By analyzing the calculated monthly AI, significant correlations can be found with meteorological and activity data (Table 7).

Table 3. Sectors affected by the emitting sources in the study zones.

Zone	Wind Impact Sector Urban waste landfill	Wind Impact Sector Industrial Park
1	SE-SSE	WSW
2	ESE-SE	WSW-W
3	ENE-E	WNW-NW
4	E	WSW-N

Table 4. Monthly input of urban solid waste at the disposal site.

Month	WVDHV (Tons)	WVD (Tons)
January	76072.35	66877.77
February	62934.45	62256.46
March	74422.77	64402.02
April	66325.45	70945.52
May	80207.41	83278.59
June	91501.92	67017.09
July	88008.23	83007.87
August	59057.36	58943.22
September	77093.15	66188.36
October	77419.96	75896.58
November	68072.45	69116.45
December	57829.71	65930.47

WVDHV: Waste Volume Deposited (historical values).
WVD: Waste Volume Deposited (odor control period).

The data analysis gives the following results:

- Values of r^2 in the range of 0.787-0.940, 0.752-0.978 and 0.660-0.902 for the average monthly AI for different participants in Zones 1, 2 and 3, respectively.
- Values of r^2 in the range of 0.800-0.920 for the monthly AI of participants in Zone 1 with respect to the impact frequency of Source 1.
- Values of r^2 in the range of 0.744-0.863 for the monthly AI of participants in Zone 2 with respect to the impact frequency of Source 1.
- Values of r^2 in the range of 0.808-0.914 for the monthly AI of participants in Zone 3 with respect to the sum of the impact frequencies of Sources 1 and 2.
- Values of r^2 equal to 0.861 for the monthly AI of the participant in Zone 4 with respect to the sum of the impact frequencies of Sources 1 and 2.
- Values of r^2 equal to 0.297 for the AI of participant A2 with respect to the impact frequency of Source 1. The low correlation for this participant is due to the difference in altitude with respect to the other participants (Figure 11), which causes a shield effect that interrupts the flow from the source to the participant's location.

- Values of r^2 equal to -0.486 and -0.420 for the AI of participant B3 with respect to the impact frequency of Sources 1 and 2, respectively. The inverse correlation for this participant with respect to the rest of the participants in this zone is due to a nearby waste treatment plant (Figure 12). The correlation (r^2) with this source is 0.934.

Table 5. Daily AI for a Zone 2 participant.

REC	Month	Day	0-8 hours	8-16 hours	16-24 hours	(AI)	Statistical Values
A2	March	1	1	1	4	25.00	
A2		2	1	1	1	0.00	MAI: 18.011
A2		3	5	1	1	33.33	
A2		4	1	1	1	0.00	DC: 0.664
A2		5	1	3	1	16.67	
A2		6	1	5	1	33.33	MEAN: 1.720
A2		7	1	1	4	25.00	
A2		8	1	1	1	0.00	STDEV: 1.370
A2		9	1	1	1	0.00	
A2		.....	......	......	......	......	
A2		.....	......	......	......	......	
A2		26	1	3	1	16.67	
A2		27	1	1	5	33.33	
A2		28	1	1	1	0.00	
A2		29	1	1	1	0.00	
A2		30	1	1	3	16.67	
A2		31	1	1	3	16.67	

(REC): Receptor code.
(AI): Annoyance Index.
(MAI): Arithmetic mean of Annoyance Index (%).
(DC): Discernment coefficient.
(MEAN): Arithmetic mean.
(STDEV): Standard deviation.

Table 6. Validation of participants in a study zone.

RECEPTOR	MONTH	NM	MPC	CPPC	DC	M	DS
A2	MARCH	93	100		0.477	2.118	1.187
	APRIL	90	100		0.453	1.678	0.910
	MAY	93	100		0.713	2.462	1.194
	JUNE	90	100		0.349	1.833	0.738
	JULY	93	100		0.645	2.656	1.068
	AUGUST	93	100		0.559	2.398	0.898
	SEPTEMBER	90	100	100	0.687	2.444	1.273
B2	MARCH	93	100		0.699	2.236	1.378
	APRIL	90	100		0.699	2.122	1.216
	MAY	93	100		0.437	2.806	0.576
	JUNE	90	100		0.443	2.089	0.895
	JULY	93	100		0.548	3.247	0.670

	AUGUST	93	100		0.436	2.097	1.074
	SEPTEMBER	90	100	100	0.458	3.156	0.598
C2	MARCH	93	100		0.438	2.258	0.721
	APRIL	90	100		0.446	2.167	0.864
	MAY	93	100		0.517	2.634	0.639
	JUNE	0	0		n.d.	n.d.	n.d.
	JULY	93	100		0.645	2.580	1.056
	AUGUST	93	100		0.576	2.043	1.052
	SEPTEMBER	90	100	85.7	0.602	2.555	0.961
D2	MARCH	0	0		n.d.	n.d.	n.d.
	APRIL	90	100		0.444	2.456	0.584
	MAY	93	100		0.410	2.978	0.570
	JUNE	90	100		0.503	2.167	0.974
	JULY	92	98.9		0.416	3.283	0.541
	AUGUST	93	100		0.546	1.742	0.943
	SEPTEMBER	90	100	85.6	0.500	3.244	0.624
E2	MARCH	93	100		0.761	1.656	1.175
	APRIL	90	100		0.671	1.522	0.963
	MAY	93	100		0.616	2.043	1.122
	JUNE	90	100		0.300	1.433	0.654
	JULY	93	100		0.543	1.817	0.966
	AUGUST	93	100		0.739	1.376	0.896
	SEPTEMBER	90	100	100	0.596	1.567	0.925
F2	MARÇ	0	0		n.d.	n.d.	n.d.
	ABRIL	36	40		0.443	2.472	0.560
	MAY	39	41.9		0.568	2.795	0.695
	JUNE	90	100		0.465	1.900	0.900
	JULY	93	100		0.587	3.280	0.743
	AUGUST	93	100		0.474	3.344	0.599
	SEPTEMBER	27	30	58.8	0.542	3.630	0.642
G2	MARCH	0	0		n.d.	n.d.	n.d.
	APRIL	0	0		n.d.	n.d.	n.d.
	MAY	0	0		n.d.	n.d.	n.d.
	JUNE	90	100		0,.86	1.622	0.869
	JULY	93	100		0.264	3.172	0.601
	AUGUST	0	0		n.d.	n.d.	n.d.
	SEPTEMBER	0	0	28.6	n.d.	n.d.	n.d.
H2	MARCH	0	0		n.d.	n.d.	n.d.
	APRIL	90	100		0.430	2.467	0.524
	MAY	0	0		n.d.	n.d.	n.d.
	JUNE	0	0		n.d.	n.d.	n.d.
	JULY	0	0		n.d.	n.d.	n.d.
	AUGUST	0	0		n.d.	n.d.	n.d.
	SEPTEMBER	0	0	14.3	n.d.	n.d.	n.d.

Eliminated receptor (participation coefficient < 70%)

Month eliminated data (participation coefficient < 70%)

Month eliminated data (discernment coefficient < 0.3)

n.d.: no data.

NM Total number of sensorial complaints values.

MPC Month participation coefficient.

CPPC Control period (March-September) participation coefficient.

DC Discernment coefficient.

M Arithmetic mean of sensorial complaints values.

SD Standard deviation of sensorial complaints values.

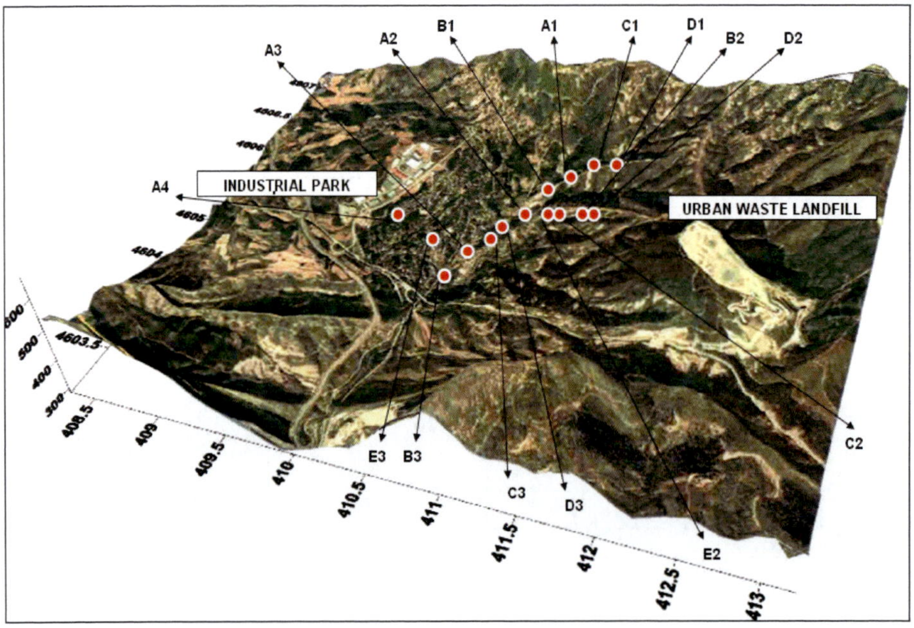

Figure 10. Location of validated participants in the study zone.

Figure 11. Shield effect caused by topography for the A2 participant.

Table 7. Linear correlations (r^2) between AI, meteorological variables and data on possible emitting activities.

ZONE 1	(AI)A1	(AI)B1	(AI)C1	(AI)D1			(AI)ZONE	ZONE 2	(AI)A2	(AI)B2	(AI)C2	(AI)D2	(AI)E2	(AI)ZONE
(AI)A1[a]		0.756	0.787	0.817			0.910	(AI)A2		0.768	0.601	0.517	0.504	0.777
(AI)B1	0.756		0.840	0.612			0.878	(AI)B2	0.768		0.911	0.937	0.610	0.978
(AI)C1	0.787	0.840		0.838			0.935	(AI)C2	0.601	0.911		0.934	0.811	0.969
(AI)D1	0.817	0.612	0.838				0.940	(AI)D2	0.517	0.937	0.934		0.707	0.936
								(AI)E2	0.504	0.610	0.811	0.707		0.752
(AI)ZONE[b]	0.910	0.878	0.935	0.940				(AI)ZONE	0.777	0.978	0.969	0.936	0.752	
IMPSOUC1	0.843	0.869	0.800	0.920			0.794	IMPSOUC1	0.297	0.552	0.744	0.822	0.863	0.672
IMPSOUC2	0.011	-0.175	0.415	0.372			0.113	IMPSOUC2	-0.535	-0.072	0.146	0.158	-0.194	-0.121
PRES	-0.762	-0.255	-0.233	-0.435			-0.516	PRES	0.176	0.205	0.055	0.206	-0.351	0.059
CALM	-0.624	-0.639	-0.924	-0.671			-0.771	CALM	-0.128	-0.619	-0.730	-0.851	-0.647	-0.673
HUM	-0.086	-0.138	0.318	-0.188			0.014	HUM	0.067	0.016	-0.337	0.103	-0.020	0.055
VEL	0.883	0.857	0.829	0.637			0.781	VEL	0.058	0.620	0.472	0.410	0.713	0.516
TEMP	-0.326	0.154	-0.614	-0.217			-0.301	TEMP	0.437	0.313	0.189	-0.068	-0.237	0.194
WASVOL	0.845	0.922	0.868	0.963			0.872	WASVOL	0.358	0.618	0.765	0.706	0.868	0.719
ZONE 3	IMOA3	IMOB3	IMOC3	IMOD3	IMOE3	S1+S2	(AI)ZONE	ZONE 4	(AI)A4					(AI)ZONE
(AI)A3		0.007	0.967	0.873	0.624	0.925	0.829	(AI)A4						
(AI)B3	0.007		0.250	0.240	0.802	-0.461	0.660							
(AI)C3	0.967	0.250		0.933	0.891	0.914	0.965							
(AI)D3	0.873	0.240	0.933		0.980	0.808	0.849							
(AI)E3	0.624	0.802	0.891	0.980		0.269	0.902							
(AI)ZONE	0.829	0.660	0.965	0.849	0.902			(AI)ZONE						

Table 7. (Continued)

IMPSOUC1	0.892	-0.486	0.930	0.852	0.628	0.988	0.586	**IMPSOUC1**	0.766					
IMPSOUC2	0.933	-0.420	0.871	0.742	0.191	0.988	0.609	**IMPSOUC2**	0.937					
S1+S2	0.925	-0.461	0.914	0.808	0.269		0.606	**S1+S2**	0.861					
PRES	0.244	0.326	0.162	0.180	0.154	0.132	0.223	**PRES**	0.525					
CALM	-0.855	0.014	-0.820	-0.815	-0.871	-0.718	-0.688	**CALM**	-0.735					
HUM	0.697	0.311	0.653	0.343	0.331	0.689	0.664	**HUM**	0.461					
VEL	0.347	-0.511	0.280	0.279	-0.393	0.352	0.106	**VEL**	0.175					
TEMP	-0.683	0.879	-0.624	-0.472	0.156	-0.903	-0.264	**TEMP**	-0.585					
WASVOL	0.128	-0.131	0.247	0.328	0.154	0.044	0.111	**WASVOL**	-0.028					

(a) Monthly Annoyance Index of the receptor A1.
(b) Monthly Annoyance Index of the zone 1.

IMPSOUC1: Monthly frequency of wind direction from Focus 1 to the receptors location (Table 3).
IMPSOUC2: Monthly frequency of wind direction from Focus 2 to the receptors location (Table 3).
S1+S2: Sum of monthly frequencies of wind directions from Focus 1 and 2 to the receptors location (Table 3).
PRES: Atmospheric Pressure.
VEL: Wind velocity.
CALM: Frequency of wind calms.
TEMP: Ambient Temperature.
HUM: Relative Humidity.
WASVOL: Volume of residues entering to the urban waste disposal site.

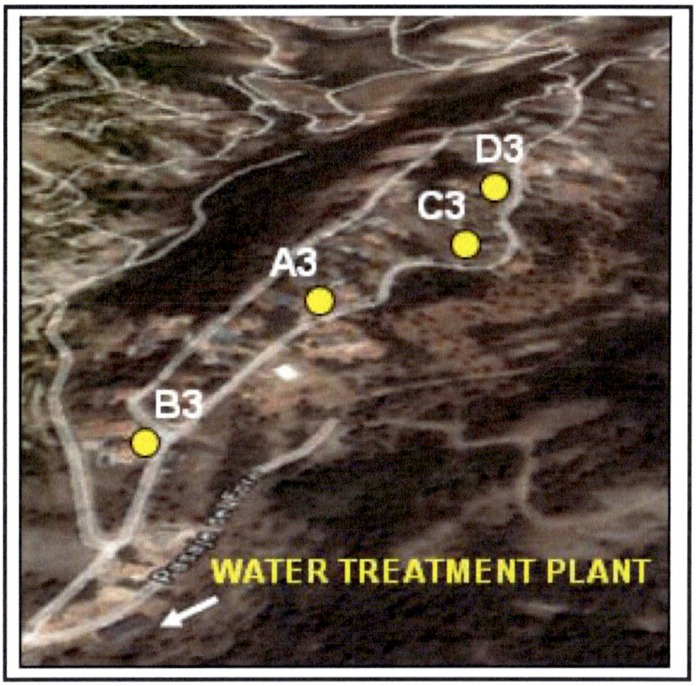

Figure 12. Impact direction (S) of the water treatment plant near participant B3.

The sensory data obtained from the participants are highly correlated to the wind impact frequencies of the sources. This makes it possible to create a protocol to assign potential sources to odor episodes in the affected zones. Two additional causes also seem to have an effect: calm percentage (percentage of time when the wind velocity is close to zero) with an inverse correlation ranging from 0.624 to 0.924 in all zones, and wind velocity, with correlations in the range of 0.620-0.883 for the points near the main source (Source 1). The participants are also highly correlated (in the range of 0.618-0.963) to the volume of waste entering the landfill site. For these reasons, the use of sensory data from people affected by odor episodes constitutes the basis of a new method for determining the contribution of each possible source in cases of a small number of sources in residential areas.

When the number of odor emission sources is greater than the number in the previous example, the sensory data obtained do not give the same good results. In such cases, the sensory data obtained can be used in a more continuous format (Figure 8, form B). The application of the data in this format, once validated using the criteria explained above, allows valid data to be obtained for more complex evaluation zones (multiple-source, long and short episodes). The contribution of the various sources to odor episodes can be determined by elaborating the following:

a) Episode roses: Frequencies of impact direction for the detected episodes and for each odor intensity, using individual analysis of all episodes with intensity values greater than 3. The impact directions are determined as specified in the German VDI 3940 standard.
b) Annoyance index maps: Elaborated with the largest AI values in the study period.

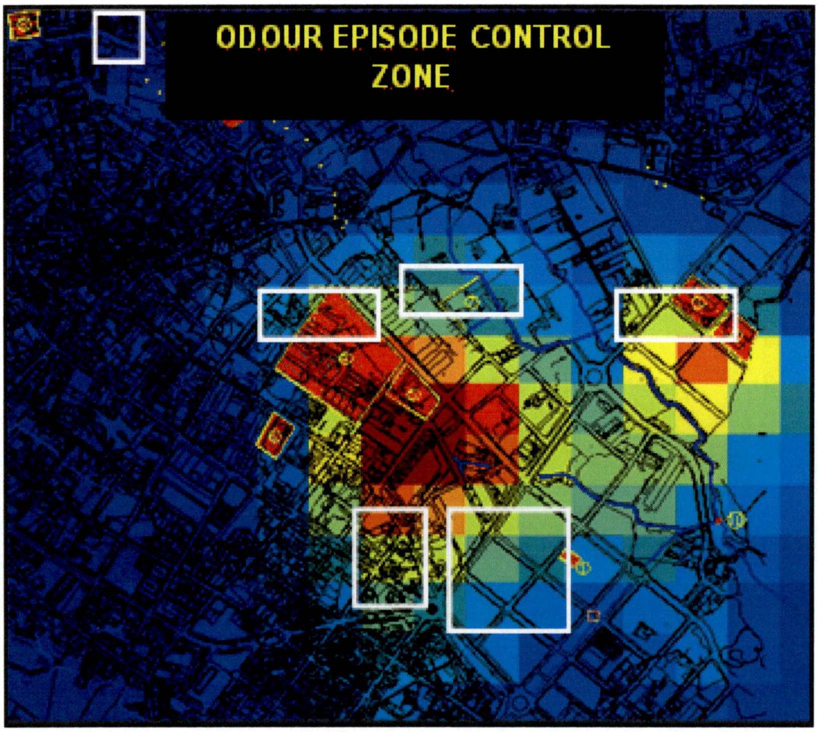

Figure 13. Study zones of odor episodes on a simulated impact map for inventoried potential sources.

Figure 13 shows the real-life application of the methodology in a city of 50,000 inhabitants, with the impact of multiple sources, mainly food industry facilities (gelatin, slaughterhouses, dairy products, etc.), leather manufacturers, and cattle and pork farms. The activities identified as potential sources are both inside the urban area and on the periphery.

An individual analysis of the registered odor episodes makes it possible to create a database and episode roses. The impact directions have already been determined based on local meteorological data as specified in the German VDI 3940 standard. From these roses, we can differentiate the episodes that have a clear source from those that do not (which must be studied on-site).

Figures 14 and 15 show episode roses based on more than 3,000 episodes recorded from June 2004 to January 2005.

From June to October, with Pasquill-Gifford atmospheric stability classes A, B and C (i.e. more unstable atmospheric conditions) predominating 26 to 42% of the time, the impact directions of the episodes were independent of the wind regime. From November to January, when stability classes D, E and F predominate 72 to 83% of the time, the impact directions were closer to the prevailing wind directions.

In the example, impact directions in Zone D always came from the same sources. In Zone F, however, they are more variable, and sources that initially were not considered were detected by on-site inspection and by using back-trajectory methods in dispersion models.

Table 8. Impact directions of episodes: Zone F.

MONTH	RECEP	INT.ODO	DATE	HOUR	WIND VEL	IMPDIR(K)	IMPDIR(U)
JUNE	4B10	3	01.06.04	19:00 a 20:00	2.4/1.4	SW/SSW	
	4B03,4B01	4	01.06.04	20:00 a 21:00	1.4/1.6		S
ZONE F	4B01	4	02.06.04	02:00 a 03:00	0.5/0.1		NNW
	4B10	3	02.06.04	04:00 a 05:00	0.2/0.9	E	
	4B10	3	03.06.04	04:00 a 05:00	0.8/0.1		S
	4B10	3	08.06.04	19:00 a 20:00	3.1/2.9	ENE	
	4B01	4	08.06.04	20:00 a 21:00	1.9/2.1	NNE	
	4B10	3	10.06.04	04:00 a 05:00	1.0/1.0	W	
	4B10	3	11.06.04	04:00 a 05:00	0.1/n.d.		NNW
	4B03	3	15.06.04	22:00 a 23:00	0.9/0.8	WNW	
	4B03	5	16.06.04	00:00 a 01:00	0.3/0.6	ENE	
	4B01	4	17.06.04	01:00 a 02:00	3.6/4.1		N
	4B01	4	17.06.04	23:00 a 24:00	0.7/1.2	NNE	
	4B03,4B01	3	21.06.04	22:00 a 23:00	3.7/3.9		S
	4B01	3	21.06.04	23:00 a 24:00	3,./1.3		S
	4B03	3	22.06.04	14:00 a 15:00	4.3/3.8	E	
	4B03	4	22.06.04	23:00 a 24:00	4.8/5.0		S
	4B03	3	25.06.04	12:00 a 13:00	1.9/2.0	E/ESE	
	4B03	4	27.06.04	20:00 a 21:00	2.5/1.9		SSE
	4B01	3	28.06.04	23:00 a 24:00	0.4/0.4	WSW	
	4B03	5	29.06.04	00:00 a 01:00	1.4/0.7		NNW
	4B01	5	29.06.04	01:00 a 02:00	1.5/0.9	W	
	4B01	4	29.06.04	05:00 a 06:00	1.6/1.9	WNW	
	4B03	4	29.06.04	23:00 a 24:00	0.4/1.2		N
	4B10	3	30.06.04	19:00 a 20:00	2.9/2.6	ENE	
	4B03	3	30.06.04	21:00 a 22:00	1.6/0.3		S
	4B01	5	30.06.04	23:00 a 24:00	0.4/0.6		S

RECEP: Participant reference.
INT.ODO: Odor Intensity.
WIND VEL: Wind velocity.
IMPDIR(K): Impact direction from known odor emission source.
IMPDIR(U): Impact direction from unknown odor emission source.

Using the graphical representation of the AI obtained from the highest odor intensity values recorded by participants in the 0-8, 8-16 and 16-24 hourly intervals, the main sources responsible for the odor episodes can be detected. Figure 16 shows the distribution of the annoyance over the urban area, with wind direction frequencies for the different monthly periods.

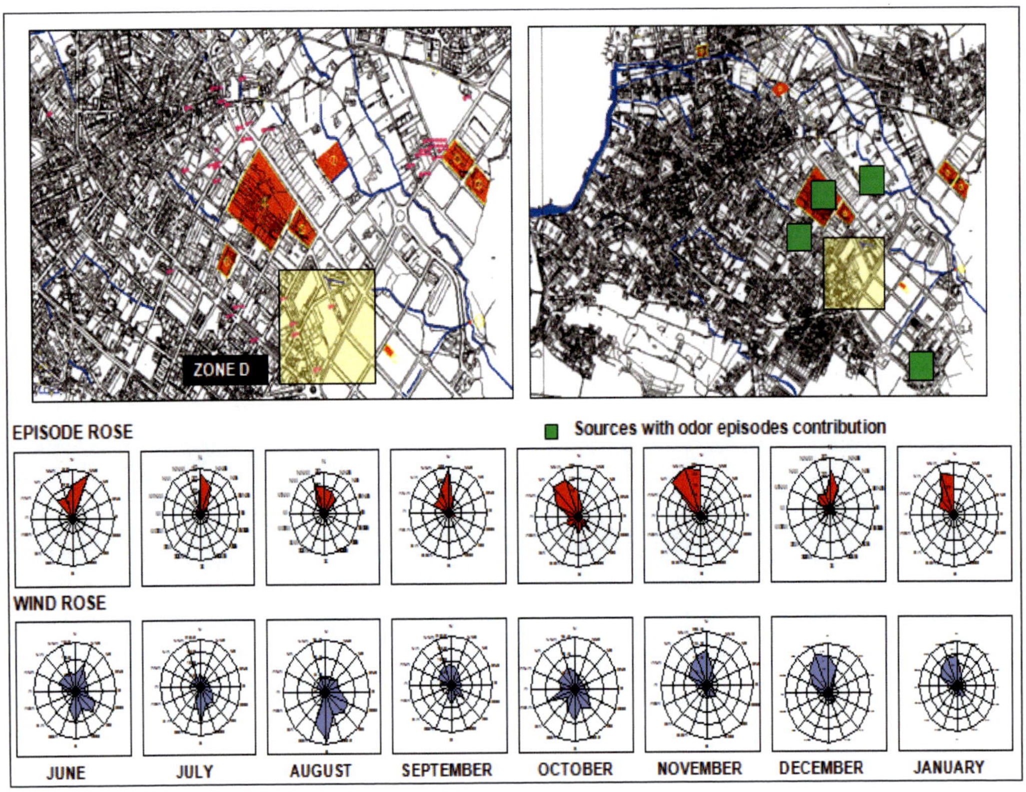

Figure 14. Episode roses. Sources contributing to odor episodes (Zone D).

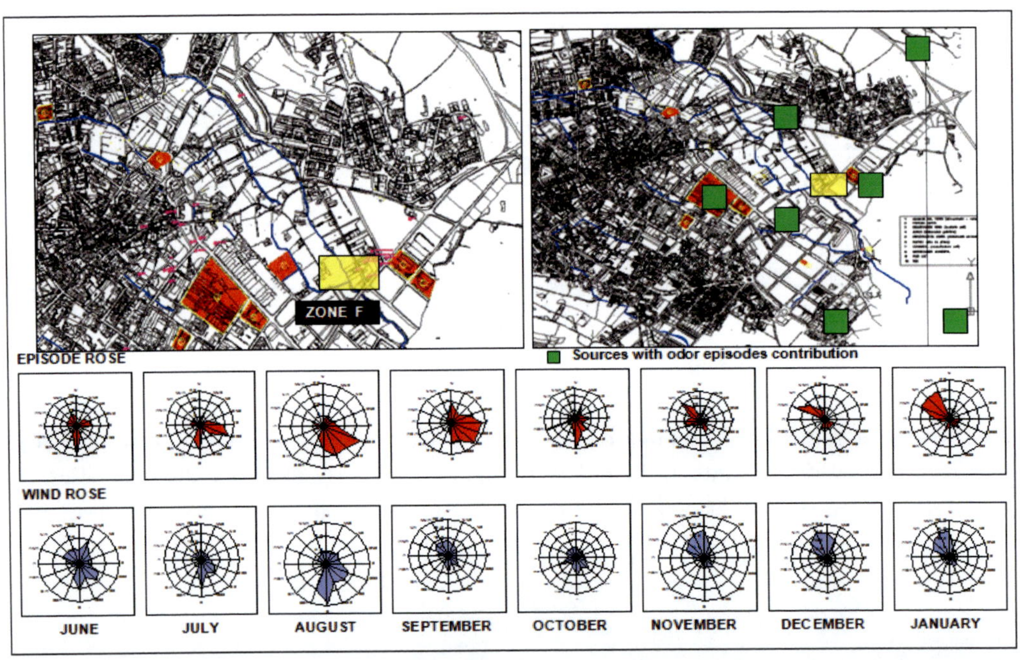

Figure 15. Episode roses. Sources contributing to odor episodes (Zone F).

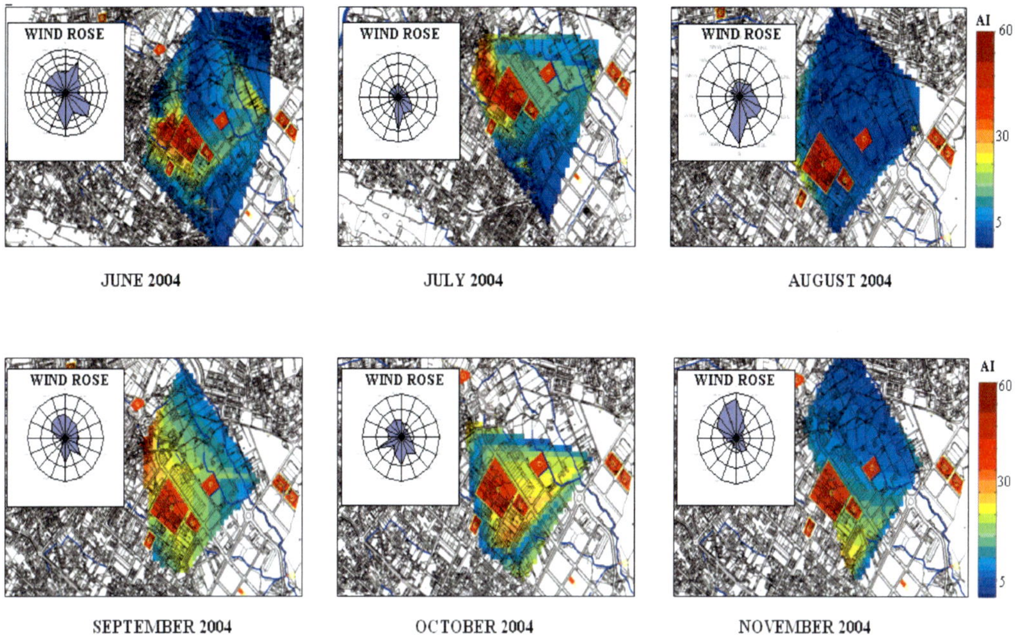

Figure 16. Odor annoyance index (AI) distribution for the urban area of Banyoles. Correspondence with wind direction frequencies.

5. Impact Maps

Air dispersion models and impact maps are used to study air quality and the impact of known probable emission sources.

The mathematical modeling of physicochemical processes allows a lot of information to be processed in a short time. The simulation uses the numerical solutions of mathematical equations, usually partial differential equations that are numerically integrated over time and space, by means of algorithms translated into computer programming languages. As tools, these models are as good as their scientific bases. New concepts are continually being introduced to improve them. Their results must be viewed critically, but they are a good option for obtaining knowledge on dispersive processes.

In the transport of pollutants from a source to the environment, many variables influence how these contaminants reach the surrounding area and affect the environment.

A pollutant released into the atmosphere (under certain conditions of gas velocity, pressure, temperature, flow and concentration at a certain point—for example, a stack with a known height and diameter) is affected by several phenomena, such as jets of air and floating due to temperature differences. The plume is then dispersed from the emitting zone according to the atmospheric conditions of wind field and stratification and is diluted, thereby reducing its concentration. A good definition of the meteorological conditions during the dispersion time is therefore required. Different atmospheric data requirements for different models will be explained below.

Many different models are in use today. A few have been accepted for regulation purposes after having been tested against sample-based field work to see how closely they estimate the real pollutant concentrations measured around a source.

Table 9. Air dispersion models for regulatory purposes.

US ENVIRONMENTAL PROTECTION AGENCY

Recommended general-use models:

- AERMOD: A Gaussian environmental model. Replaces the Industrial Source Complex (ISC3) model.
- CALPUFF: A non-steady-state Gaussian-Lagrangian model. Instead of a continuous plume, it uses a puff of gas released to the atmosphere.

Specific-use models:

- Buoyant Line- and Point-Source Dispersion Model (BLP): A Gaussian plume model for aluminum reduction plants and other sources with plume elevation and building effects (downwash).
- CALINE3: For line-source emitters such as roads.
- CAL3QHC: An adapted CALINE3 model for studying intersections, etc.
- Complex Terrain Dispersion Model Plus Algorithms for Unstable Situations (CTDMPLUS).

Gaussian model:

- Offshore and Coastal Dispersion (OCD) model.

AUSTRALIA

- AUSPLUME: A Gaussian regulatory model approved by the Australian EPA.
- The Air Pollution Model (TAPM): A very complete Eulerian atmospheric diagnostic model.

EUROPE

Great Britain

- Atmospheric Dispersion Modeling System (ADMS): Includes URBAN, ROADS and SCREEN versions.
- Urban Dispersion Model (UDM): A Gaussian model for urban areas developed by the Ministry of Defense.

Germany

- AUSTAL2000: A Germany Lagrangian regulatory model based on the 2002 TA Luft (Technische Anleitung zur Reinhaltung der Luft).

Impact maps are the graphical result of the calculation, using dispersion models, of the distribution of a pollutant at ground-level concentration. In this application, there is no difference between odorous and non-odorous compounds. Impact maps provide information on the area that must be studied in the field through sampling, chemical analysis and, in particular, social participation. Impact maps from several studies on both urban and rural environments are shown as examples below.

There are two basic methods for calculating impact maps, long-term and short-term, depending on the period of time studied. Short-term methods run the model for time periods ranging from half an hour to several years, and give the sequential evolution of the ground-level concentration for each time step (usually half an hour or one hour, depending on the meteorological data interval). The average concentrations for the entire period, maps of areas with the highest concentrations, etc., can also be calculated using short-term methods.

Long-term methods calculate concentrations using meteorological data in a frequency distribution form, as shown in Table 1, in which the variables of wind velocity, wind direction and atmospheric stability class are grouped by frequency.

The biggest difference between long-term and short-term methods, therefore, is that the latter only calculate at steady state for situations described by each of the terms of the frequency array. After that, they find the average concentrations using the frequency data. The calculation time is much shorter than that needed for short-term methods, which require that each time step be calculated.

The model developed at the LCMA can be used for both short-term and long-term calculations. Short-term calculations are usually used to simulate short episodes, while long-term calculations are used to simulate monthly or yearly impact maps.

The LCMA model has been used in cases where surface meteorological station data was available for the study area. In situations in which there were no such stations, meteorological data from mesoscale models were processed with a diagnostic model and the local meteorological data were obtained through successive nesting.

An odor dispersion model is basically an application of a general air dispersion model with the following characteristics:

- The scale is local and calculations are normally done for relatively short distances from the emission sources (on the order of several kilometers).
- The relation between odor intensity and concentration is exponential rather than linear and depends on the particular compounds distributed in the air.
- The number of odor units at ground level and the zones where odor will be detected can be determined if the emission sources are well-defined and the odor thresholds for the compounds are known.

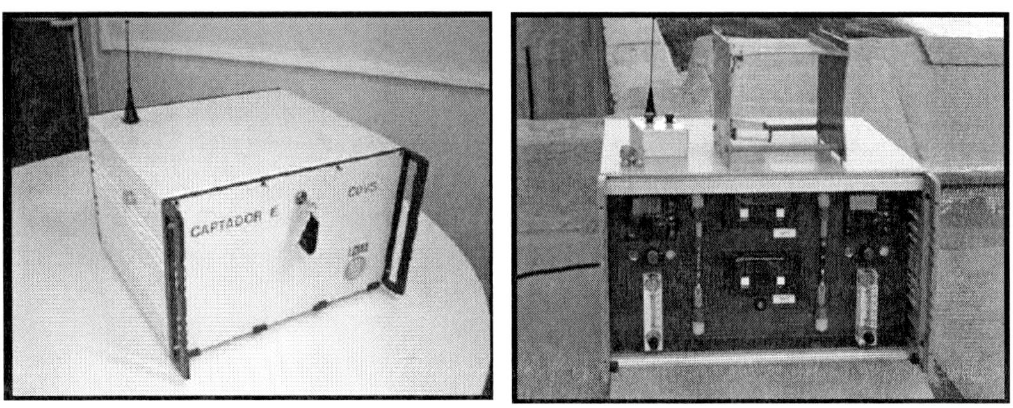

Figure 17. Air VOC sampler.

6. Chemical Control

Interest in determining VOC concentrations in both indoor and outdoor air has increased over the last several decades, since VOCs are known precursors of photochemical smog formation [Derwent et al., 1996, Chang et al., 1999, Atkinson 2000, Peng et al. 2006, Vautard et al. 2007] as well as can cause short- and long-term adverse effects on human health (irritation of mucous membranes, psychological stress and long-term toxic reactions) [ECA-IAQ 1997, Hutter et al. 2006, Liang and Liao 2007] and comfort (bad odours) [Wolkolff and Nielsen 2001, Zuraimi et al. 2006]. Chemical control is used for the individual determination of a wide range of odor nuisance and air-quality VOC present in air samples. A sorbent-based method (the air is pulled through glass tubes filled with appropriate sorbent) has been developed to collect successfully VOC in air, both target and wide-range screening compounds, including alkanes, aromatic hydrocarbons, aldehydes, alcohols, chlorides, esters, ketones, terpenes, amides, carbon disulfide and isocyanates. A good combination of different sorbents allows the qualitative determination of a wide range of C_2-C_{14} VOC, around 200-300 in a standard urban sample [Ribes et al. 2007].

6.1. Air Sampler Development

A collector pump sampler has been specially designed to collect dynamically air samples in sorbent tubes, both outdoors and indoors (determination of indoor air quality in cases of sick building syndrome evaluation) (Figure 17). The VOC sampler is equipped with inert captation line and high precision total volume measurement. Other characteristics include 10 calibration flow levels, high flow stability, very low breakthrough values and inexistent tube contamination during pre-activation processes. The airtihghtness of the system was evaluated coupling a sorbent tube to the collector pump sampler during a month. The analysis of the sorbent tube revealed no VOC contamination. The operating flow ranges between 40 and 200 ml min^{-1}. Samples can be taken dynamically during 24-hour controls or during odor episodes. By means of remote control (both radio frequency and mobile phone), the air sampler can be activated during odor episodes when medium and high odor intensity and nuisance is percept. The potentially affected people and municipality technicians can activate the sampler. In the same way, a mobile unit may also be handed to the local police in order to follow directionally changing episodes. Collected ambient air samples are further analyzed by thermal desorption and gas chromatography-mass spectrometry (TD-GC/MS) [Ribes et al. 2007].

Outdoors, the sampling system can be used in situations involving permanent and discontinuous odor episodes in air, either in urban, industrial, waste treatment facilities or rural areas. To determine permanent odors, the areas of major potential impact are delimited by site inspection and auditing of real activities, as well as through the creation of impact maps, where the probable impacts of the different emission activities are represented in the study zone. Furthermore, collector pump samplers are located in several fixed sites in order to be representative of the different areas of the urban and/or industrial zone. A thorough analysis of the meteorological conditions in the area during the sampling period in combination with the chemical information obtained through the individual VOC determination allow an irrefutable tracking down of the permanent odor sources and to draw up a list of potential odor episodic sources.

In order to check the system effectiveness during discontinuous odor episodes sampling, the collector pump sampler was tested by identifying individual VOC in ambient air collected in the vicinity of municipal solid waste landfill sites, chemical plants and industrial locations.

6.2. Analytical Methodology

A versatile validated analytical method, based on thermal desorption (TD) coupled to gas chromatography (GC) and mass spectrometry (MS), is used for the determination of VOC in indoor and outdoor air [Ribes et al. 2007]. The VOC analyzed quantitatively are selected on the basis of their occurrence and adverse effects on environment and human health and comfort.

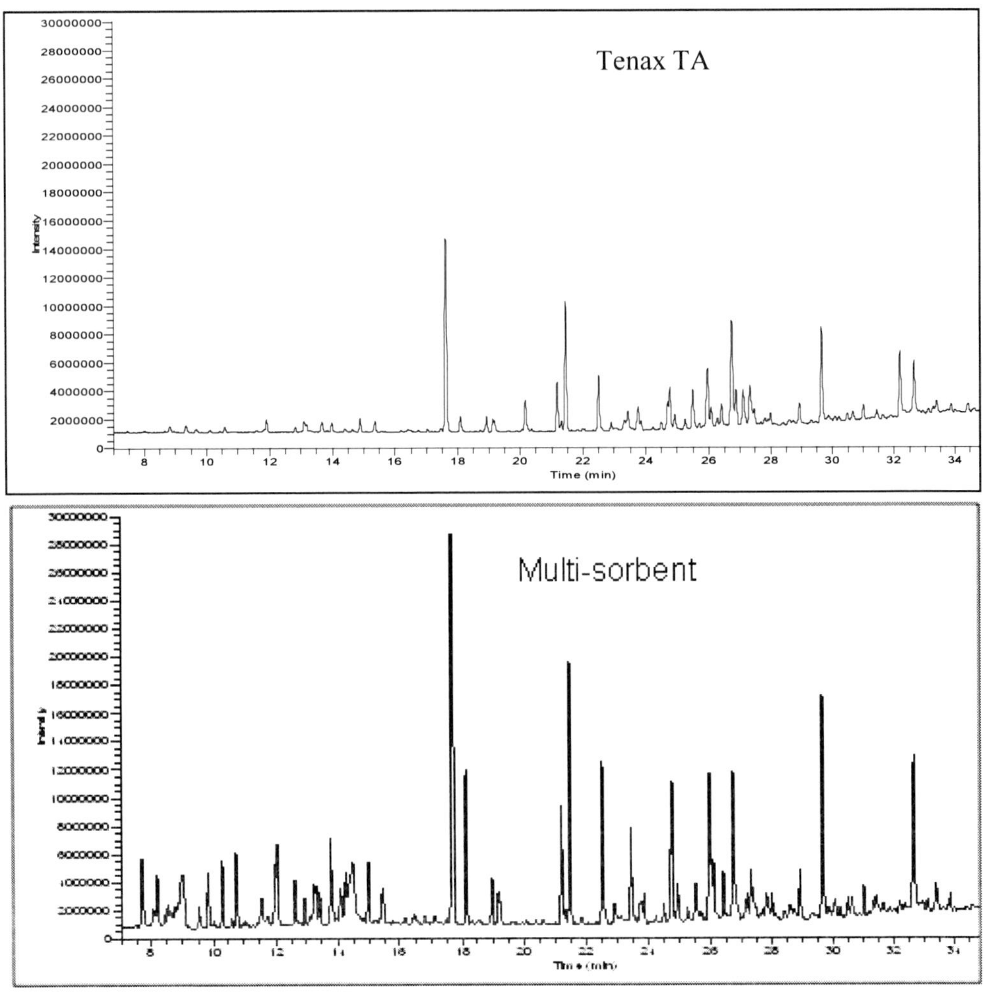

Figure 18. TD-GC/MS chromatograms of a simultaneous sample using Tenax TA and multi-sorbent (Carbotrap, Carbopack X and Carboxen 569) adsorbents for its sampling.

6.2.1. Multi-sorbent Tubes

A multi-sorbent tube filled with Carbotrap (activated graphitised black carbon, 20/40 mesh, weak sorption strength, hydrophobic), Carbopack X (activated graphitised black carbon, 40/60 mesh, medium sorption strength) and Carboxen 569 (spherical carbon molecular sieve, 20/45 mesh, high sorption strength) has been optimized to analyze successfully both polar and non-polar VOC. Adsorbents are placed in glass tubes ordered from weak to strong sorption strength (air stream introduction way) [Ribes et al. 2007]

Before the first use, tubes are conditioned by thermal cleaning (250°C for 20 min, 300°C for 20 min, 330°C for 20 min, 350°C for 20 min and 400°C for 20 min) under a flow rate of helium of 70 ml min^{-1}. For subsequent uses, pre-conditioning at 400°C for 20 min is applied. After conditioning, they are immediately sealed with Swagelock end caps fitted with PTFE ferrules and stored in boxes filled with desiccant material. Samples are stored at 4°C and analyzed as soon as possible (1 week at the latest).

A comparative study was done between the generally used adsorbent Tenax TA and the above-mentioned multisorbent-tube, containing Carbotrap, Carbopack X and Carboxen 569. The performance evaluation between the two kinds of sampling adsorbent methods was done in real samples. Important differences were observed between the sampling efficiency for the most volatile fraction of VOC, giving better qualitatively and quantitatively responses the multisorbent-tube compared to Tenax TA (Figure 18) [Berenguer et al. 2002].

6.2.2. Analytical Instrumentation

The analysis of VOC is performed by Automatic Thermal Desorption coupled with capillary Gas Chromatography/Mass Spectrometry Detector. The thermal desorption of the sampling tubes is carried out at 300°C with a flow rate of 50 ml min^{-1} for 10 min (primary desorption), during that time the eluted compounds are swept from the tube to a cryofocusing trap (containing approximately 15 mg of Tenax TA and 15 mg of Carbotrap) maintained at –30°C, applying a flow split of 4 ml min^{-1}. After a primary desorption, the cold trap is rapidly heated from –30°C to 300°C (secondary desorption) and then maintained at this temperature for 10 min. During the secondary desorption, the compounds are submitted to a flow split of 7 ml min^{-1} and are injected onto the capillary column (DB-624, 60 m x 0.25 mm x 1.4 μm) via a transfer line heated at 200°C. The column oven temperature starts at 40°C for 1 min, increases to 230°C at a rate of 6°C min^{-1} and then it is maintained at 230°C for 5 min. Helium (99.999%) carrier gas flow in the analytical column is approximately 1 ml min^{-1} (1.4 bar).

Mass spectral data are acquired over a mass range of 20-300 amu. A 6 min solvent delay time is applied for standards analysis to avoid saturation of mass spectrometer detector. Qualitative identification of target compounds is based on the match of the retention times and the ion ratios of the target quantification ions and the qualifier ions. Quantification of field samples is conducted by the external standard method. Standards are prepared in methanol and injected at 30°C on the multisorbent-tubes under an inert Helium gas flow (100 ml min^{-1}) using a conventional gas chromatograph packed column injector. Tube loading lasts not less than 5 min. The large variability of VOC occurrence in air samples, in terms of abundance, lets to the necessity of working with two different concentration ranges. Hence, the quantification method is validated for two different quantification ions for each VOC to allow a reliable quantification in both very diluted (major characteristic ion in the VOC spectra) and very concentrated samples (minor characteristic ion, normally the molecular ion) [Ribes et al. 2007].

The chromatographic separation is satisfactory for most of VOC (Figure 19, Table 10). Mainly all co-eluting VOC are satisfactory quantified using characteristic ions. Only *m*-xylene and *p*-xylene are quantified together as they exhibit identical mass spectra. Limits of detection (LOD), determined applying a signal-to-noise ratio of 3, range form 0.001 to 10 ng. However, *NN*-Dimethylformamide and *N*-Methylformamide present values of 14 and 97 ng, respectively. The studied compounds exhibit linearity ranges (ng) from 3 to 4 orders of magnitude and most of them show repeatibilities (% relative standard deviation values) lower than 25%, accomplishing the EPA performance criteria [U.S. EPA 1999]. No significant breakthrough values are observed for those VOC with boiling points >70°C. Extreme precautions are established for quality assurance, injecting periodically blank samples and a known concentration of toluene [Ribes et al. 2007].

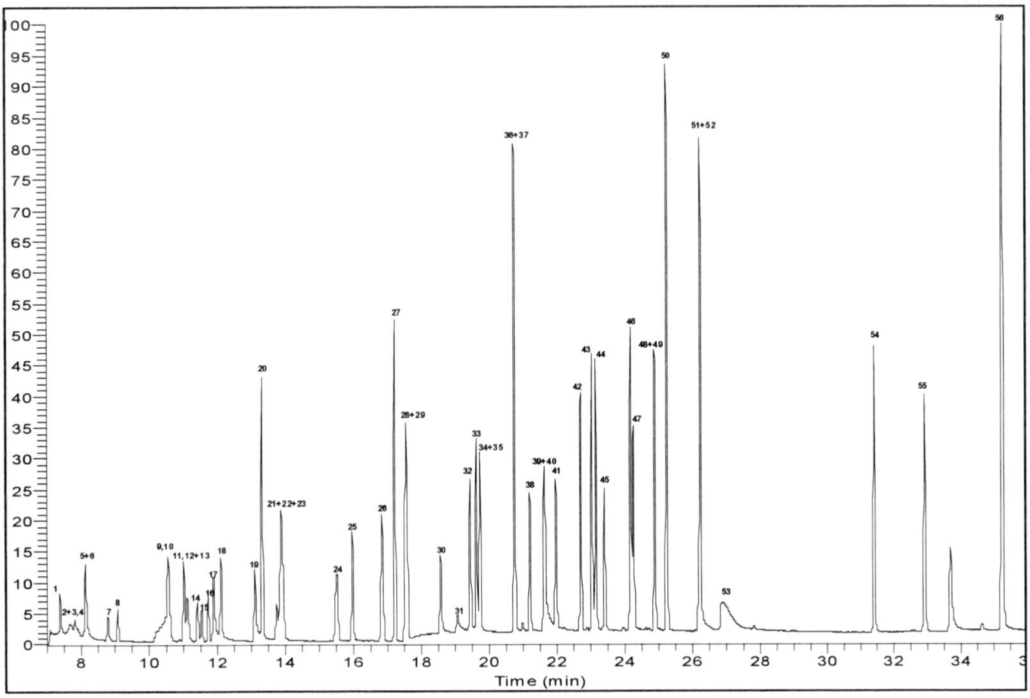

Figure 19. GC chromatogram for stock standard solution. Table 10 shows corresponding VOC reference numbers.

6.3. Concentration Maps

To exemplify the results obtained form chemical control (VOC concentrations, both 24-hour average and maximum episodic values), concentration maps are represented in a study area (Figure 20). These maps use interpolation between concentrations at the air sampling points and can show individual VOC, families of VOC (sum of the concentrations of individual VOC from the same family) or Total VOC (TVOC) concentrations. Concentration maps are important tools for determining the origin of VOC emissions in a concrete study area, as they show clearly the focuses with higher concentrations of the studied compounds.

Besides, they can be useful to establish the contribution of each emission activity to individual or familiar VOC concentrations in a concrete area, showing the most important emission activities in respect to the air quality of the zone. These maps validate the predictive impact maps mentioned in chapter 5, as the most important impacted zones in an area (where higher concentrations of VOC are found) generally fit well with the predicted impact areas (Figure 21).

Table 10. Method target VOC analytes

Target VOCs	VOC Ref. No.	CAS No.	MW (g/mol)	BP (ºC)	RT (min)	Quantification ions	
						m/z 1	m/z 2
Ethanol	1	64-17-5	46	79	7.4	45	46
Propanal	2	123-38-6	58	49	7.5	58	58
Acetone	3	67-64-1	58	56	7.56	43	44
Carbon disulfide	4	75-15-0	76	46	7.84	76[a]	76
Methyl acetate	5	79-20-9	74	57	8.14	74[a]	74
Isopropanol	6	67-63-0	60	82	8.16	45	59
Tert-butylmethylether	7	1634-04-4	88	55	8.81	73	57
n-Hexane	8	110-54-3	86	69	9.11	57	86
Butanal	9	123-72-8	72	75	10.16	44	72
Ethyl acetate	10	141-78-6	88	77	10.48	61	88
Chloroform	11	67-66-3	117	61	11.02	83	87
Methylethylketone	12	78-93-3	72	80	11.12	72	57
Tetrahydrofuran	13	109-99-9	72	66	11.12	42	72
1,1,1-Trichloroethane	14	71-55-6	131	74	11.42	97	117
Cyclohexane	15	110-82-7	84	81	11.54	56	84
Carbon tetrachloride	16	56-23-5	151	77	11.74	117	121
Isobutanol	17	78-83-1	74	108	11.9	43	74
Benzene	18	71-43-2	78	80	12.12	78	51
1-Butanol	19	71-36-3	74	118	13.11	56	31
Trichloroethylene	20	79-01-6	129	87	13.13	130	134
Methylcyclohexane	21	108-87-2	98	101	13.77	55	98
Pentanal	22	110-62-3	86	103	13.78	44	86
Methyl methacrylate	23	80-62-6	46	100	13.83	100[a]	100
Methylisobutylketone	24	108-10-1	100	117	15.53	43	100
Toluene	25	108-88-3	92	111	15.97	92	65
1,1,2-Trichloroethane	26	79-00-5	132	114	16.83	166	168
Tetrachloroethylene	27	127-18-4	163	121	17.22	166	168
Butyl acetate	28	123-86-4	116	126	17.55	73[a]	73
Hexanal	29	66-25-1	100	131	17.59	44	72
N,N-Dimethylformamide	30	68-12-2	73	153	18.56	73	58
N-Methylformamida	31	123-39-7	59	183	19.06	59	30
Ethylbenzene	32	100-41-4	106	137	19.45	106	65
n-Nonane	33	111-84-2	128	150	19.63	57	128
m-Xylene	34	108-38-3	106	139	19.73	106	77
p-Xylene	35	106-42-3	106	138	19.73	106	77
o-Xylene	36	95-47-6	106	145	20.75	91[a]	91
Styrene	37	100-42-5	104	145	20.75	104[a]	104
Heptanal	38	111-71-7	114	150	21.2	44	86
2-Butoxyethanol	39	111-76-2	118	171	21.64	57	87

α-Pinene	40	7785-70-8	136	157	21.65	93	136
Cyclohexanone	41	108-94-1	98	155	21.97	98	83
Propylbenzene	42	103-65-1	120	159	22.71	91	120
n-Decane	43	124-18-5	142	174	23.04	71	142
1,3,5-Trimethylbenzene	44	108-67-8	120	165	23.15	105	120
β-Pinene	45	127-91-3	136	167	23.4	93	136
1,2,4-Trimethylbenzene	46	95-63-6	120	168	24.16	105	120
Benzaldehyde	47	100-52-7	106	178	24.25	77	106
Isocyanatocyclohexane	48	3173-53-3	125	169	24.88	82	125
Limonene	49	5989-27-5	176	177	24.88	93	136
p-Dichlorobenzene	50	106-46-7	147	174	25.24	140	75
n-Undecane	51	1120-21-4	156	195	26.22	57	156
Phenol	52	108-95-2	94	182	26.32	94	66
1-Octanol	53	111-87-5	130	194	26.88	41	84
Naphthalene	54	91-20-3	128	218	31.4	128	102
Isothiocyanatocyclohexane	55	1122-82-3	141	219	32.9	55	141
2-Methylnaphthalene	56	91-57-6	142	242	35.34	142	115
1-Methylnaphthalene	57	90-12-0	142	245	35.9	142	115

CAS number (CAS No.), molecular weight (MW, g mol^{-1}), boiling point (BP, °C, at 760 mmHg), retention time (RT, min) and quantification ions m/z 1 (low concentration range) and m/z 2 (high concentration range).

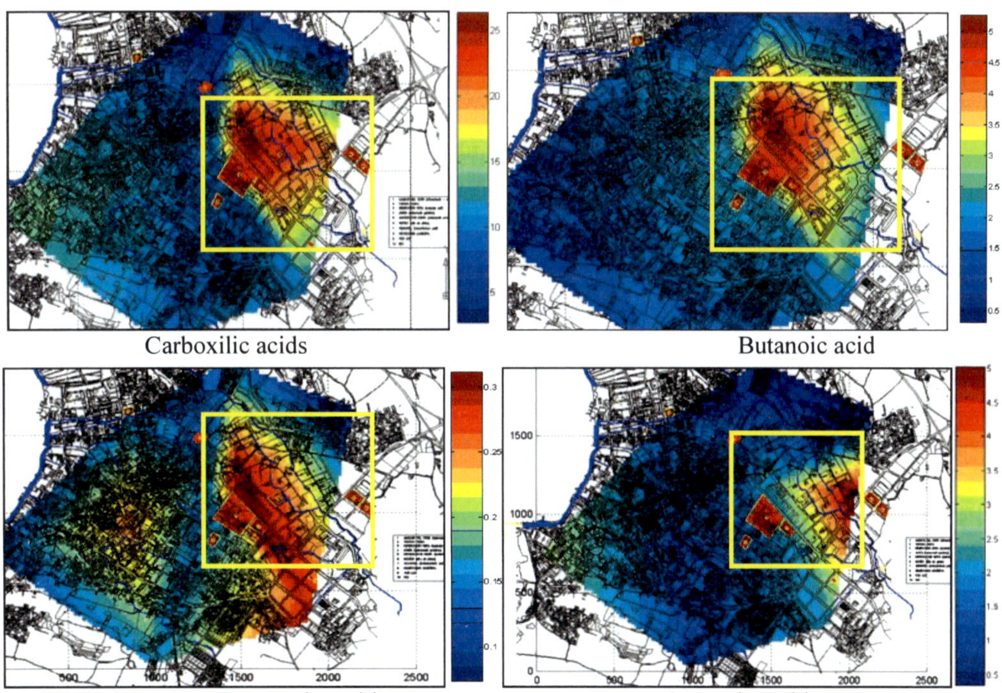

Figure 20. Maps of concentrations (μg Nm^{-3}) of carboxylic acids in Banyoles (Catalonia, Spain).

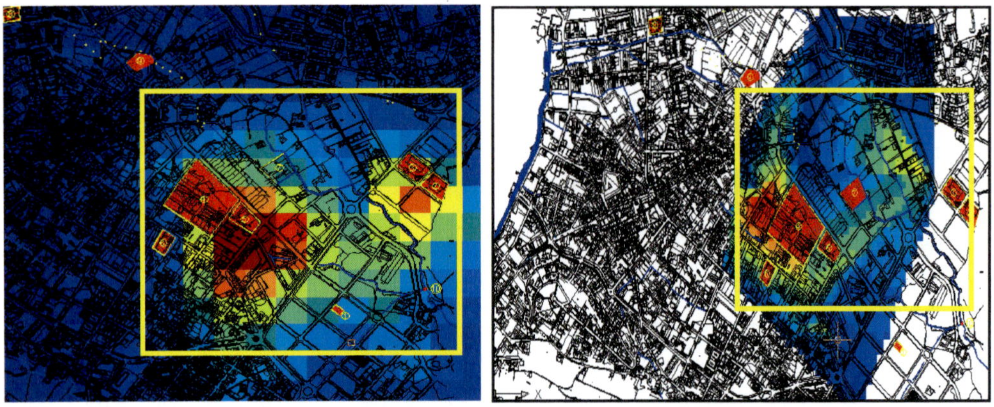

Impact prediction map Concentration map

Figure 21. Impact prediction map and concentration map for Banyoles (Catalonia, Spain).

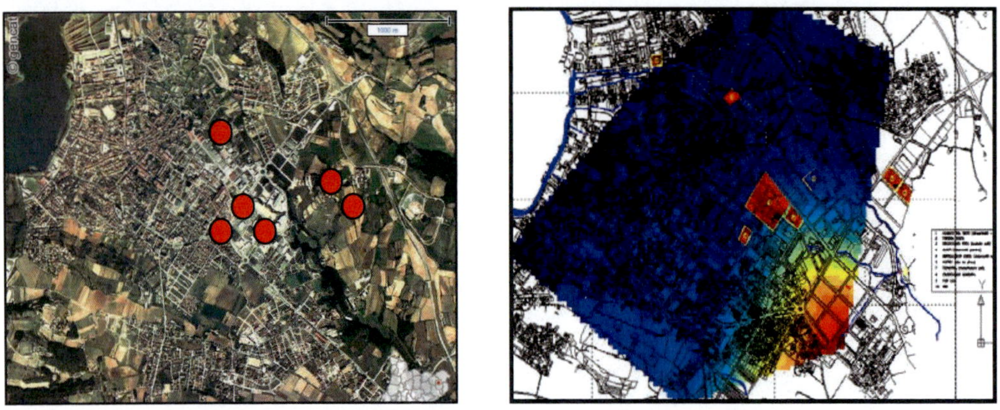

Figure 22. Initial inventory of emission activities. Carbon disulfide concentration map.

6.4. Odor Maps

Most VOCs are odorous substances capable of eliciting an olfactory response in people. The odor sensation is caused when an odorant stimulates the olfactory organs [Powers, 2004]. Not all odorants make people react the same way; it depends on the odor threshold, which is defined as the lowest concentration of a particular compound at which 50% of a human panel can perceive the odor. The threshold of a chemical compound is determined in part by its shape, polarity, partial charges and molecular weight (Table 11). The perception of a mixture of odorants is very different from how each chemical would be perceived independently. Odorants can act as additive, masking or synergistic agents [Powers, 2004]. The detection of an odor does not tell us what kind of compounds are interacting in the odor. The specific chemical substances producing the odor can be determined by applying chemical control.

Table 11. Odor thresholds of several VOCs.

Chemical compound	Odor threshold (μg m^{-3})	Odor description
Acetic acid	43	Vinegar
Butanoic acid	0.35-86	Rancid
Pentanoic acid	8-12,000	Sweat
Hydrogen sulphide	0.76	Rotten egg
Methyl mercaptan	0.003-38	Cabbage
Ethyl mercaptan	0.043	Decayed cabbage
Dimethyl sulphide	0.34-1.1	Rotten vegetable
Dimethyl disulphide	1.1-46	Putrefaction
Ammonia	100-11,600	Pungent and irritating
Dimethylamine	47-160	Fish
Formaldehyde	490	Acrid
Acetaldehyde	0.01-4	Fruit, apple
Pentanal	2.5-34	Fruit, apple
Butanone	870	Green apple

Odor maps can be represented by means of odor units (OUs) (Figure 23). The OU value is the amount of odorant, over the threshold limit, present in a cubic meter of odorous gas. It is calculated by dividing the concentration of a specific compound by its threshold limit. This calculation indicates how many times the threshold limit has been exceeded. Like concentration maps, odor maps can help us determine the source of VOC emissions.

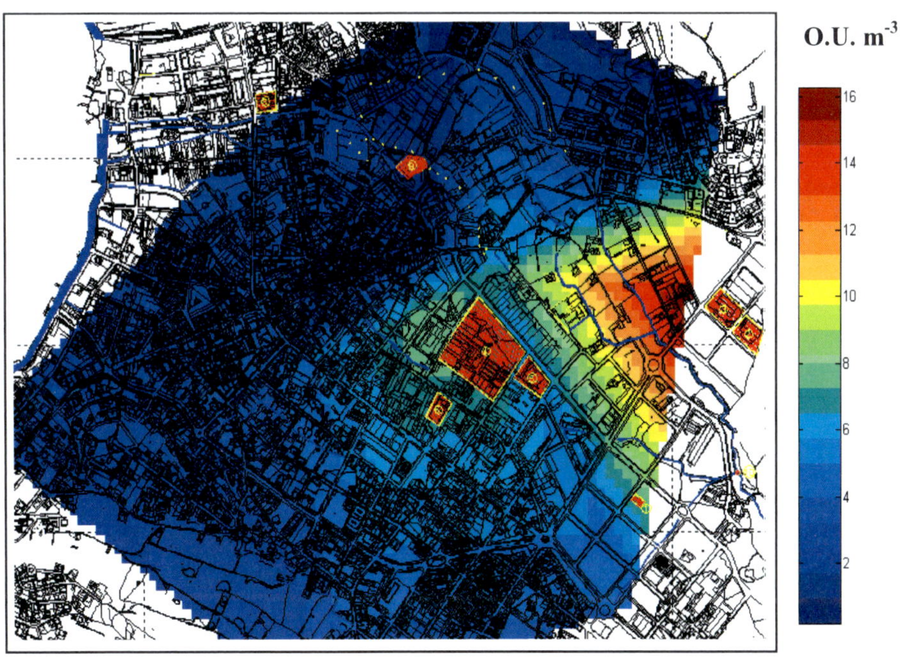

Figure 23. Odor map (OU m^{-3}) of Banyoles (Catalonia, Spain).

6.5. Air Quality Determination

Air quality can be defined in a variety of ways, but the most realistic way is to define it in terms of a measurable atmospheric condition. Generally, specific atmospheric pollutants are established and measured, physically or chemically, in real time or as average concentrations [Godish, 1997]. Several criteria based on quality guidelines and standards have been established for determining air quality. Some of them, proposed by the US Environmental Protection Agency, refer to outdoor air and are intended to protect the population from adverse health effects caused by exposure to toxic substances. Other countries have determined similar reference values for evaluating outdoor air quality (Table 12). Other types of standards, such as the threshold limit values (TLVs) developed by the American Conference of Governmental Industrial Hygienists (ACGIH), are used as professional exposure limits, applicable to the working population. These values are not intended to be used directly on the general population, as they have been established with specific exposure conditions. The World Health Organization (WHO) determines ambient air quality reference levels for CO, NO_2, SO_2, O_3, lead, particulate matter and several carcinogenic and non-carcinogenic compounds [WHO, 2005, Bartual et al., 2008].

The outdoor air quality of a specific place and period of time depends on several variables, such as chemical emissions to the atmosphere, emission source density, the topography and orography of the area, meteorological conditions and regional planning (industrial facilities are often located very close to inhabited areas). Due to the specific meteorology of an area, changes in wind regimes can cause variability in air quality over the course of a year, over 24-hour periods and from one location to another [Godish, 1997].

Indoor air quality (in homes, commercial establishments, offices and public buildings) depends on emissions from a variety of sources, including building materials, appliances of various types, consumer products, environmental tobacco smoke, the entrance of outdoor air and ventilation rates [Maroni et al., 1995, Godish, 2001].

6.5.1. Outdoor Air

As mentioned above, different countries have different guideline values for evaluating outdoor atmospheric pollutants (Table 12). These values are generally based on indoor professional exposure TLVs. TLVs cannot be applied directly to general population; they must be corrected by a factor because they are established for specific exposure conditions. The factor applied to the TLVs must correct several points that are characteristic of working exposure. TLVs are defined for an average exposure, sometimes discontinuous, of 40 hours per week. However, the general population is sometimes exposed continually to outdoor air for more than 40 hours per week. In addition, TLVs are applied to the working population (healthy people aged 16 to 65), whereas the general population includes small children, the elderly and ill people. Finally, in working environments, regulation and control targets the compounds derived from professional exposure, and not the other compounds present in the air [Bartual et al., 2008].

Table 12. Ambient air quality regulations.

Country	State	Regulation	Information
World Health Organization (WHO)		**NOAEL**: No Observed Adverse Effect Level **LOAEL**: Lowest Observed Adverse Effect Level	www.who.int
United States (US EPA)		**AEGL**: Acute Exposure Guideline Level	www.epa.gov
United States (HIOSH)		**IDLH**: Immediately Dangerous to Life Health concentrations	www.cdc.gov/niosh
United States		**ACL**: Ambient Air Concentration Limits	
United States	California	Odor threshold and irritant characteristics	www.dhs.ca.gov
United States	Massachusetts	**TEL**: Threshold Effects Exposure Limits	www.mass.gov
United States	Massachusetts	**AAL**: Allowable Ambient Limits for Ambient Air	www.mass.gov
United States	Michigan	**IRSL**: Initial Risk Screening Level **SRSL**: Secondary Risk Screening Level	www.michigan.gov
United States	Texas	**ESL**: Effects Screening Levels	www.tceq.state.tx.us
United States	Vermont	Hazardous Ambient Air Standards	www.anr.state.vt.us
Canada	Ontario	**POI**: Point of Impingement	www.ene.gov.on.ca
Canada	Ontario	**AAQC**: Ambient Air Quality Criteria	www.canlii.org
Spain (Environment Ministry)		TLV/30	www.mma.es

6.5.2. Indoor Air

Several inorganic, organic and biological contaminants can be found in indoor environments, such as asbestos, radon, CO, CO_2, NO_x, SO_2, O_3, VOC, bacteria, viruses, allergens and fungi [WHO 1982, Jones 1999, Godish 2001]. As it has been said in previous chapters, VOC are a highly diverse class of chemical contaminants, and between 50 and 300 compounds may be found in non-industrial indoor air environments [Mølhave 1992, Johansson 1999]. VOC have been generally less studied than other indoor air contaminants, and the knowledge of the adverse effects on health associated to indoor air VOC concentrations is nowadays very limited (Johansson, 1999; Edwards etal., 2001; Venn et al., 2003). Several human exposure studies have been carried out [Wolkoff 1995, Andersson et al. 1997, Pappas et al. 2000, Mølhave 2001], however, the VOC concentrations experimented were relatively high and far from the usual dwelling concentrations, making difficult the extrapolation between the results obtained and the real effects of indoor air VOC concentrations [Holcomb and Seabrook 1995, Wolkoff et al. 1997]. The evaluation of health issues caused by complex VOC mixtures is also difficult because their effects may be additive, synergistic, antagonic or even independent for each other, making hardly predictable the toxicological consequences of VOC exposition to human wellness [Mølhave et al. 1997, Johansson 1999]. On the other hand, the wide range of sampling strategies and Total VOC (TVOC) concentration calculation methods used in the literature often make TVOC published data not comparable [Mølhave 1992, ECA-IAQ 1997, Mølhave et al. 1997, Wolkoff et al. 1997, Johansson 1999]. TVOC should be determined following the established standardized method (ECA-IAQ 1997 and ISO 16000-6), analyzing the main indoor VOC families, such as aliphatic hydrocarbons, aromatic hydrocarbons, terpenes, alcohols, aldehydes, ketones, halocarbons and esters (Table 13). TVOC should be used as a screening tool, as has not biological relevance and is not recommended for making definitive conclusions about indoor air quality [Andersson et al. 1997]. However, it has some useful applications as an indicator of the presence of VOC indoors [Wolkoff and Nielsen 2001, Mølhave 2003].

Table 13. Minimum number of compounds to include in TVOC analysis[1]

Chemical compound	Cas No.
Aromatic Hydrocarbons	
Benzene	71-43-2
Toluene	108-88-3
Ethylbenzene	100-41-4
m+*p*-Xylene	108-38-3/106-42-3
o-Xylene	95-47-6
n-Propylbenzene	103-65-1
1,2,4-Trimethylbenzene	95-63-6
1,3,5-Trimethylbenzene	108-67-8
2-Ethyltoluene	611-14-3
Styrene	100-42-5
Naphthalene	91-20-3
4-Phenylcyclohexene	31017-40-0
Aliphatic Hydrocarbons	
n-Hexane	110-54-3
n-Heptane	142-82-5
n-Octane	111-65-9
n-Nonane	111-84-2
n-Decane	124-18-5
n-Undecane	1120-21-4
n-Dodecane	112-40-3
n-Tridecane	629-50-5
n-Tetradecane	64036-86-3
n-Pentadecane	629-62-9
n-Hexadecane	544-76-3
2-Methylpentane	107-83-5
3-Methylpentane	96-14-0
1-Octene	111-66-0
1-Decene	872-05-9
Cycloalkanes	
Methylcyclopentane	96-37-7
Cyclohexane	100-82-7
Methylcyclohexane	108-87-2

Terpenes	
3-Carene	13466-78-9
α-Pinene	80-56-8
β-Pinene	181172-67-3
Limonene	138-86-3
Alcohols	
2-Propanol	67-63-0
1-Butanol	71-36-3
2-Ethyl-1-hexanol	104-76-7
Glycols/Glycolethers	
2-Methoxyethanol	109-86-4
2-Ethoxyethanol	110-80-5
2-Butoxyethanol	111-76-2
1-Methoxy-2-propanol	107-98-2
2-Butoxyethoxyethanol	112-34-5
Aldehydes	
Butanal	123-72-8
Pentanal	110-62-3
Hexanal	66-25-1
Nonanal	124-19-6
Benzaldehyde	100-52-7
Ketones	
Methylethylketone	78-93-3
Methylisobutylketone	108-10-1
Cyclohexanone	108-94-1
Acetophenone	98-86-2
Halocarbons	
Trichloroethene	79-01-6
Tetrachloroethene	127-18-4
1,1,1-Trichloroethane	71-55-6
p-Dichlorobenzene	106-46-7
Acids	
Hexanoic acid	142-62-1
Esters	
Ethylacetate	141-78-6
Butylacetate	123-86-4
Isopropylacetate	108-21-4
2-Ethoxyethylacetate	111-15-9
Texanolisobutyrate	6846-50-0
Other	
2-Pentylfuran	3777-69-3
Tetrahydrofuran	109-99-9

[1] ECA-IAQ, 1997.

Table 14. TVOC percentage for different chemical families.[1]

Family	% TVOC
Alkanes	33
Aromatic hydrocarbons	17
Terpenes	10
Halocarbons	10
Esters	7
Aldehydes and Ketones[2]	7
Other	16

[1]Seifert, 1990; [2]Except formaldehyde

Two possible approaches for determining indoor air quality through TVOC calculation have been proposed by Mølhave [Mølhave, 1991] and Seifert [Seifert, 1990]. Mølhave [Mølhave, 1991] suggested four exposure ranges of TVOC: comfort range (<0.2 mg m^{-3}), multifactorial exposure range (0.2-3 mg m^{-3}), discomfort range (3-25 mg m^{-3}) and toxic range (>25 mg m^{-3}). On the other hand, [Seifert (1990) estimated a target indoor air quality guideline value based on two empirical field studies in German and Dutch homes [WHO, 1989; Lebret, et al., 1986; Krause et al., 1987; Seifert and Abraham, 1982], suggesting that TVOC concentrations in indoor air should not exceed 300 μg m^{-3}. On the basis of the above mentioned empirical field studies, a standard classification of the dwelling indoor air VOC concentrations was proposed depending on the different chemical families present indoors (Table 14). Alkanes, aromatic hydrocarbons, terpenes, halocarbons, esters, aldehydes (except formaldehyde) and ketones and other compounds, should contribute with 33%, 17%, 10%, 10%, 7%, 7%, and 16% of the TVOC, respectively. In addition to that, two indoor air quality criterions were established. Criterion 1: no individual compound should exceed 50% of the average value of its class and Criterion 2: no individual compound should exceed 10% of the TVOC value.

It has been observed that people repetitively exposed to long-term relatively high levels of VOC can modulate their physiological response to a given VOC [Wolkoff and Nielsen 2001, Hummel et al. 2000]. For this reason, legislated guideline recommendations should be established for VOC concentrations in indoor air. In the interim period, VOC indoor concentrations should be maintained as low as reasonably achievable (ALARA), to diminish the organic pollution load in indoor environments [ECA-IAQ 1997, Mølhave, L. 2003].

6.6. Identification of Recently Identified Pollutants and Understudied Compounds

Indoor and outdoor air quality controls and odor episode studies based on TD-GC/MS air analysis can detect a series of notorious compounds with high odor thresholds. Depending on the concentrations, the odor of these compounds sometimes cannot be perceived. These potentially toxic compounds (e.g. isocyanatocyclohexane and isothiocyanatocyclohexane) are generally recently identified pollutants or understudied compounds. In some cases, no information is available about recognized or suspected health hazards of long-term exposure to these compounds [Gallego et al., 2007].

Several of these compounds are not currently included in the lists of chemical compounds in urban air quality analysis protocols and have only been recorded in indoor workplaces. Due to their toxic nature, compounds of this sort must be taken into account in determining air quality, despite the fact that their effects on humans are still under evaluation.

7. Case Studies

7.1. Back-Trajectory Application

The back trajectory of an air mass shows the path a particular air mass has followed to reach a final position at a certain time.

When a well-defined wind field is obtained for the area in question during the study period, the trajectory of an air mass during this time interval can be calculated. This calculation consists in the numerical or graphical integration of air velocity vectors over time, beginning at a certain position using calculated contour values for the wind field and making reverse time steps: t-dt, t-2dt, t-3dt and so on.

Back-trajectory calculation is used, for example, to find possible emitters when there is no clear relation between an episode and the known emission sources in a certain area.

Figure 24. Graphical description of the study area.

The back-trajectory application was used to study an odor episode in the city of Terrassa that originated at a municipal solid waste landfill located several kilometers from the urban area (Figure 24). Until this study was carried out, residents did not know that the landfill was causing the odor episodes they had been enduring for several years. On 22 June 2003, several people reported an odor episode between 5:30 and 6:30 a.m. GMT. A chemical analysis (Figure 25) of air samples revealed the presence of chemical compounds that suggested that a waste disposal site was a possible source.

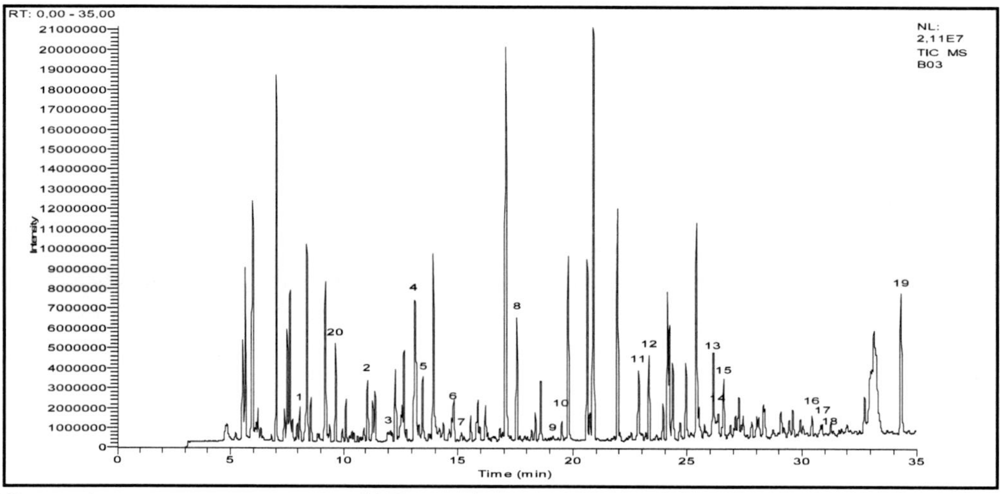

1 Propanal	11 Heptanal
2 Trimethyl silanol	12 Octamethylcyclotrisiloxane
3 Butanal	13 Cyclohexylisocyanate
4 Formic acid	14 Pentanoic acid
5 Acetic acid	15 Hexanoic acid
6 Butanol	16 Ethylheptanoic acid
7 Pentanal	17 Octanoic acid
8 Hexamethylcyclotrisiloxane	18 Nonanoic acid
9 Hexanal	19 Cyclohexylisothiocyanate
10 Butyric acid	20 Carbon disulphide

Figure 25. Chromatogram of an air sample from Terrassa.

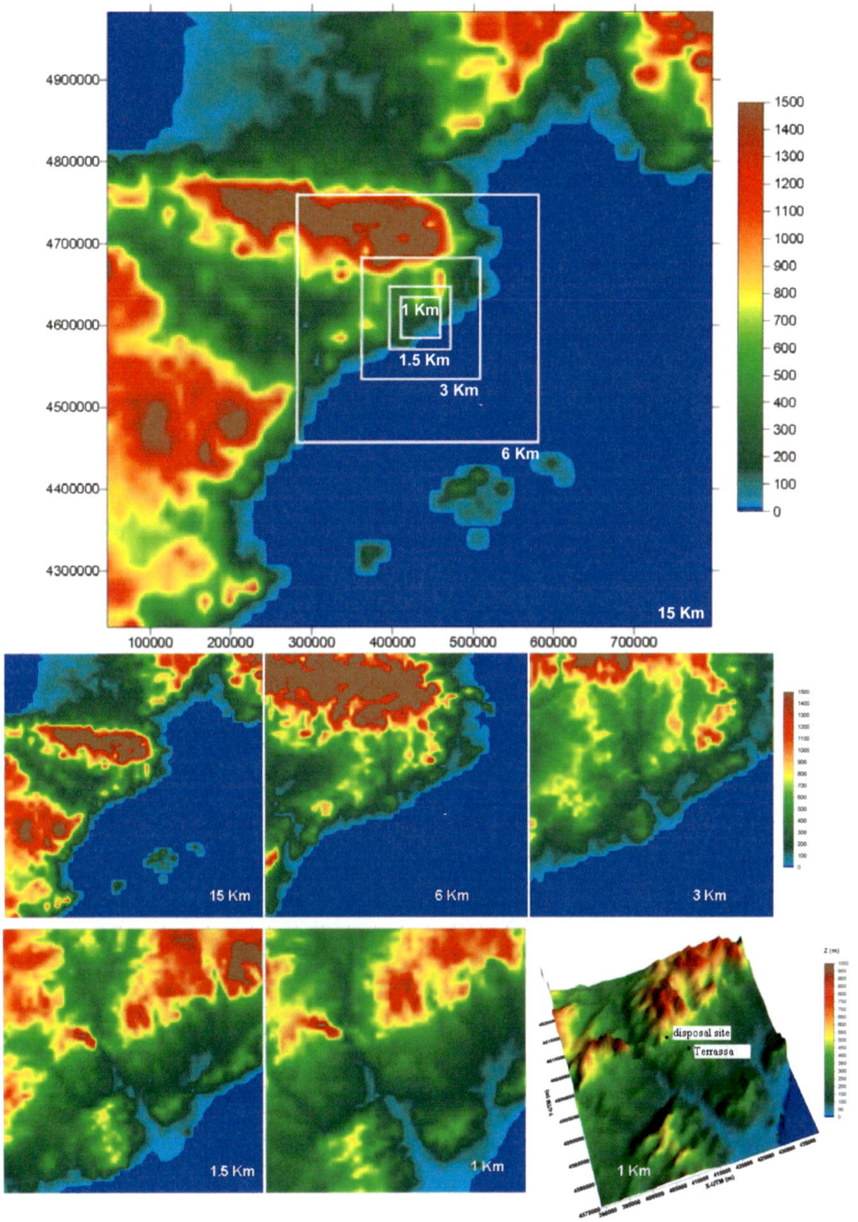

Figure 26. Dominions and nesting zones in the meteorological model simulation.

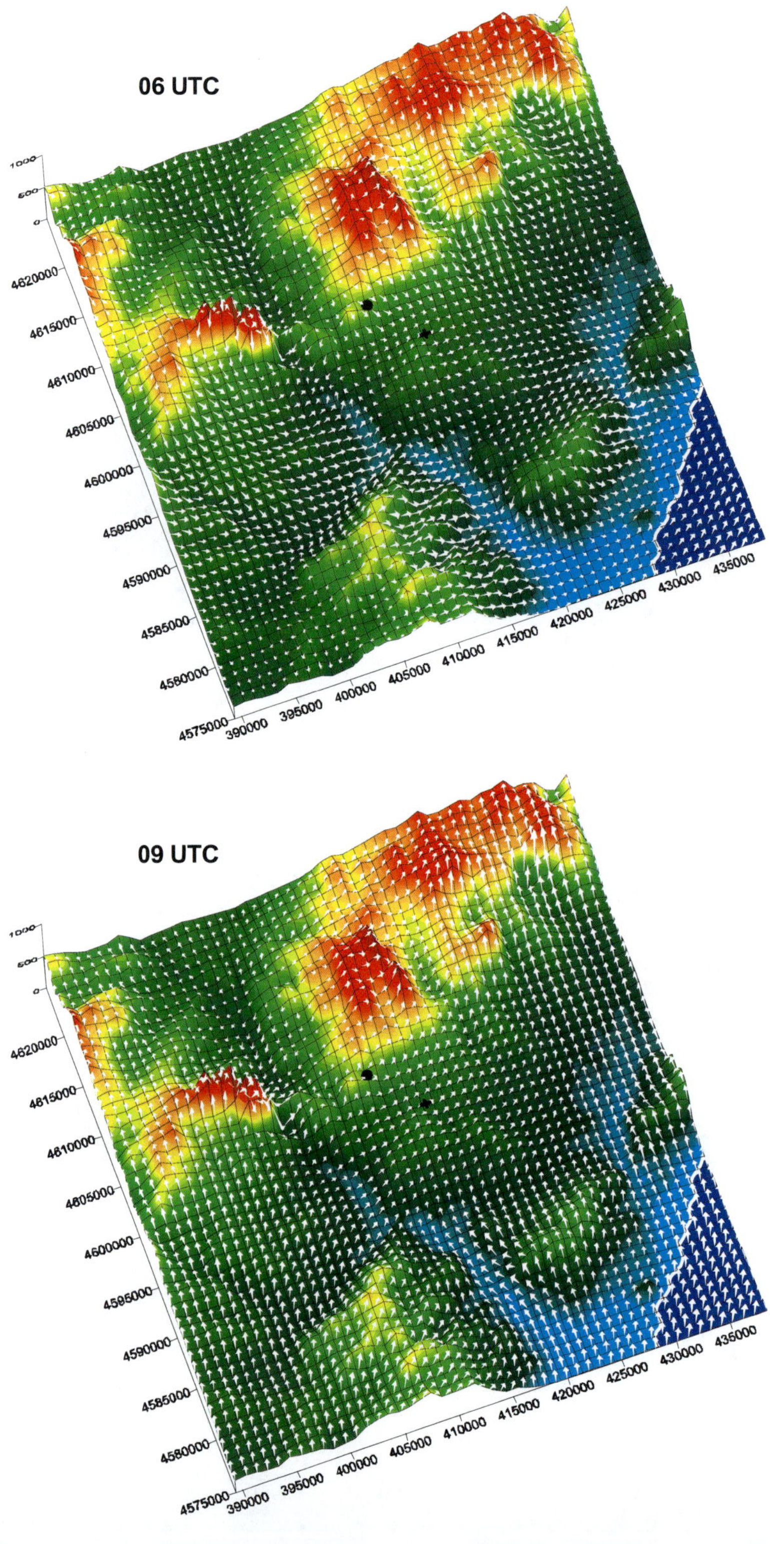

Figure 27. Samples of hourly wind fields obtained at the innermost nesting zone.

The meteorological model was applied to a very complex topographical zone that affected the wind regimes and required a very high resolution. This was achieved by using consecutive nesting, as shown in Figure 26 [Soriano et al., 2004]. Figure 27 shows the hourly wind field calculated based on the numerical simulation, which features high variability of wind directions.

Figure 28 shows the back trajectories calculated for different times of day on 22 June 2003. For the location of the air mass over Terrassa at each specific time, the path it took to arrive there is shown. When the residents detected the odor episode (from 5:00 to 6:00 a.m., approximately), they were breathing an air mass that had passed over the municipal waste landfill site. This points to the landfill as the source of the odor episode. From 7:00 a.m. on, the air over the city did not come from the landfill, which explains the end of the odor episode.

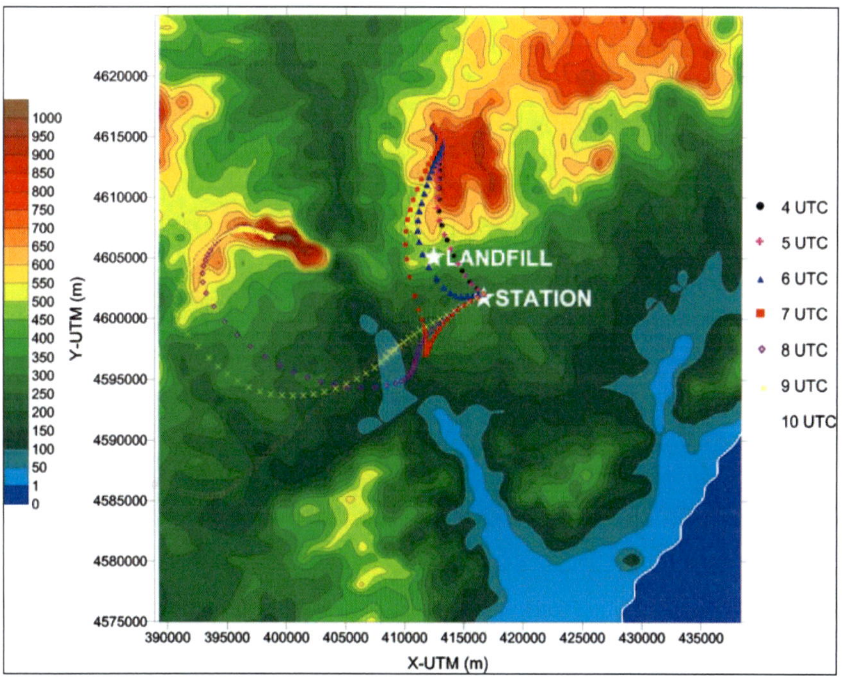

Figure 28. Back trajectories calculated for different times.

7.2. Application of Integrated Methodology (Social Participation, Chemical Control and Numerical Modeling)

The application of chemical control is illustrated by a study conducted in Benicarló (Autonomous Community of Valencia, Spain). One of its main objectives was to determine the chemical composition and source of odor episodes that had affected much of the city.

7.2.1. Description of the Study Area

Benicarló is a city in the Autonomous Community of Valencia (Spain) with approximately 25,000 inhabitants. It is located at the shore of the Mediterranean Sea and is surrounded by irrigated orchards. The surrounding orography generally does not exceed a height of 150 meters above sea level. Benicarló is characterized by its Mediterranean climate. Winters are mild, with average temperatures ranging from 7 to 10°C. Summers are cool, with average temperatures ranging from 20 to 26°C, and are dominated by high pressure areas and varying humidity levels, making rainfall unlikely except for the odd thunderstorm.

The most important VOC emission sources in Benicarló are furniture manufacturers, polyester resin plants and fragrance producers (Figure 29).

The available information on the emitting activities lacked data on VOC emissions. These activities have multiple emission sources from both the industrial process and combustion (including waste incineration in activities S1 and S2). The emission heights are lower than 20 meters.

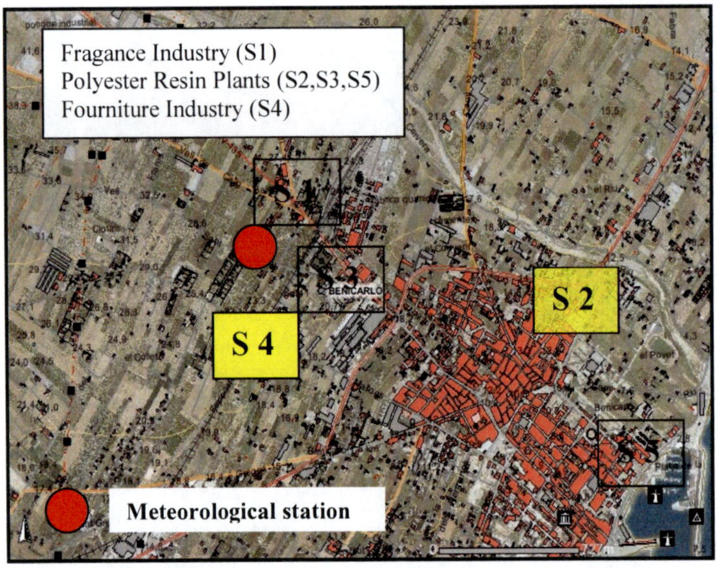

Figure 29. VOC emission sources related to industrial activities.

7.2.2. Meteorological Analysis

The wind direction frequencies are those typical of coastal zones: prevailing winds from the NW sector in the interval 0-8 hours and from the SE in the interval 8-16 hours. In the period of maximum solar radiation (8-16 hours), the wind direction frequency is divided between the two aforementioned wind sectors (Figure 30), with NW and SSE being the most frequent directions.

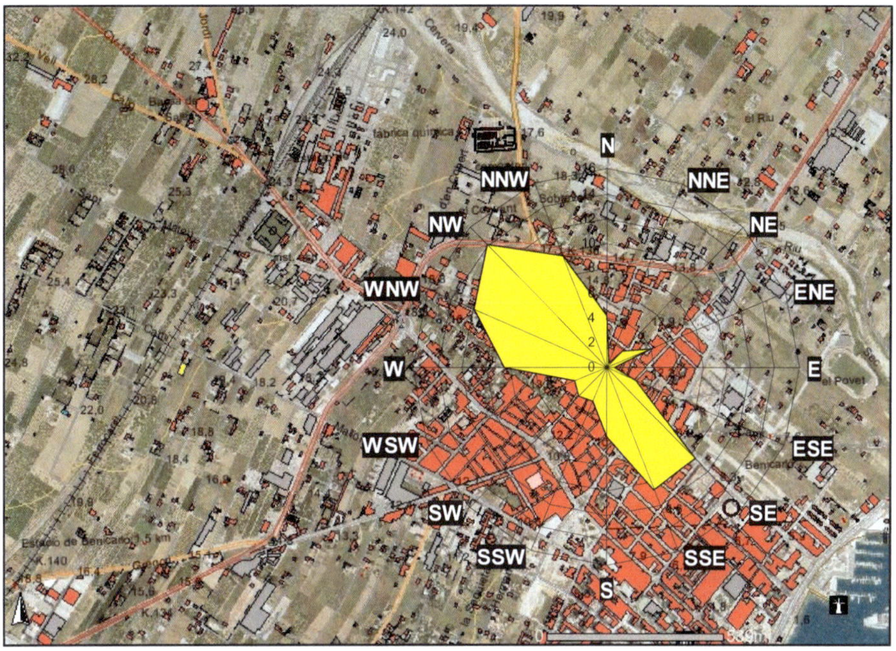

Figure 30. Annual wind rose.

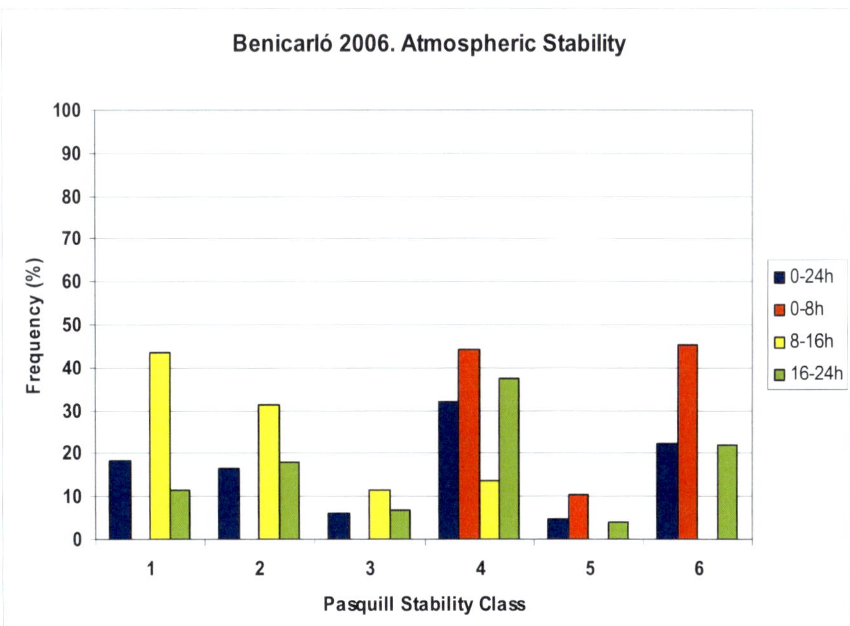

Figure 31. Annual atmospheric stability in Benicarló at daily and hourly intervals.

The atmospheric stability classes are plotted in Figure 31. In the interval 0-8 hours, the more stable classes (4, 5 and 6 or D, E and F) account for 100% of the time. For the interval 8-16 hours, the more unstable classes account for 87% of the time. Between 0 and 8 hours, therefore, there is a greater risk of strong odor episodes and high concentration of air pollutants. The potential emitting activities are located in what were found to be the prevailing wind directions (NW, ENE), which is consistent with the large number of odor episodes recorded.

7.2.3. Impact Maps

Using the meteorological database generated by the initial meteorological study, in conjunction with the emission parameters (location of the source, contaminant emission flow, emission height, etc.) of the known emitters that are potential generators of episodes, we can elaborate impact maps (contaminant concentration distribution according to meteorological conditions during the study period) that can be used to design a chemical control program. The information provided by impact maps can be used to optimize chemical control samplings (i.e. by ensuring that measurements are only taken during maximum and minimum impact periods) and to identify the most appropriate locations for future emitting facilities in order to minimize the impact on urban areas.

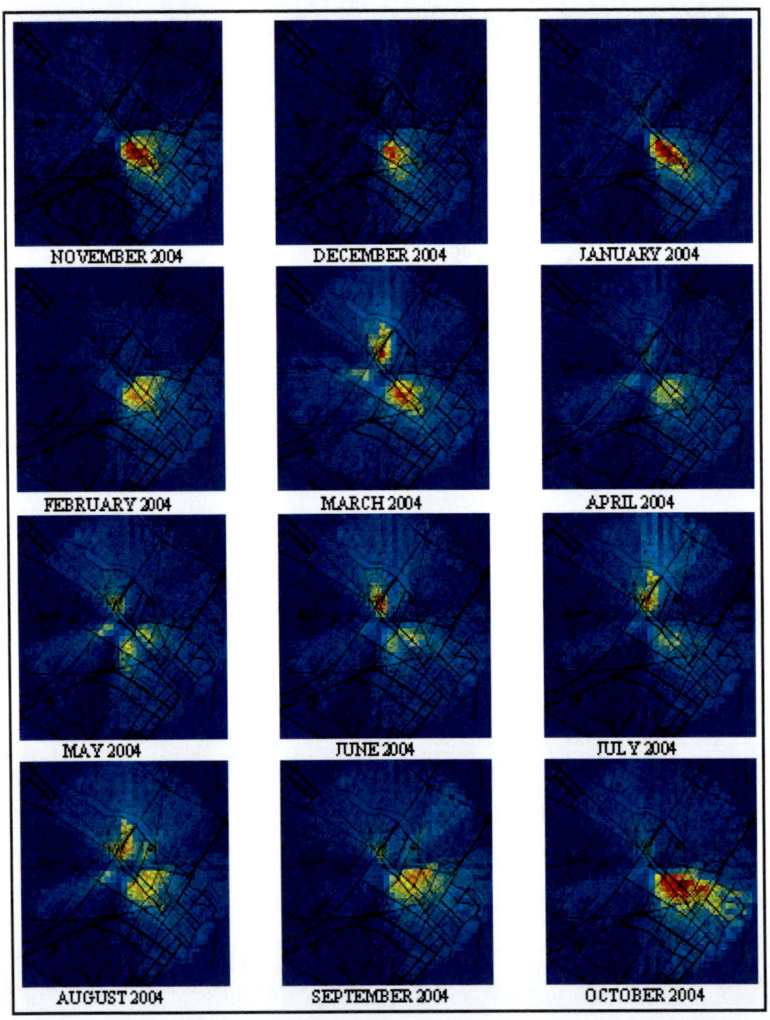

Figure 32. Impact maps (November 2004 to October 2005).

When no real emission data is available, published emitting factors for similar activities can be used. In the total absence of information, unitary emission values are used (for example, 1 g s^{-1}). The latter approach would obtain relative concentration maps, which can also be used to design a chemical control program with validation of the experimental emission concentration results obtained from chemical analysis in the urban area. Figure 32 shows monthly impact maps based on the available data on the sources initially considered (those shown in Figure 29).

7.2.4. Social Control (Social Participation)

The informational sessions on the proposed air quality evaluation process led to the participation of 23 residents Figure 33 in the social control process in the various study zones Figure 34. Of these participants, 14 recorded odor episodes on a daily basis for more than 3 months. In this study, due to the low level of information about the potential emitting activities, the type B form was used. The study lasted 9 months.

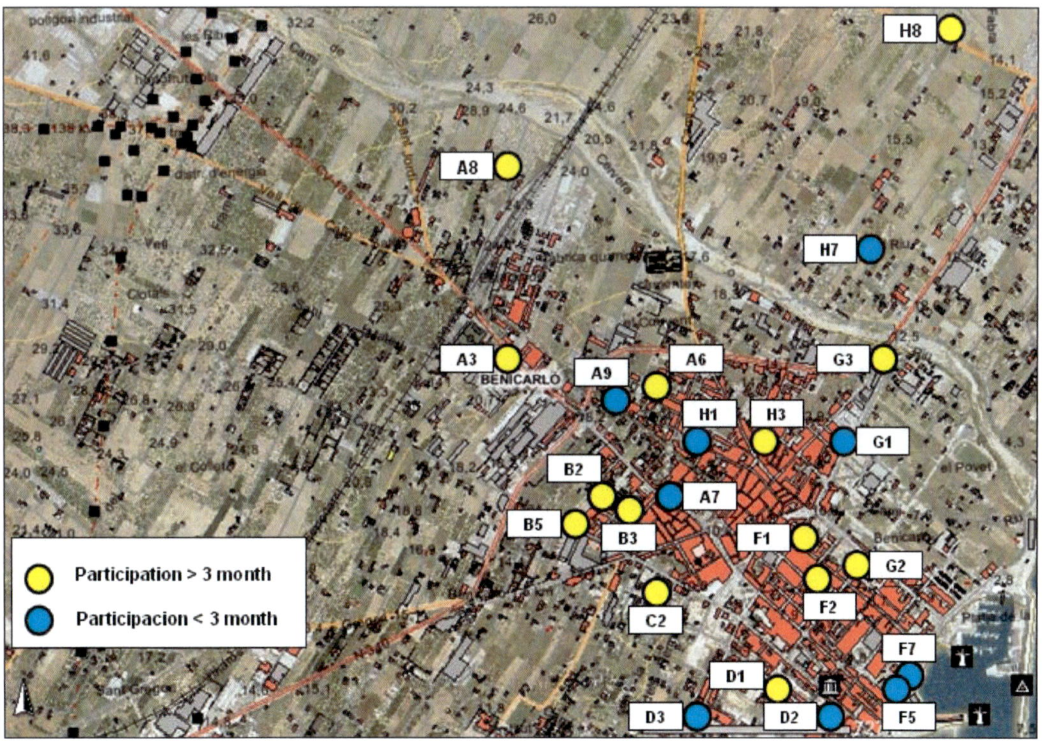

Figure 33. Distribution of participants in the urban area.

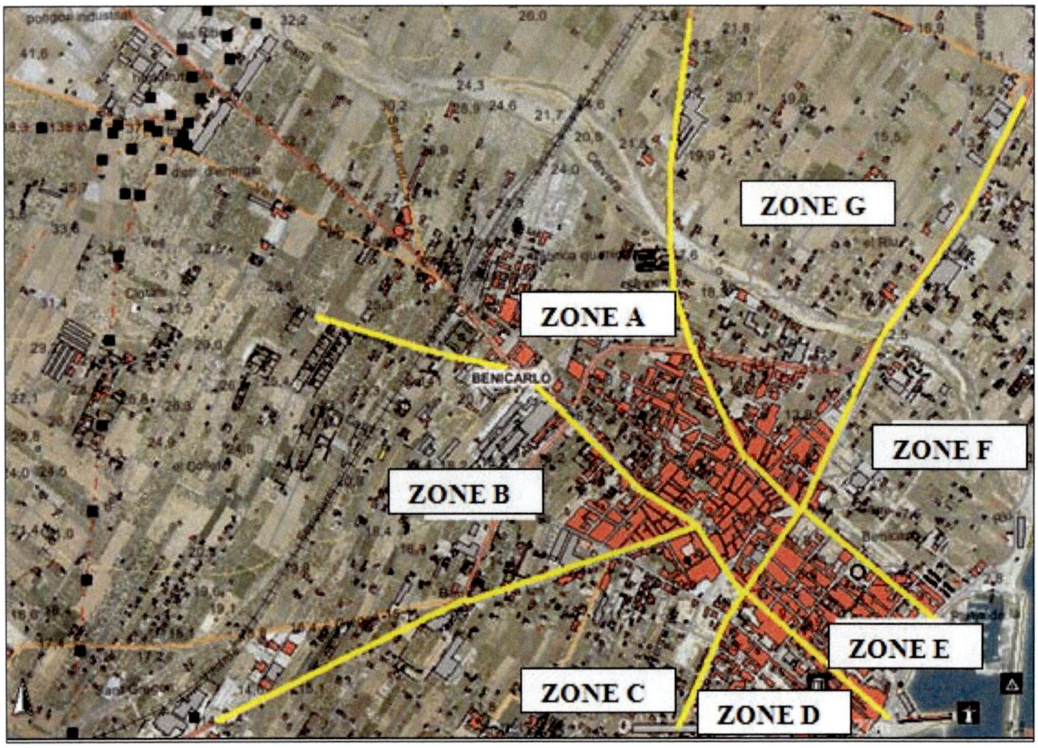

Figure 34 Distribution of control zones in the urban area of Benicarló.

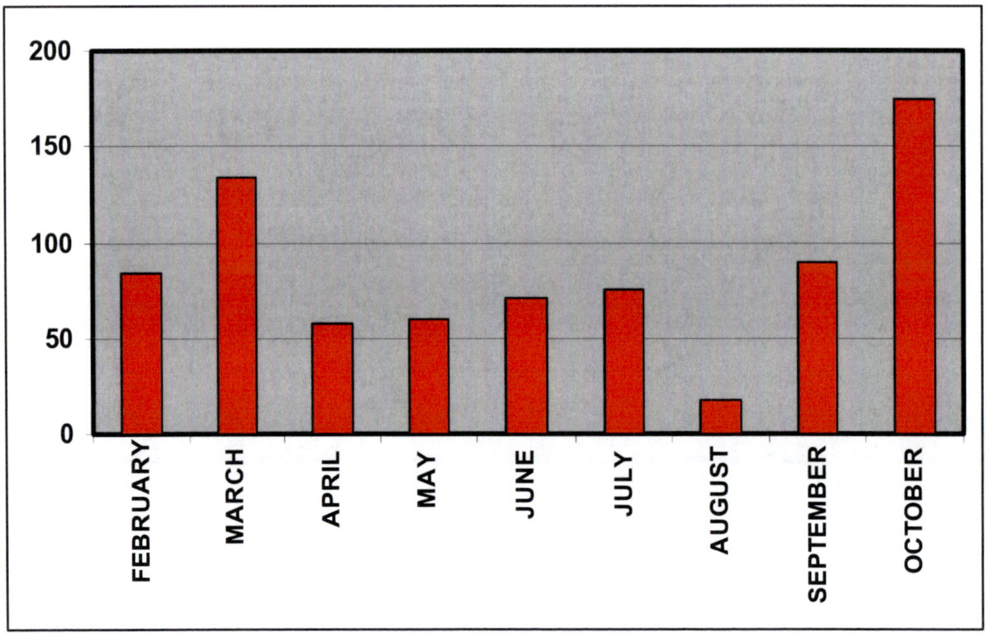

Figure 35. Monthly evolution of total odor episodes in all study zones.

Figure 35 shows the evolution of odor episodes in the different zones over the study period.

Episode roses (Figure 36) for each study zone were created based on individual analysis of the various recorded episodes with an intensity of 3, 4 or 5 (with exception of zone D, where the only episodes recorded had an intensity of 2).

Zone B is used to illustrate the methodology used. Known emission sources contributed to the episodes of odor and low air quality recorded in the city area (Table 15). Based on databases of recorded episodes, new sources that complete the final inventory of emitters were detected, as some episodes did not correspond with any known emission sources (Figure 37). Almost all of the newly detected sources were small industrial emitters that were not initially considered potential polluters. The exception was the emitting source located in the SSE-SSW sector of the city that could only be attributed to the municipal sewer (Figure 38) The path of the sewer corresponded to the intersection of the impact directions of unknown origin. The use of chemical control and the elaboration of concentration maps confirmed this hypothesis.

Table 15.

Zone B	**Impact frequency (%)**									
RECEPTOR	S1	S2	S3	S4	S5	S10	S11	S12	S13	US
B1,B3	20.0	4.3	20.0	24.3	2.9	4.3	1.4	1.4	2.9	18.6
B2	15.1	5.9	15.1	26.9	5.0	5.9	1.7	1.7	4.2	18.5
B5	12.5	6.3	12.5	37.5	6.3					25.0

US: Unknown source.

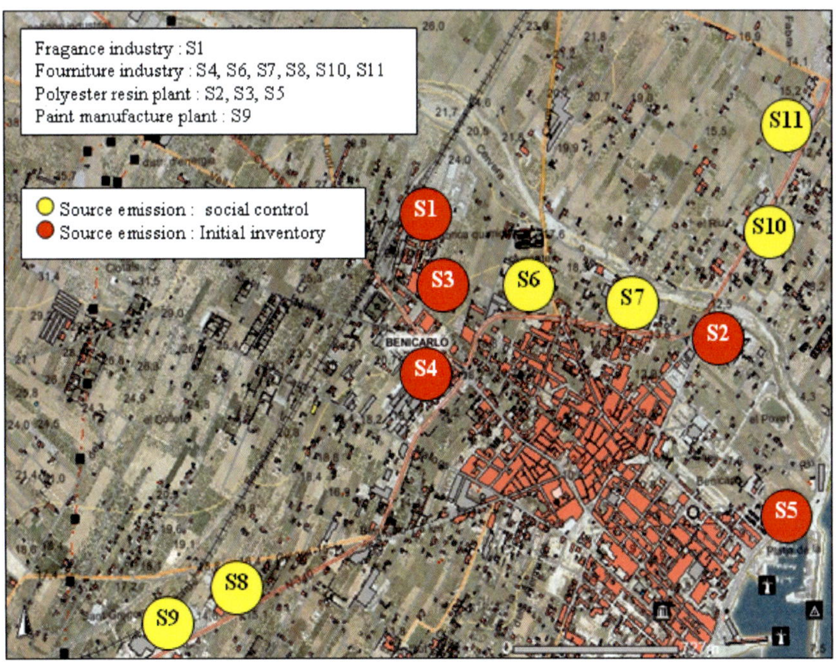

Figure 37. Inventory of emitters identified by social control.

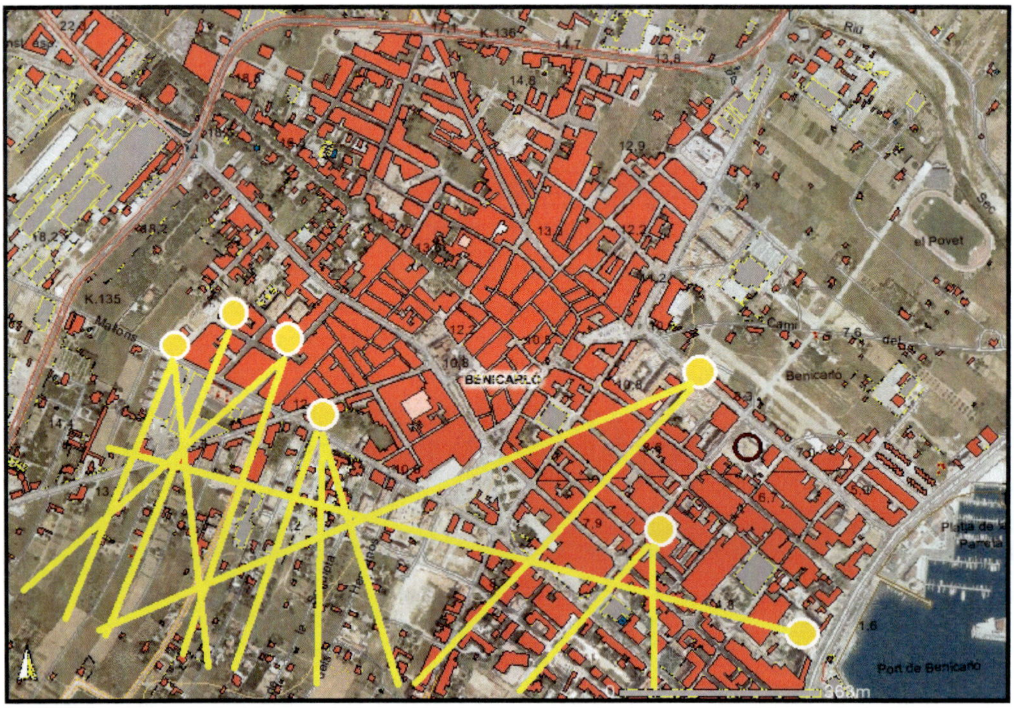

Figure 38. Area of the episode sources identified using impact directions determined based on the contributions of several social control participants.

7.2.5. Chemical Control

The chemical control was designed using cost-optimization criteria based on information from the impact prediction maps (Figure 32). It comprised two study phases:

Phase I. April: Emitters' minimum impact period. The affected people monitored the odor and discomfort episodes by activating air sampling equipment. 24-hour air sampling was planned.

Phase II. October: Maximum impact period. Episodic and 24-hour planned air sampling was carried out.

7.2.5.1. Qualitative VOC Determination

The abovementioned analytical technology qualitatively determined 233 chemical compounds during odor episodes and in the 24-hour control samples. Table 16 shows all of the determined compounds, along with their CAS number, odor threshold, TLVs and associated R-phrases (risk phrases). TLVs are occupational exposure values, but they are commonly used as advisable outdoor air limits, generally by dividing the TLV by a factor of 420, as explained in Section 6.5.1.

Table 17 shows the definitions of the R-phrases associated with each compound [EU, 1967, EU, 2001]. R-phrases are warnings about the nature of special risks attributed to dangerous substances and preparations.

7.2.5.2. Quantitative VOC Determination

Of the 233 compounds determined qualitatively, 75 of them were quantified as individual compounds. These compounds were selected based on several criteria, such as their probable or possible toxicity on humans, their frequency of appearance in samples, and their nature, as some of them are emitted from specific industrial activities. The sampling points were distributed throughout the city, as shown in Figure 39. Table 18 shows VOC chemical family concentrations in Phases I and II for odor episodes and 24-hour air quality controls.

Figure 36. Episode roses for the study zones. Emitting source located along the impact direction.

Table 16. Compounds determined qualitatively in Benicarló (Autonomous Community of Valencia, Spain). Odor threshold (µg m^{-3}), TLV (mg m^{-3}), TLV/420 (mg m^{-3}) and associated R-phrases.

Compound	CAS No.	Odor threshold	TLV	TLV/420	R-phrases
(1-methylpropyl)benzene	135-98-8				
(dimethylamino)difluorophosphine	814-97-1				
1,1,1-trichloroethane	71-55-6	5,300	1,925	4,583	20
1,1,2-trimethylcyclohexane	7094-26-0				
1,1,3-trimethylcyclohexane	3073-66-3				
1,1-dichloro-1-fluoroethane	1717-00-6				
1,1-dimethylcyclopentane	1638-26-2				
1,1-dimethyllcyclopropane	1630-94-0				
1,2,3-trichlorobenzene	87-61-6				
1,2,4-trimethylbenzene	95-63-6	140	125	298	20-36/37/38
1,2,4-trimethylcyclohexane	2234-75-5				
1,2.3-trimethylbenzene	526-73-8		125	298	
1,2-diethylbenzene	135-01-3				20/21-38
1,2-dimethylcyclohexane	583-57-3				
1,2-dimethylcyclopentane	2452-99-5				
1,2-pentadiene	591-95-7				
1,3,5-trimethylbenzene	108-67-8	180	125	298	37
1,3,5-trimethylcyclohexane	1839-63-0				
1,3-dimethylcyclohexane	591-21-9				
1,3-dimethylcyclopentane	2453-00-1				
1,3-pentadiene	504-60-9				
1,4-dimethyl-5-(1-methylethyl)-cyclopentene	61142-33-4				
1,4-dimethylcyclohexane	589-90-2				
1,4-dioxane	123-91-1	270,000	74	176	36/37-40-66
1,4-epoxy-p-menthane	470-67-7				
1,4-pentadiene	591-93-5				
1,8-cienole	470-82-6	7.5			
1-ethyl-2-methylcyclohexane	3728-54-9				
1-fluoro-1,2-dichloro-ethane	75-69-4				
1-hexene	592-41-6		175	417	
1-isopropyl-2,3-dimethyl-cyclopentene	7712-73-4				
1-methoxy-2-propanol	107-98-2	37,000	275	655	
1-methoxy-2-propyl acetate	108-65-6				
1-methyl-4-(2-methylpropyl)-benze	5161-04-6				
1-methyl-4-propyl-benze	1074-55-1				
1-methylcyclopentene	693-89-0				
1-methylindene	767-59-9				
1-methylpyrrole	96-54-8				

1-octanol	111-87-5	140			
1-pentene	109-67-1				
2,2,3-trimethylbutane	464-06-2				
2,2,3-trimethylpentane	564-02-3				
2,2,4-trimethyl-1,3-dioxolane	1193-11-9				
2,2,4-trimethylpentane	540-84-1		1,425	3,393	38-65-67
2,2-dimethylbutane	75-83-2				
2,3,4-trimetihypentane	565-75-3				
2,3-dimethylbutane	79-29-8				
2,4,4-trimethyl-1-pentene	107-39-1				
2,4-dimethyl-1-heptene	19549-87-2				
2,4-dimethyl-2-pentene	625-65-0				
2,4-dimethylheptane	2213-23-2				
2,4-dimethylpentane	108-08-7	870,000			
2,4-hexadiene	592-46-1				
2,5-dimethyl-1,4-dioxane	15176-21-3				
2,5-dimethylheptane	2216-30-0				
2,5-dimethylhexane	592-13-2				
2,6-dimethylheptane	1072-05-5				
2,6-dimethyloctane	2051-30-1				
2,7-dimethyl-1,6-octadiene	40195-09-3				
2-[(trimethylsilyl)oxy]- trimethylsilyl ester of benzoic acid	3789-85-3				
2-butanol	78-92-2	285,000	308	733	36/37-67
2-butenal	4170-30-3	1,700	0.9[1]	2	24/25-26-37/38-41-48/22-68
2-butene	107-01-7	4,800			
2-ethyl-1-hexene	1632-16-2				
2-ethyl-4-methyl-1,3dioxolane	4359-46-0				
2-ethyl-4-methyl-1,3-dioxolane	4359-46-0				
2-ethylbutanol	97-95-0	300			
2-ethylhexanoic acid	149-57-5				63
2-ethylhexanol	104-76-7	400			
2-ethyl-m-xylene	2870-04-4				
2-ethylpyridine	100-71-0	57			
2-heptene	592-77-8				
2-hexene	592-43-8				
2-methyl-1,3-pentadiene	1118-58-7				
2-methyl-1-butene	563-46-2				
2-methyl-2-butene	513-35-9				
2-methyl-2-hexene	2738-19-4				
2-methyl-2-pentene	625-27-4				
2-methyl-2-propanol	75-65-0	2,900,000	308	733	20
2-methyl-2-propenal	78-85-3				
2-methylheptane	592-27-8		1,425[2]	3,393	38-65-67
2-methylhexane	591-76-4				38-65-67
2-methyloctane	3221-61-2				

Table 16. Continued

Compound	CAS No.	Odor threshold	TLV	TLV/420	R-phrases
2-methylpentane	107-83-5		1,792[3]	4,267	
2-methylpropanal	78-84-2				
2-methylpropanoic acid	79-31-2	240			21/22
2-methylpropenal	78-85-3				
2-nonanone	821-55-6	5,500			
2-octene	111-67-1	2,000			
3,3-dimethylpentane	562-49-2				
3,7-dimethyl-1,6-octadien-3-ol	78-70-6	9			
3-carene	13466-78-9	8,600			
3-ethylheptane	15869-80-4				
3-ethylhexane	619-99-8				
3-ethyl-o-xylene	933-98-2				
3-ethylpentane	617-78-7				
3-heptene	592-78-9				
3-hexene	592-47-2				
3-methyl-1-pentene	760-20-3				
3-methyl-2-pentanol	565-60-6				
3-methyl-2-penten-4-one	565-62-8	140			
3-methyl-2-pentene	616-12-6				
3-methylheptane	589-81-1		1,425	3,393	
3-methylhexane	589-34-4				
3-methylnonane	5911-04-6				
3-methyloctane	2216-33-3				
3-methylpentane	96-14-0		1,792	4,267	
4-ethyl-o-xylene	934-80-5				
4-methyl-1,3-heptadiene	17603-57-5				
4-methyl-1,3-pentadiene	926-56-7				
4-methyl-1-hexene	3769-23-1				
4-methyl-2-hexene	3404-55-5				
4-methyl-2-pentanone	108-10-1	140	208	495	20-36/37-66
4-methyl-2-pentene	4461-48-7				
4-methylheptane	589-53-7		1,425	3,393	
4-methylnonane	17301-94-9				
4-methyloctane	2216-34-4				
5,5-dimethyl-2-hexene	39761-61-0				
Acetic acid	64-19-7	90	25	60	35
Acetic acid pentyl ester	628-63-7				
Acetone	67-64-1	1,000	1,208	2,876	36-66-67
Acetonitrile	75-05-8	285,000	34	81	20/21/22-36
Acetophenone	98-86-2	10	50	119	22-36
Acetylcyclobutane	3019-25-8				
Alcanfor	464-49-3	490			
a-pinene	80-56-8	230			

Benzaldehyde	100-52-7	10			22
Benzene	71-43-2	1,500	1.6	4	45-46-36/38-48/23/24/25-65
Benzothiazole	95-16-9				
Benzyl acetate	140-11-4	85,000	62	148	
b-myrcene	123-35-3	130			
b-pinene	127-91-3	8,900			
Butanal	123-72-8	0.84			
Butanoic acid	107-92-6	0.35			34
Butyl acetate	123-86-4	7,700	725	1,726	66-67
Camphene	79-92-5	26,000			
Caprolactam	105-60-2	300	24	57	20/22-36/37/38
Carbon disulfide	75-15-0	180	3	7	36/38-48/23-62-63
Carbon tetrachloride	56-23-5	884,000	32	76	23/24/25-40-48/23
Chloroform	67-66-3	650,000	49	117	22-38-40-48/20/22
Cyclohexane	110-82-7	35,600	350	833	38-65-67
Cyclohexanone	108-94-1	880	82	195	20
Cyclohexyl isocyanate	3173-53-3				
Cyclohexyl isothiocyanato	1122-82-3				
Cyclopentane	287-92-3		1,750	4,167	
Cyclopentanone	120-92-3	31,000			36/38
Cyclopropane	75-19-4				
Dichloromethane	75-09-2	4,100	177	421	40-48/20
Dihydromyrcenol	18479-58-8				
Diisopropylnaphthalene	38640-62-9				
Dimethylmalonic acid	595-46-0				
Dimethylnitrosamine	62-75-9				45-25-48/25
Dimethylstyrene	27576-03-0				
DL-limonene	7705-14-8	1,700			38-43
Ethanol	64-17-5	154,000	1,917	4,564	
Ethyl acetate	141-78-6	623,000	1,467	3,493	36-66-67
Ethylbenzene	100-41-4	10	442	1,052	20
Ethylcyclohexane	1678-91-7				
Ethylcyclopentane	1640-89-7				
Ethylcyclopropane	1191-96-4				
Formic acid	64-18-6	2,000	10	24	35
Formic acid butyl ester	592-84-7				36/37
Heptanal	111-71-7	64			
Heptanoic acid	111-14-8	22			34
Hexamethylcyclotrisiloxane	541-05-9				
Hexamethylethane	594-82-1				
Hexanal	66-25-1	34			

Table 16. Continued

Compound	CAS No.	Odor threshold	TLV	TLV/420	R-phrases
Hexanoic acid	142-62-1	20			
Isobutanol	78-83-1		154	367	37/38-41-67
Isobutyric acid	79-31-2	240			21/22
Isocineole	470-67-7				
Isohexane	107-83-5		1,792	4,267	
Isopentane	78-78-4		3,000	7,143	65-66-67
Isopropanol	67-63-0	8,000	500	1,190	36-67
Isopropylbenzene	98-82-8	650	250	595	37-65
Linalool	78-70-6	1.3			
m+p-xylene	108-38-3/106-42-3	540/520	442	1,052	20/21-38
Methyl acetate	79-20-9	22,000	617	1,469	36-66-67
Methyl ethyl ketone	78-93-3	2,900	600	1,429	36-66-67
Methylacetophenone	26444-19-9	2			
Methylcyclobutane	598-61-8				
Methylcyclopentane	96-37-7				
m-ethyltoluene	620-14-4				
methyl-vinyl-cetone	78-94-4	500			
m-propyltoluene	1074-43-7				
N-acetylmorpholine	1696-20-4				
Naphthalene	91-20-3	7	53	126	22-40
n-butanol	71-36-3	480	62	148	22-37/38-41-67
n-decane	124-18-5	11,300			
n-dodecane	112-40-3	11,800			
N-formylmorpholine	4394-85-8				
n-heptane	142-82-5	165,000	1,667	3,969	38-65-67
n-hexane	110-54-3	107,000	179	426	38-48/20-62-65-67
Nitrocyclohexane	1122-660-7				
Nitromethane	75-52-5		51	121	22
Nitrosotrimethylurea	3475-63-6				
NN-dimethylacetamide	127-19-5	170,000	36	86	61-20/21
NN-dimethylformamide	68-12-2	140	30	71	61-20/21-36
n-nonane	111-84-2	60,000	1,067	2,540	
Nonanal	124-19-9	20			
n-pentane	109-66-0	92	1,800	4,286	65-66-67
n-pentadeca	629-62-9				
n-pentanal	110-62-3	92	179	426	
n-tetradecane	629-59-4	5,000			
n-tridecane	629-50-5	42,000			
n-undecane	1120-21-4	9,600			
Octamethylcyclotetrasiloxane	556-67-2				62

Octane	111-65-9	71,000	1,425	3,393	38-65-67
Octanoic acid	124-07-2	600			
o-cymene	527-84-4				
o-ethyltoluene	611-14-3				
o-xylene	95-47-6	770	442	1,052	20/21-38
p-dichlorobenzene	106-46-7	730	122	290	36-40
p-diethylbenzene	105-05-5				
p-ethyltoluene	622-96-8				
Phenylether	101-84-8		7	17	
Pinane	473-55-2				
p-propyltoluene	1074-55-1				
Propanal	123-38-6	3.6	48	114	36/37/38
Propionic acid	79-09-4	5.1	31	74	34
Propylbenzene	103-65-1	14,400			
Styrene	100-42-5	12	86	205	20-36/38
tert-butyl methyl ether	1634-04-4		183	436	38
tert-butyltrichlorosilane	18171-74-9				
Tetrachloroethylene	127-18-4	8,300	171	407	40
Tetrahydrofuran	109-99-9	270	150	357	36/37
Tetramethylurea	632-22-4				
Toluene	108-88-3	3,800	77	183	38-48/20-63-65-67
Trichloroethylene	79-01-6	3,900	54	129	45-36/38-67
Trichlorofluoromethane	75-69-4	28,000			
Trimethylsilanol	1066-40-6				
Tripropylene glycol	1638-16-0				
Tripropylene glycol methyl ether	20324-33-8				

[1]Ceiling value; [2]As octanes; [3]As hexane isomers.

Table 17. R-phrases

R-phrase	Definition
20	Harmful by inhalation
22	Harmful if swallowed
25	Toxic if swallowed
26	Very toxic by inhalation
34	Causes burns
35	Causes severe burns
36	Irritating to eyes
37	Irritating to respiratory system
38	Irritating to skin
40	Possible risk of cancer
41	Risk of serious damage to eyes
43	May cause sensitization by skin contact
45	May cause cancer
46	May cause heritable genetic damage
61	May cause harm to the unborn child

Table 17. R-phrases (Continued)

R-phrase	Definition
62	Possible risk of impaired fertility
63	Possible risk of harm to the unborn child
64	May cause harm to breastfed babies
65	Harmful: may cause lung damage if swallowed
66	Repeated exposure may cause skin dryness or cracking
67	Vapors may cause drowsiness and dizziness
20/21	Harmful by inhalation and in contact with skin
20/21/22	Harmful by inhalation, in contact with skin and if swallowed
20/22	Harmful by inhalation and if swallowed
21/22	Harmful in contact with skin and if swallowed
23/24/25	Toxic by inhalation, in contact with skin and if swallowed
24/25	Toxic in contact with skin and if swallowed
36/37	Irritating to eyes and respiratory system
36/37/38	Irritating to eyes, respiratory system and skin
36/38	Irritating to eyes and skin
37/38	Irritating to respiratory system and skin
48/20	Harmful: danger of serious damage to health by prolonged exposure through inhalation
48/20/22	Harmful: danger of serious damage to health by prolonged exposure through inhalation and if swallowed
48/22	Harmful: danger of serious damage to health by prolonged exposure if swallowed
48/23	Toxic: danger of serious damage to health by prolonged exposure through inhalation
48/23/24/25	Toxic: danger of serious damage to health by prolonged exposure through inhalation, in contact with skin and if swallowed
48/25	Toxic: danger of serious damage to health by prolonged exposure if swallowed

The compounds that were not quantified individually, were quantified with the response factor of toluene, and their summarized concentrations appear as "Other compounds" in Table 18.

Table 18. VOC family concentrations (μg m^{-3}) in 24-hour controls and odor episodes at the various sampling points shown in Figure 39 during Phases I and II.

Sampling point	1	1	2	2	3	3	4	4	5	6	6	7	8	9	10	11
Type	Ep	24-h	Ep	24-h	Ep	24-h	Ep	24-h	Ep	Ep	24-h	Ep	24-h	24-h	24-h	24-h
Aromatic Hydrocarbons	121.8	129.0	8.0-35.1	80.4	230.7-251.1	18.3	8.0	111.4	6.4	7.4	37.0	17.9	8.0-9.8	6.8-41.7	18.5-121.1	38.9-65.4
Terpenes	0.9	108.0	0.5-46.5	4.9	n.d.	n.d.	10.2	3.6	8.2	4.7	5.9	11.1	1.3-5.3	0.4-9.5	2.9-51.9	0.8-9.1
Chlorines	0.6	1.4	0.1-19.3	3.2	1.2-2.1	0.8	0.6	0.9	19.5	0.3	3.0	1.8	2.7-8.3	1.1-3.4	5.6-31.2	4.2-13.9
Esters	4.1	12.7	2.1-22.8	36.7	0.7-2.9	8.2	10.1	1.5	16.6	0.5	2.1	2.1	3.9-4.7	1.1-6.8	4.6-27.0	8.5-22.5
Aldehydes	n.d.[1]	9.1	n.d.-49.7	2.5	n.d.-105.9	10.2	n.d.	n.d.	n.d.	n.d.	n.d.	n.d.	1.9-3.0	1.2-4.8	7.6-31.5	6.6-22.0
Carboxylic acids	26.2	21.0	10.4-152.5	30.6	10.7-44.3	12.9	15.3	16.7	55.8	11.1	n.d.	13.3	9.5-23.2	2.0-15.3	23.5-128.8	19.5-45.4
Ketones	166.5	141.3	0.5-173.6	37.5	152.1-175.6	43.7	47.2	13.9	49.8	26.4	14.6	8.6	4.0-11.4	3.7-9.3	19.1-78.8	15.3-35.1
Alcohols	21.1	126.1	n.d.-970.9	61.5	12.3-41.0	13.4	157.5	50.9	n.d.	15.2	73.2	15.1	27.8-52.5	3.4-24.2	102.5-269.6	5.9-23.4
Furanes	n.d.	4.3	n.d.	0.8	n.d.-0.2	n.d.	2.2	0.8	2.9	1.3	0.4	0.6	0.2-0.9	0.1-2.2	1.2-19.2	6.5-25.2
Nitrogen compounds	29.8	42.1	0.9-106.1	5.2	4.3-18.7	6.2	5.8	4.6	0.9	0.8	18.5	n.d.	13.0-19.8	2.2-10.0	42.4-46.0	0.5-4.0
Sulfur compounds	0.7	39.2	0.1-107.8	0.4	0.1-0.9	0.3	2.2	3.1	23.1	25.5	27.1	0.1	0.1-17.2	0.1-17.2	30.8-38.8	0.3-14.2
Ethers	2.7	n.d.	n.d.	n.d.	n.d.	n.d.	11.8	9.3	n.d.	n.d.	5.1	0.8	1.2-1.9	1-0-4.0	4.7-12.4	5.5-9.9
Other compounds[2]	581.0	288.2	389.6-786.9	151.6	2198.1-2227.6	72.6	399.6	411.8	485.8	721.1	224.4	449.4	n.d.-0.3	n.d.-0.4	0.3-2.7	0.2-1.1
TVOC	954.7	922.8	412.3-2475.5	415.7	2633.9-2846.8	186.6	669.4	633.2	675.1	819.3	412.0	520.6	227.2-1751.9	226.9-3579.2	912.9-5655.7	966.9-3260.9

[1]Not detected; [2]Evaluated with the response factor of toluene.

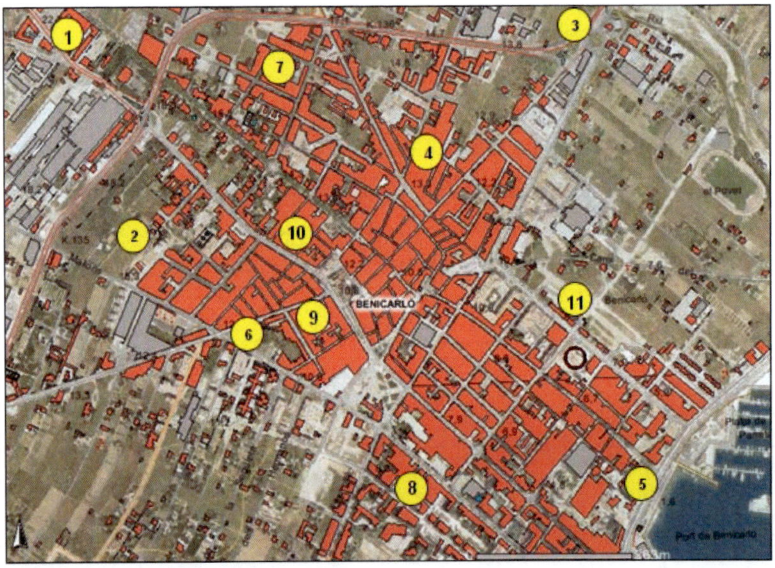

Figure 39. Sampling points in the urban area of Benicarló.

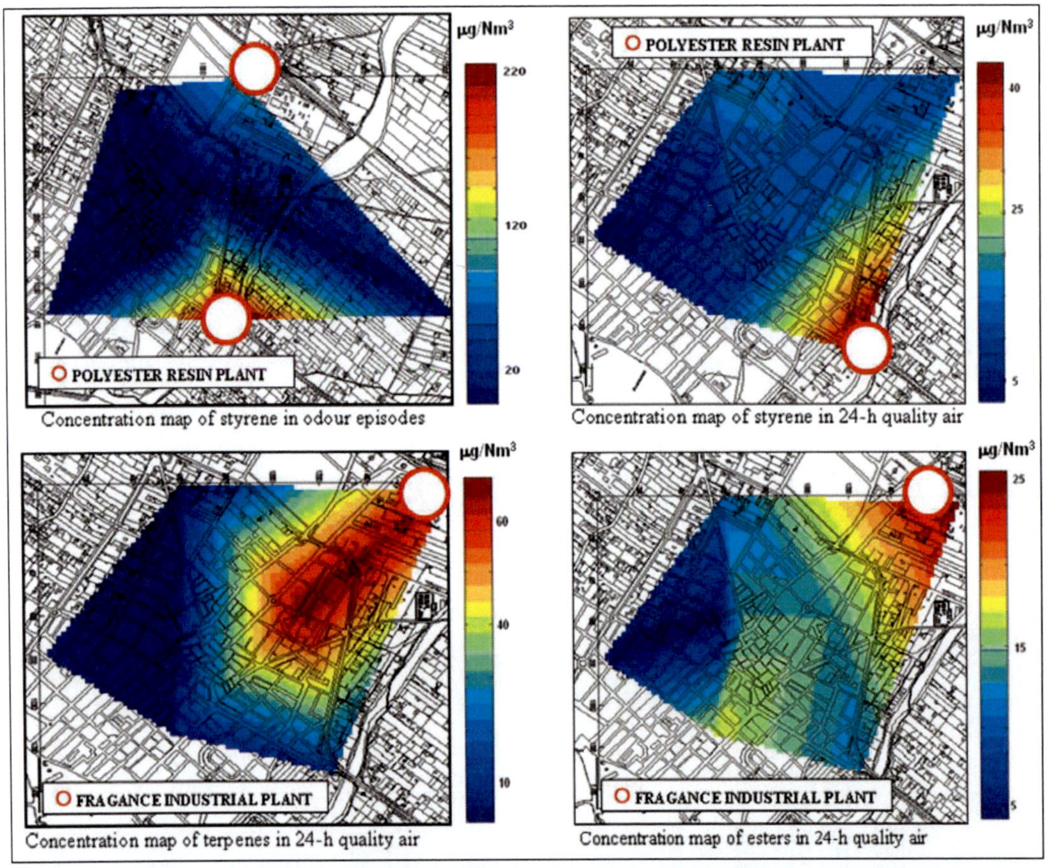

Figure 40. Concentration maps of individual VOCs and VOC compound families in the urban area of Benicarló.

7.2.5.3. Concentration Maps

A graphical representation of VOC concentrations on concentration maps, based on quantitative chemical analysis, clearly shows the distribution in the different zones of the urban area (Figure 40). The analysis of the total VOC (TVOC) concentration ranges by study zone, in odor episodes and in 24-hour controls, validates the previously calculated impact maps due to the correspondence of concentrations and the areas of maximum and minimum impact (Figure 41).

In our example, the emission points of most emitters are quite low (<20 m). This causes the concentration distribution of the polluted plume to begin near the emitters. It also allows us to observe the source of the emissions in terms of individual chemical compound levels.

Concentration maps show the impact of certain industrial activities on the urban area, as in the case of styrene (emitted mainly by polyester synthesis plants) and terpenes (emitted by the fragrance plant).

No emitter was assigned for the concentration maps of carbon disulfide and dichloromethane. Nevertheless, the distribution of these compounds coincided with the path of the main municipal sewer (Figure 42). This confirmed the source of the odor episodes recorded by several social control participants (Figure 38).

The results of the chemical and social controls made it possible to the detect illicit disposal of chemical contaminants from several activities that use dichloromethane in their production processes (such as furniture factories).

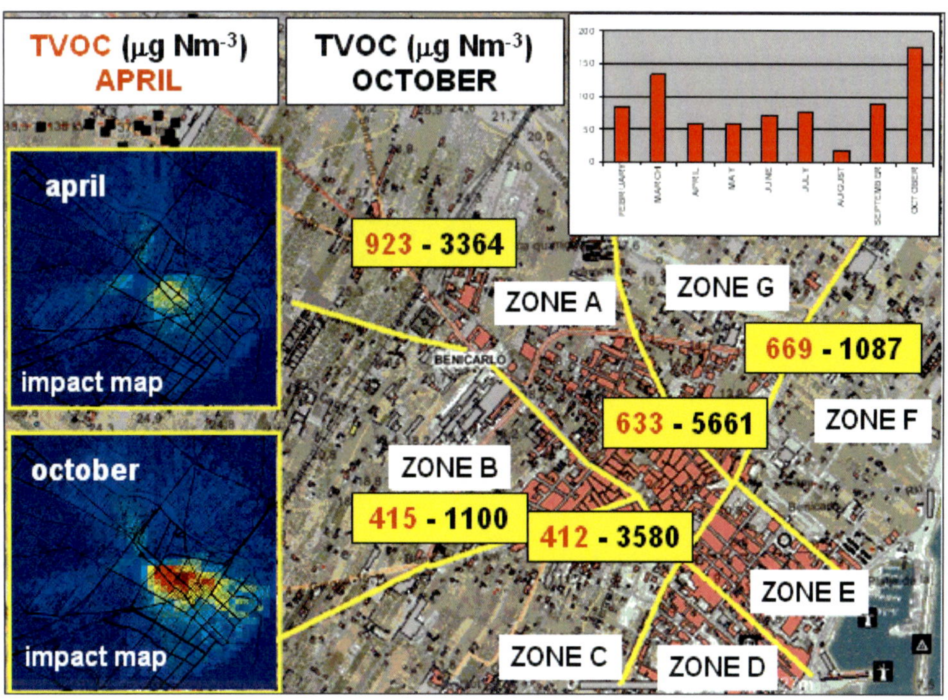

Figure 41. Comparison of the maximum and minimum impact periods of the various study zones in the city of Benicarló.

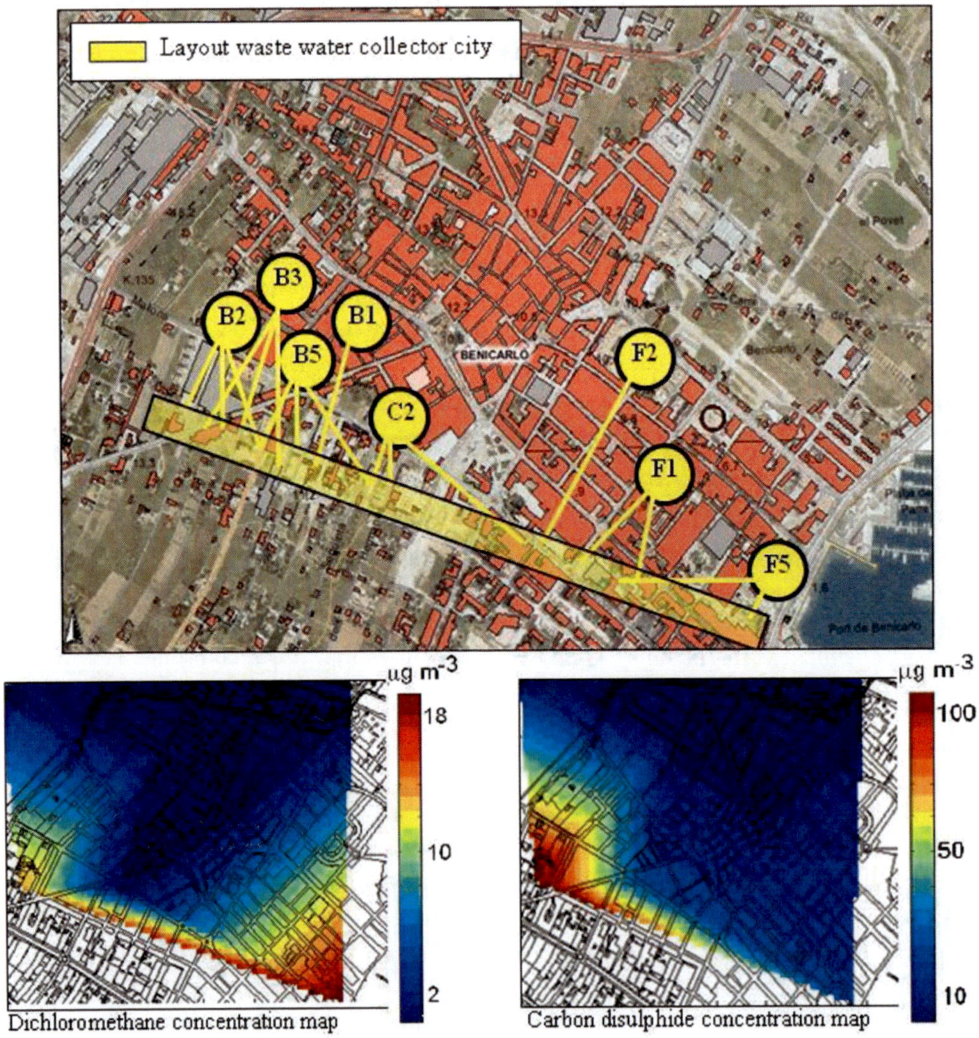

Figure 42. Area where episode sources determined by social control coincide with the concentration maps for dichloromethane and carbon disulfide.

8. CONCLUSIONS

The combined application of meteorological analysis, social participation, chemical control and numerical modeling provides very complete results in air quality evaluations, especially during odor episodes. This technique can identify the chemical compounds that cause the episodes and their sources, and can be applied to 24-hour air quality analysis. A real inventory of all potential emitting activities can be obtained, including not only nearby emitters but also sources that are several kilometers away from the study zone, for example in neighboring towns.

The combined application of data from meteorological analysis, emissions inventory and social participation can be used in regional planning through the elaboration of impact maps, which are useful in designing chemical control programs.

Results from the various applied techniques intervalidate the entire air quality control process, as shown in Figure 41, because the chemical compound concentrations correlate well with the maximum and minimum impact periods (April and October, respectively) determined by both numerical simulation and social control, which provided the total number of episodes recorded in these months. In our example, it was shown that TVOC concentrations in the maximum impact period (October), ranged from 3,364 to 5,661 μgNm^{-3}. In the minimum impact period (April), TVOC values ranged from 412 to 923 $\mu g\ Nm^{-3}$. The same correspondence is seen in the evolution of the episodes recorded by means of social participation. There were 181 episodes in October and 55 in April, the latter being the month with the fewest episodes (except for August, due to the annual vacation period).

Air quality criteria can be applied to the chemical compound concentrations obtained with the presented methodology. Since most of the analyzed compounds are determined individually, they can be classified by their potential impact on human health by using risk phrases (chemical security charts, ICSC), listings, and international legislation on toxic compounds, ozone precursors and the potential effects of new compounds. In each urban zone, several VOCs of interest can be studied in order to evaluate air quality. No international guidelines have been established for individual compounds, so different quality criteria are in place in different countries. In workplace settings, the most commonly used criterion is the TLV, divided by a factor ranging from 30 to 1000. The value obtained depends on the toxic potential of the compound. The TLV/420 values of individual compounds are similar to those published by the World Health Organization (WHO) for several compounds.

The application of these criteria presents certain difficulties. TLV values are not available for all identified compounds, although the concentration/TLV ratio can be high (70-90%) for the compounds of interest. Moreover, it is not possible to evaluate the individual effects of all compounds, especially those present in low concentrations. In assessing air quality, it would be useful to have a global value, such as total volatile organic compounds (TVOC). This value can be obtained by adding the concentrations of individually quantified compounds of interest with the concentrations of other quantified compounds, using the response factor of toluene. A preliminary criterion based on the values established for indoor air, corrected by the factor of average time exposure, could then be considered. In urban areas of developed countries, people spend between 75 and 85% of their time indoors, compared with just 15 to 25% outdoors. Alternatively, another criterion can be applied considering the same discomfort observed for TVOC concentration levels indoors (3,000 $\mu g\ Nm^{-3}$ for outdoor air), without any correction factor for exposure time.

In our example, the application of the preliminary criterion would surpass the TLV/420 criterion for two individual compounds: carbon disulfide at control point 9, with a value of 56 $\mu g\ Nm^{-3}$ (TLV/420 = 50 $\mu g\ Nm^{-3}$), and 2-butenal at most control points, with values ranging from 2.5 to 11.6 $\mu g\ Nm^{-3}$ (TLV/420 = 2 $\mu g\ Nm^{-3}$), taking into account that the 81.8% of VOCs of interest have been assigned a TLV/420 value. In the low-impact period, TVOC had a ratio of 0.075 to 0.3 with respect to the overall concentration 3,000 $\mu g\ Nm^{-3}$ (threshold of discomfort for indoor air), whereas for the high-impact period this ratio ranged from 0.4 to 1.9, with the highest value in Zone A.

Another application of the proposed air quality methodology is odor evaluation. Odor units can be determined for individual compounds using published odor threshold values. This makes it possible to quantify odor strength and determine the main chemical compounds responsible for odor episodes, which gives an indication of the emission source. This level of information greatly surpasses that obtained by applying olfactometric techniques, especially in scenarios with multiple emitting sources.

REFERENCES

Andersson, K., Bakke, J.V., Bjørseth, O., Bornehag, C-G., Clausen, G., Hongslo, J.K., Kjellman, M., Kjaergaard, S., Levy, F., Mølhave, L., Skerfving, S., and Sundell, J. *Indoor air*, 1997, *7*, 78-91.

Atkinson, R. *Atm. Env.*, 2000, *34*, 2063-2101.

Aitken, M.D i Okun, M.F. Quantification of wastewater odours by the affected public, *Water Environmental Research*, 1992, *64*, 720-727.

Bartual, J., Berenguer, M.J., Bernal, F., Freixa, A., Guardino, X., Hernández, A., Martí, M.C., Rosell, M.G., Solans, X. *Calidad de aire interior*, Instituto Nacional de Seguridad e Higiene en el Trabajo, Centro Nacional de Condiciones de Trabajo: Barcelona, 2008 p. 200.

Berenguer, M.J., Cid, F.J., Guardino, X., Roca F.X. *Optimization of a Thermal Desorption-GC-MS method for a wide range of volatile organic compounds in outdoor/indoor air monitoring.* Proceedings of the 10th Jornadas de Análisis Instrumental, Barcelona, 2002, p. 246.

Chang, T.Y., Nance, B.I., Kelly, N.A. *Atm. Env.*, 1999, *33*, 4695-4708.

Department of Environment and Conservation NSW, Assessment and management of odour from stationary sources in NSW, Sydeny, 2006.

Derwent, R.G., Jenkin, M.E., Saunders, S.M. *Atm. Env.*, 1996, *30*, 181-199.

ECA-IAQ, *Indoor Air Quality and its Impact on Man. Total Volatile Organic Compounds in Indoor Air Quality Investigations*. EUR 17675 EN, Office for Official Publications of the European Communities, Luxemburg, 1997, Report No. 19, pp. 48.

Edwards, R.D., Jurvelin, J., Koinstinen, K., Saarela, K., and Jantunen, M. *Atm. Env.*, 2001, *35*, 4829-4841.

EU (European Union) *Council Directive 2001/59/EC*. Brussels, Belgium, 2001.

EU (European Union) *Council Directive 67/548/EEC*. Brussels, Belgium, 1967.

Gallego, E., Roca, F.X., Perales, F., Ribes, A., Carrera, G., Guardino, X., Berenguer, M.J. *Atm. Env.*, 2007, *41*, 8228-8240.

Godish, T. *Air Quality*; Lewis Publishers: Boca Raton, FL, 1997; p. 448.

Godish, T. *Indoor Environmental Quality*, Lewis Publishers: Boca Raton, FL, 2001; p. 461.

Holcomb, L.C. and Seabrook, B.S. *Indoor Environ.*, 1995, *4*, 7-26.

Hummel, T., Dalton, P., and Dilks, D.D. *Chem. Sens.*, 2000, *25*, 788.

Hutter, H.-P., Moshammer, H., Wallner, P., Damberger, B., Tappler, P., Kundi, M. *Int. J. Hyg. Environ. Health*, 2006, *209*, 65-68.

Johansson, I. *The role of volatile organic compounds in the assessment of indoor air quality*, Thesis, Department of Chemistry, Division of Analytical Chemistry, Royal Institute of Technology, Stockholm, Sweden, 1999.

Jones, A.P. *Atm. Env.*, 1999, *33*, 4535-4564.

Krause, C., Mailahn, W., Nagel, R., Schulz, C., Seifert, B., and Ullrich, D. *Occurrence of volatile organic compounds in the air of 500 homes in the Federal Republic of Germany*. Proceedings of indoor Air '87, Institute of Water, Soil and Air Hygiene, Berlin-Dahlem, 1987, pp. 102-106.

Lebret, E., van del Wiel, H.J., Bos, H., Moij, D., and Boleij, J.S.M. *Env. Int.*, 1986, *12*, 323-332.

Liang, H.-M., Liao, C.-M. *Chemosphere*, 2006, *68*, 781-789.

Maroni, M., Seifert, B., Lindvall, T. *Indoor Air Quality. A Comprehensive Reference Book*, Elsevier: Amsterdam, 1995, p. 989.

McGinley, M.A., Quantifying Public Perception of odours in a Community utilitzing Telemarketing Protocol, Proceedings of Air and Waste Management Association International Specialty Conference: Odour and Environmental Air, 310-322, Bloomington, 1995.

McKenzie, J.R., Mann, J., Use of odour annoyance index surveys for the confirmation of computer based dispersion modelling, Proceedings of Air Quality Modelling Conference: Air Quality Modelling. New methods for a new reality, Air & Waste Management Association, 2004, 285-299.

Ministry for the Environment, Good Practice Guide for Assessing and Managing Odour in New Zeeland. Air Quality Report 36, Wellington, 2003.

Mølhave, L. *Indoor Air*, 1991, *1*, 357-376.

Mølhave, L. *Indoor Air*, 1992, *2*, 65-77.

Mølhave, L. *Indoor Air*, 2003, *13*, 12-19.

Mølhave, L., Clausen, G., Berglund, B., De Ceauirriz, J., Kettrup, A., Lindvall, T., Maroni, M., Pickering, A.C., Risse, U., Rothweiler, H. Seifert, B., and Younes, M. *Indoor Air*, 1997, *7*, 225-240.

Mølhave, L."Sensory irritation in humans caused by volatile organic compounds (VOCs) as indoor air pollutants: a summary of 12 exposure experiments". In: Spengler et al. (eds.), *Indoor Air Quality Handbook*, McGraw-Hill: New York, NY, 2001; pp. 25.1-25.28.

Nicolas, J., Denne, P., Romain, A.C., Delva, J., Estimation de nuisances olfactives. Campagne de mesures des odeurs sur le CET de Malvoisin, Département des Sciences et de Gestion de l'Environment-Université de Liège, 1-40, Arlon, 2007.

Pappas, G.P., Herbert, R.J., Henderson, W., Koeing, J., Stover, B., and Barthart, S. *Int. J. Occ. Environ. Health*, 2000, *6*, 1-8.

Peng, Y.P., Chen, K.S., Lai, C.H., Lu, P.J., Kao, J.H. *Atm. Env.*, 2006, *40*, 6741-6751.

Powers, W. *The Science of the Smell Part 3: Odour detection an measurement*, United States Department of Agriculture, Washington, 2004.

Ribes, A., Carrera, G., Gallego, E., Roca, X., Berenguer, M.J., Guardino, X. *J. Chrom. A*, 2007, *1140*, 44-55.

Roca, F.J. Disseny d'un sistema de control d'olors i de la qualitat de l'aire. Caracterització, quantificació i assignació de l'origen dels compostos orgànics volàtils, Doctoral Thesis, Departament d'Enginyeria Minera i Recursos Naturals, Universitat Politècnica de Catalunya, 2006

Roca, F.J. *Disseny d'un sistema de control d'olors i de qualitat de l'aire. Caracterització, quantificació i assignació de l'origen dels compostos orgànics volàtils.* Thesis, Departament d'Enginyeria Minera i Recursos Naturals, Manresa, Spain, 2006.

Roca, F.J., Gallego, E., Perales, J.F., Guardino, X., Projecte de realització de control d'episodis d'olor i qualitat de l'aire a l'àrea urbana de Benicarló, Laboratori del Centre de Medi Ambient-Universitat Politècnica de Catalunya, 1- 70, Barcelona, 2006

Roca, F.J., Ribes, A., Perales, J.F., Carrera, G., Projecte de caracterització d'episodis d'olor i de la qualitat de l'aire a les zones d'impacte de l'àrea urbana de Banyoles – Laboratori del Centre de Medi Ambient- Universitat Politècnica de Catalunya, 1-239, Banyoles, Spain, 2005.

Roca, F.X., Folch, J., Corruchaga, A., Cid, J., Puigjaner, L. *Design and development of a control system for odours and volatile organic compounds (VOC) from industrial and waste treatment processes.* Proceedings of the 9th Mediterranean Congress of Chenical Engineering, Barcelona, 2003, p. 609.

Seifert, B. *Regulating indoor air.* Proceedings of the 5th International Conference on Indoor Air Quality and Climate, Toronto, Canada, 1990, pp. 95-49.

Seifert, B., Abraham, H.J. *Ecotox. Environ. Safety*, 1982, *6*, 190-192.

Soriano, C., Roca, F.J., Alarcón, M. *Study of odor episodes using analytical and modelling approaches*, 27th NATO/CCMS International Technical Meeting on Air Pollution Modeling and its Application, /Banff (Canada), 2004

U.S. EPA, *Compendium of Methods for the Determination of Toxic Organic Compounds in Ambient Air*, Method TO-15, EPA/625/R-96/010b, Center for Environmental Research Information, Office of Research and Development, U.S. EPA, Cincinnati, OH, 1999.

Vautard, R., Szopa, S., Beekmann, M., Menut, L., Haulustaine, D.A., Laurence, R. *Develop. Env. Sci.*, 2007, *6*, 210-219.

Venn, A.J., Cooper, M., Antoniak, M., Laughlin, C., Britton, J., and Lewis, S.A. *Thorax*, 2003, *58*, 955-960.

Verein Deutscher Ingenieure. Norma VDI-3940: Determination of odourants in ambient air field inspections. Düsseldorf , 1993.

Verein Deutscher Ingenieure. VDI 3883 Part 2: Determination of annoyance parameters by questioning. Repeated brief questioning of neighbour panellists. Düsseldorf, 1993

WHO (World Health Organization). *Indoor air pollutants: exposure and health effects*, EURO Reports and Studies 78, WHO Regional Office for Europe, Copenhagen, Denmark, 1982.

WHO (World Health Organization). *Indoor air quality: organic pollutants*, Euro Reports and Studies No. 111, WHO, Regional Office for Europe, Copenhagen, Denmark, 1989.

WHO (World Health Organization). *WHO Air Quality Guidelines. Global update*, WHO Regional Office for Europe, Copenhagen, Denmark, 2005.

Wolkoff, P., Nielsen, G.D. *Atm. Env.*, 2001, *35*, 4407-4417.

Workolff, P. *Indoor Air Suppl.*, 1995, *3*, 9-73.

Workolff, P., Clausen, P.A., Jensen, B., Nielsen, G.D. and Wilkins, C.K. *Indoor Air*, 1997, *7*, 92-106.

Zuraimi, M.S., Roulet, C.-A., Tham, K.W., Sekhar, S.C., David Cheong, K.W., Wong, N.H., Lee, K.H. *Build. Env.*, 2006, *41*, 316-329.

In: Air Quality in the 21st Century
Editors: G. C. Romano and A. G. Conti, pp. 211-260
ISBN 978-1-60456-793-9

Chapter 5

LICHEN BIOMONITORING OF AIR POLLUTION: ISSUES FOR APPLICATIONS IN COMPLEX ENVIRONMENTS

***G. Brunialti*[a,*]*, L. Frati*[a]*, G. Incerti*[b]*, G Rizzi*[c]*, M. Vinci*[b] *and P. Giordani*[c]

[a]TerraData environmetrics, Department of Environmental Science, University of Siena, Via PA Mattioli 4, I-53100 Siena, Italy
[b]Department of Biology, University of Trieste, via L. Giorgieri 10, I-34127 Trieste, Italy
[c]Dip.Te.Ris., Polo Botanico Hanbury, University of Genova, Corso Dogali 1M, I-16136 Genova, Italy

ABSTRACT

Lichens have been widely used as biomonitoring of atmospheric pollution, because these organisms respond to phytotoxic gases (especially SO_2 and NO_x) at cellular, individual and community level. Nevertheless, due to the biological and ecological variability of the organisms, scientists often find difficult to discern the effects of pollution against a natural background noise. On the other hand, statistically supported responses are needed if we aim to include these techniques in decisional processes of environmental management. In this chapter we try to investigate this complex trade-off, by focusing the attention on bioindication techniques, based on the observation of a decreased richness of lichen communities, related to increasing concentration of atmospheric pollutants. In particular we present worked examples of biomonitoring applications in heterogeneous environments, to provide an overview of main sampling, analytical and statistical procedures, which can lead scientists and stakeholders to a better knowledge of the relationship between lichen diversity and air pollution. We take into account three main factors of variability for lichen communities across spatial and temporal scales.

In the first case, we investigate the different responses of lichens to atmospheric pollutants in relation to macroclimatic predictors. Because of close dependence on the atmosphere for their metabolic processes, lichens are strongly influenced by climate-related variables (e.g. rainfall

* Corresponding author: Giorgio Brunialti, TerraData environmetrics, Department of Environmental Science, University of Siena, Via PA Mattioli 4, I-53100 Siena, Italy; e-mail: brunialti@terradata.it

and temperature). It was shown how, under the same level of air pollution, lichen diversity is higher in areas with higher rainfall. We present results from dry vs. humid bioclimatic areas in Italy and we discuss how synergistic and antagonistic relations among pollutants, lichen and climate may significantly vary, together with the predictive power of biomonitoring methods.
Modern approach to environmental assessments includes an evaluation of the pollution impact not only on human health, but also on natural heritage. Hence, lichen biomonitoring was used to detect the effects of gaseous pollutants on natural ecosystems. In the second example, we show that, in the framework of lichen biomonitoring in natural forests, habitat-related variables may play a major role on lichen diversity and more caution should be taken when interpreting these data in terms of direct effects of pollution.
At a more detailed spatial scale, the variability of lichen diversity can be very high and consequently the level of uncertainty in the interpretation of data may arise considerably. This should be taken into account when higher scale surveys are planned, by considering an adequate sampling intensity at within-site scale. In the third working example we discuss this issue, showing data from intensive experiments and comparing the variability unrelated to atmospheric pollution at local scale in natural environments with that observed at higher scale in anthropized ones.

INTRODUCTION

Monitoring and assessment of natural systems are increasingly focusing on the application of indicators of ecosystem status, and substantial efforts are currently devoted to the identification and development of suitable indicators (National Research Council, 1986; Noss, 1990; Messer *et al.* 1991; Bruns *et al.*, 1991; 1997; Kurtz *et al.*, 2001). However, given the variability and diversity of ecological systems, the assessment of their condition is further complicated by the vast diversity in structure, extent and composition of these ecosystems (White, 2004).

Due to this variability, study of environmental properties with known cause–effect relationship is to be preferred. Indeed, the higher the potential to isolate the cause, the lesser is the error in interpreting the data. For some methodologies the cause–effect relationship is more obvious, for example, in the case of ecotoxicological experiments or in the case of ozone–sensitive tobacco, in which noise is easily identified since the cultivar Bel W3 is preferentially sensitive to ozone–related injuries (Brunialti *et al.*, 2004). For other types of biomonitoring, it is more difficult to detect the influence of a single variable.

Also considering that populations of organisms underlie to dynamic processes in a constant state of flux (Olsen *et al.*, 1999), the challenge is to select biomonitoring methods that well answer the pertinent monitoring question (i.e. pollutant impact) in the most suitable, unbiased and repeatable manner.

In this context, lichens are widely used as biomonitors, both as bioaccumulators of trace elements, due to their resistance to heavy metals (Garty, 2001), and as bioindicators, due to their sensitivity to phytotoxic gases (Hawksworth & Rose, 1970; Herzig *et al.*, 1989). Because of these features and since lichens are slow-growing organisms they have been used as long-term biomonitors of air pollution, both for small-scale (Nimis *et al.*, 1990; Loppi *et al.*, 2002a; Loppi & Corsini, 2003) and large-scale surveys (Nimis *et al.*, 1991; van Dobben & DeBakker, 1996; Bennet & Wetmore, 1999; Frati & Brunialti, 2006).

For a better use of these features, bioindication techniques were developed in order to correlate lichen community data and increasing concentration of pollutants. For this reason the results of several lichen biomonitoring surveys were used to integrate instrumental data of atmospheric pollution (Nimis *et al.*, 1990; Giordani *et al.*, 2002; Pinho *et al.*, 2004), and for developing forecasts in connection with human health (Cislaghi & Nimis, 1997).

Nevertheless, given the variability of biological data, lichen diversity depends on many factors other than air pollution, such as climate, substratum ecology and other types of anthropogenic interference, i.e. land use activities (Barkman, 1958; Giordani, 2005; Brunialti & Giordani, 2003; Frati *et al.*, 2007). These aspects stress on the necessity to standardise all the main phases of biomonitoring techniques beginning from sampling design and including data elaboration and interpretation. In particular, there is a clear need to standardize methods so that they adapt to environmental variability. Concerning lichen biomonitoring methods, in recent years a long process of improvement was developed, in order to understand how these organisms respond to air pollutants and, at the same time, how the effects of variables that may cause a bias in this phenomenon can be highlighted.

For this reason, pilot studies have been carried out in heterogeneous situations in order to i) identify major limiting factors that influence lichen diversity (i.e. latitude, longitude, altitude, temperature, rainfall, etc.) ii) find out the suitable sampling methods to obtain statistically valid and unbiased data.

In this chapter we try to investigate this complex trade–off focusing on bioindication techniques, based on lichen diversity changes in relation to different concentrations of atmospheric pollution. The aim of this chapter is to provide an overview of main sampling, analytical and statistical procedures, that can help scientists and stakeholders for a better interpretation of the data.

Lichen Diversity Value (LDV) Method

Several works using lichen diversity have been done worldwide, assuming that sensitive species decline in polluted areas, and tolerant species remain (Martin & Coughtrey, 1982; Galun & Ronen, 1988; Showman, 1988; Wetmore, 1988; Nimis *et al.*, 1991, Vokou, 1999; Geebelen & Hoffmann, 2001; Giordani *et al.*, 2002; Loppi *et al.*, 2002b; Pirintsos & Loppi, 2003).

Actually, the most used lichen bioindication method in Europe has been developed as the result of a standardisation process carried out in the last few years starting from the Index of Atmospheric Purity (IAP) approach (Hawksworth & Rose, 1970; Amman *et al.*, 1987; Nimis *et al.*, 2002) and upgrading it with VDI guidelines (2005) and with field experience of several European researchers in the field. The experimental evidence at the basis of the assumption of lichen diversity as a valuable tool in the assessment of air pollution derives from several studies showing a good correlation among this response variable and the impact of the main atmospheric pollutants, above all SO_2 and NO_x (Hawksworth & Rose, 1970; Amman *et al.*, 1987; Nimis *et al.*, 2002; Giordani *et al.*, 2002) and recently the effects of nitrogen compounds, mainly NH_3 (Søchting, 1995; Van Herk, 1999; Ruoss, 1999; van Dobben & ter Braak, 1998; Frati *et al.*, 2007).

The method proposed by Asta *et al.* (2002) has been recently adopted in Europe with several application on the field (Castello & Skert, 2004, Pinho *et al.*, 2004 Frati & Brunialti, 2006; Giordani & Incerti, 2007) and several standardisation processes (Ferretti *et al.*, 2004;

Brunialti & Giordani, 2002). Despite this, much work still remains to do, since several environmental factors other than air pollution affect the forecasting precision of lichen diversity measures.

MONITORING IN A VARIABLE ENVIRONMENT

According to Yoccoz *et al.* (2001) we can define monitoring as the process of gathering information about some system-state variables at different points in time for the purpose of assessing the state of the system and making inferences about changes in state over time. If our focus is on the monitoring of biological diversity, the systems of interest to us are typically ecosystems or components of such systems (e.g., communities and populations), and the variables of interest include quantities such as species richness, species diversity, biomass, and population size.

In the assessment of environmental quality by means of biomonitors, it is important to control the variability of biological data, which often affects the forecasting precision of these techniques (Laskowsky & Kutz, 1998).

Environmental factors such as geomorphology, climatic variables, and substrate could have a great impact on the ecosystem property studied to assess environmental quality, such as the rate of indicator species, the biodiversity of a community, or the presence of injuries on organisms. For this reason, it is important to understand the environmental processes and the interactions underlying the changes they undergo (Brunialti *et al.*, 2004).

In lichen biomonitoring context the research of the last years focused on several aspects of variability in ecosystems of interest, which affect lichen diversity and may introduce problems in data interpretation. In fact, besides pollution, lichen biodiversity is also sensitive to abiotic variables related to macro- and micro-climatic variations (Pinho *et al.*, 2004). Macroclimatic variations, e.g. changes in temperature, precipitation, geomorphology, and soil chemistry cannot be avoided when studying lichen biodiversity for biomonitoring purposes, particularly, if the study area is large enough to enclose different climatic conditions (Brunalti & Giordani, 2003). However, these variations must be considered in data elaboration (Giordani *et al.*, 2001; Brunalti & Giordani, 2003).

Further, microclimatic parameters, particularly light, water and nutrients may also cause variations in lichen biodiversity. These alterations may be driven by local sources of disturbance, such as roads or farms, different land uses or habitat fragmentation (Jonsson & Jonsell, 1999; Sillett & Goslin, 1999; Moen & Jonsson, 2003). Bark tree characteristics such as tree age or tree species and correspondent bark pH are also important limiting factors for biodiversity of corticolous lichens (Kuusinen, 1996; Herk, 2001). Therefore, in atmospheric pollution biomonitoring studies using lichens, microclimatic variations that are able to influence biodiversity data must be reduced or, ideally, completely removed (Pinho *et al.*, 2004).

Objective and Approach

The purpose of this chapter is to describe criteria for selecting lichen biomonitoring tools to detect the effects of air pollution on ecosystems in different environments. In order to discuss this subject reference will be made to several worked examples dealing with three main topics, which will be discussed in detail in each paragraph:

i) Macroclimatic influence. Because of close dependence on the atmosphere for their metabolic processes, lichens are strongly influenced by climate-related variables (e.g. rainfall and temperature). It was shown how, under the same level of air pollution, lichen diversity is higher in areas with higher rainfall. We present results from dry vs. humid bioclimatic regions in Italy and we discuss how synergistic and antagonistic relations among pollutants, lichens and climate may significantly vary, together with the predictive power of biomonitoring methods.

ii) Monitoring air pollution in forest ecosystems. We show how, in this context, habitat-related variables may play a major role on lichen diversity and more caution should be taken when interpreting these data in terms of direct effects of pollution.

iii) Spatial scale influence. At a more detailed spatial scale, the variation of lichen diversity is usually very high and therefore the level of uncertainty in the interpretation of data may arise considerably. In the third case, we analyse this aspect, showing data from intensive experiments that investigated the variability unrelated to atmospheric pollution at local scale in natural vs. anthropized environments.

Worked Example I. Pollution-related Signal and Climate-related Noise in Lichen Biomonitoring at Regional Scale: A Case-study from Liguria (NW Italy)

It is largely assumed that epiphytic lichens respond to combined effect of several gaseous pollutants (Nimis *et al.*, 2002) and this aspect is often considered as one of the major advantages of this biomonitoring technique. Nevertheless, research on this topic has shown how comparable lichen diversity may be driven by different relative relationship among the main predictive factors (Giordani, 2006, 2007; Brunialti & Giordani 2003).

This is particularly the case of the changing pollution scenario we are observing after the relevant decreasing of SO_2 concentrations (Greenstone, 2004), due to recent policies for pollution reduction. Several authors (e.g. VDI, 2005; Sparrius, 2007) showed the increasing relevance of nitrophytic lichen species vs. acidophytes, as a consequence of the arising of concentrations of various nitrogen compounds (van Dobben & ter Braak, 1998; Ruoss, 1999; van Herk, 1999; Wolseley & James, 2002; Frati *et al.*, 2007) lacking a prevailing limiting influence of SO_2 concentrations.

These ecological evidences suggest to re-discuss the operational approaches for biomonitoring, including probabilistic sampling design for collecting data, quality assurance procedures for evaluating them and more flexible tools for interpretation, in terms of "effects of atmospheric pollution on organisms" and/or "deviations from background reference conditions of epiphytic lichen communities in absence of pollution sources" (Loppi *et al.*, 2002b; Brunialti *et al.*, 2004; VDI, 2005).

At the same time, progress in biomonitoring research pointed out the need of a regionalization for interpreting lichen diversity data, due to the high positive influence of atmospheric humidity (or, in general, water-related factors). That means: looking at the same lichen diversity in different climatic regions, we cannot deduce the same effect of atmospheric pollutant, because lichens are naturally more abundant in humid climate. But even this simple observation does not take into account that climate also interacts with pollutants. Depending on bio-climate, factors may co-vary differently and at various extent, so that the effects of atmospheric pollutants on organisms may be consistently different. E.g. Loppi & De Dominicis (1996) pointed out that in agricultural landscape dust and nitrogen compound may strongly alter the natural composition of epiphytic community. Hence, in an artificially dryer environment, the physiological stress for lichens is significantly higher and consequently their sensitive to atmospheric pollution.

The changing environmental scenario also implies to change operational questions: the story is no more if lichens are sensitive to atmospheric pollution, but if we can still observe it. Assuming that the physiological cause-effect relationship between pollution and lichen vitality is constant, can be still we confident that the signal-to-noise ratio of lichen diversity vs. environmental variability is still detectable? Can we also expect that it will be reliable in the next years?

Taking into account the variation of several environmental factors other than pollution, the model described in this worked example gives a contribution to the previous questions.

MATERIALS AND METHODS

Sampling Design

All the 5,420 kilometric UTM intersections within the survey area (Liguria, NW-Italy) were systematically selected and classified into nine strata, based on habitat type and altitude. A number of intersections proportional to the surface occupied by each stratum within the survey area were selected by means of a stratified random sampling design, performed using the 'random sampling' option implemented within the software package Statistica 6.0 (StatSoft Italia, 2001). The sample included 165 intersections, corresponding to about 3% of the statistical population. In the field, each intersection selected was positioned using a GPS and a 30 m radius plot was installed, centred on the selected coordinates.

Lichen Sampling

Within each plot, all trees conforming to minimum standards (circumference > 60 cm; inclination of the bole < 10°; absence of damage and of decorticated areas on the trunk) within the 30 m radius plots were considered.

For each tree, the abundance of each lichen species was relevėd using a sampling grid consisting of a 10×50 cm ladder divided into 5 10×10 cm quadrants. This ladder grid was placed systematically on the N, E, S and W side of the bole of each tree (4 per tree), with the top edge 1.5 m above ground, following the standards suggested by Asta *et al.* (2002). Summary measurements of species richness and abundance were calculated for each plot: mean number of species per tree; total number of species releved within the plot; mean Lichen Diversity Value (LDV – Asta *et al.*, 2002) calculated as the sum of the abundance of each species within the sampling grids on a tree, averaged for all trees within a plot.

Data Analysis

Two matrices were considered in analysing the data: (1) a matrix of sampling plots × species abundances and (2) a matrix of sampling plots × environmental factors. To detect and exclude possible outliers from the analysis, an exploratory multivariate analysis was carried out using PC-ORD (McCune & Mefford 1999). In particular, we used the Species Occurrence (SpOcc) –NPMR model (McCune *et al.*, 2003), which gives equal weight (1) to all sampling points within the window while all observations outside the window are given zero weight. Outlier sampled plots, plots with < 3 sampled trees, rare species (occurring in < 5% of sampling plots) and plots containing more than one tree species were excluded. For each model, a list of correlated variables, with a given tolerance interval and a cross-R^2 are given. Model quality is evaluated with a cross-R^2 in terms of the size of the residual sum of squares in relationship to the total sum of squares. For further details on NPMR technique see McCune (2006).

Environmental predictors were selected, basing on previous studies (Giordani, 2006; 2007). Furthermore, other factors were added, that were not considered in the previous works. Details of data collection are reported in Giordani (2006).

To reduce the effect of the high variability and the different orders of magnitude within variables on the final model, quantitative variables in matrix 2 were log-transformed (McCune *et al.*, 2002).

RESULTS

Table 1.1 shows the best SpOcc model for Lichen Diversity Value (LDV) in Liguria, in relation to environmental predictors: it includes 15 variables, associated to different ecological subsets (substrate, climate, pollution and other anthropogenic sources of alterations) and produce a cross-R^2 = 0.62, statistically significant ($p < 0.05$) when compared with 10,000 randomized iterative interactions with Monte Carlo test. All pollution-related factors (namely SO_2, NO_x, CO) seem to have a relevant effect on lichen diversity. On the basis of this model, it is possible to investigate synergistic and/or antagonistic relationships among the most important predictors of lichen diversity, with particular reference to climate- and pollution-related factors.

Table 1.1. Best SpOcc model for Lichen Diversity Value (LDV) in Liguria, in relation to environmental predictors

Evaluated cross R^2	Average size of neighbours	Subset of variables	Variables (tolerance)
0.626 *	1.1	Substrate	Circumference of the trunk (20 cm)
			Water retention (categorical)
			Exfoliation (categorical)
			Bark texture (categorical)
			Bark pH (categorical)
		Pollution	Hourly av. SO_2 (18 μg/m^3 per hour)
			Hourly av. NO_x (60 μg/m^3)
			Hourly av. CO (333 μg/m^3)
		Climate (direct)	Yearly av. temperature (2.7 °C)
			Yearly av. rainfall (135 mm)
		Climate (indirect)	Latitude (40,050 m)
			Longitude (59,400 m)
			Elevation (461.4 m)
		Other anthropogenic disturbances	Harvesting impact (categorical)
			Agricultural practices (categorical)

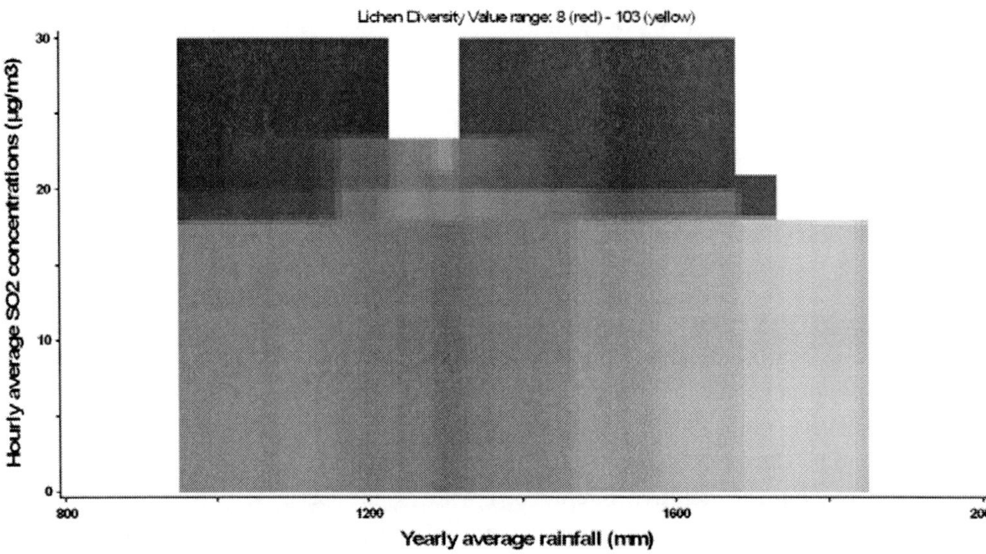

Figure 1.1. LDV response to SO_2 and rainfall, as estimated by SpOcc-NPMR model. Higher LDV corresponds to light and grey, and vice versa.

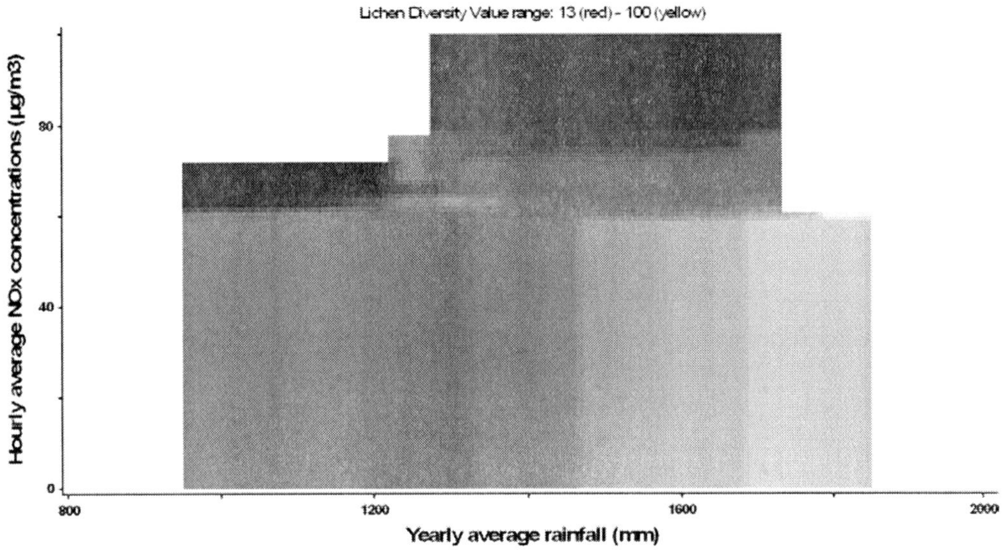

Figure 1.2. LDV response to NO_x and rainfall, as estimated by SpOcc-NPMR model. Higher LDV corresponds to light and grey, and vice versa.

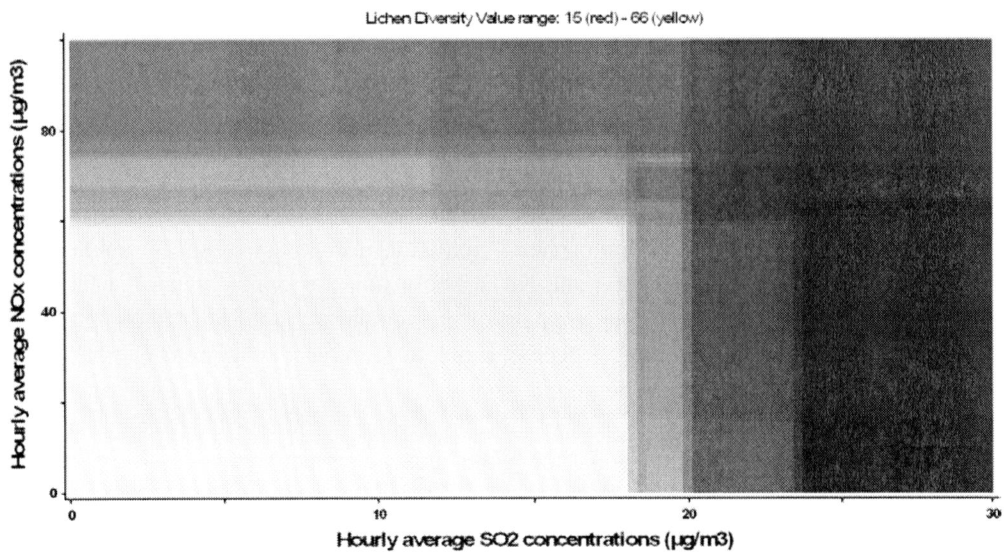

Figure 1.3. LDV response to NO_x and SO_2, as estimated by SpOcc-NPMR model. Higher LDV corresponds to light and grey, and vice versa.

Considering the interaction between precipitations and SO_2 concentrations, the model predicts a LDV range scoring from 8 to 103 (Figure 1.1). Modelled SO_2 concentrations > 20 $\mu g/m^3$ are always strongly limiting for lichen diversity, independently from the rainfall amount. For lower pollutant concentrations, we predict an increasing gradient of LDV co-varying with increasing precipitations. In particular, LDV < 60 is expected for areas with precipitations < 1,300 mm. A second relevant step was found for precipitations = 1,500 mm, whereas LDV considerably arises for mean annual rainfall > 1,700 mm. Similar trends were founds for NO_x, in relation to the mean annual amount of precipitations (Figure 1.2): lowest

LDV (= 13) was shown for NO_x concentrations > 80 μg/m^3 for medium to high mean annual rainfall, but a strong decreasing of lichen diversity was also found for lower pollutant concentration (60 μg/m^3) in dryer conditions (< 1,200 mm rainfall per year).

Figure 1.3 evidences the bivariate interactions between lichen diversity, SO_2 and NO_x. The model predicted very low LDV (< 20), regardless from the NO_x level, for SO_2 > 24 μg/m^3 (but note that additional important steps were also detected for concentrations > 20 and 18 μg/m^3). NO_x caused a decreasing below the limit of LDV = 40 for concentrations > 60 μg/m^3 and this is predicted also for low SO_2 concentrations. On the other hand, medium to high LDV values (> 60) were clearly expected for SO_2 < 12 μg/m^3 and NO_x < 18 μg/m^3.

DISCUSSION

In this contribution we provide an example of a conceptual flow that will allow to detect better the effect of phytotoxic gases against the noise of other ecological predictors, interacting at different spatial and temporal scales. The process includes three main steps that would lead to a better interpretation of lichen biomonitoring data: i) to define homogeneous bioclimatic regions with comparable response of lichens (in terms of distribution and/or abundance); ii) to identify for each region the influence of each atmospheric pollutant on the lichen biodiversity and to analyse trends of biodiversity indices; iii) to elaborate proper scales for interpreting lichen biomonitoring data in terms of alteration from background conditions.

i) To Define Homogeneous Bioclimatic Regions

Concentration thresholds of atmospheric pollutants affecting lichen communities are probably related to a regional context and we can only suppose similar trends for larger and ecologically different areas. The spatial extent of the comparability of the effects is far to be constant, being strictly depending on geomorphologic features of the areas (Ferretti *et al.*, 2004) and, consequently, on bioclimatic characteristics (Brunialti & Giordani, 2003). Nevertheless, good results may be obtained, basing on the definition of homogeneous areas with similar biological traits. Among the wide array of bioclimatic classifications (e.g. Emberger, 1955; Rivas-Martinez, 1995), more specific approaches taking into account the ecological characteristics of lichens should be suggested to draft a regionalization for interpreting lichen biomonitoring data. Some authors proposed a similar approach respectively for large areas of North America (Goward & Spribille, 2005) and for Italy (Nimis & Martellos, 2002). Giordani & Incerti (2007) refined this method with statistical tools at regional scale in Liguria, the survey area of the present worked example, selecting subsets of epiphytic lichens as bioclimatic indicators for humid sub-Mediterranean and Mediterranean units. Then, these taxonomic and bioclimatic considerations are introductory to the analysis of the trends of synthetic indices of biodiversity (e.g. LDV). In the elaborations of Figures 1.1 and 1.2, a distinction between the two bioclimatic regions may be drawn for 1,200 mm yearly average precipitation.

ii) To Identify the Influence of Each Atmospheric Pollutant for Each Bioclimatic Region and to Analyse Trends of Biodiversity Indices

As an important output, using NPMR, we show modelled background LDV along an increasing gradient of mean annual rainfall (and generally speaking of climate-related variables), under comparable load of atmospheric pollution.

We confirm the well-known phytotoxic effect of the two pollutants, but we also provide quantitative estimations on the synergistic effects on lichen diversity after direct observations in the field. The interactions among these pollutants have been largely reported in the past (e.g. Hawksworth & Rose, 1970), but also more recently for critical environmental conditions (van Dobben *et al.*, 2001; Giordani *et al.*, 2002; Purvis *et al.*, 2003; Kapusta *et al.*, 2004; Cristofolini *et al.*, 2008). Our data suggest a prevailing negative effect of SO_2 on lichen diversity. Contrarily to what reported by other authors (e.g. Isocrono *et al.*, 2007), we still observed low lichen diversity even for relatively low SO_2 concentrations (i.e. = 20 $\mu g/m^3$) and this level still strongly limits the re-colonization process also under very low NO_x concentrations. According to several authors (Bates *et al.*, 2001; Giordani *et al.*, 2002; Purvis *et al.*, 2003), this pollutant, mainly due to road traffic, is presently the main limiting factor for lichen colonization in urban areas (van Dobben *et al.*, 2001). However, the question is still on debate: Frati *et al.* (2006) found no association between NO_2 and the diversity of epiphytic lichens very close to a rural highway in Italy, probably because of low concentration of the pollutant, whereas Davies *et al.* (2007) showed how in London lichen diversity declined where NO_x exceeded yearly average concentrations = 70 $\mu g/m^3$, but the pollutant seems to be rapidly dispersed away from roads, decreasing by ca. 70% within 20-30 m of the roadside.

On the other hand, we have not yet enough informations about the time frame of the signal-to-noise ratio and we can also hypothesize a prevalent contribution of habitat-related parameters to the scarce re-colonization (e.g. the dramatic drop out of vegetative propagules in large areas) (Will-Wolf & Scheidegger, 2002).

The model gives a good schematization of the relationship among lichens, climate and SO_2. Two sub-niches of low lichen diversity were predicted for mean annual rainfall < 1,200 and SO_2 > 25 $\mu g/m^3$. Over these limits, lichen diversity showed a continuous and progressive increasing, which is probably at the basis of the fast re-colonization processes recently reported for various areas of Europe (Davies *et al.*, 2007; Isocrono *et al.*, 2007). Hourly average NO_x concentrations > 80 $\mu g/m^3$ seem to represent an important threshold limiting colonization, but the bivariate response of lichens to NO_x and climate showed a higher level of variability: comparable LDV were predicted for concentrations = 70 $\mu g/m^3$ and 1,600 mm rainfall per year vs. 20 $\mu g/m^3$ and 1,200 mm rainfall per year.

iii) To Elaborate Proper Scales for Interpreting lichen Biomonitoring Data

The final step is represented by the elaboration of interpretative scales, which allow to couple lichen diversity to a given level of alteration caused by atmospheric pollutants. These scales are generally developed in terms of deviations from reference conditions within a certain homogeneous ecological context. Loppi *et al.* (2002b) suggested reference values of lichen diversity in background undisturbed areas of Thyrrenian Italy, calculating a potential maximum and, therefore, progressive classes of alteration from it. Similar ideas come from the biomonitoring of hydrosystems: the E.U. Water Framework Directive (European Parliament, 2000) suggested qualitative reference for evaluating the ecological status of water bodies, assessing the highest potential quality, based on the composition of aquatic communities. Tison *et al.* (2007) were able to predict how the diatom communities should be like out of anthropogenic pressure in each French hydro-ecoregion, designed for a topology of running waters and validated with benthic macro-invertebrates from reference sites (Wasson *et al.*, 2002)

Coming back to lichens, other authors have recently proposed to disaggregate the ecological composition of lichen communities, according to particular ecological requirements, e.g. for nitrogen compounds. The interpretative scale of German guidelines for lichen bioindication (VDI, 2005) represents a good step toward this goal, where a different weight is given to comparable LDV obtained with a different species composition (namely nitrophytic vs. acidophytic species). Similar approaches in this sense were also reported for The Netherlands (van Herk, 2003) and France (Gombert *et al.*, 2005). Nevertheless, Jovan and McCune (2006) showed that these species were probably also responding to climatic variability and bark pH, this latter being one of the main factors affecting changes in lichen community structure (Barkman, 1958; van Herk, 2003; Loppi & Frati, 2004; Sparrius, 2007; Larsen *et al.*, 2007). Hence, a step beyond is in the estimation of LDV in dynamic systems, such as changing pollution scenarios, in order to give informations about the responses of lichen diversity under ecological, spatial and temporal interactions of pollutant concentrations. Additionally, in our model, we link real pollution data to the response of lichens, in a multivariate environmental space, where the significance and the amplitude of the effects of each variable is weighted. The iterative and simultaneous evaluation of different predictors considerably increases the model quality, which allows predicting the response of LDV along a broader gradient. This is to say that the model translates in quantitative terms and statistic probability the interactive and dynamic response of lichen diversity to environment.

Worked Example II. Biomonitoring of Air Pollution in Forest Ecosystems: Problems and Perspectives

One advantage of biomonitoring techniques over traditional instrumental recording is the possibility of estimating the effects of atmospheric pollution not only in urban areas, but also in remote environments, such as forested landscapes (Giordani, 2006).

Lichen biomonitoring surveys have been carried out to estimate the pollutant load in forest ecosystems (see e.g. McCune, 2000; Kapusta *et al.* 2004; Jovan & McCune, 2005), but care should be taken when considering factors affecting the variability of lichen diversity in such habitats. In fact, while standardising a large scale extensive monitoring, such as an Italian wide network, we must consider that the territory is highly heterogeneous and that a mosaic of different landscapes is represented, with forested and non-forested areas.

In this context, the interactions between environmental factors other than air pollution and lichen communities may affect their response and the forecasting precision of this ecological approach.

The use of lichens in mapping studies is based on the assumption that ecological parameters other than air pollution have a constant effect on these organisms (Richardson, 1988). Nevertheless, it is widely accepted that climate and other factors affect lichen communities (Barkman, 1958) and, as a consequence, it is difficult to discriminate the effect of air pollution on the frequency and distribution of lichen species from the effects of other environmental parameters (De Wit, 1976). To overcome this problem, lichen diversity method for air pollution monitoring was developed in non-forested areas, and it was diffused mainly in urban and sub-urban districts (Nimis *et al.*, 2001). In fact, the great correlation among lichen diversity and the main atmospheric pollutants (SO_2, NO_x, PM10, etc.) has been observed in studies developed in these situations and with relevés mainly carried out on isolated *Tilia* and *Quercus* trees. However, we must consider that forested landscape covers a great portion of the Italian territory (35.4%; 10673589 ha) and if we exclude it from monitoring programs we may lose important information dealing with pollution impact. For this reason, it is necessary to understand which kind of influence forest management and structure have on lichen diversity variability. In fact, we must underline that highly diversified silvicultural practises contribute to a high macro- and micro-climatic variability within forested habitats. In these habitats several variables may have an effect on lichen diversity stronger than the atmospheric pollution one, with the possibility of an incorrect interpretation of LDV scores.

A lot of studies have been carried out on this topic, mainly focused on finding a sustainable forestry management (see e.g. Lesica *et al.*, 1991; Neitlich & McCune, 1997; Hilmo & Sastad, 2001; Humphrey *et al.*, 2002; McCune *et al.*, 2003). The results of great part of these works confirm a high influence of forest structure and habitat characteristics on lichen communities (Lesica *et al.*, 1991; Sillett, 1995; Neitlich & McCune, 1997; McCune *et al.*, 2003). In particular, a lot of these studies strongly suggest that forest management contributes to the reduction of lichen biomass and diversity, by changing habitat conditions and influencing the dynamics of managed forests over a vast range of spatial and temporal scales (Dettki & Esseen, 2002; McCune *et al.*, 2003; Lemhkhul, 2004; Jovan & McCune, 2005).

This paragraph reports the results of two experimental studies carried out with different approaches with the aim of distinguishing the main limiting factors for lichen diversity in heterogeneous forests not subjected to air pollution and in forested – non-forested landscapes subject to air pollution respectively. The first study was carried out in a non-polluted area characterized by a great heterogeneity from the landscape point of view (Casentino National Park, Tuscany, Italy). Six forest types were considered in which different forest managements were adopted. The second study was developed along a gradient of air pollution. The area (Genova Province, Liguria, Italy) is characterized by a mosaic of forested and non-forested landscapes. Forest type and incidence of environmental pollution were considered to develop a predictive model to explain lichen diversity response to limiting factors.

Our primary objective was to develop a model that related epiphytic lichen community composition to air quality within forested landscapes under different ecological conditions.

The results of these studies can explain some aspects of lichen diversity behaviour and response important to find out a standardized and repeatable monitoring protocol.

MATERIALS AND METHODS

The Study in Tuscany

Study area: Casentino National Park (17,730 ha)

Sampling design: 22 sampling plots were randomly selected proportionally to the surface occupied by forests in the area. The sites had been previously investigated to analyse the influence of forest structure and chemical and physical characteristics of forest types on bryophytes and vascular plants biodiversity (Chiarucci *et al.*, 1996; Ferretti *et al.*, 1996; Chiarucci & Bonini, 2005). Sampling stations include 6 forest types: Beech wood (9 plots), Turkey oak wood (2), Broadleaf mixed wood (2), Silver fir wood (5), Pine wood (2) and Douglas fir wood (2). A total of 183 trees were releved, belonging to 12 tree species (*Fagus sylvatica, Pseudotsuga menziesii, Alnus glutinosa, Quercus cerris, Pinus nigra, Abies alba, Fraxinus ornus, Prunus avium, Ostrya carpinifolia, Acer pseudoplatanus, A. platanoides, A. obtusatum*). At each plot, consisting in a 200 m^2 square, all suitable trees (circumference > 50 cm) were investigated. For each tree LDV was calculated, as reported in worked example I (see lichen sampling in Material and Methods section). For each plot the following variables were considered to test their predictivity on LDV:

i) tree species (categorical): 8 categories (*Quercus cerris, Fagus sylvatica, Abies alba, Pseudotsuga menziesii, Pinus nigra, Alnus glutinosa, Prunus avium, Ostrya carpinifolia*);
ii) tree type (categorical): 2 categories (hardwood, conifer);
iii) forest type (categorical): 6 categories (beech wood, turkey oak wood, broadleaf mixed wood, silver fir wood, pine wood and douglas fir wood);
iv) forest management (categorical): 3 categories (coppice-woodlands, old coppice-woodlands, high forests).

See Brunialti (2006) for further details on data achievement and elaboration.

Data analysis: one-way non parametric Kruskal Wallis ANOVA (Kruskal & Wallis, 1952) was used to test the effects of environmental categorical variables on LDV. The analysis were performed with STATISTICA 6.0 (StatSoft Inc.,Tulsa, Oklahoma, USA).

The Study in Liguria

Study area: The Genova Province (1,835 km^2) is characterized by a high variability in elevation (0-1,800 m a.s.l.) and related climatic parameters, such as mean annual temperature (10-14°C) and rainfall (900-1,800 mm). A sub-Mediterranean climate-type can be found along the coast, more humid on the Thyrrenian side, dryer on the Padanian one. A montane climate-type is present inland at higher elevations. Population, consisting of about 1 million inhabitants, is concentrated in the capital town and in coastal areas (~2000 inhab/km^2), where high levels of atmospheric pollutants occur, mainly from traffic and industry. On the other hand, many mountainous areas are scarcely populated (~10 inhab/km^2) and without local sources of air pollution.

Sampling design: 53 sampling sites were selected by means of a stratified random sampling design, based on habitat type and altitude, proportionally to the surface occupied by each stratum within the survey area.

At each site, consisting of a 30 m radius plot, all suitable trees (circumference > 60 cm; inclination of the bole < 10°, absence of damage and decorticated areas on the trunk) were investigated.

For each tree LDV was calculated, as reported in worked example I (see lichen sampling in Material and Methods section).

The following environmental data were considered:

i) atmospheric pollutants: SO_2, NO_x, and CO concentrations (quantitative). Data were obtained from the Regional Inventory (Liguria Regional Council, 1999) and from the Coordination Territorial Plan (Genova Provincial Council, 2002).
ii) Forest management (categorical): 3 categories (coppice-woodlands, old coppice-woodlands, old-growth forests).
iii) Occurrence of forest fires in the last 20 years (categorical): 2 categories (burnt, unburnt).
iv) Agricultural practices (categorical): 2 categories (treated, non-treated).

See Giordani (2006) for further details on data achievement and elaboration.

Data analysis: to select the best models describing the relationships between LDV, pollutant concentrations and other environmental variables, the transformed matrix of the whole dataset (log-transformation, outlier analysis) was analysed by NPMR (Bowmann & Azzalini, 1997; Peterson, 2000; McCune *et al.*, 2003) using the software Hyperniche 1.0 (McCune & Mefford, 2004). In particular, the Species Occurrence (SpOcc-NPMR) model was used (Peterson, 2000; McCune *et al.*, 2003).

RESULTS

Casentino National Park

A significant effect of tree species was found on lichen diversity (Table 2.1). In particular, broadleaved trees such as *Quercus cerris*, *Alnus glutinosa*, *Prunus avium*, *Fagus sylvatica* and *Ostrya carpinifolia* had the highest lichen diversity, with a mean LDV range from 44 to 71. On the other hand, an evident gap in species richness was found on conifer trees, which showed a highly lower gradient of biodiversity (LDV from 14 to 21). At the same time a considerable variability in LDVs was observed for *Fagus* (range 17 – 112) and *Ostrya* (range 15 – 120) with respect to *Prunus avium* (range 43 – 50), *Pinus nigra* (range 0 – 33) and *Pseudotsuga menziesii* (range 0 – 24).

Subdividing the dataset of LDVs detected respectively in hardwood or in conifer tree landscapes a significant difference was found (Table 2.1, $p < 0.001$). In particular, lichen diversity was distinctly higher in hardwood (mean LDV = 58 ± 22) than in conifer stands (mean LDV = 23 ± 20).

Table 2.1. Relationship between Lichen Diversity Values and predictive variables. ANOVA Kruskal Wallis test

Predictive variables	Categories	N	Lichen Diversity (mean ± st. dev.)	Lichen Diversity (min – max)	K-W ANOVA test
Tree species	*Quercus cerris*	34	71.56 ± 16.98	33 – 112	H (6, 174) = 113.4 p < 0.001
	Alnus glutinosa	3	65.33 ± 24.13	45 – 92	
	Prunus avium	3	46.33 ± 3.51	43 – 50	
	Fagus sylvatica	61	49.39 ± 20.44	17 – 112	
	Ostrya carpinifolia	5	44.00 ± 42.90	15 – 120	
	Abies alba	36	21.42 ± 17.90	0 – 64	
	Pseudotsuga menziesii	13	14.46 ± 8.37	0 – 24	
	Pinus nigra	24	14.29 ± 9.71	0 – 33	
Tree type	Hardwood	93	58.14 ± 22.18	17 – 120	H (1, 183) = 85.8 p < 0.001
	Conifer	90	22.77 ± 19.95	0 – 92	
Forest type	Beech wood	62	53.05 ± 21.76	17 – 120	H (5, 183) = 98.2 p < 0.001
	Turkey oak wood	18	66.61 ± 16.45	33 – 102	
	Broadleaf mixed wood	13	70.69 ± 23.89	32 – 112	
	Silver fir wood	49	28.94 ± 24.05	0 – 92	
	Pine wood	20	15.25 ± 10.02	0 – 33	
	Douglas fir wood	21	15.52 ± 9.10	0 – 30	
Forest management	Coppice-woodlands	72	62.18 ± 22.07	22 – 120	H (2,183) = 94.9 p < 0.0001
	Old coppice-woodlands	10	60.70 ± 22.05	32 – 88	
	High forests	101	23.49 ± 17.83	0 – 92	

The pattern of biodiversity in relation to forest type reflects the influence observed for tree species. In fact, lichen diversity is significantly lower for pine and for silver and Douglas fir (mean LDV from 15 to 29) than for hardwood forest types (mean LDV from 53 to 71).

LDVs strictly depend also by forest management (Table 2.1) with an increasing gradient of biodiversity from high forests (monospecific conifer stands, LDV = 23 ± 18) to old coppice-woodlands (LDV = 61 ± 22) and coppice-woodlands (LDV = 62 ± 22).

Liguria

Once established that habitat characteristics of different forest types and management influence lichen diversity in non-polluted areas, we may ask if the same limiting factors are predominant also under different ecological conditions (namely forested and non-forested landscapes) in order to construct a model to describe lichen diversity changes in relation to one or more predictors. Data collected in Liguria gave us the opportunity to find out this relationship. Figure 2.1 shows the best SpOcc-NPMR models describing the LDV for *N* predictors (with *N* ranging from 1 to 7) in forested areas. A model with a significant ($p<0.05$) cross-R^2 = 0.702 was found, including 5 variables, both climatic (mean annual rainfall and temperature) and related to habitat characteristics and to forest management (e.g. trunk circumference and occurrence of recent forest fires). On the other hand, pollutant concentrations were relevant factors for non-forested areas (Figure 2.2), where a significant ($p<0.05$) cross-R^2 = 0.957 was found for a model giving high weight to the effects of atmospheric pollutants, mainly SO_2 total load and CO daily median concentrations. Also in this case climatic variables, such as yearly average rainfall and temperature, play a relevant role in affecting lichen diversity.

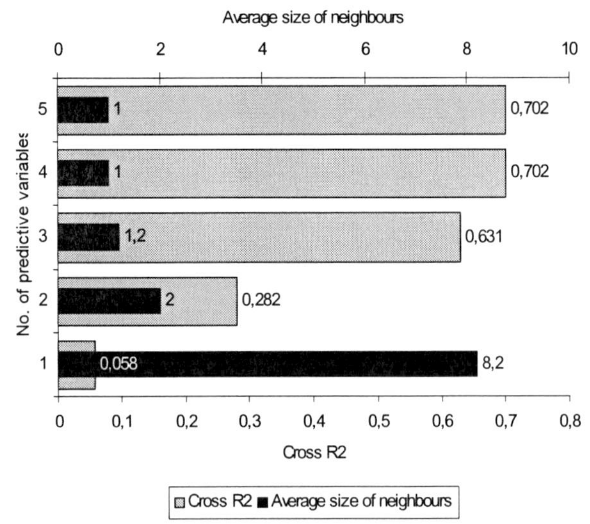

Predictive variables	Variables (tolerance)
1	Circumference of the trunk (7.6 cm)
2	Altitude (146.7 m) Circumference of the trunk (7.6 cm)
3	Yearly average temperature (0.4 O°) Altitude (220 m) Circumference of the trunk (7.6 cm)
4	Yearly average rainfall (270 mm) Yearly average temperature (0.3 O°) Altitude (220 m) Circumference of the trunk (7.6 cm)
5	Yearly average rainfall (270 mm) Yearly average temperature (0.3 O°) Forest fires in the last six years Altitude (220 m) Circumference of the trunk (7.6 cm)

Figure 2.1. Forested areas: best SpOcc-NPMR models from 1 to 7 predictors for the response variable LDV.

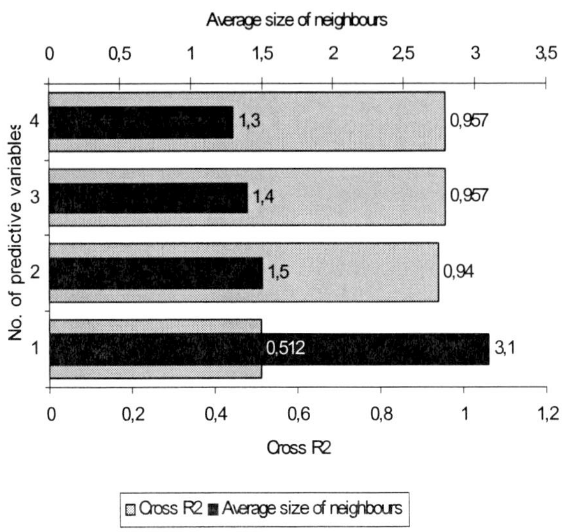

Predictive variables	Variables (tolerance)
1	Yearly average rainfall (22.5 mm)
2	Yearly average rainfall (22.5 mm) Yearly average temperature (0.3 O°)
3	SO_2 total load (3,75 tons/year) Yearly average rainfull (22.5 mm) Yearly average temperature (0.15 O°)
4	SO_2 total load (3,75 tons/year) CO average daily emission (1875 μ/m^3) Yearly average rainfall (22.5 mm) Yearly average temperature (0.15 O°)

Figure 2.2. Non-forested areas: best SpOcc-NPMR models from 1 to 7 predictors for the response variable LDV.

DISCUSSION

Casentino National Park

The results of the first study show that in forested areas lacking air pollution lichen diversity strictly depends on structural characteristics of the managed forests, mainly due to tree substrate, forest type and management. In particular, a great influence of tree species and forest type was observed, which may be considered the main factors affecting LDVs and promoting a great variability in lichen biodiversity among different sites. This variability could be interpreted not only with substrate features but also with the great environmental variability of the forest types that characterizes the study area. Significant differences between the tree species in their epiphytic lichen and bryophyte flora and diversity have been previously demonstrated in a large number of studies in various temperate and boreal areas (for a review see Kuusinen, 1996). As a matter of fact, substrate suitability for lichen colonisation is, in turn, determined by a number of interrelated factors. According to Adams & Risser (1971) the species of the host tree is important because the physical and chemical characteristics of bark strongly vary among tree species. The influence on lichen diversity is likely to be found on physical and chemical properties of the bark such as pH, water retention and buffer capacity, bark roughness, all parameters which proved to be important for lichen colonization (Armstrong, 1990; Hyvärinen *et al.*, 1992). For example the scaly, peeling bark of young pines, with a marked exfoliation, supports fewer lichens than the bark of deciduous trees (Hyvärinen *et al.*, 1992; Kuusinen, 1996). Another important parameter that may explain lichen diversity variability observed in this study is the too smooth bark of *Fagus* and *Ostrya* that may represent a limiting factor for vegetative diaspores establishment and dispersal (Armstrong, 1990; Sillett *et al.*, 2000; Hilmo, 2002). This parameter may explain the high range of variability of LDVs releved on these substrates and may be a confounding factor to data interpretation with respect to air pollution monitoring.

As regards the difference between hardwood and conifer stands the results confirm that tree substrate and forest structure are important environmental factors for lichen colonization in areas not affected from air pollution impact. Similar results were reported also by several other authors (Kuusinen, 1996; Nietlich & McCune, 1997; Humphrey *et al.*, 2000). For example Nietlich & McCune (1997) report that hardwood gaps within conifer-dominated systems host the greater lichen biomass and diversity. Also in the present work hardwood forest types (beech, turkey oak and broadleaf mixed woods) showed to be better habitats for lichen colonization with respect to pine, silver fir and Douglas fir woods. These results strictly depend on the different structural features of these forest types. Different forest structure implies, in fact, a different canopy openness, that strongly affects light and moisture regime, but also substrate and habitat suitability (for a review see Will-Wolf *et al.*, 2002).

At last, forest management affects lichen biodiversity with a strong reduction in correspondence of high forests, mainly characterised by monospecific conifer woods with a low structural diversity and, as a consequence, with homogenous ecological conditions.

Coppice-woodlands, and in particular woods with standard trees retention, show higher LDVs under the same conditions of air pollution. These results confirm an important influence of land use and forest management on LDV that may represent an important confounding effect on data interpretation in the context of atmospheric pollution assessment, due to the fact that to higher values of LD a low air pollution is normally associated and *vice versa*. For this reason the rule to exclude conifer trees from extensive LD monitoring of air pollution, stated in Italian sampling protocol (Nimis *et al.*, 2001), can be justified.

On the other hand, an interpretative approach is necessary to include also forested areas in air pollution surveys in order to include a broad portion of territory that actually is not considered.

Liguria

A high heterogeneity of LDV scores, both among different habitats and within the same ecological conditions, highlighted the co-existence of several factors affecting lichen diversity in terms of both anthropogenic sources and natural site characteristics. In particular, climatic variables related to humidity and temperature strongly affect LDV in both the models, denoting a great influence of climate on lichen colonization on bark, as previously reported in worked example I (see Discussion section).

On the contrary, air pollutants, such as CO and SO_2, are the main limiting factors for the development of epiphytic lichen communities in non-forested areas, where the relevés are mainly carried out on isolated trees reducing the confounding effect of microclimatic variables, which often affect species richness in forested habitats.

Will-Wolf *et al.* (2006) obtained similar results, concluding that in most cases ecological indicator species and groups are reliable only over finite, often relatively narrow geographic and ecological scales, where the effects of climatic variables can be controlled.

Several authors observed that direct SO_2 damage to lichens was probably confined to major urban areas and also that ambient concentrations of NO_x often correlate with SO_2, making it difficult to separate sulphur dioxide effects on lichen communities from nitrogen oxides ones (van Dobben *et al.*, 2001; Geiser & Neitlich, 2007). In fact, direct effects of this latter pollutant on lichen diversity are still poorly known or absolutely unknown, calling for studies of NO_x effects on lichens at urban ambient concentrations (Jovan & McCune, 2005; Geiser & Neitlich, 2007).

According to our results, the relation between lichen diversity and environmental variables changes under different ecological conditions: air pollutants are the main limiting factors in non-forested areas, whereas other factors, such as harvesting of fires and trunk circumference, predominate in forested areas. Furthermore, the strong influence of climatic and forest variables highlights the problem of applicability of the method for monitoring air pollution over vast spatial areas. For this reason, the method used till now must be standardized for forested areas in order to include this land use category in the sampling design at Italian nation-wide scale.

Bioindication with lichen communities allows flexible, intensive sampling across a landscape without the costs of installing and maintaining permanent instrumentation, above all in remote areas. Notwithstanding, a suitable model to explain lichen diversity response in relationship to environmental variables other than air pollution is necessary.

One of the driving questions behind these studies was whether forest management could affect lichen diversity more than by air pollution effects. The aim of distinguishing these two effects is important for the fact that misleading results are possible if we do not consider lichen diversity responses.

Our results have important implications for releving LD in forest ecosystems with the aim of assessing air pollution because of the strong influence of habitat structure on our response variable, i.e. lichen communities on trees.

As a consequence, in the first study, we found that these organisms, in the context of heterogeneous situations, are influenced by limiting factors other than air pollution variables, but, if we consider forested and non-forested landscapes within the same survey area (2nd study), the predictivity of several environmental anthropogenic variables can be isolated and explained.

In order to obtain repeatable and unbiased data on air pollution using this approach we can conclude that the high variability in lichen diversity due to the effect of noisy environmental data with respect to air pollution effect must be explained and controlled.

In this context, the device of excluding conifer forests from the sampling design of Italian monitoring program (Nimis *et al.*, 2001) revealed a good choice, but the problem of representativeness of great part of Italian territory (alpine and montane areas) still remains and a standardized approach is necessary.

Several critical issues must be considered in the design of effective monitoring programs in order to provide a conceptual framework and general recommendations (Will-Wolf *et al.*, 2002; Ferretti & Erhardt, 2002; Noss, 1999; Vos *et al.*, 2000). In particular, progress towards the development of lichen community-based methods for monitoring air pollution in forested landscapes will be hampered until several devices are adopted:

i) *using subsets and stratification:* due to the observed influence of land use characteristics on LDV, the introduction of a stratified sampling design can be recommended in the sampling protocol of lichen biomonitoring survey. In this way it is possible to control the variability in lichen communities related to land use properties (above all forested and non-forested areas). For example, in the case of Geiser & Neitlich (2007) study, modelling a data subset containing equal numbers of polluted and clean plots across regional ranges of variability was the key to isolating lichen pollution response from potentially confounding environmental variables.

ii) *Adopting an integrated monitoring network:* another important tool, which can be important to introduce in the sampling protocol, is to plan large scale monitoring on the basis of an integrated monitoring network based not only on lichens but including also passive samplers of the main air pollutants co-located in the same sampling stations.

    This approach should be based on pilot studies in which both the monitoring approaches are carried out at the same time in a percentage of the same sampling stations. In this way a suitable model of air pollution trends and effects can be obtained. A similar approach allowed Jovan & McCune (2005) to construct a model that relates lichen communities composition to air quality and climatic patterns throughout the greater Central Valley region of California, USA.

iii) *Using lichen guilds or indicator species distribution:* besides using total lichen diversity as a response variable it is possible to consider indicator species or lichen guilds response as suggested by several authors (Will-Wolf *et al.*,2002; Giordani, 2006; Geiser & Neitlich, 2007). In fact, the use of air-pollution sensitive species, with a known response to pollutant concentrations, has also proved to be a valid tool under conditions of decreasing pollution (Bates *et al.*, 2001).

iv) *Adopting interpretative scales for LDV scores in different ambient conditions:* collecting data in a large dataset of forested and non-forested landscapes on a nation-wide scale, will permit to elaborate interpretative scales for lichen diversity values as previously reported by Loppi *et al.* (2002a; 2002b) and by Brunialti *et al.* (2004) to solve the problem of the comparability of lichen diversity data sampled in different bioclimatic zones and on several tree species.

WORKED EXAMPLE III. NON POLLUTION-RELATED NOISE IN LICHEN BIOMONITORING AT WITHIN-SITE SCALE: A CASE STUDY FROM ITALIAN-SLOVENIAN BORDERLINE

Lichens have long been used as environmental monitors. The use of epiphytic lichens as indicators of atmospheric quality dates to the 1860's when independent observations in Europe showed that lichen declines were caused by air pollution. Studies done since have established that lichens respond in predictable ways to changes in air quality, and over 1500 papers have been published (see reviews by Nash, 1989; Richardson, 1992; Hyvärinen *et al.*, 1993; Stolte *et al.*, 1993; Gries, 1996; Garty, 2000, 2001; Nash & Gries, 2002; Nimis *et al.*, 2002). Lichen sensibility to atmospheric pollutant such as sulphur dioxide (SO_2) and nitrogen oxides (NO_X) is well known, since in last decades, many authors have shown that increasing airborne levels of several gaseous pollutants, negatively affect both individuals and communities, in terms of physiological stress and diversity decline, respectively (Hawksworth & Rose, 1970; Ferry *et al.*, 1973; Nash & Wirth, 1988; Nimis *et al.*, 1989; Purvis, 2000; van Dobben *et al.*, 2001).

Several methodological approaches have been proposed in the past for estimating the variation of epiphytic lichen diversity after air pollution (De Sloover & Le Blanc, 1968; Le Blanc & De Sloover, 1970; Liebendoerfer, 1988; Nimis, 1999, for a review see Kricke & Loppi, 2002); nowadays a consolidated approach has widespread established, considering species frequencies inside a grid, having standardized shape and size, overlapping the trunk of a given number of trees in each sampling site (e.g. ANPA, 2001; Asta *et al.*, 2002; VDI, 2005).

The methodology, with minimal case by case differences, is currently employed at almost global scale, both in biodiversity research field studies and in air-pollution monitoring surveys at institutional level (Castello & Skert, 2005; Frati & Brunialti, 2006; Walker *et al.*, 2006; Sparrius, 2007; Giordani, 2007; Frati *et al.*, 2007; Nali *et al.*, 2007; Svoboda, 2007; Poličnik *et al.*, 2008). Several aspects have been recently discussed, with particular attention on data variability and predictivity at different scale, due to both a changing pollution scenario (Bartok, 1999; van Herk 2001, 2003; Lambley *et al.*, 2003; Loppi, 2003; Davies *et al.*, 2003; Kricke & Feige, 2003; Stapper & Franzen-Reuter, 2003; Frati *et al.*, 2006), and to non-pollution related factors, such as climate (Giordani & Incerti, 2007), landcover (Pinho *et al.*, 2008), substrate chemistry (Hauck & Spribille, 2005) and habitat, foliage and tree morphology (Cristofolini *et al.*, 2008).

One of the focus points in the methodological approach is the appropriate number of trees to be selected for sampling. The effect of sampling intensity (i.e.: the proportion of individuals sampled out of the reference population) on data variability is a main topic in biodiversity research field (e.g. Gotelli & Colwell, 2001). Some authors have studied this topic focusing their attention on sampling density at between-site scale (i.e. number of sampling sites per km^2, see Ferretti *et al.*, 2004) whereas, to our knowledge, the role of sampling intensity at within-site scale, notwithstanding it could affect lichen diversity estimates even at larger scale, is far from being clarified. However, to estimate the variability of lichen diversity at within-site scale, a high sampling intensity should be taken into account, planning *ad hoc* case studies, in sites where an adequate number of suitable trees are present.

In most of the literature on lichen diversity a high sampling density at between-sites scale is usually favored with respect to the sampling intensity at within-site scale, also due to cost-benefit reasons.

Scaling of biodiversity is a core topic discussed by modern ecology (e.g. Purvis & Hector, 2000; Gaston, 2004; Tylianakis *et al.*, 2005). Relationships such as distance-decay (i.e. how community composition changes with geographic distance) and species-area (i.e. species richness, abundance and patch occupancy) are basic in the spatial scaling of biodiversity (Green & Bohannan, 2006) and some authors demonstrated that species-area response curves could be sensitive indicators of community-level changes in biodiversity (Tittensor *et al.*, 2006).

There are only few studies focusing on the scaling of epiphytic lichens diversity (Dietrich & Scheidegger, 1997; Kuusinen & Siitonen, 1998; Ellis & Coppins, 2007); in a recent one by Pinho *et al.* (2008) the importance of a multiple spatial scale analysis was shown, to interpret the impact of neighbourhood land-cover on epiphytic lichen diversity. This worked example deals with a case study concerning the scale-dependency of lichen diversity variability. It aims at estimating lichen diversity variability at within-site scale, by means of an adequate sampling intensity. Further, it aims to show how non-pollution related factors impact lichen diversity at within-site scale, with respect to what observed for air pollutant in studies at between-sites scale.

Materials and Methods

Survey Area

Field data were collected in a protected parkland in Lipica, close to the Italian-Slovenian borderline, ca. 3 km from Trieste (Figure 3.1). Local climate is generally submediterranean, conditioned by geomorphology, the mild effects of the sea and the peculiar cold and dry wind (Figure 3.2), with an annual run of about 40,000 km and gust maximum speed over 40 m/s every year (Stravisi, 2001).

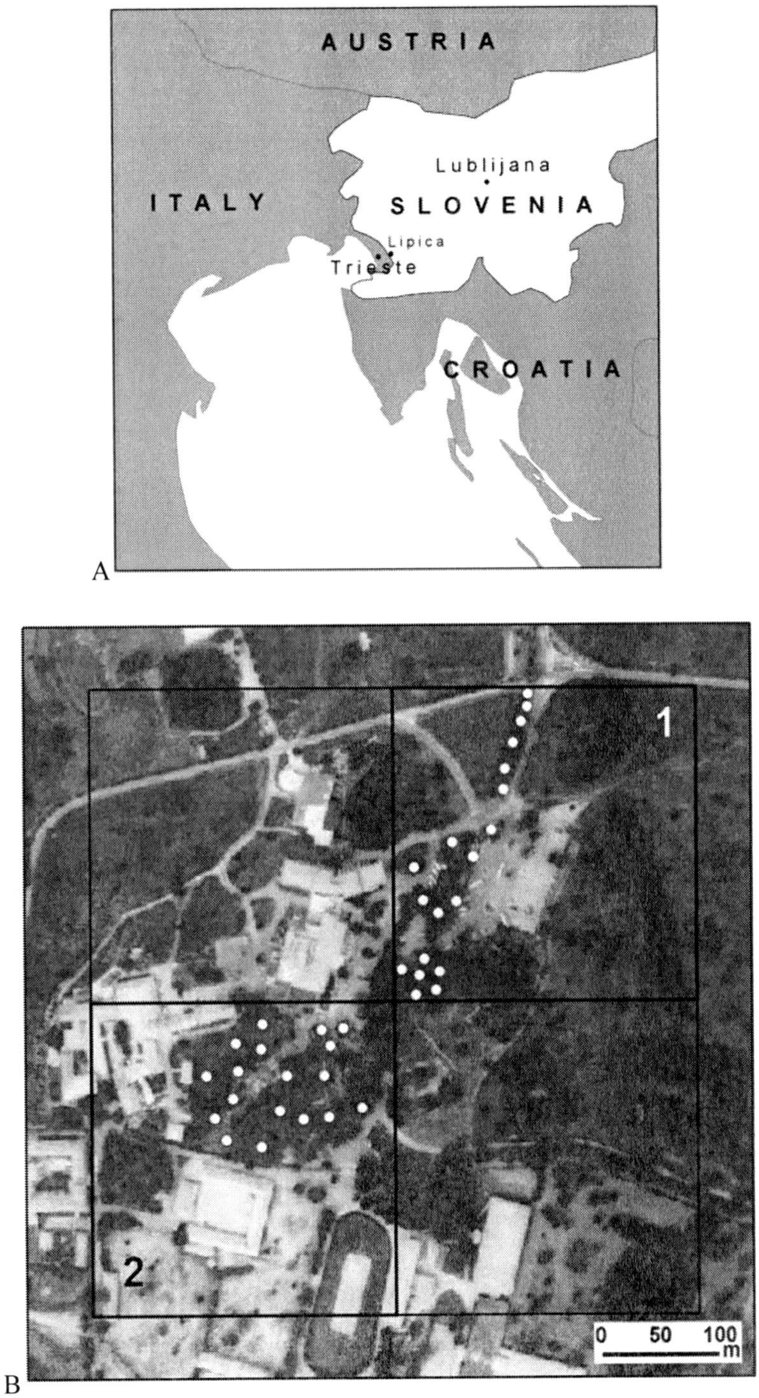

Figure 3.1. Location of Lipica parkland (A) and aerial photograph of the survey area (B). Plot limits (black), subplot numbers and tree locations (○) are shown.

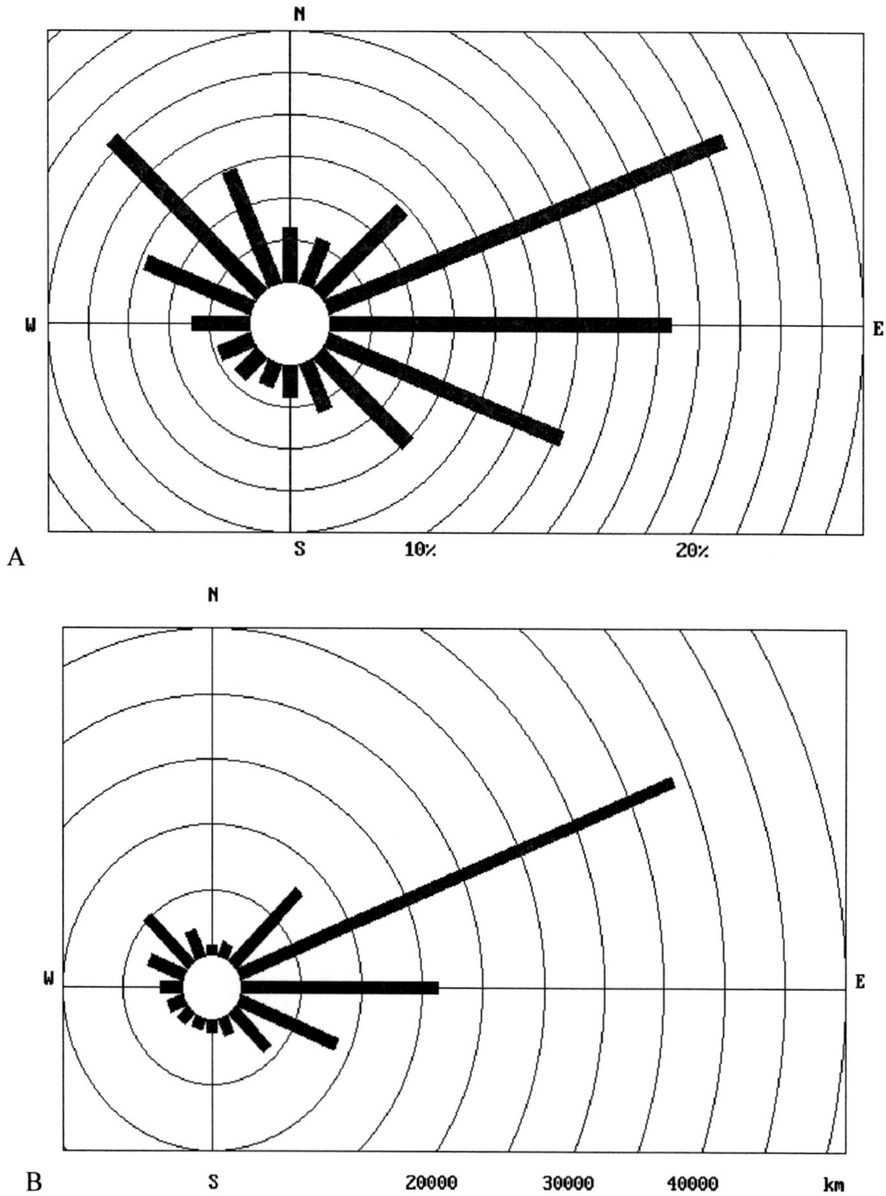

Figure 3.2. Wind frequency (A) and running (B) direction, as yearly average from 1990-2001 in Trieste, ca. 3 km away from Lipica (Stravisi, 2001).

Natural tree vegetation consists of close stands of deciduous oaks (mainly *Quercus pubescens* Willd.), whereas several ornamental lime-trees (*Tilia europaea* L.), maples (*Acer pseudoplatanus* L.) and horse-chestnuts (*Aesculus hippocastanum* L.) are planted along the park pathways.

Sampling Strategy

Following operative protocols standardized at international level to estimate lichen diversity for monitoring purposes (ANPA, 2001; Asta *et al.*, 2002; VDI, 2005), a systematic sampling design was adopted. It prescribes to select sampling sites, consisting of squared or circular sampling units (SU), at the intersections of a grid overlapping the survey area. Distance between adjacent sites and sampling unit size are chosen according to the scale of observation (local, regional, national) and generally vary the range 1-20 km and 1-100 ha, respectively. In most cases SU is typically a 1×1 km squared area, subdivided into subplots, and trees are systematically selected within each subplot according to their proximity to the subplot centre.

In this paper a SU of 500×500 m was considered as the survey area, geo-referenced with a GPS (Geoexplorer 3, Trimble Inc., Sunnyvale, CA) to a UTM intersection of the national grid (ANPA, 2001) and subdivided into two subplots of 250×250 m.

Within each subplot, only lime-trees were considered for sampling, due to their prevalence in the site, to *Tilia* morphological suitability in terms of bole inclination, branching height and bark roughness, and to avoid data variability consequent to the selection of different tree species (e.g. Jüriado *et al.*, 2003; Eckhardt *et al.*, 2003; Loppi & Frati, 2004). All the lime-trees conforming to minimum standards (bole circumference at breast height > 60 cm; inclination of the bole < 10°; absence of damage and of decorticated areas on the trunk) were considered.

For each tree, lichen abundance and LDV were acquired, as reported in worked example I (see lichen sampling in Material and Methods section). Lichen species were identified and noted in the field where possible or collected for herbarium examination. Identification of specimens was undertaken using standard light microscopy at ×6–32 (Makroskop M420, Wild, CH), ×100, ×400, and under oil at ×1000 (Axioplan, C. Zeiss GmbH, D) and standard chemical spot tests with K, C, and Pd (Orange *et al.*, 2001). Taxonomic nomenclature follows Nimis (2003).

Non-pollution Related Factors

Several variables where recorded at either grid ladder- or tree-scale (Table 3.1). Some of these, such as stand density, or distance from the closest tree, being related to foliage shading, can indirectly estimate the amount and duration of sunlight reaching the trunk, affecting lichen growth (e.g. Coxson & Stevenson, 2007). Analogously bole circumference, bark roughness and fissures can indirectly estimate tree age, and, if associated to bole inclination, indicate preferential pathways for water drainage and eutrophication dusts leaching (Purvis *et al.*, 2008); bryophytes and algae coverage are related to humidity, whereas grazing by gastropods, coleopterans and harvestmen negatively affect lichen growth (Benesperi & Tretiach, 2004; Gauslaa, 2005; Nimis & Skert, 2006).

Table 3.1. List of variables tested as predictors for Lichen Diversity Values. For each variable (rows) observation scale, variable type and unit of measure are reported

Code	Description	Observation Scale	Type	Unit of Measure
S.Slo	Site slope	Tree	Quantitative	degree
S.Exp	Site Exposition	Tree	Quantitative	degree
B.Cfr	Bole circumference	Tree	Quantitative	cm
B.Inc	Bole inclination	Tree	Quantitative	degree
SDen	Stand density (radius 20 m^2)	Tree	Quantitative	trees/ha
DCT	Distance from the closest tree	Tree	Quantitative	m
DPW	Distance from pathway	Tree	Categorical	m*
Dist	Disturbance	Tree	Binary	-
BTR	Bark texture roughness	Grid ladder	Quantitative	mm
BKF	Bark knots frequency	Grid ladder	Quantitative	n
BFF	Bark fissures frequency	Grid ladder	Quantitative	n
MC	Bryophytes percent coverage	Grid ladder	Quantitative	%
AC	Algae percent coverage	Grid ladder	Quantitative	%
GC	Grazing percent coverage	Grid ladder	Quantitative	%
Herb	Herbivore class	Grid ladder	Categorical	see note†

* four distance classes (A: 0 - 10 m., B: 10 - 50 m., C: 50 - 100 m., D: > 100 m.).
† three taxa: *Insecta*, *Opiliones* and *Gastropoda*, or 'none'.

Data Analysis

Summary measurements of species richness and abundance were calculated for each subplot, tree and grid ladder: total number of species and Lichen Diversity Value (LDV, Asta *et al.*, 2002), calculated as the sum of the abundance of each species both within each grid ladder (LDVg, i.e. number of 10×10 cm squares where a species is present) and for each tree (LDVt). Lichen diversity dependency on subplot location was evaluated, by using Mann-Whitney non-parametric U test (Mann & Whitney, 1947). Significant differences in lichen communities and in non pollution-related factors distributions were tested, between grid ladder expositions and subplots, respectively, by using Kruskall-Wallis non-parametric ANOVA statistics (Kruskall & Wallis, 1952).

Following McCune *et al.* (2002), a multivariate modeling approach was used to evaluate LDV variability at tree-scale related to independent predictive factors. Two matrices were considered: (1) a predictor matrix (trees × non pollution-related factors), and (2) a response matrix (trees × LDV and species richness). To detect and exclude possible outliers, an explorative multivariate analysis was carried out using PC-ORD method (McCune & Mefford, 1999). The whole dataset was analysed by Non Parametric Multiplicative

Regression (NPMR, McCune *et al.*, 2003), using the software Hyperniche 1.0 (McCune & Mefford, 2004), in order to calculate best predictive models for LDV with an increasing number of independent variables. Local mean estimator method (LM) was selected to fit values for specific points in the space defined by the independent variables. For details on NPMR-LM method see Bowman & Azzalini (1997).

The best NPMR-LM models were evaluated by leave-one-out cross validation method, by calculating cross cross-R^2 as suggested by Guisan & Zimmermann (2000). This parameter is based on the size of the residual sum of squares (RSS) in relation to the total sum of squares (TSS), and ranges between $-\infty$ and 1, higher values corresponding to higher accuracy in model predictions.

A further model evaluation was carried out with a Monte Carlo permutation test. The procedure tests the null hypothesis that the fit of the selected model is no better than could be obtained by chance, given an equal number of predictor variables. By shuffling the LDV observed values, the best NPMR-LM models were fitted for each of 10,000 randomized permutation sets; then cross-R^2 for the selected model was compared with values calculated for the best models of randomized datasets. The probability associated with the Monte Carlo test was calculated as:

$$p = (1 + f)/(1 + N)$$

where f is the number of models for which cross-R^2 higher than that calculated for the selected model are found, and N is total number of permutations.

RESULTS

Local Epiphytic Vegetation

A total of 50 epiphytic lichen species were recorded in 148 grid ladders (4 expositions × 37 lime-trees, Table 3.2). The number of species, with reference to the spatial extent of survey area, is rather high, particularly when compared to studies carried out in similar environmental conditions (e.g. see other case studies in the text). Most frequent lichen species are characteristic of *Xanthorion* s.l. (e.g. *Xanthoria parietina, Phaeophyscia orbicularis, Physcia adscendens, Candelaria concolor, Candelariella reflexa*) and *Parmelion* s. l. (*Parmelina tiliacea, Punctelia subrudecta, Flavoparmelia caperata, Melanelixia fuliginosa, Melanohalea elegantula, Parmelia sulcata*) alliances (Barkman, 1958). Past studies nearby the survey area (e.g. Nimis, 1985; 1986) have proved that these two guilds are differently distributed in southerly and northerly tree faces, principally according to their light and humidity requirements. Our study partially confirms those observations (Figure 3.3); anyway details on lichen vegetation in the survey area are beyond the purpose of this paper, and will be addressed elsewhere.

Table 3.2. List of species recorded in the 148 grid ladders. Species abundances in sampled trees are shown, as average frequency (± standard deviation) in the 4 grid ladders

Lichen species / Tree number	1	2	3	4	5	6	7	8	9	10	11	12	13	14	15	16	17	18	19	20
Amandinea punctata																		0.8±1.5		0.8±1.5
Arthonia radiata					0.5±1.0															
Caloplaca cerina var. *cerina*					0.5±0.6															
Caloplaca ferruginea					0.3±0.5															
Caloplaca pyracea																				
Candelaria concolor	4.8±0.5	4.3±1.5	4.3±1.5	1.5±1.0	4.8±0.5	1.5±1.9	1.3±1.0	1.3±2.5	3.8±1.5	1.3±2.5	1.0±1.4	2.8±2.6	4.8±0.5	2.5±1.3	3.5±1.9	2.3±2.6	2.8±2.1	4.3±1.0	2.5±2.4	2.0±2.2
Candelariella reflexa				2.8±1.7	0.5±1.0		1.5±1.7	1.5±1.9	1.3±1.5	2.5±2.9	2.0±2.4						0.8±1.5	1.0±2.0		2.8±2.1
Candelariella xanthostigma			0.8±1.5		1.0±2.0									0.5±0.6					1.3±2.5	0.5±1.0
Collema flaccidum				0.3±0.5		0.3±0.5						0.3±0.5		0.3±0.5					1.0±2.0	
Degelia plumbea												0.5±0.6								
Evernia prunastri																				0.5±1.0
Flavoparmelia caperata	0.8±1.0	1.0±1.4	1.3±1.3	1.3±1.0	1.0±0.8		2.3±2.6	2.0±2.2	3.3±2.2	1.3±2.5	2.0±2.4			0.5±1.0				0.3±0.5	0.8±1.5	2.3±1.5
Hyperphyscia adglutinata			2.0±2.3		1.5±1.7	1.5±1.7	1.3±2.5	0.8±1.5	3.8±1.9	4.3±1.5	2.3±2.6		0.8±1.5		2.0±2.4	2.0±2.3	2.3±2.6	2.0±2.4	0.5±1.0	1.5±2.4
Hypogymnia physodes										0.3±0.5	0.3±0.5									1.5±1.9
Lecanora carpinea			2.3±2.1		1.5±1.3													1.0±1.4		
Lecanora chlarotera			2.8±1.3		0.8±1.0					0.3±0.5								0.3±0.5		
Lecanora expallens					0.3±0.5															0.5±1.0
Lecanora hagenii					0.5±0.6															
Lecanora symmicta					0.5±1.0															
Lecanora umbrina					0.3±0.5															
Lecidella elaeochroma	0.8±1.5	1.3±2.5	2.3±1.7		4.8±0.5		1.0±0.8			0.8±1.0				0.5±1.0	0.5±1.0	0.5±1.0		1.0±2.0		0.3±0.5
Lepraria incana																				1.3±2.5
Leptogium cianescens												0.3±0.5								
Melanelixia fuliginosa					1.5±1.9				2.3±2.1	3.3±2.4	1.0±1.4							0.5±0.6	1.0±2.0	0.5±1.0

Melanelixia subaurifera									0.5±1.0											
Melanohalea elegantula																				
Opegra atra																		0.5±0.6		
Opegrapha rufescens									1.5±2.4	1.3±2.5	1.3±2.5									
Parmelia sulcata					0.5±1.0		0.8±1.0		2.5±2.9	2.5±0.6	2.3±2.2									1.5±1.9
Parmelina quercina											0.3±0.5									
Parmelina tiliacea	2.5±2.9	1.3±2.5	5.0±0.0	2.0±2.4	2.0±2.4	0.8±1.5	5.0±0.0	2.5±1.9	4.5±0.6	4.8±0.5	3.8±2.5	2.0±2.4	0.3±0.5	0.8±1.5		2.5±2.1	0.8±1.5	1.8±2.4	1.5±2.4	4.8±0.5
Pertusaria albescens			0.8±1.0	0.3±0.5	0.8±1.5					0.5±0.6			1.3±2.5	1.8±2.4	0.5±0.6	0.8±1.5	1.8±1.3	0.3±0.5	1.3±1.5	
Pertusaria amara																		0.3±0.5		
Phaeophyscia chloantha		1.5±1.7										1.0±2.0		0.8±1.0						
Phaeophyscia hirsuta	0.3±0.5																		0.5±1.0	
Phaeophyscia orbicularis	1.3±2.5	1.0±2.0	0.5±1.0		1.8±2.4	1.0±2.0	0.8±1.5						1.3±1.5	1.3±2.5	2.0±2.2	1.0±1.4	3.0±1.8	1.5±1.7	0.8±1.5	
Physcia adscendens	2.5±1.9	1.8±2.1	2.0±2.4	3.3±1.0	3.5±1.0		1.5±1.3	0.8±1.0	1.3±1.5	1.8±2.4	1.5±2.4	3.5±1.3	1.0±2.0	0.5±0.6	1.3±1.3	1.5±1.3	2.5±0.6		2.8±2.1	
Physcia aipolia														1.3±1.5						
Physcia biziana		0.8±1.5	2.8±2.2		3.8±1.9													0.5±1.0		
Physcia tenella																0.5±1.0				
Physconia distorta																				
Physconia grisea		1.3±2.5				2.0±2.2	0.3±0.5	0.3±0.5												
Pseudovernia furfuracea																				0.3±0.5
Punctelia jeckeri									2.8±2.6		0.5±1.0									
Punctelia subrudecta	1.0±1.4	1.3±1.5	2.0±2.0	0.5±1.0	1.8±2.2	1.0±2.0	3.0±2.2	3.0±1.6	2.3±2.6	3.5±2.4	2.0±1.4	1.0±1.4	0.5±0.6		0.8±1.5	0.5±1.0		5.0±0.0	1.3±1.5	3.0±2.2
Rinodina exigua			0.5±1.0						1.3±2.5	1.3±2.5	1.3±2.5									
Rinodina pyrina																				
Rinodina sophodes																				
Xanthoria fallax				0.5±0.6		2.3±2.2		0.3±0.5						0.3±0.5		2.3±2.6				
Xanthoria parietina			2.8±2.6		1.8±2.4									0.3±0.5					1.3±2.5	

Table 3.2. Continued

Lichen species / Tree number	21	22	23	24	25	26	27	28	29	30	31	32	33	34	35	36	37
Amandinea punctata									0.3±0.5	2.5±2.9				0.3±0.5			0.3±0.5
Arthonia radiata																	
Caloplaca cerina var. cerina																	
Caloplaca ferruginea	0.3±0.5																
Caloplaca pyracea										0.3±0.5							
Candelaria concolor	2.3±2.6	1.0±1.2	3.8±1.5	4.5±0.6	2.5±2.9	3.8±1.5	3.0±0.8	5.0±0.0	5.0±0.0	4.5±1.0	5.0±0.0	4.5±1.0	2.3±1.5	3.3±2.2	1.5±2.4	5.0±0.0	2.3±2.2
Candelariella reflexa	1.5±1.9	2.3±1.7		3.5±2.4	0.5±1.0	0.5±1.0	4.0±1.4		1.0±1.4	4.5±1.0			4.5±0.6	3.0±1.4	1.5±2.4	1.3±1.5	4.0±1.4
Candelariella xanthostigma									0.5±1.0							2.3±2.2	
Collema flaccidum																	
Degelia plumbea																	
Evernia prunastri																	0.3±0.5
Flavoparmelia caperata	1.0±1.2	1.5±1.7	0.5±1.0	1.3±1.9		0.3±0.5	2.0±1.8		1.0±1.4	2.3±1.5			3.8±1.5	3.0±2.3	2.0±1.4		5.0±0.0
Hyperphyscia adglutinata		1.0±1.4	3.5±1.3	3.3±2.4		1.3±1.5		5.0±0.0		2.5±2.4	0.5±1.0		0.3±0.5	0.5±1.0	0.5±1.0	1.5±1.9	
Hypogymnia physodes																	0.8±1.0
Lecanora carpinea	0.3±0.5			0.3±0.5			0.5±0.6	0.5±1.0	0.8±1.0		2.3±1.7	0.5±1.0			0.3±0.5	2.3±2.1	0.3±0.5
Lecanora chlarotera	0.5±0.6			0.8±1.0				1.3±1.3	0.5±0.6	0.3±0.5	0.5±1.0			0.3±0.5		2.0±1.4	
Lecanora expallens	0.8±1.5	1.0±2.0					0.3±0.5						0.8±1.0				
Lecanora hagenii																	
Lecanora symmicta				0.3±0.5													
Lecanora umbrina																	
Lecidella elaeochroma	0.8±1.5			0.3±0.5				4.8±0.5	2.8±2.1		1.8±1.7	0.8±1.5		1.0±0.8		5.0±0.0	
Lepraria incana						0.5±1.0	1.3±2.5										
Leptogium cianescens																	
Melanelixia fuliginosa	1.5±2.4	1.3±1.3	2.0±0.8	2.0±1.6		0.3±0.5	4.3±1.5	0.3±0.5	1.3±1.5	0.8±1.0	3.0±2.4	1.5±2.4	1.8±1.5	1.0±1.2	1.0±1.2		3.0±0.8
Melanelixia subaurifera																	
Melanohalea elegantula					1.0±2.0												
Opegra atra																	

Opegrapha rufescens																	
Parmelia sulcata	0.3±0.5									1.3±1.5			2.3±1.0	1.8±1.5			1.0±1.4
Parmelina quercina																	
Parmelina tiliacea	4.3±1.5	5.0±0.0	5.0±0.0	4.5±1.0	0.5±0.6	0.3±0.5	5.0±0.0	1.3±2.5	2.8±2.2	4.3±1.5	1.8±2.2	3.8±1.5	5.0±0.0	4.3±1.0	3.3±2.4	0.8±1.0	5.0±0.0
Pertusaria albescens	0.8±1.0			1.0±0.8	0.8±1.0	2.8±1.3	2.3±1.9		0.5±1.0	1.8±1.3	1.0±1.4	1.3±2.5		0.3±0.5	1.3±1.0		0.3±0.5
Pertusaria amara																	
Phaeophyscia chloantha											2.0±1.4				0.8±1.5		
Phaeophyscia hirsuta																	
Phaeophyscia orbicularis	0.8±1.0	0.5±1.0		0.8±1.5			1.0±2.0	0.5±1.0	3.3±2.2	1.3±2.5	3.5±1.9			2.0±2.4	1.3±1.9	4.8±0.5	1.0±0.8
Physcia adscendens	2.3±1.0	0.3±0.5	1.5±1.3	1.5±1.9		0.8±1.5	0.3±0.5	1.8±2.4		0.8±1.0	1.3±1.5			1.0±1.2		0.5±0.6	
Physcia aipolia																	
Physcia biziana	0.8±1.5		0.8±1.0	0.3±0.5				2.8±1.3	0.8±1.5		1.0±1.2		0.3±0.5				
Physcia tenella																	
Physconia distorta			0.3±0.5		1.3±2.5												
Physconia grisea				0.5±1.0	3.0±1.4	0.5±1.0			1.3±1.5		0.5±1.0						
Pseudovernia furfuracea																	
Punctelia jeckeri																	
Punctelia subrudecta	3.5±1.7	4.0±1.4	4.3±1.0	4.3±0.5	1.0±1.4	3.0±1.8	5.0±0.0	1.5±2.4	3.3±1.0	4.0±1.4	4.5±0.6	4.8±0.5	4.5±1.0	2.3±1.5	4.5±1.0	0.3±0.5	3.3±1.0
Rinodina exigua										0.8±1.5			3.5±2.4				
Rinodina pyrina										0.8±1.5							
Rinodina sophodes	1.3±1.5			0.8±1.5					0.3±0.5	1.3±2.5				0.5±0.6			2.8±1.7
Xanthoria fallax					0.5±0.6												
Xanthoria parietina									0.3±0.5								

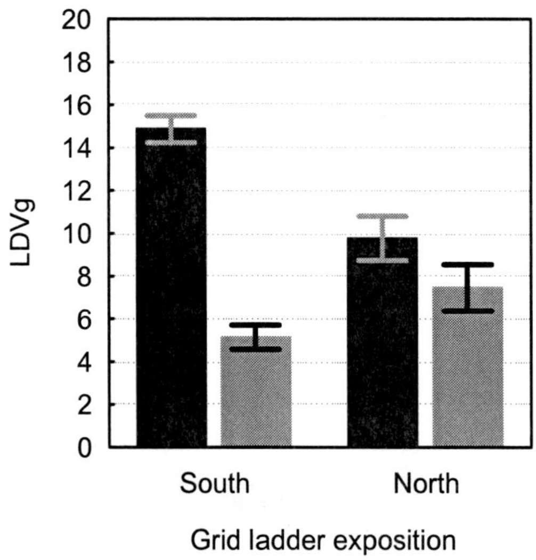

Figure 3.3. Differential distribution of *Xanthorion* (black bars) and *Parmelion* (grey bars) species in south and north expositions. Data are expressed as average ± standard error of respective contributions to lichen diversity values at grid ladder scale (LDVg), species attributed to alliances following Barkman (1958).

Lichen Diversity Variability at within-Plot Scale

Average species richness at different sampling scale (Table 3.3) depicts a typical, not asymptotic increasing in the species-area curve (Figure 3.4).

LDV values are distributed following unimodal curves at ladder scale, both considering the whole dataset of 148 grid ladders and three out of the four expositions (Figure 3.5), whereas the East exposed ladders show a bimodal distribution of LDV. This could be eventually explained by the Bora effect: this predominant cold and dry wind, blowing from ENE at very high speed for more than 80 days every year (Stravisi, 2001), could negatively affect lichen growth on the most exposed trunks, in terms of hydric, thermic and mechanical stress, whereas the higher modal value could correspond to much wind-protected conditions.

Table 3.3. Species richness at different within-site observation scales

Species richness/observation scale	Grid Ladder	Tree	Subplot	Plot
Average number of species	6.7	12.1	38.0	50.0
Standard deviation	3.0	3.8	4.2	-
Coefficient of Variation (%)	44.4%	31.4%	11.1%	-
Number of observations	148	37	2	1
Sampled bark area (m2)	0.05	0.2	3.7	7.4

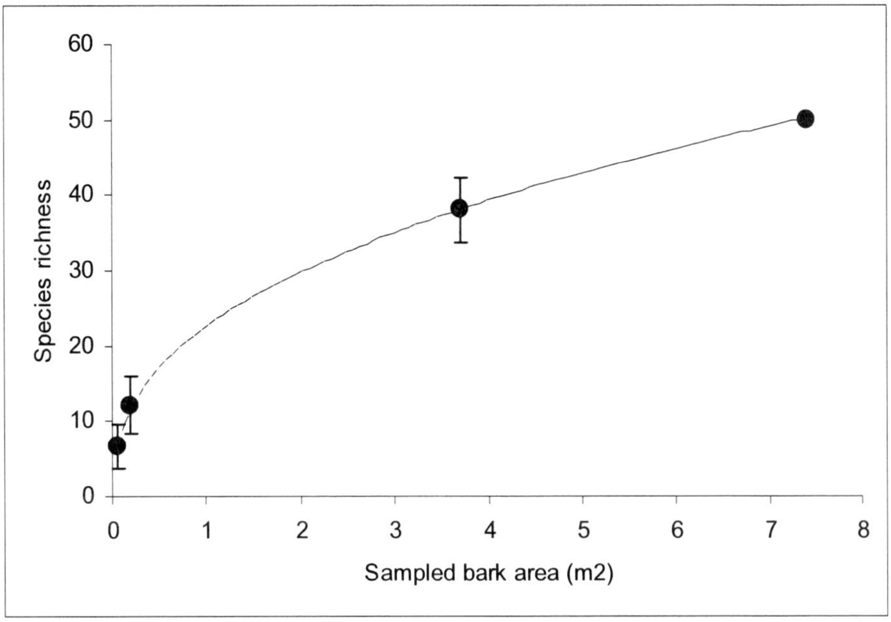

Figure 3.4. Species-area curve in survey area, based on data in Table 3.3.

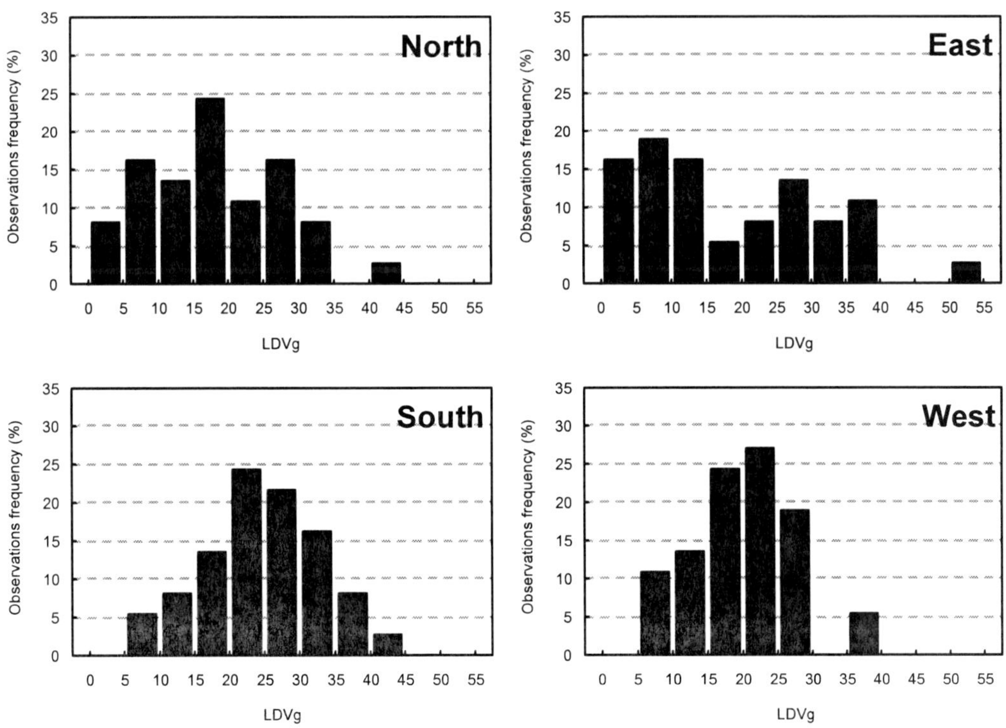

Figure 3.5. Frequency distributions of lichen diversity values at grid ladder scale (LDVg), in main expositions.

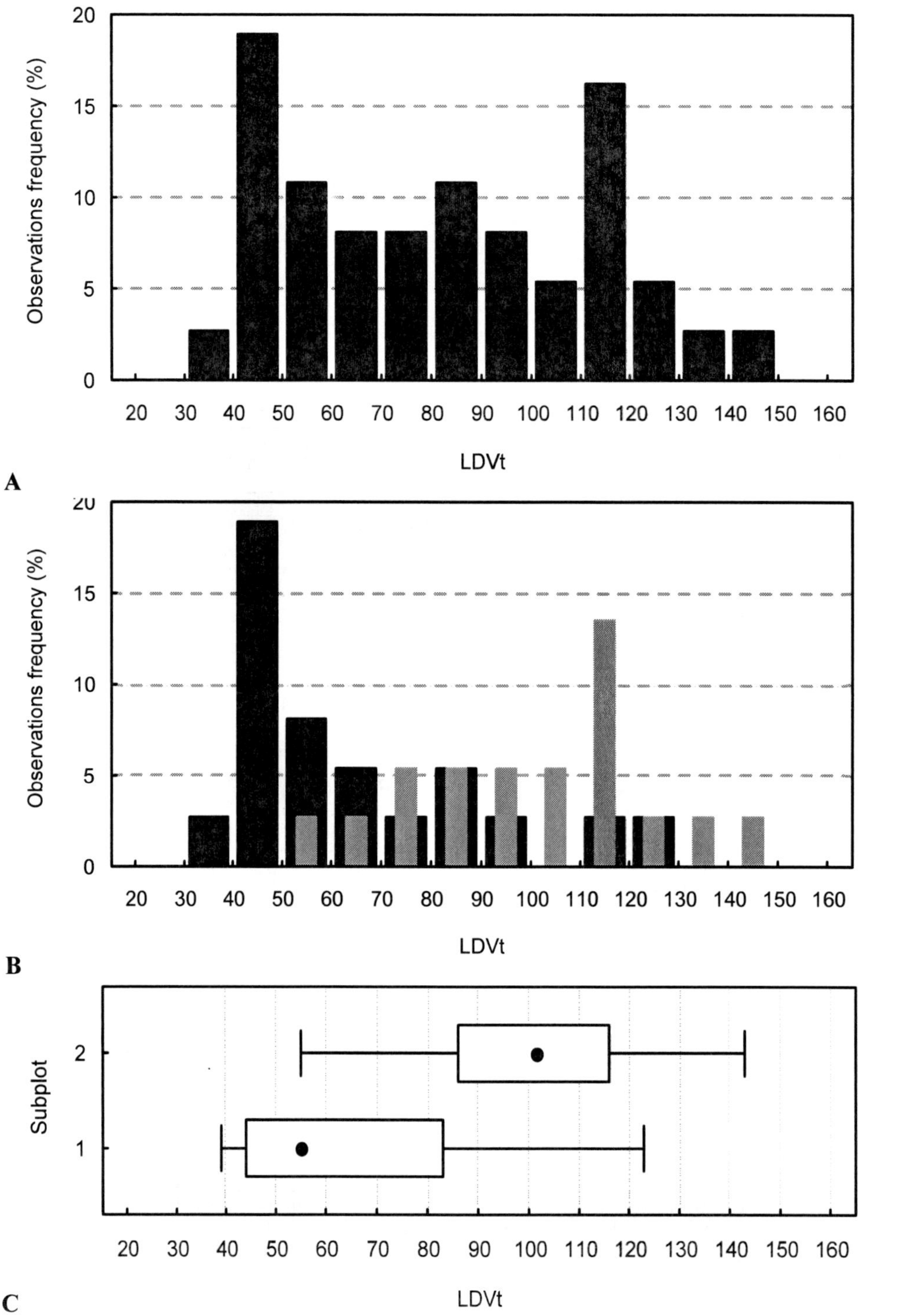

Figure 3.6. Frequency distributions of lichen diversity values at tree scale (LDVt; A: all sampled trees; B: sub-samples corresponding to subplots 1 [■] and 2 [■]). Box plot for significant differences in LDVt between subplots is also shown (C): extremes (whiskers), quartiles (boxes) and median values (dots).

Also at tree scale LDV values follow a bimodal distribution. Two corresponding modal values can be observed if the whole dataset is subdivided into two sub-samples based on the two subplots (Figure 3.6 A and B). Differences in LDV values at tree scale between the two subplots samples are statistically significant (U Mann-Whitney =50.5, $p < 0.001$, Figure 3.6 C), which means that they cannot be considered as a unique population, and that some predictive factor should exist, related to such a difference, which will be discussed in the following.

LDV Variability and Non Pollution-related Factors

The best NPMR-LM model for LDV in the survey area includes 5 predictive variables, describing different situations in terms of substrate competitors (mosses and algae bark coverage), stand density (number of surrounding trees, distance from the closest tree) and tree age (bole circumference, bark texture roughness). The model produces a cross-R^2 = 0.90; the evaluation of model fitness (Table 3.4, Figure 3.7) indicates a highly significant relationship between LDV and the predictors, according to Monte Carlo permutations test ($p < 0.001$).

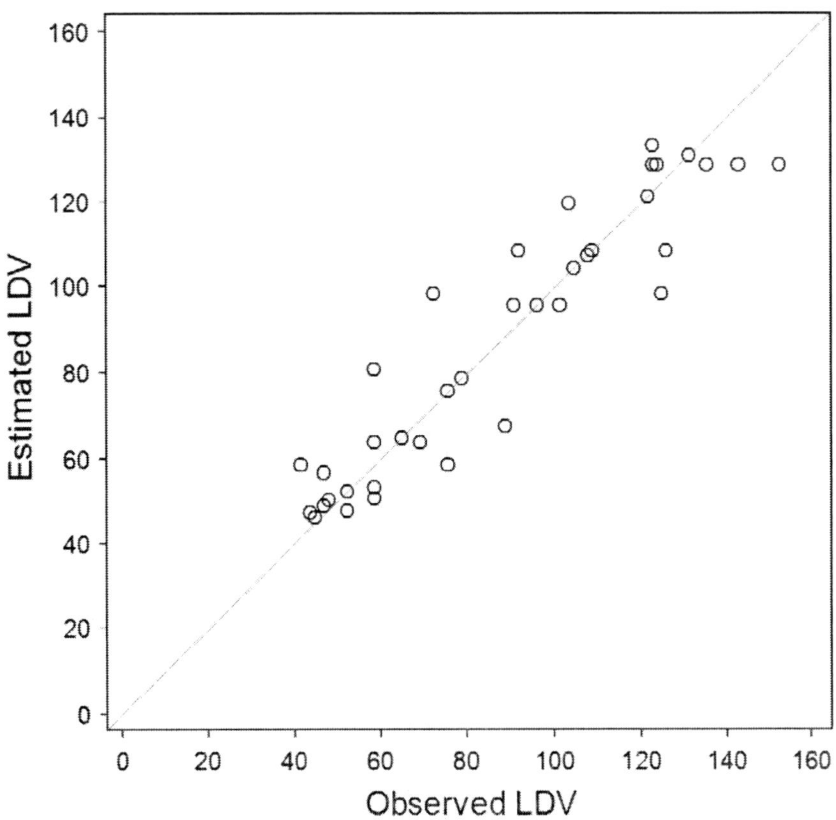

Figure 3.7. Model fitness: scatter plot of estimated and observed LDV values.

On the basis of the model, it is possible to investigate synergistic and/or antagonistic relationships between the predictors of lichen diversity at within-site scale, with particular reference to substrate competitors, stand density and tree age. Univariate modelled responses to the predictors (Figure 3.8) show that LDV at tree-scale depicts a complex pattern along the ecological gradients: a progressive increase in bryophyte percent coverage of tree bark (MC) is firstly coupled with an increase in LDV, probably related to favourable microclimatic conditions for both lichens and mosses; then LDV rapidly decreases, indicating that bryophytes are likely more competitive than lichens in substrate colonization. The minimum LDV can be observed at about 50% of MC, followed by a slight increase, corresponding to higher frequency of muscicolous lichen species (e.g. *Leptogium cyanescens, Physconia spp.*) at higher moss coverage values.

The distance from the closest tree (DCT) seems to indirectly estimate the canopy effect on light diffusions on the trunk: the predictor is positively related to LDV up to a mean threshold value at ca. 6 m., likely indicating a progressive increase in direct light radiation penetrating throughout the foliage; At higher DCT values lichen diversity has an asymptotic trend, which likely does not depend on light effects.

The bole circumference has more or less no evident effect on LDV in the range between 60 and 140 cm; higher values correspond to a slight and almost linear decrease in lichen diversity. This could be explained by considering that younger trees (bole circumference < 140 cm) are more colonized than older ones (bole circumference > 150 cm) by pioneer crustose species, whose small thalli can contribute to LDV by covering a smaller bark area with respect to macro-foliose. On the other hand, a combined effect of increasing moss coverage and bark texture roughness on older trees (at bole circumference > 150 cm) could contribute to limit lichen growth.

Table 3.4. Best LM-NPMR predictive models for Lichen Diversity Value at tree scale (LDVt)

Model	× R^2	p	Predictor count	Predictor 1 (tolerance)	Predictor 2 (tolerance)	Predictor 3 (tolerance)	Predictor 4 (tolerance)	Predictor 5 (tolerance)	Predictor 6 (tolerance)
1	0.3199	0.59334	2	MC (15.1%)	DCT (0.5 m)				
2	0.5640	0.33985	3	MC (15.1%)	DCT (0.5 m)	BCfr (7 cm)			
3	0.7071	0.04776	4	MC (15.1%)	DCT (0.5 m)	BCfr (7 cm)	SDen (64 tr/ha)		
4	*0.8211*	*0.00321*	5	MC (15.1%)	DCT (0.5 m)	BCfr (7 cm)	SDen (64 tr/ha)	BTR (1 class)	
5	*0.9004*	*0.00009*	6	MC (15.1%)	DCT (0.5 m)	BCfr (7 cm)	SDen (64 tr/ha)	BTR (1 class)	AC (8.7%)

Type, fitness (cross-R^2), evaluation result (p and LogB), predictor count, variables and tolerances are shown for each model (see Table 3.1 for variables description). Highly significant results are in italic.

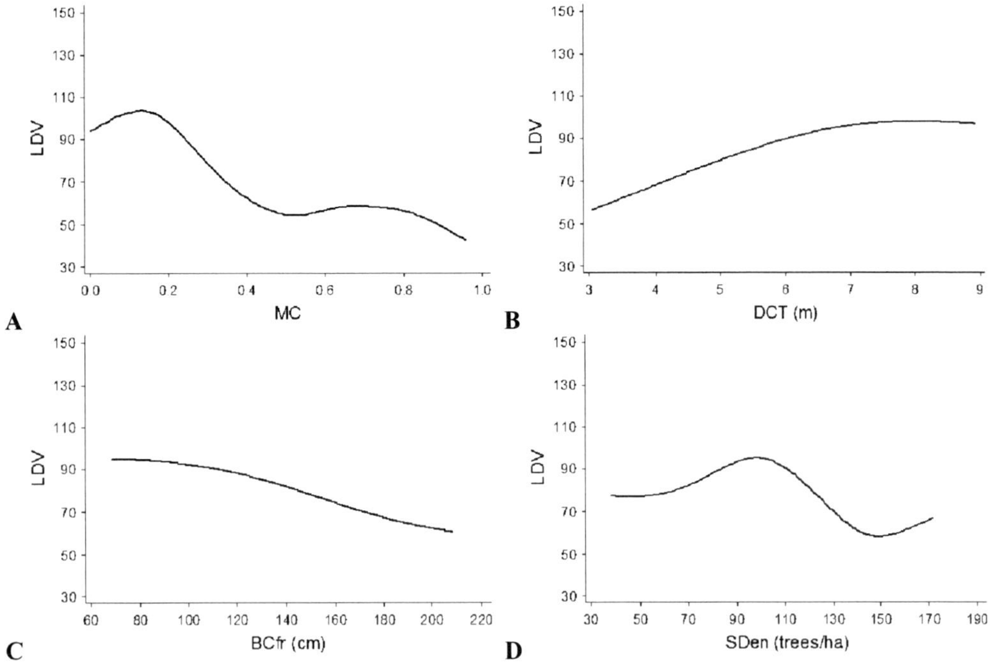

Figure 3.8. Fitted LDV response curves to the predictors, estimated by model 6 in Table 3.4: bryophytes percent coverage (A), distance from closest tree (B), bole circumference (C), stand density (D).

Stand density (SDen) appears to affect LDV in a way hard to explain. If we consider that SDen could indirectly estimate the light regimen, and that most of the recorded lichen species belong to two communities requiring different light conditions (more sciaphytic, stenoic *Parmelion* s.l. and more eliophytic, eurioic *Xanthorion* s.l.), then it would be possible that the variability in LDV values along the SDen gradient, with a relative maximum and minimum, could correspond to a change in the relative contribution of the two communities to the overall LDV values.

Table 3.5. Non parametric univariate statistics for differences in LDV predictors between subplots at tree- and grid ladder-scale

code	Predictor description	Trees					Grid Ladders				
		N1	N2	U	Z	p	N1	N2	U	Z	p
MC	moss % coverage	19	18	*98.00*	*2.22710*	*0.025941*	76	72	*1649.50*	*4.16825*	*0.000019*
DCT	distance from closest tree	19	18	131.00	-1.21605	0.223966	76	72	*2708.00*	*-0.10742*	*0.013999*
BCfr	bole circumference	19	18	*78.50*	*2.81162*	*0.004930*	76	72	1392.00	5.15613	*0.000000*
SDen	stand density	19	18	*87.00*	*2.55842*	*0.010515*	76	72	*2096.00*	*-2.45530*	*0.000000*
BTR	bark texture roughness	19	18	160.50	0.39714	0.691264	76	72	2590.00	0.56012	0.507575
AC	algae % coverage	19	18	142.50	-1.07421	0.282729	76	72	1256.00	5.67788	0.841855

Significant results are in italic.

The predictors selected by the LM-NPMR model at tree level are likely to explain most of lichen diversity variability also at subplot scale. Notwithstanding our dataset does not allow fitting a suitable model for LDV at subplot scale, due to the insufficient number of observations (only two subplots), significant differences between the subplots in terms of LDVt (Figure 3.6) correspond to significant differences also in the predictors, when a non-parametric univariate approach is followed (Table 3.5).

DISCUSSION

LDV variability at within- and between-sites scale: pollution and non pollution-related predictors

Our results show that non-pollution related predictors explain the variability of lichen diversity, at within-site scale. In several papers focused on the application of epiphytic lichen biomonitoring in larger areas, the estimation of within-site variability was often neglected, or reported as LDV standard deviation (Castello & Skert, 2005; Paoli *et al.*, 2006), standard error (Larsen *et al.*, 2007; Pinho *et al.*, 2008) or coefficient of variation (Frati & Brunialti, 2006). Moreover these estimations are mostly based on a scarce number of trees per site, so that a true discussion on the role of within-site non pollution-related factors, or a quantitative comparison of between- and within-sites variability, can hardly be found in literature.

Notwithstanding this scarce knowledge of lichen diversity scaling, a profusion of interpretative tools have been proposed at international level, such as land classification systems of environmental alteration (e.g. Castello & Skert, 2005). These, relating lichen diversity values to pre-defined levels of environmental "naturality", are mostly based on the assumption that the relationships between lichen diversity and target environmental parameters are the strongest at ecosystem and regional scale (e.g.: since high air levels of SO_2 have been proved to be the causal factor of a decrease in lichen diversity, then LDV would be a major proxy for SO_2 pollution).

Table 3.6. LDV scale proposed for land classification and lichen diversity interpretation in terms of environmental alteration in North Adriatic submediterranean region (from Castello & Skert, 2005)

Class	LDV	Environmental alteration	Mapping color
A	0 (lichen desert)	Very high alteration	Magenta
B	1 to 15	High alteration	Red
C	16 to 30	Medium alteration	Orange
D	31 to 45	Low alteration / low naturality	Yellow
E	46 to 60	Medium naturality	Light green
F	61 to 75	High naturality	Dark green
G	> 75	Very high naturaliy	Blue

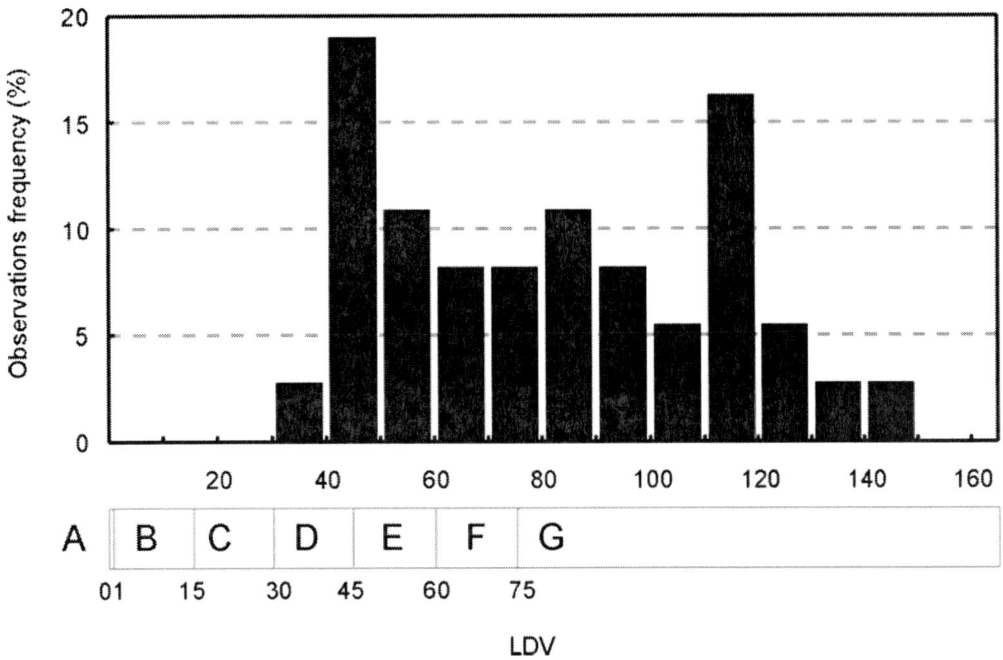

Figure 3.9. Frequency distributions of lichen diversity values at tree level in survey area (LDVt), compared to scale of environmental alteration proposed to interpret LDV in North Adriatic submediterranean region (See Table 3.6).

But what happens if LDV variation at large scale (Table 3.6), which is related to corresponding changes in pollution-related factors (or even climate-related factors, e.g. Giordani & Incerti, 2007, Giordani, 2007), has the same order of magnitude within a plot of 500×500 m (Figure 3.9)? Moreover, what to say if *within-site* variability can be explained by non pollution-related factors, such as substrate competitors, stand density and ecological succession stage?

Possible answers to these questions should be given after considering that: (1) a changing pollution scenario is ongoing, characterized by the decreasing of gaseous well-known pollutants (mostly SO_2, and NO_X, see WHO, 2000) and the increasing of emerging pollutants, such as PM10 and PM2.5, heavy elements, VOCs, ozone, etc (WHO, 2006); (2) changes in air quality policy framework and the consequent decrease in acidificating pollutant were evidenced in a relatively short time, at least when compared to extremely low lichen growth rates (e.g. Hale, 1959, 1970; Hedenås & Ericson, 2003); (3) one of the most reported reactions of epiphytic lichens, since recently, is the re-colonization of urban areas throughout European and American countries (Davies *et al.*, 2003; Jovan & McCune, 2005; Munzi *et al.*, 2007; Frati & Brunialti, 2006); (4) the recolonization process is still in act (Vokou *et al.*, 1999; Hultengren *et al.*, 2004; Isocrono *et al.*, 2007; Gombert *et al.*, 2005), preventing from acritical usage of standardized protocols including generalized interpretative tools, although they were largely useful just until the very recent past; (5) an up to date approach, based on both a probabilistic sampling design and a statistically rigorous data analysis, should be taken into account in order to interpret LDV values, since the traditional simplification of the system, based on the predominance of the LDV-SO_2 relationship above all the possible others, is not exhaustive anymore.

The above considerations put the lichen ecologist, working on lichen and air pollution, in face of two alternatives, both of them aimed at improving current interpretation of lichen diversity variability in space and time: a) furthering the basic eco-physiologic research on phytotoxic effects of emerging pollutants (PM2.5, VOCs, ozone, etc.) since the physiologic responses at species level are indispensable to better understand ecological processes at community and ecosystem level, b) improving knowledge on lichen ecology at community scale, with particular regard to potential predictors unrelated to air pollution, since the effect of air pollutants on the dependent variable (the signal, *sensu* Erhardt & Ferretti, 2002) would be clearer once we will be able to isolate it from the multivariate noise at within-site scale.

General Conclusions

In this chapter three worked examples on lichen biomonitoring tools have been reported, to detect the effects of air pollution on ecosystems in different environments.

All the results clearly showed that standardized sampling, analytical and statistical procedures are badly needed, in order to help scientists and stakeholders for a better interpretation of lichen biomonitoring data.

We showed that lichens respond to air pollutants at large scale, and at the same time, to other environmental predictors whose respective role both depends on the considered ecosystem and on the observation scale.

Synergistic and antagonistic relations among pollutants, lichens and other macro- and micro-environmental factors are current research topics in landscape, ecosystem and community ecology, whereas eco-physiology and population ecology studies could help to better clarify causal effects of single pollutants at individual- and species-scale.

At institutional level the use of lichens as biomonitors is currently widespread, and the standardization of bioindication techniques is continuously updating based on the improvements of scientific research. In this context, our contribution represent a step further, on the way of a better knowledge of lichen response variability in a complex environment.

References

[1] Ammann, K., Herzig, R., Liebendoerfer, L., & Urech, M. (1987). Multivariate correlation of deposition data of 8 different air pollutants to lichen data in a small town in Switzerland. *Advanced Aerobiolology, 51,* 401–406.

[2] ANPA (2001). I.B.L.: Indice di Biodiversità Lichenica. Serie Manuali e Linee Guida, 2/2001. ANPA, Roma.

[3] Asta, J., Erhardt, W., Ferretti, M., Fornasier, F., Kirschbaum, U., Nimis, P.L., Purvis, W., Pirintsos, S., Scheidegger, C., Van Haluwyn, C. & Wirth, V. (2002). Mapping lichen diversity as an indicator of environmental quality. In P. L. Nimis, C. Scheidegger & P. Wolseley (eds.), *Monitoring with Lichens–Monitoring Lichens,* (273-281). Dordrecht: Netherlans.

[4] Barkman, J.J. (1958). *Phytosociology and Ecology of Cryptogamic Epiphytes*. Assen, Netherlands: Van Gorcum.

[5] Bartók, K. (1999). Pesticide usage and epiphytic lichen diversity in Romanian orchards. *The Lichenologist, 31,*21-25.

[6] Bates, J.W., Bell, J.N.B., Massara, A.C. (2001). Loss of *Lecanora conizaeoides* and other fluctuations of epiphytes on oak in S.E. England over 21 years with declining SO_2 concentrations. *Atmospheric Environment,* 35, 2557-2568.

[7] Benesperi, R. & Tretiach, M. (2004). Differential land snail damage to selected species of the lichen genus *Peltigera. Biochemical Systematics and Ecology, 32,* 127-138.

[8] Bennet, J.P. & Wetmore, C.M. (1999). Changes in element contents of selected lichens over 11 years in northern Minnesota, USA. *Environ. Exp. Bot., 41*, 75-82.

[9] Bowman, A.W. & Azzalini, A. (1997). *Applied smoothing techniques for data analysis.* Oxford: Clarendon Press.

[10] Brunialti, G. & Giordani, P. (2002). Applicabilità del nuovo protocollo di campionamento del metodo di Biodiversità Lichenica (BL). In: Verso una rete nazionale per il rilevamento della qualità dell'aria mediante l'indice di biodiversità lichenica. Una valutazione preliminare per la progettazione e le procedure di assicurazione di qualità. *APAT 2002, internal report.*

[11] Brunialti, G. & Giordani, P. (2003). Variability of lichen diversity in a climatically heterogeneous area (Liguria, NW Italy). *Lichenologist, 35*, 55-69.

[12] Brunialti, G., Giordani, P. & Ferretti, M. (2004). Discriminating between the good and the bad: quality assurance is central in biomonitoring studies. In: Wiersma G.B. (Ed.), *Environmental Monitoring,* (443-464). Boca Raton: CRC Press LLC.

[13] Bruns, D.A., Wiersma G.B. & Rykiel Jr. E.J. (1991). Ecosystem monitoring at global baseline sites, *Environ. Monit. Assess., 17*, 3-31.

[14] Bruns, D.A., Wiersma G.B. & White G.J. (1997). Testing and application of ecosystem monitoring parameters. *Toxicol. Environ. Chem., 62*, 169-196.

[15] Castello, M. & Skert, N. (2004). Evaluation of lichen diversity as an indicator of environmental quality in the North Adriatic submediterranean region. *Sci. Tot. Env., 336,* 201-214.

[16] Castello, M. & Skert, N. (2005). Evaluation of lichen diversity as an indicator of environmental quality in the North Adriatic submediterranean region. *Science of The Total Environment, 336,* 201-214.

[17] Cislaghi, C. & Nimis, P.L. (1997). Lichens, air pollution and lung cancer. *Nature, 387*, 463-464.

[18] Conti, M.E. & Cecchetti G. (2001). Biological monitoring: Lichens as bioindicators of air pollution assessment-A review. *Environmental Pollution, 114,* 471–492.

[19] Coxson, D.S. & Stevenson, S.K. (2007). Growth rate responses of *Lobaria pulmonaria* to canopy structure in even-aged and old-growth cedar–hemlock forests of central-interior British Columbia, Canada. *Forest Ecology and Management, 242,* 5-16.

[20] Cristofolini, F., Giordani, P., Gottardini, E. & Modenesi, P. (2008). The response of epiphytic lichens to air pollution and subsets of ecological predictors: A case study from the Italian Prealps. *Environmental Pollution, 151,* 308-317.

[21] Davies, L., Purvis, W. & James, P. (2003). A study of epiphytes in London under a climate of low sulphur dioxide. In: P. Lambley & P. Wolseley (eds), *Lichens in a changing pollution environment.* English Nature Research Reports, (pp. 21-26).

[22] Davies, L. Bates, J.W., Bell, J.N.B., James, P.W., Purvis, O.W. (2007). Diversity and sensitivity of epiphytes to oxides of nitrogen in London. *Environmental Pollution*, 146, 299-310.

[23] De Sloover, J. & Le Blanc, F. (1968). Mapping of atmospheric pollution on the basis of lichen sensitivity. In: R. Misra e B. Golpal (eds.), *Proc. Symp. on Recent Adv. Trop. Ecol.* (42-56). Varanasi, India.

[24] Dietrich, M. & Scheidegger, C. (1997). Frequency, diversity and ecological strategies of epiphytic lichens in the Swiss central plateau and the pre-Alps. *Lichenologist, 29,* 237-258.

[25] Eckhardt, M., Günzl, B., Schmidt, M. & Kriebitzsch, W.U. (2003). Welche Faktoren beeinflussen die Artenvielfalt baumbewohnender Flechten in Laubwäldern? *AFZ – Der Wald, 21,* 1083-1085.

[26] Ellis, C.J. & Coppins, B.J. (2007). 19th century woodland structure controls stand-scale epiphyte diversity in present-day Scotland. *Diversity and Distributions, 13,* 84-91.

[27] Emberger, L. (1955). Une classification biogéographique des climats. *Travail du Laboratoire de Botanique et Zoologie de la Faculté de Science Serveur Botanique de Montpellier,* 7, 3-43.

[28] European Parliament (2000). Directive 2000/60/EC of the European Parliament and of the Council establishing a framework for Community action in the field of water policy. O.J.L327, p. 72.

[29] Ferretti, M., Brambilla, E., Brunialti, G., Fornasier, F., Mazzali, C., Giordani, P. & Nimis, P.L. (2004). Reliability of different sampling densities for estimating and mapping lichen diversity in biomonitoring studies. *Enviromental Pollution, 127,*249-256.

[30] Ferry, B.W., Baddeley, M.S. & Hawksworth, D.L. (1973). *Air pollution and lichens.* London: Athlone Press.

[31] Frati, L. & Brunialti G. (2006). Long-term biomonitoring with lichens: comparing data from different sampling procedures. *Environmental Monitoring and Assessment, 119,* 391-404.

[32] Frati, L., Caprasecca, E., Santoni, S., Gaggi, C., Guttova, A., Gaudino, S., Pati, A, Rosamilia, S., Pirintsos, S.A. & Loppi, S. (2006). Effects of NO2 and NH3 from road traffic on epiphytic lichens. *Environmental Pollution, 142(1),* 58-64.

[33] Frati, L., Santoni, S., Nicolardi, V., Gaggi, C., Brunialti, G., Guttova, A., Gaudino, S., Pati, A., Pirintsos, S.A. & Loppi, S. (2007). Lichen biomonitoring of ammonia emission and nitrogen deposition around a pig stockfarm, *Environmental Pollution, 146,* 311-316.

[34] Galun, M. & Ronen, R. (1988). Interaction of lichens with pollutants, in M. Galun (ed.), *Handbook of Lichenology* (55-72), Boca Raton: CRC Press.

[35] Garty, J. (2000). Environment and elemental content of lichens. In: Markert, B. and Friese, K. (eds.). Trace elements: their distribution and effects in the environment. Elsevier Science B.V., (245-276).

[36] Garty, J. (2001). Biomonitoring atmospheric heavy metals with lichens: Theory and application. *Crit. Rev. Plant Sci., 20,* 309-371.

[37] Gaston, K.J. (2004). Macroecology and people. *Basic and Applied Ecology, 5(4),* 303-307.

[38] Gauslaa, Y. (2005). Lichen palatability depends on investments in herbivore defence. *Oecologia, 143,* 94-105.
[39] Geebelen,W. & Hoffman, M. (2001). Evaluation of bio-indication methods using epiphytic correlations with SO2 pollution parameters. *Lichenologist, 33,* 249-260.
[40] Giordani, P. & Incerti, G. (2007). The influence of climate on the distribution of lichens: a case study in a borderline area (Liguria, NW Italy). *Plant Ecology, in press, DOI 10.1007/s11258-007-9324-7*
[41] Giordani, P. (2006).Variables influencing the distribution of epiphytic lichens in heterogeneous areas: A case study for Liguria, NW Italy. *Journal of Vegetation Science*, 17, 195-206.
[42] Giordani, P. (2007). Is the diversity of epiphytic lichens a reliable indicator of air pollution? A case study from Italy. *Environmental Pollution, 146,* 317-323.
[43] Giordani, P., Brunialti, G. & Alleteo, D. (2002). Effects of atmospheric pollution on Lichen Biodiversity (LB) in a Mediterranean region (Liguria, NW-Italy). *Environmental Pollution, 118,* 53-64.
[44] Giordani, P., Brunialti, G. & Modenesi, P. (2001). Applicability of the lichen biodiversity method (L.B.) to a Mediterranean area (Liguria, NW Italy). *Cryptogamie Mycol., 22,* 193.
[45] Gombert, S. Asta, J. & Seaward, M.R.D. (2005). The use of autecological and environmental parameters for establishing the status of lichen vegetation in a baseline study for a long-term monitoring survey. *Environmental Pollution, 135,* 501-514.
[46] Gotelli N.J. & Colwell R.K. (2001). Quantifying biodiversity: procedures and pitfalls in the measurement and comparison of species richness. *Ecology Letters, 4,* 379-391.
[47] Goward, T. & Spribille, T. (2005). Lichenological evidence for the recognition of inland rain forests in western North America. *J. Biogeogr.,* 32, 1209-1219.
[48] Green, J. & Bohannan, B.J.M. (2006). Spatial scaling of microbial biodiversity . *Trends in Ecology and Evolution, 21,* 501-507.
[49] Greenstone, M. (2004). Did the Clean Air Act cause the remarkable decline in sulfur dioxide concentrations? *Journal of Environmental Economics and Management 47,* 585-611.
[50] Gries, C. (1996). Lichens as indicators of air pollution. In: T.H. Nash (Ed.), *Lichen Biology* (240-254). Cambridge: University Press
[51] Guisan, A. & Zimmermann, N.E. (2000). Predictive habitat distribution models in ecology. *Ecological Modelling, 135,* 147-186.
[52] Hale, M.E. Jr (1959). Studies on Lichen Growth Rate and Succession. *Bulletin of the Torrey Botanical Club, 86(2),* 126-129.
[53] Hale, M.E. Jr. (1970). Single-lobe growth-rate patterns in the lichen Parmelia caperata. *Bryologist 73,* 72-81.
[54] Hauck, M. & Spribille, T. (2005). The significance of precipitation and substrate chemistry for epiphytic lichen diversity in spruce-fir forests of the Salish Mountains, northwestern Montana. *Flora, 200,* 547-562.
[55] Hawksworth, D.L. & Rose, L. (1970). Qualitative scale for estimating sulphur dioxide air pollution in England and Wales using epiphytic lichens. *Nature, 227,* 145-148.
[56] Hedenås, H., Ericson, L. (2003). Response of epiphytic lichens on *Populus tremula* in a selective cutting experiment. *Ecological Applications, 13(4),* 1124-1134.

[57] Herzig, R., Liebendoerfer, L., Urech, M., Amman, K., Cuecheva, M. & Landolt, W. (1989). Passive biomonitoring with lichens as a part of an integrated biological measuring system for monitoring air pollution in Switzerland. *Int. J. Environ. An. Ch., 35,* 43-57.

[58] Hultengren, S., Gralén, H. & Pleijel, H. (2004). Recovery of the Epiphytic Lichen Flora Following Air Quality Improvement in South-West Sweden. *Water, Air, & Soil Pollution, 154(1),* 203-211.

[59] Hyvärinen, M., Soppela, P., Halonen, P. & Kauppi, M. (1993). A review of fumigation experiments on lichens. *Aquilo Ser. Bot. 32,* 21-31.

[60] Isocrono, D., Matteucci, E., Ferrarese, A., Pensi, E. & Piervittori, R., (2007). Lichen colonization in the city of Turin (North Italy) based on current and historical data. *Environmental Pollution, 145,* 258–265.

[61] Jonsson, B. G. & Jonsell, M. (1999). Exploring potential biodiversity indicators in boreal forests. *Biodiversity and Conservation, 8,* 417-1433.

[62] Jovan, S. & McCune, B. (2005). Air-quality bioindication in the greater central valley of California, with epiphytic macrolichen communities. *Ecological Applications, 15(5),* 1712–1726.

[63] Jovan, S. & McCune, B. (2006). Using epiphytic macrolichen communities for biomonitoring ammonia in forests of the greater Sierra Nevada, California. *Water, Air, and Soil Pollution*, 170, 69–93.

[64] Jüriado, I., Paal, J. & Liira J. (2003). Epiphytic and epixylic lichen species diversity in Estonian natural forests. *Biodiversity and Conservation, 12(8),* 1587-1607.

[65] Kapusta, P., Szarek-Lukaszewska, G. & Kiszka, J., 2004. Spatial analysis of lichen species richness in a disturbed ecosystem (Niepolomice Forest, S Poland). *Lichenologist*, 36, 249-260.

[66] Kricke, R. & Feige, G.B. (2003). Changes in epiphytic lichen flora in urban environments. In: P. Lambley & P. Wolseley (eds), *Lichens in a changing pollution environment. English Nature Research Reports, 525* (pp. 27-30).

[67] Kricke, R., Loppi, S. (2002). Bioindication: the I.A.P. approach. In: P.L. Nimis, C. Scheidegger e P.A. Wolseley (eds.), *Monitoring with Lichens - Monitoring Lichens, NATO Science Series 7* (21-38). Kluwer, Dordrecht, The Netherlans.

[68] Kruskal W.H. e Wallis W.A., 1952. Use of ranks in one-criterion variance analysis. *Journal of the American Statistical Association, 47,* 583-621.

[69] Kurtz, J.C., Jackson, L.E. & Fisher, W.S. (2001). Strategies for evaluating indicators based on guidelines from the Environmental Protection Agency's Office of Research and Development. *Ecol. Indicat., 1,* 49-60.

[70] Kuusinen, M. & Siitonen, J. (1998). Epiphytic lichen diversity in old-growth and managed Picea abies stands in southern Finland. *Journal of Vegetation Science, 9,* 283-292.

[71] Kuusinen, M. (1996). Epiphyte flora and diversity on basal trunks of six old-growth forest tree species in southern and middle boreal Finland. *Lichenologist, 28,* 443-463.

[72] Lambley, P., Wolseley, P. & James P. (2003). The impact of a changing pollution climate on the conservation of lichen-rich sites in England and the UK. In: P. Lambley & P. Wolseley (eds) *Lichens in a changing pollution environment. English Nature Research Reports, 525* (pp. 107-116).

[73] Larsen, R.S., Bell, J.N.B., James, P.W., Chimonides, P.J., Rumsey, F.J., Tremper, A. & Purvis, O.W. (2007). Lichen and bryophyte distribution on oak in London in relation to air pollution and bark acidity. *Environmental Pollution, 146(2),* 332-340.

[74] Laskowsky, S.L. & Kutz, F.W. (1998). Environmental data in decision making in EPA regional offices, *Environ. Monit. Assess., 51,* 15.

[75] Le Blanc, F. & De Sloover, J. (1970). Relation between industrialization and the distribution and growth of epiphytic lichens and mosses in Montreal. *Canadian Journal of Botany, 48,* 1485-1496.

[76] Liebendoerfer, L., Herzig, R., Urech, M. & Ammann, K. (1988). Evaluation and calibration of the Swiss lichen-indication-method with important air pollutants. *Staub Reinhaltung der Luft ,48(6),* 233-238.

[77] Loppi, S. & Corsini, A. (2003). Diversity of epiphytic lichens and metal contents of Parmelia caperata thalli as monitors of air pollution in the town of Pistoia (C Italy). *Environ. Monit. Assess., 86,* 289-301.

[78] Loppi, S. & De Dominicis, V. (1996). Effects of agriculture on epiphytic lichen vegetation in central Italy. *Isr. J. Plant Sci, 44,* 297-307.

[79] Loppi, S. & Frati, L. (2004). Influence of tree substrate on the diversity of epiphytic lichens: comparison between *Tilia platophyllos* and *Quercus ilex* (Central Italy). *Bryologist 107(3),* 340-344.

[80] Loppi, S. (2003). Mapping the effects of air pollution, nitrogen deposition, agriculture and dust by the diversity of epipytic lichens in central Italy. In: P. Lambley & P. Wolseley (eds), *Lichens in a changing pollution environment. English Nature Research Reports, 525,* (pp. 37-41).

[81] Loppi, S., Destito, G., Pirintsos, S.A. & De Dominicis, V. (2002a). Temporal variation of air pollution in a geothermal area of central Italy: Assessment by the biodiversity of epiphytic lichens. *Israel J. Plant Sci., 50,* 45-50.

[82] Loppi, S., Giordani, P., Brunialti, G., Isocrono, D. & Piervittori, R. (2002b). A new scale for interpretation of lichen biodiversity values in the Thyrrenian side of Italy, in Progress and Problems in Lichenology at the Turn of the Millenium-IAL-4, *Bibliotheca Lichenologica, 82,* 237-243.

[83] Mann, H.B. & Whitney, D.R. (1947). On a test of whether one of two random variables is stochastically larger than the other. *Annals of Mathematical Statistics, 18,* 50-60.

[84] Martin, M.H. & Coughtrey, P.J. (1982). *Biological Monitoring of Heavy Metal Pollution*, London: Applied Science Publishers.

[85] McCune, B. (2006). Non-parametric habitat models with automatic interactions. *Journal of Vegetation Science*, 17, 819-830.

[86] McCune, B. & M.J. Mefford. (1999). *PC-ORD. Multivariate Analysis of Ecological Data, Version 4.0.*. Gleneden Beach, Oregon: MjM Software Design.

[87] McCune, B. & Mefford, M.J. (2004). *Nonparametric Multiplicative Habitat Modeling.* Gleneden Beach, Oregon: MjM Software Design.

[88] McCune, B., Berryman, S.D., Cissel, J.H. & Gitelman A.I. (2003). Use of a smoother to forecast occurrence of epiphytic lichens under alternative forest management plans. *Ecol Appl, 13,* 1110-1123.

[89] McCune, B., Grace, J.B. & Urban, D.L. (2002). *Analysis of ecological communities.* Gleneden Beach, Oregon: MjM Software Design.

[90] Messer J.J., Linthurst R.A. & Overton W.S. (1991). An EPA program for monitoring ecological status and trends. *Environ. Monit. Assess., 17,* 67-78.
[91] Moen, J. & Jonsson, B.G. (2003). Edge effects on liverworts and lichens in forest patches in a mosaic of boreal forest and wetland, *Conservation Biol., 17,* 380-388.
[92] Munzi, S., Ravera, S. & Caneva, G. (2007). Epiphytic lichens as indicators of environmental quality in Rome. *Environmental Pollution, 146(2),* 350-358.
[93] Nali, C., Balducci, E., Frati, L., Paoli, L., Loppi, S. & Lorenzini, G. (2007). Integrated biomonitoring of air quality with plants and lichens: A case study on ambient ozone from central Italy. *Chemosphere, 67(11),* 2169-2176.
[94] Nash III T.H. & Wirth V. (1988). Lichens, Bryophytes and Air Quality. *Biblio. Lichenol., 30,* 1-297.
[95] Nash III, T.H. & Gries, C. (2002). Lichens as bioindicators of sulfur dioxide. *Symbiosis, 33,* 1-21.
[96] Nash III, T.H. (1989). Metal tolerance in lichens. In: Shaw, J. (Ed.), *Metal Tolerance in Plants: Evolutionary Aspects* (119-131). Boca Raton, Florida: CRC Press.
[97] National Research Council, (1986). *Indicator species and biological monitoring, in Ecological Knowledge and Environmental Problem-Solving: Concepts and Case Studies* (81-87)., Washington, D.C.: National Academy Press.
[98] Nimis, P.L., Castello, M. & Perotti, M. (1990). Lichens as biomonitors of sulphur dioxide pollution in La Spezia (Northern Italy), *Lichenologist, 22,* 333-344.
[99] Nimis, P.L., Lazzarin, A., Lazzarin, G. & Gasparo, D. (1991). Lichens as bioindicators of air pollution by SO2 in the Veneto region (NE Italy), *Studia Geobotanica ,11,* 3-76.
[100] Nimis, P.L., Scheidegger, C. & Wolseley, P.A. (2002). Monitoring with lichens-monitoring lichens: An Introduction, in P.L. Nimis, C. Scheidegger, and P.A. Wolseley (eds.), *Monitoring with Lichens-Monitoring Lichens* (pp. 408). Nato Science Program-IV, Vol. VII, Kluwer Academic Publisher, The Netherlands.
[101] Nimis, P.L. & Martellos, S. (2002). ITALIC – The information system on Italian lichens. *Bibliotheca Lichenologica,* 82, 271-283.
[102] Nimis, P.L. & Skert, N. (2006). Lichen chemistry and selective grazing by the coleopteran *Lasioderma serricorne*. *Environmental and Experimental Botany, 55,* 175-182.
[103] Nimis, P.L. (1985). Urban Lichen Studies in Italy. 1st : the town of Trieste. *Studia Geobotanica, 5,* 49-74.
[104] Nimis, P.L. (1986). Urban Lichen Studies in Italy. 2nd : the town of Udine. *Gortania, 7,* 147-172.
[105] Nimis, P.L. (1999). Linee-guida per la bioindicazione degli effetti dell'inquinamento tramite la biodiversità dei licheni epifiti. In: C. Piccini e S. Salvati (eds.), *Atti del workshop "Biomonitoraggio della qualità dell'aria sul territorio nazionale".* A.N.P.A. (pp. 97-122).
[106] Nimis, P.L. (2003). Checklist of the Lichens of Italy 3.0, University of Trieste, Dept. of Biology, IN3.0/02 *(http://dbiodbs.univ.trieste.it/).*
[107] Nimis, P.L., Ciccarelli, A., Lazzarin, G., Bargagli, R., Benedet, A., Castello, M., Gasparo, D., Lausi, D., Olivieri, S. & Tretiach, M. (1989). I licheni come bioindicatori di inquinamento atmosferico nell'area di Schio-Thiene-Breganze (VI). *Atti Mus. Civ. St. Nat. Verona, 16,* 1-154.

[108] Noss R.F. (1990). Indicators for monitoring biodiversity: a hierarchical approach, *Conserv. Biol., 4,* 355-364.

[109] Olsen, A.R., Sedransk, J., Edwards, D., Gotway, C.A., Liggett, W., Rathbun, S., Reckhow, K.H. & Young, L.J. (1999). Statistical issues for monitoring ecological and natural resources in the United States, *Environ. Monit. Assess., 54,* 1.

[110] Orange, A., James, P.W. & White, F.J. (2001). *Microchemical methods for the identification of lichens*. London: British Lichen Society.

[111] Paoli, L., Guttová, A. & Loppi, S. (2006). Assessment of environmental quality by the diversity of epiphytic lichens in a semi-arid mediterranean area (Val Basento, South Italy). *Biologia, 61(4),* 353-359. *DOI 10.2478/s11756-006-0064-2*

[112] Pinho, P., Augusto, S., Branquinho, C., Bio, A., Pereira, M. J., Soares, A. & Catarino, F. (2004). Mapping lichen diversity as a first step for air quality assessment. *Journal of Atmospheric Chemistry, 49,* 377-389.

[113] Pinho, P., Augusto, S., Máguas, C., Pereira, M.J., Soares, A. & Branquinho, C. (2008). Impact of neighbourhood land-cover in epiphytic lichen diversity: Analysis of multiple factors working at different spatial scales. *Environmental Pollution, 151(2),* 414-422.

[114] Pirintsos, S.A. & Loppi, S. (2003). Lichen as bioindicators of environmental quality in dry Mediterranean areas: A case study from northern Greece. *Israel J. Plant Sci., 51,* 143-151.

[115] Poličnik, H., Simončič, P. & Batič, F. (2008). Monitoring air quality with lichens: A comparison between mapping in forest sites and in open areas. *Environmental Pollution, 151,* 395-400.

[116] Purvis, A. & Hector, A. (2000). Getting the measure of biodiversity. *Nature, 405,* 212-219.

[117] Purvis, O.W., Chimonides, J., Din, V., Erotokritou, L., Jeffries, T., Jones, G.C., Louwhoff, S., Read H., Spiro B., 2003. Which factors are responsible for the changing lichen floras of London? *The Science of The Total Environment,* 310, 179-189.

[118] Purvis, O.W., Dubbin, W., Chimonides, P.D.J., Jones, G.C. & Read, H. (2008). The multi-element content of the lichen *Parmelia sulcata*, soil, and oak bark in relation to acidification and climate. *Science of the Total Environment, 390(2-3),* 558-568.

[119] Purvis, W. (2000). *Lichens*. Natural History Museum, London & Smithsonian Institution. (pp.112). Washington DC.,

[120] Richardson, D.H.S. (1992). *Pollution monitoring with lichens*. Naturalists' Handbooks, 19. Ltd. Slough, England: Richmond Publishing Co.

[121] Rivas-Martínez, S. (1995). Clasificación bioclimatica de la tierra. *Folia Botanica Madridensis, 16,* 1-32.

[122] Ruoss, E. (1999). How agriculture affects lichen vegetation in central Switzerland. *Lichenologist, 31,* 63-73.

[123] Showman, R.E. (1988). Mapping air quality with lichens: The North American experience. *Bibliotheca Lichenologica, 30,* 67-89.

[124] Sillet, S.C. (1999). Distribution of epiphytic macrolichens in relation to remnant trees in a multiple-age Douglas-fir forest, *Canadian J. Forest Res., 29,* 1204-1215.

[125] Sochting, U. (1995). Lichens as monitors of nitrogen deposition. *Cryptogamic Botany, 5,* 264-269.

[126] Sparrius, L.B. (2007). Response of epiphytic lichen communities to decreasing ammonia air concentrations in a moderately polluted area of The Netherlands. *Environmental Pollution, 146,* 375–379.

[127] Stapper, N.J. & Franzen-Reuter, I. (2003). Mapping aerial hypertrophication with epiphytic lichens as biomonitors in North Rhine-Westphalia (NRW Germany). In: P. Lambley & P. Wolseley (eds). Lichens in a changing pollution environment. *English Nature Research Reports, 525,* 31-36.

[128] StatSoft Italia (2001). STATISTICA (sistema software di analisi dei dati), versione 6. www.statsoft.it.

[129] Stolte, K., Mangis, D., Doty, R., Tonnessen, K. & Huckaby, L. (1993). *Lichens as bioindicators of air quality*. General Technical Report RM-224. USDA Forest Service, Rocky Mountain Forest and Range Experiment Station, Fort Collins, CO.

[130] Stravisi, F. (2001). La Bora a Trieste (The borawind in Trieste). Unione Meteorologica Friuli-Venezia Giulia, *Atti I Conv,* (23-34).

[131] Svoboda, D. (2007). Evaluation of the European method for mapping lichen diversity (LDV) as an indicator of environmental stress in the Czech Republic. *Biologia, 62(4),* 424-431.

[132] Tison, J., Park, Y.-S., Coste, M., Wasson, J.G., Rimet, F., Ector L., & Delmas F. (2007). Predicting diatom reference communities at the French hydrosystem scale: A first step towards the definition of the good ecological status. *Ecological Modelling,* 203, 99-108.

[133] Tittensor, D.P., Micheli, F., Nyström, M. & Worm, B. (2006). Human impacts on the species-area relationship in reef fish assemblages. *Trends in Ecology & Evolution, 21(9),* 501-507.

[134] Tylianakis, J.M., Klein, A.-M. & Tscharntke, T. (2005). Spatiotemporal variation in the diversity of Hymenoptera across a tropical habitat gradient. *Ecology, 86,* 3296-3302.

[135] van Dobben, F.H., Wolterbeek, Th.H., Wamelink, W.W. G. & ter Braak, F.J.C. (2001). Relationship between epiphytic lichens, trace elements and gaseous atmospheric pollutants, *Environmental Pollution, 112(2),* 163-169.

[136] van Dobben, H.F & DeBakker, A.Y. (1996). Re-mapping epiphytic lichen biodiversity in the Netherlands: effects of decreasing SO2 and increasing NH3. *Acta Botanica Neerlandica, 45,* 55-71.

[137] van Dobben, H.F. & ter Braak, C.J.F. (1998). Effects of atmospheric NH3 on epiphytic lichens in the Netherlands: the pitfalls of biological monitoring. *Atmospheric Environment, 32,* 551-557.

[138] van Dobben, H.F., Wolterbeek, H.T., Wamelink, G.W.W. & ter Braak, C.J.F. (2001). Relationship between epiphytic lichens, trace elements and gaseous atmospheric pollutants. *Environmental Pollution, 112,* 163-169.

[139] van Herk, C.M. (1999). Mapping of ammonia pollution with epiphytic lichens in The Netherlands. *Lichenologist, 31,* 9-20.

[140] van Herk, C.M. (2001). Bark pH and susceptibility to toxic air pollutants as independent causes of changes in epiphytic lichen composition in space and time. *The Lichenologist, 33(5),* 419-441.

[141] van Herk, C.M. (2003). The effects of short and long distance nitrogen deposition on epiphytic lichens. In: P. Lambley & P. Wolseley (eds). *Lichens in a changing pollution environment. English Nature Research Reports, 525,* 13-20.

[142] VDI (2005). *Biological measurement measure procedures for determining and evaluating the effects of ambient air pollutions by means of lichens (bioindication).* Verein Deutscher Ingenieure – Richt Linien VDI3957-blatt 13. Düsseldorf.

[143] VDI, (2005). VDI 3957 Blatt 11: *Probenahme von Blättern und Nadeln zum Biomonitoring von Immissionsbelastungen (passives Biomonitoring)*, Entwurf, 14 S.

[144] Vokou, D. (1999). Lichens as bioindicators of temporal variations in air quality around Thessaloniki, northern Greece, *Ecol. Res., 14,* 89-96.

[145] Vokou, D., Pirintsos, S.A. & Loppi S. (1999). Lichens as bioindicators of temporal variations in air quality around Thessaloniki, northern Greece. *Ecological Research, 14(2),* 89-96. *DOI 10.1046/j.1440-1703.1999.00294.x*

[146] Walker, T.R., Crittenden, P.D., Young, S.D. & Prystina, T. (2006). An assessment of pollution impacts due to the oil and gas industries in the Pechora basin, north-eastern European Russia. *Ecological Indicators, 6(2),* 369-387.

[147] Wasson, J.G., Chandesris, A., Pella, H. & Blanc, L. (2002). Typology and reference conditions for surface water bodies in France: the hydro-ecoregion approach. In: Ruoppa, M., Karttunen, K. (Eds.), *Typology and Classification of Lakes and Rivers, Symposium on Typology and Classification of Lakes and Rivers.* Helsinki, Finland, October 24–26, 2002. Tema Nord 566, pp. 37–41.

[148] Wetmore, C.M. (1988). Lichen floristics and air quality, *Bibliotheca Lichenologica, 30,* 55-65.

[149] White, G.J. (2004). Selection of ecological indicators for monitoring terrestrial systems. In: Wiersma G.B. (Ed.), *Environmental Monitoring* (263-282). Boca Raton: CRC Press LLC.

[150] Will-Wolf, S. & Scheidegger, C. (2002). Monitoring lichen diversity and ecosystem function. An Introduction. In: Nimis, P.L., Scheidegger, C. & Wolseley P. (eds.) *Monitoring with Lichens – Monitoring Lichens,* pp. 143-145. Kluwer, Dordrecht.

[151] Wolseley, P. & James, P. (2002). Assessing the role of biological monitoring using lichens to map excessive ammonia (NH3) deposition in the UK. In: *Report Produced by Imperial College & The Natural History Museum. Effects of NOX and NH3 on Lichen Communities and Urban Ecosystems. A Pilot Study.* London.

[152] World Health Organization (2000). *Air Quality guidelines for Europe* (pp. 91-273). WHO Reg. Pub., European Series,

[153] World Health Organization (2006). *Air quality guidelines. Global update 2005. Particulate matter, ozone, nitrogen dioxide and sulfur dioxide.* WHO Pub.

[154] Yoccoz, N.G., Nichols, J.D. & Boulinier, T. (2001). Monitoring of biological diversity in space and time, *Trends Ecol. Evol., 16,* 446.

In: Air Quality in the 21st Century
Editors: G. C. Romano and A. G. Conti, pp. 261-285
ISBN 978-1-60456-793-9

Chapter 6

HYDROCARBON CONTAMINATION AND ENVIRONMENTAL HEALTH QUALITY: AN OVERVIEW

***Leo C. Osuji*[1] *and Adaobi E. Osuji*[2]**
[1]Department of Pure and Industrial Chemistry, University of Port Harcourt, PMB 5323, Choba, Port Harcourt, Nigeria
[2]University of Port Harcourt Teaching Hospital, Port Harcourt, Nigeria

ABSTRACT

Health, ecological, and socio-economic impact of hydrocarbon contaminants (crude oils, natural gas, lubricating oils, and so on) in soil, water, and air have been reviewed for purposes of self-risk recognition and assessment. Over 56 hydrocarbons are involved in environmental contamination. Daily exposure doses of 0.1 to 5.0 milligrams per kilogram (oral ingestion) and 0.1 to 6.3 milligrams per kilogram (inhalation) of gasoline to kerosene range hydrocarbons (C_6 to C_{16}) can cause kidney, liver and hematological malfunctions. Leaded gasoline is the single largest source of lead exposure in urban areas. Urban congestion, worn-out vehicles, low quality fuels (having sulfur greater than 0.1 percent), and odor-emitting wastes, which provide 'maternity wards' and 'free lunch counters' for flies and rats, are some of the despairs of poverty in developing nations. A careful husbandry of the scientific, political and socio-economic measures suggested herein might enhance the quality and stability of the environment.

Keywords: hydrocarbon; health, environmental quality; toxicity; contamination

INTRODUCTION

The growing concern for environmental quality within the past two decades has prompted scientific enquiry from various academic disciplines into contemporary ecological problems. Developing nations of Africa, Asia, and Latin America, for instance, have benefited immensely from industrial development (labor absorption, economic emancipation and national esteem), but they have also witnessed an unprecedented breakdown of ecological balance and community structure. The resulting socio-economic pressures when proffered compensations for environmental spoils are deemed inadequate, and the peculiar mix of politics associated with them, are believed to be some of the major causes of communal crisis, youth restiveness and the recent cases of expatriate hostage-taking in some regions. More disturbing, however, is the far-reaching implications on the health of humans who are directly or indirectly exposed to the principal environmental hazard. All in all, the hydrocarbon industry has been the centerpiece of controversy.

Hydrocarbon contaminants are the major effluents from the petroleum industry. They include spilled oils, flared gases, produced water, spent lubricating oils, flue gases (vehicle emissions), diesel oils, et cetera. Hydrocarbon contamination derives mainly from its hydrophobic (water-despising) nature and complex chemical composition. These characteristics predispose it as a primary contaminant over many other secondary contaminants such as heavy metals and oxygen-demanding substances [1].

The three major components of the environment, the soil, water and air media, have been direct targets of discharged hydrocarbons. The effects of such released hydrocarbons on the receiving environment range from massive destruction of ecological plants and animals, to the impairment of public health. Intangible effects include the desecration of socio-cultural values and urban congestions arising from resettlements. The psychological depressions and nostalgia that follow the release of such hydrocarbons into the environment cannot be easily quantified [2][3].

Our group and many other authors have investigated typical, on-site cases of hydrocarbon contaminations and their concomitant impact on environmental quality. In 2005, we reviewed the various tradeoffs of atmospheric pollution in industrial areas of Nigeria and gave an insight into the phenomenal concentrations of discharged effluents. And recently [4], gave a general overview of petroleum industry effluents and other oxygen-demanding wastes. For purposes of self-risk reconnaissance, we have decided to refocus our attention on the increasing insurgency of hydrocarbons in the environment, with a view to exposing their contaminable levels and health implications. The present article is an attempt in this direction.

Concept of Environmental Quality

Definition of the Environment

The environment can be broadly defined as the sum total of all the conditions, agencies (i.e. chemical, biological, physical and material) and influences that affect development, growth, life and death of an organism, species or race. With reference to highly evolved animals, it includes, in its widest sense, mental influences [5]. When any of these conditions is altered, the environment is said to be degraded. A man's environment is his ecological niche, his mode of life in a community, particularly in relation to soil, water and air.

Major Components of the Environment

1. Soil

The soil is the outermost layer of the earth's crust, a layer distinctively different from the underlying bedrock. The soil, as it were, is not a lifeless zone, but an active system with inputs and outputs of energy and matter; it is a dynamic layer in which various chemical, physical and biological activities are constantly going on. Through time, the soil adjusts itself to prevailing conditions of the environment and may change internally when such conditions either change or become subsumed in anthropogenic machinations [6]. Naturally, soil-borne plants require sixteen essential elements for growth, three of which come from the air (i.e., hydrogen, oxygen and carbon), and the other thirteen from the soil. And of the mineral nutrients (i.e., the macronutrients and micronutrients), the primary macronutrients (nitrogen, phosphorus, and potassium) are the most significant because plants use them in large amounts for growth and productivity. Acceptable ranges for these nutrients (nitrogen, phosphorus, and potassium) are 15,000, 2,000 and 10,000 milligram per kilogram respectively [7].

2. Water

Water is a valuable natural resource, a dominant part of the physical environment essential for sustenance of life. About 70% of the human body, by weight, consists of water and many of the body functions depend on it. Thus, humans can rarely survive for more than three days without water. Water is also essential for the sustenance of fisheries and aquatic resource, personal and environmental hygiene, irrigation, transport, power generation, sports and recreation. When water is contaminated, there is a reduction in available oxygen, an obligate requirement for a healthy living. Therefore, major indices of organic contamination in water are its biochemical oxygen demand (BOD) and dissolved oxygen (DO). The decomposition process of hydrocarbon contaminants utilizes oxygen; the more the organic matter in water, the greater the quantity of dissolved oxygen required. The minimum concentration of dissolved oxygen in freshwater necessary for aquatic fauna like fish to live is about five milligrams per liter. Table 1 contains some selected water quality indicators (dissolved oxygen and fecal coliform) for various rivers in the world.

3. Atmosphere

The atmosphere can be described as a natural laboratory in which various gases, solid particles and sunlight are involved in many chemical processes [3]. The natural state of the atmosphere, as the by-products of these processes, is however being altered by man through the injection of various contaminants into the air. Unfortunately, the air above us is not as unlimited as it seems, because the mixing layer in which the majority of the contaminants spread and are chemically converted varies in height between a few hundred meters and two kilometers. In effect, all we have is a limited film of air to live in and share between us [8]. The atmosphere, therefore, is of prime significance and consideration for a healthy environment [9]. Table 2 contains some selected air quality indicators for various cities in the world while Table 3 gives an insight into the trend in population growth and projections from 1965 to 2030. Combined indicators of ambient air quality with the numbers of people exposed to such levels of contaminants shows the severity of unhealthy urban air. This is worsened by the "unrelenting" population pressure in the urban areas which exposes a larger percentage of the populace to discharged effluents [3][10-12].

Table 1. Selected water quality indicators (Dissolved oxygen and fecal coliform) for various world cities and rivers

Country	*River, city*	**Dissolved oxygen (DO)**					**Fecal coliform**				
		Annual mean concentration (milligrams per litre)			*Average annual growth rate (percent)*		*Annual mean concentration (number/100 milliliter sample)*			*Average annual growth rate for series (percent)*	
		1979-82	1983-86	1987-90			1979-82	1983-86	1987-90		
Low-income											
Bangladesh	Karnaphuli	5.7	6.1	..	-1.1	(5)	3,133	..	..	..	(3)
Bangladesh	Meghna	6.5	7.0	..	2.6	(5)	519	700	..	-35.1	(5)
China	Pearl, HongKong	7.6	7.8	7.8	0.4	(11)	316	563	174	-14.4	(10)
China	Yangtze, Shanghai	8.3	8.3	8.2	-0.1	(11)	711	464	731	10.6	(11)
China	Yellow, Beijing	9.8	9.7	9.8	-0.1	(11)	51	1,337	1,539	9.8	(11)
India	Cauveri, Satyagalam	7.0	7.3	7.5	1.1	(9)	..	684	920	121.8	(9)
India	Godavari, Polavaram	7.2	7.2	6.9	0.0	(8)	248	2	4	-3.8	(7)
India	Sabarmati, Dharoi	9.4	9.1	8.9	0.0	(9)	659	222	220	15.4	(8)
India	Subarmarekha	8.0	7.9	7.5	-0.2	(9)	1,239	4,513	2,800	89.0	(9)
India	Tapti, Nepanagar	7.5	6.9	6.1	-2.3	(9)	..	110	130	-23.2	(4)
Parkistan	Indus, Kotri	6.2	6.8	7.1	1.8	(10)	436	463	446	-1.7	(10)
Parkistan	Ravi, d/s from Lahore	7.6	7.2	2.6	-13.6	(11)	105	121	78	-3.4	(11)
Parkistan	Blue Nile	7.2	6.7	7.0	-0.8	(12)	275	392	249	-6.6	(10)
Sundan		7.3	8.2	..	3.3	(7)	..	..	..	..	(0)
Middle-income											
Argentina	de la Plata, Buenos Aires	7.6	7.5	..	0.0	(8)	828	230	..	-23.1	(8)
Argentina	Parana' Corrientes	8.1	8.0	8.1	0.1	(10)	185	146	111	-6.6	(10)
Brazil	Guandu, Tomada d'Agua	8.1	7.8	7.7	-0.7	(11)	1,202	2,452	6	-47.0	(8)
Brazil	Paraiba, Aparecida	6.0	6.1	6.0	-0.4	(7)	13,950	9,800	6,075	-11.5	(7)
Chile	Maipo, Los Almendros	12.9	13.2	10.8	-1.4	(10)	871	705	775	5.3	(8)
Chile	Mapocho	11.8	12.1	10.0	-17	(10	2	2	5	8.0	(8)
Colombia	Cauca Juanchito	..	5.2	4.8	1.0	(5)	..	10,000	10,000	0.0	(4)
Ecuador	San Pedro	7.7	7.8	..	-0.1	(5)	80,000	30,603	..	-31.5	(4)
Fiji	Waimanu	7.6	7.8	8.0	0.5	(9)	600	1,605	..	8.1	(7)
Hungary	Danube	9.4	10.4	9.9	1.7	(10)	3,416	3,075	3,750	1.2	(10)

Korea	Han	..	10.5	10.4	-0.2	(8)	..	8	12	14.4	(8)
Malaysia	Kinta	6.8	7.5	8.3	2.9	(7)	..	..	..	..	(0)
Malaysia	Muda	7.3	7.2	6.3	-1.3	(8)	..	..	..	..	(0)
Mexico	Atoyac	3.5	1.7	0.3	-47.5	(9)	157,500	105,000	916,667	23.9	(7)
Mexico	Balsas	7.6	6.3	6.8	-1.9	(10)	1,558	26,333	130,000	95.4	(8)
Mexico	Blanco	5.0	3.4	4.1	-3.7	(9)	21,717	39500	12,150	1.8	(8)
Mexico	Colorado	7.9	8.7	8.2	1.4	(9)	277	58	37	-28.7	(7)
Mexico	Lerma	0.3	0.4	0.5	-18.6	(10)	192,250	165,000	67	5.7	(7)
Panama	Aguas Claras	7.9	8.2	..	0.4	(7)	290	143	..	-14.4	(6)
Panama	San Felix	8.2	8.0	..	-1.0	(7)	850	753	..	-6.2	(6)
Philippines	Cagayan	7.8	7.9	8.1	0.3	(11)	..	..	..	..	(3)
Portugal	Tejo, Santarem	8.9	8.6	8.4	-0.7	(9)	2,252	4,163	4,225	24.6	(9)
Thailand	Chao Phrya	6.3	6.3	..	0.2	(8)	1,093	1,745	..	47.7	(7)
Thailand	Prasak, Kaeng Khoi	6.6	7.7	..	8.0	(5)	596	2,724	..	9.9	(8)
Turkey	Porsuk, Agackoy	9.0	9.1	9.2	0.7	(9)	..	..	..	..	(1)
Turkey	Sekarya, Adetepe	9.2	8.7	8.9	-0.3	(8)	..	..	..	..	(1)
Uruguay	de la Plata, Colonia	..	..	..	7.1	(3)	..	453	93	54.6	(4)
Uruguay	Uruguay Bella Union	..	7.9	8.4	-1.4	(4)	..	200	1,100	66.9	(4)
High-income											
Australia	Murray, Mannum	7.1	8.2	8.6	2.4	(8)	33	103	80	15.8	(8)
Belgium	Escaut, Bleharies	5.7	6.2	5.9	1.1	(11)	76	579	867	40.8	(11)
Belgium	Meuse, Heer/Agimont	10.5	10.8	11.3	0.8	(11)	30	1,391	1,700	69.7	(11)
Japan	Kiso, Inuyama	10.8	10.5	10.8	-0.2	(10)	610	491	600	-2.0	(10)
Japan	Shinano, Zuiun Bridge	10.1	10.3	10.3	0.2	(10)	290	346	193	-3.0	(10)
Japan	Tone, Tone-Ozeki	10.0	9.9	10.4	0.5	(10)	521	593	618	3.7	(10)
Japan	Yodo, Hirakata Bridge	8.7	8.4	8.4	-0.4	(11)	72,000	70,333	..	9.3	(7)
Netherlands	Ijssel, (arm of Rhine)	8.7	7.9	..	-3.3	(6)	9,833	2,050	..	-43.0	(5)
Netherlands	Rhine, German frontier	8.5	8.0	..	-2.6	(6)	17,633	10,500	..	-11.8	(5)
United Kingdom	Thames	9.9	10.3	9.1	0.2	(8)	..	..	..	..	(0)
United States	Delaware, Trenton, NJ.	11.1	10.6	..	-2.5	(7)	74	197	..	-4.0	(7)
United States	Hudson, New York	9.8	12.1	..	4.2	(7)	941	792	..	-7.4	(7)
Untied States	Mississippi, Vickburg	8.4	8.3	..	-0.2	(7)	435	1,473	..	40.2	(7)

Note: d/s, downstream; u/s, upstream. a. Numbers in parentheses denote the number of years of observations. Data have been presented only when they are available for four or more years.

Table 2. Selected ambient air quality indicators for various cities around the world

Country group	City	Type of site	Sulphur dioxide [SO_2] Annuall mean concentration (micrograms / cubicmeter) 1979-82	1983-86	1987-90	Average annual growth rate for series (percent)	Suspended particulate matter [SPM] Annuall mean concentration (micrograms per cubic meter) 1979-82	1983-86	1987-90	Average annual growth rate for series (percent)
Low-income										
China	Beijing	CCC	77	119	107	3.5 (8)	475	500	413	-2.7 (8)
China	Guangzhou	CCR	59	107	95	7.7 (9)	146	209	234	7.4 (9)
China	Shanghai	CCC	66	59	69	2.5 (9)	224	214	253	2.5 (9)
China	Shanghai	CCR	57	84	104	9.2 (9)	240	230	290	3.8 (9)
China	Xian	CCR	116	111	100	1.4 (9)	401	485	580	6.7 (9)
Egypt,	Cairo	CCC	5	101	18	-11.0 (8)	…	…	…	… (0)
Ghana	Accra	SI	..	…	…	… (1)	119	109	144	2.4 (9)
India	Bombay	CCC	23	23	…	1.8 (4)	154	140	…	-1.1 (6)
India	Calcutta	CCC	71	54	…	4.6 (7)	410	393	…	-1.0 (13)
India	Delhi	CCC	42	86	…	12.0 (6)	460	460	…	-0.3 (7)
Indonesia	Jakarta	CCR	…	…	…	… (2)	254	271	…	2.2 (6)
Pakistan	Lahore		…	…	…	… (1)	745	…	496	-5.1 (6)
Middle-income										
Brazil	Sao Paulo	…	78	46	41	-7.5 (12)	134	98	..	-9.1
Chile	Santiago	CCC	69	85	..	2.5 (10)	..	..	..	..
Greece	Athens	CCC	57	34	..	-4.8 (9)	224	178	..	-6.0
Iran	Tehran	CCC	130	115	165	6.9 (14)	226	248	261	-2.4
Iran	Tehran	SR	114	61	64	-2.7 (14)	215	251	238	-1.3
Malaysia	Kuala Lumpur	SC	…	..	..	.. (2)	172	135	119	-3.9
Philippines	Davao	SI	…	..	..	.. (3)	163	205	..	-2.4
Philippines	Manila	SI	73	34	..	-12.0 (9)	90	..	..	0.8
Poland	Warsaw	CCC	42	35	23	-6.4 (13)	..	..	..	..
Portugal	Lisbon	CCR	32	21	27	-3.0 (10)	99	97	99	0.4
Thailand	Bangkok	SR	15	15	14	-1.7 (10)	136	163	105	-2.4
Venezuela	Caracas	CCC	32	27	21	-0.5 (13)	..	..	..	..
Yugoslavia	Zagreb	CCC	79	107	92	-4.3 (19)	114	127	135	-1.7

Yugoslavia	Zagreb	SR	33	66	47	-0.9 (19)	129	117	91	-2.6
High-income										
Australia	Melbourne	CCC	7	6	..	-14.3 (10)	71	58	..	-4.5 (9)
Australia	Sydney	SI	31	15	..	-7.3 (11)	76	58	..	-8.5 (11)
Belgium	Brussels	CCC	74	42	..	-11.5 (11)	24	22	..	-3.3 (11)
Belgium	Brussels	SR	60	37	..	-9.1 (15)	..	..	..	.. (2)
Canada	Hamilton	CCC	..	..	..	.. (3)	102	89	89	-2.8 (13)
Canada	Montreal	SR	27	20	..	0.7 (11)	58	39	35	-8.3 (13)
Canada	Toronto	CCC	..	11	14	4.0 (5)	60	60	61	-0.5 (8)
Canada	Vancouver	CCC	21	..	..	-7.0 (5)	70	50	42	-4.5 (14)
Denmark	Copenhagen	SI	33	27	..	-5.7 (7)	53	55	..	3.0 (9)
Finland	Helsinki	CCC	24	..	..	-2.8 (5)	72	79	81	2.0 (11)
Germany	Frankfurt	CCC	71	56	36	-7.2 (17)	24	39	42	0.5 (17)
Hong Kong	Hong Kong	..	..	25	46	47.3 (4)	..	99	132	14.9 (4)
Ireland	Dublin	CCI	40	34	32	-3.2 (12)	..	..	..	.. (0)
Israel	Tel Aviv	CCC	16	30	..	-7.1 (11)	..	..	..	.. (0)
Italy	Milan	CCC	160	90	..	-14.5 (7)	..	..	..	.. (0)
Italy	Milan	CCR	259	114	..	-11.4 (7)	..	..	..	.. (0)
Japan	Osaka	CCC	37	28	28	-8.4 (14)	51	41	42	-6.3 (13)
Japan	Tokyo	SR	42	30	20	-5.7 (17)	54	51	..	-4.5(14)
Netherlands	Amsterdam	CCC	33	24	..	-6.7 (15)	..	..	..	.. (0)
New Zealand	Auckland	CCR	8	3	..	-37.2(6)	..	..	..	.. (0)
New Zealand	Christchurch	SR	20	18	19	-3.5(12)	..	..	..	.. (0)
Spain	Madrid	CCC	105	54	36	-9.8(17)	..	..	..	.. (0)
United Kingdom	Glasgow	CCC	73	52	..	-8.8(8)	..	..	..	.. (0)
United Kingdom	London	CCC	66	44	..	-11.4(13)	..	..	..	.. (0)
United States	Birmingham	CCC	..	..	..	.. (3)	83	75	..	-3.0 (11)
United States	Chicago	CCI	..	..	..	.. (2)	121	99	..	-6.2 (10)
United States	Fairfield	SI	..	..	..	.. (3)	71	53	..	-5.6 (11)
United States	Houston	CCC	..	..	..	.. (3)	82	62	..	-7.3 (11)
United States	New York City	SR	38	31	..	-5.9(9)	49	46	..	-2.7 (11)

CCC is city centre commercial; CCI, city centre industrial; CCR, city center residential; SI, suburban industrial; SR, suburban residential; SC, suburban commercial. Numbers in parenthesis denote the number of years of observations. Data have been presented only when they are available for more than four or more years.

Table 3. Population (midyear) and average annual growth

Country group	Population (millions)							Average annual growth (percent)				
Low-and middle-income	1965	1973	1980	1990	1991	2000	2030	1965 to 1973	1973 to 1980	1980 to 1990	1990 to 2000	2000 to 2030
Low-income	2,403	2,923	3,383	4,146	4,226	4,981	7,441	2.5	2.1	2.0	1.9	1.4
Middle-income	1,776	2,168	2,501	3,058	3,117	3,670	5,430	2.5	2.0	2.0	1.8	1.3
Severely indebted	627	755	883	1,088	1,109	1,311	2,011	2.3	2.3	2.0	1.9	1.4
Sub-Saharan Africa	258	314	370	455	464	546	794	2.5	2.3	2.1	1.8	1.3
East Asia and the Pacific	245	302	366	495	510	668	1,346	2.7	2.8	3.1	3.0	2.4
South Asia	792	1,195	1,347	1,577	1,602	1,818	2,378	2.6	1.7	1.6	1.4	0.9
Europe	645	781	919	1,148	1,170	1,377	1,978	2.4	2.4	2.2	1.8	1.1
Middle East and North African	154	167	182	200	195	217	258	1.1	1.2	1.0	0.8	0.6
Middle East and North African	125	154	189	256	264	341	674	2.7	3.0	3.1	2.9	2.3
Latin America and the Caribbean	243	299	352	433	441	516	731	2.6	2.4	2.1	1.8	1.2
Other economies	252	275	294	321	323	345	..	1.1	1.0	0.9	0.7	..
High-income	679	726	766	816	821	859	919	1.0	0.8	0.6	0.5	0.2
OECD members	649	698	733	777	781	814	863	0.9	0.7	0.6	0.5	0.2
World	3,326	3,924	4,443	5,284	5,370	6,185	8,869	2.1	1.8	1.7	1.6	1.2

OECD is organization for Economic Cooperation and Development (Australia, Austria, Belgium, Caanada, Denmark, Finland, France, Germany, Greece, Iceland, Ireland, Italy, Japan, Luxembourg, Netherlnds, New Zealand, Norway, Portugal, Spain, Sweden, Switzerland, Turkey, United Kingdom, and United States of America).

Environmental Impact Assessment

One of the ways of knowing the state of environmental quality is through a standard procedure known as environmental impact assessment (EIA). Environmental impact assessment can be defined or described as a systematic investigation, examination, measurement, interpretation and prediction of the chemical, biological, physical, and socio-economic consequences of projects, policies and programs. The whole concept of EIA began in the United States of America in 1969 with the passage of the US Environmental Protection Act. EIA has since been adopted by a host of other countries like Canada in 1973, Australia in 1974, the Netherlands in 1981, the United Kingdom in 1988 and Nigeria in 1992 (with the enactment of Decree 88). On June 27 1985, the European community adopted a directive that "the assessment of certain public and private projects on the environment must be conducted before embarking on the project". Major components of EIA include baseline studies and post–impact assessment. A baseline study is the first series of surveys usually conducted before an area is polluted; the study provides baseline data from where the direction and extent of change and possible ecological interactions can be predicted or detected and quantified by direct reference or comparison to subsequent monitoring programs. The latter, post-impact assessment is the systematic auditing or monitoring of the environmental impact of a project that has already been executed.

CONCEPT OF HYDROCARBONS

Hydrocarbons are chemical compounds that contain hydrogen and carbon atoms. They are usually present in three physical states (gaseous, liquid and solid) at room temperature. Those containing one to four carbon atoms are gases, while those with five to sixteen are liquids and above sixteen are solids. In general, the tendency to exist in the solid state is greater with increasing number of carbon atoms. On the basis of molecular structure, hydrocarbons can be classed as acyclic (aliphatic) e.g. propane, aromatic e.g. benzene, or alicyclic e.g. cyclohexane. Incomplete oxidation of hydrocarbons results in the formation of soot (unburnt carbon), unburnt hydrocarbons and ammonia, in addition to particulates and oxides of sulfur, carbon and nitrogen. Specific reference doses for gasoline to kerosene range hydrocarbons are given in Table 4.

Table 4. Specific reference doses for gasoline to kerosene range hydrocarbons

Carbon range	Oral RfD (mg/kg –day)	Inhalation RfD (mg/kg – day)	Target organ
C6 – C8	5.0	5.3	Kidney
C8 – C10	0.1	6.3	Liver, haematological system
C10 – C12	0.1	0.3	Liver, haematological system
C12 – C16	0.1	0.3	Liver, haematological system

Primary Sources of Hydrocarbon Contaminants and Environmental/Health Impact

Crude Oils

Crude oils contain hundreds of different hydrocarbons, some nitrogen, sulfur and oxygen compounds, and metals such as nickel, vanadium and cadmium. So basically, crude oils contain the paraffin, naphthene, and aromatic hydrocarbons and the asphaltics (non hydrocarbon compounds). Much of the sulfur in crude oils stays with the heavier fractions such as fuel/automotive oil, residual oil, and asphalt, and less in the volatile fractions such as petrol/gasoline (motor spirit). Heavy fuel oil, which is used in oil fired boilers in industry, contains fairly high sulfur content (2 to 4 percent). Crude oils are classified as "sour" or "sweet", for sulfur-containing and sulfur-free crude oils respectively. The bulk differences between the crude oil in the reservoir and the one exposed to the environment, otherwise called the "weathered" or "degraded" crude oil, is that the latter decreases in the paraffin content, especially from the gasoline to kerosene range (C_6 to C_{15}), and increases in density and viscosity, nitrogen and sulfur contents as well as in vanadium and nickel contents. World's proven reserves for crude oil in 1950 stood at 30 billion tons. Today, they exceed 250 billion tons, notwithstanding a total world oil consumption of over 120 billion tons over the period of 50 years (see Figure 1).

Oil Spill

Crude oil impacts the environment through uncontrolled releases (spills), resulting from technical mishaps (of equipment and operations), and "unsighted fingers" of sabotage. When in contact with the soil, the oil binds to soil particulate matter. Partial coating of the soil surfaces by the hydrophobic hydrocarbons in crude oil reduces the water-holding capacity of the soil due to a significant reduction in the binding properties of clay. Hydrocarbons also tend to accumulate in the pores between soil particles, which results in reduced oxygen and water permeability through the soil. Several authors, including Ndes [13] and Osuji et al [14] have reported the presence of high molecular weight hydrocarbons, such as the acyclic isoprenoids and polycyclic aromatic hydrocarbons (PAHs), in surface and subsurface soils long after spillage. Spilled oils have been known to also induce fire outbreaks thereby causing more harm on the environment. Osuji and Ukale [6] assessed the state of the environment after an oil spill inferno. Contrary to the 'celebrated' opinion of some that the fires improvised bush fallowing for cropping, the site had witnessed severe impoverishment.

In the aquatic (water) media, oil spills have been known to have deleterious effects on fishes, crustaceans and other aquatic animals. Short term (acute) and long term (chronic) toxicity have been recorded among aquatic organisms. Data obtained from research on humans and animals indicate that acyclic and aliphatic hydrocarbons can cause systemic injury; high molecular weight polycyclic aromatic hydrocarbons (PAHs) are known to induce cancer in experimental animals. Laboratory experiments indicate that concentrations of between 1,000 and 10,000 parts per billion of oil give measurable effects on marine life. Between 1.7 and 8.8 million metric tons of oil enter the oceans annually, with the dominant figure being 3.2 metric tons per annum, representing 0.1 percent of the approximately 3 billion tons of oil produced in the world every year. Out of 6 million tones of petroleum being discharged into Nigeria's offshore waters annually, about 600,000 tonnes come from oil spillage. On March 18, 1967, when the oil-carrying tanker, Torrey Canyon, grounded off the coast of England, spilling 117,000 tons of Kuwait oil, the fish industry in the area suffered a decline of more than 60 percent. It was reported that French housewives boycotted the fish market; cost of compensation and rehabilitation was estimated at about 32.5 million US Dollars [15]. Thus oil spillage is a major health and environmental hazard. Osuji and Uwakwe [4] have reviewed the environmental impact major world oil spills from 1967 to 2005.

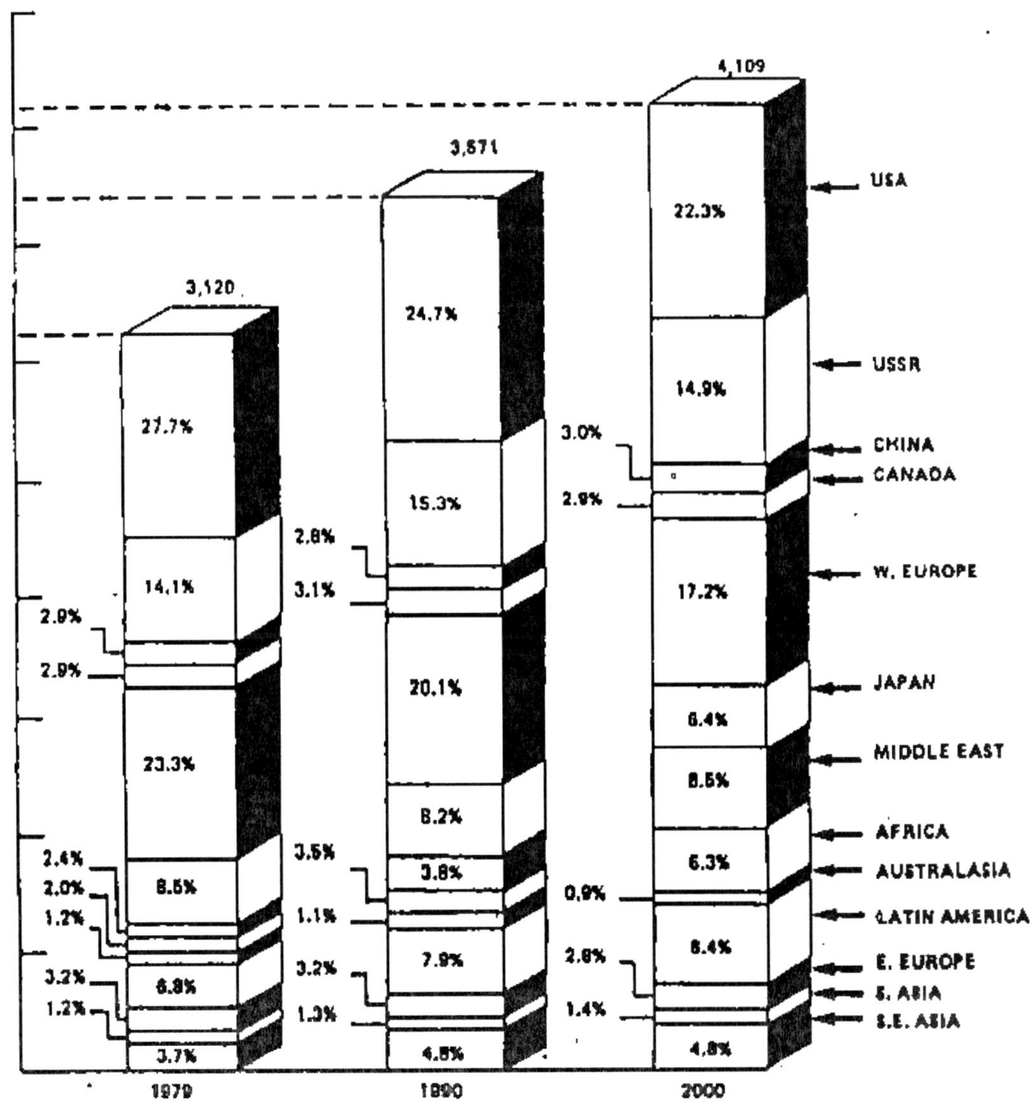

Figure 1. The rising trends of world crude oil production and consumption from 1979 to 2000.

Natural Gas

Natural gas, the gaseous phase of petroleum, is a naturally occurring mixture of hydrocarbon and non-hydrocarbon gases found in underground reservoirs either as a free gas or in association with crude oil. The hydrocarbon gases normally found in natural gas are low molecular weight paraffins (alkanes) namely methane (CH_4), which is the most predominant component, ethane (C_2H_6), propane (C_3H_8), butane (C_4H_{10}), pentane (C_5H_{12}), and small amounts of hexane (C_6H_{14}), heptane (C_7H_{16}), and traces of the heavier gases, up to nonane (C_9H_{20}). The non-hydrocarbon constituents are nitrogen (N_2), hydrogen sulphide (H_2S), carbon dioxide (CO_2), and water vapor (H_2O). Altogether, fifty-six (56) hydrocarbon compounds or more are involved in air pollution.

Gas Flaring

When crude oil is separated from its associated gas and produced water from the production or flow station, the gas is ignited at the well head and flared. The flares are massive and permanent. In 1993, it was reported that 4.8 percent of the world gross natural gas production was being flared. Organization of Petroleum Exporting Countries (OPEC) nations were flaring a total of 18 percent of their gross production, with Nigeria ranking highest with 76 percent of the nations gross natural gas production [3]. In 1989, 617 billion cubic feet of associated gas was flared, releasing 30 million tones of carbon monoxide. By the end of 1999, cumulative gas production in Nigeria amounted to approximately 27,795 barrels per cubic feet of which about 23,005 barrels per cubic feet was flared representing 82.8% of the net gas produced.

Gas flare emissions include suspended particulate matter (SPM), nitrogen oxide (NO_x) sulfur dioxide (SO_2), volatile organic compounds (VOC), carbon monoxide (CO), and carbon dioxide (CO_2). Gas flares cause noise and elevated temperatures in the immediate neighborhood of the flares; literature sources give flare temperatures as ranging from 1,300 to 1,400 degrees Celsius [13]. The heat associated with the emissions has been reported to kill vegetation around the flare stack, suppress the growth of plants, and reduce agricultural production. The noise drives the animals away and is unpleasant to humans who work and live near the flares. The 'perpetual' light produced by the flares is aesthetically unappealing and harsh to nocturnal animals. Snails, slow moving animals and insects (moths and termites) attracted to the flares are killed. The gaseous emittance from the flares mixes with the humid air around to produce contaminated 'rain' termed acid rain. Figure 2 contains a photographic plate of a typical gas flare in Niger Delta, Nigeria.

Estimates of environmental health risks in developing countries still rely on cautious extrapolations from population-response. Combining indicators of ambient air quality with the numbers of people exposed to such levels shows the severity of unhealthy urban air. An extrapolation from GEMS (Global Environment Monitoring System) data on airborne particulates for a sample of about fifty cities indicates that in the mid-1980s, about 1.3 billion people, mostly in developing nations like Nigeria, lived in towns or cities (of more than 250,000 population) which did not meet the World Health Organization (WHO) standards for suspended particulate matter (SPM). In the early 1980s cities such as Bangkok, Beijing, Calcutta, New Delhi, and Tehran exceeded on more than 200 days a year; the SPM concentrations that WHO guidelines indicate should not be exceeded more than seven days a year [11].

Gas flaring (Osuji and Avwiri, [3]).

A recent oil spill site.

Figure 2. Typical gas flaring and oil spillage sites in the hydrocarbon-rich Niger Delta region of southern Nigeria in the West African coast.

Produced Water

Produced water is the water derived in extracting crude oil from fluids emanating from a petroleum reservoir. The water contains several hundreds to perhaps one thousand or more parts per million of oil and grease. In addition, produced water may be high in total dissolved solids (TDS), and oxygen demanding organic materials. On a general note however, the characteristics of produced water vary from one formation to another, and are affected by factors such as the type of crude oil produced, total hydrocarbon concentration, salinity, temperature, oxygen content, and size of solid particles. In petroleum production, produced water represents the largest volume of polluted effluent. The effect of effluent discharged offshore may be diluted and dissipated by currents, reducing the hazards to the environment. Effluents discharged on land or into brackish waters may undergo comparatively little dilution, especially during the dry season, and may pose a hazard to aquatic organisms. Produced waters can have significantly high levels of trace metals, phenolic compounds and other toxic materials. Existing technologies for the on-site removal of oil and grease from produced water include gas flotation, parallel or tilted plate coalescers, loose or fibrous media filtration, gravity separation, and addition of chemicals to assist oil-water separation. Three groups of chemicals *viz* surface active agents (e.g. surfactants, demulsifiers), coagulating agents and polyelectrolytes are used, but unfortunately these chemicals are themselves potential

Lubricating Oils

Lubricating oils as well as sludge and bitumen slops from tank-cleaning operations are commonly discharged onto the landscape. Motor mechanics characteristically throw used lubricating oil carelessly into any available space. These wastes which have high biochemical oxygen demand (BOD) and chemical oxygen demand (COD) often find their way through runoff and the erosion of contaminated soil into water bodies, thus polluting them. Oil and grease used in lubricating machines and machine parts in industries also constitute high BOD and COD waste. Though these oil and grease are cleaned periodically with solvents, degreasers, chemicals and detergents, much fuel oil is still lost through leaks in the machines, pipes and others. The factory floors contain oil and grease which are also cleaned with degreasers and detergents. Spilled machine/engine oils and floor oil and grease also find their way through gutters or sewers into floodplains, streams, rivers, to lagoons and estuaries. In the Niger Delta area where oil pollution of land and water are most serious, the effects have been quite disturbing. The formation of a film of oil on water bodies, for instance, effectively prevents natural aeration, leading to death of the organisms trapped below by asphyxiation (oxygen deprivation).

Coal

Coal is a hydrocarbon from terrestrial plants. It contains carbon, hydrogen, oxygen and small amount of sulfur and nitrogen, as well as traces of mineral impurities. Typically, a kilogram of coal contains 150 to 200 cubic centimeters of methane (CH_4). Coal-fired stations are currently the main source of emissions from power because they make up more than half of the total thermal generating capacity and because of the high sulfur content of coal in many regions. Combustion efficiencies are often poor, and modern emission control technologies are not widely deployed; this gives rise to high emission rates of particulates and sulfur dioxide. In 1950, the world's proven reserve of coal stood at 450 billion tons. Today, its proven reserves exceed 570 billion tons. The trend in coal production from 1869 to 2000 is shown in Figure 3.

SECONDARY SOURCES OF HYDROCARBON CONTAMINANTS AND ENVIRONMENTAL / HEALTH IMPACT

Secondary hydrocarbon contaminants are those contaminants sourced from the major contaminants mentioned above.

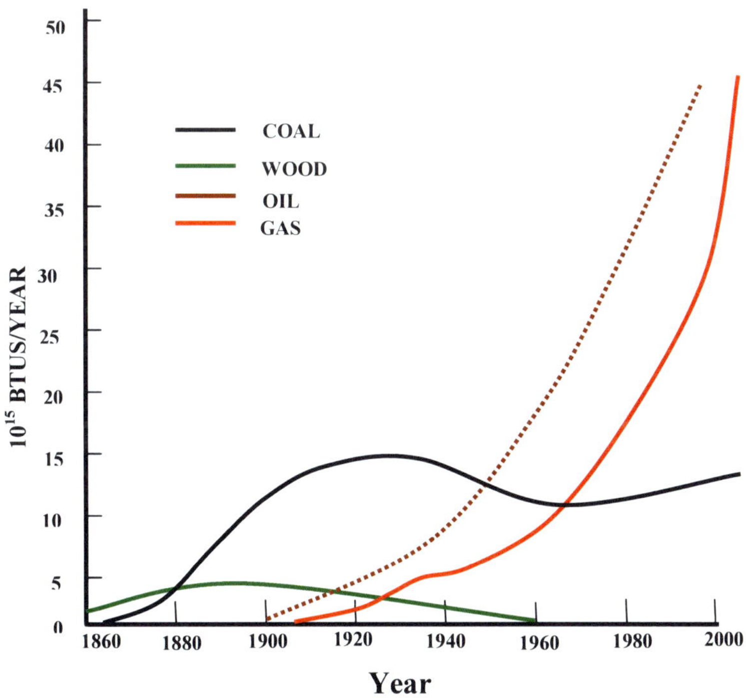

Figure 3. Trends in various hydrocarbon-source energy production and utilization from 1860 to 2000 (after Odiete [18]).

Oil Refinery

The crude oil refinery sector is another point source of hydrocarbon discharge. Effluents from refineries include benzene, toluene, ethyl benzene, and xylene, commonly called BTEX hydrocarbons. Others are polycyclic aromatic hydrocarbons, combustion gases, oil and grease, phenol, cyanide, sulphides, surfactants, suspended solids, heavy metals, and so on.

Quite a number of serious cases of pollution of well-water and streams with petroleum refinery effluents have been reported around oil refineries. Refinery effluents have also been found to affect the physiology and genetic machinery of plants [16].

Petrochemicals / Petrochemical Plants

Petrochemicals are chemicals produced mainly from crude oil and natural gas. They include benzene, toluene, ethylene, propylene, and vinyl chloride.

i) *Benzene*

Benzene is widely used as a solvent and as an intermediate in the production of aviation gasoline and other chemicals. However, exposure to 750 parts per million (750 milliliters per litre) of benzene for 30 minutes can be fatal, while exposure to concentrations larger than 300 parts per million (300 milliliters per liter) can cause nausea, drowsiness, headache and coma. Acute toxic effects of benzene is the depression of the central nervous system and chronic exposure can result in very serious toxic effects, the most significant being an insidious and unpredictable injury to the bone marrow [2][17].

ii) *Toluene (methyl benzene)*

Toluene (another name for methyl benzene) is a petrochemical solvent. Exposure to 800 parts per million (800 milliliters per liter) can lead severe fatigue and ataxia; 10,000 parts per million (10,000 milliliters per liter) can produce rapid loss of consciousness. Like benzene, toluene is also a central nervous depressant.

iii) *Vinyl chloride*

Vinyl chloride (with the chemical formula $CH_2 = CHCl$) is formed when ethylene is heated with chlorine at 350 to 450 degrees Celsius. It is used for making polyvinyl chloride (PVC), a plastic used for many purposes such as wrapping food, water pipes and bottles for beverages. It has also been used as a propellant for hair sprays and insecticides. However, minute amounts of vinyl chloride contaminate the finished products and pass to food and drinks. Vinyl chloride causes angiosarcoma, cancer in the blood vessels of the liver. Polyvinyl chloride is also carcinogenic and causes contact dermatitis.

Flue Gases / Vehicle Emissions

It has long been known and well documented that motor cars, lorries, buses and other automobiles and industrial machines emit contaminable gases [10]. These automobile and industrial machines run mostly on gasoline (petrol), diesel, and other crude oil derivatives. Through combustion, the engines release tones of exhaust (flue) gases, thus contaminating the atmosphere. The exhaust from diesel-burning vehicles contains oxides of nitrogen. Such vehicles also emit black smoke if the engine is poorly maintained. Petrol engines give out a mixture of hydrocarbons, carbon monoxide, and oxides of nitrogen and lead; diesel exhausts emit relatively high amounts of benzo-(a)-pyrene, a carcinogenic polycyclic aromatic hydrocarbon. Osuji and Avwiri [3] observed that gas-emitting generators used as standby electricity sources, "standby" for areas lucky to have epileptic power supply, have become a near–constant means of supply in some developing countries.

Three factors make contamination from vehicles more serious in the less industrialized countries than the industrialized. First, many vehicles are in poor condition, and lower-quality fuels are used. Second, motor vehicles are concentrated in a few large cities. In Mexico and Thailand, for example, about half of the vehicle fleet operates in the capital city, and in Brazil a quarter of the fleet operates in São Paulo. A larger percentage of the population moves and lives in the open air and thus more exposed to automotive contaminants. In certain circumstances, complete photochemical reactions can occur between the components of exhaust gases, giving rise to toxic compounds [1]. Where air is stagnant, exhaust gas from cars and generators will accumulate, leading to eye irritation, plant damage, and even fatalities [18]. The recommended threshold limit values for these gases are contained in Table 5.

Table 5. Threshold limit values (TLV) of some primary and secondary hydrocarbon contaminants

Compound	TLV (parts per million)	
	TWA[1]	STEL[2]
Benzene	0.5	2.5
Carbon monoxide	25	NA
Carbon tetrachloride	5	10
Chloroform	10	NA
Nitrogen dioxide	3	5
Ozone	0.05 to 0.1	0.1
Sulphur dioxide	2	5
Tetrachloroethane	25	100
Toluene	50	NA
1,1,1-Trichloroethane	350	450
Trichloroethylene	50	100

[1] TLV-TWA is the concentration for a normal 8-hour workday or 40-hour workweek to which workers may be repeatedly exposed without adverse effects.

[2] TLV-STEL is the maximum concentration that should not be exceeded at any time during a 15-minute exposure period.

NA = none assigned.

i) Carbon Monoxide (CO)

Carbon monoxide (CO) is a colorless, tasteless, odorless, and non-irritating gas. The average concentration of CO in the atmosphere is about 0.1 parts per million. During the hours of peak traffic flow and traffic congestion in the heat of the day, the carbon monoxide (CO) content of the air in urban and industrial rises to at least 100 parts per million. This decrease is equivalent to the loss of at least half a liter of blood. Incidentally along the world's busiest streets the CO level exceeds 400 parts per million at peak traffic hours (in dramatic contrast to the said average of about 0.1parts per million in the normal atmosphere) [3]. The affinity of hemoglobin for carbon monoxide is 200 times that for oxygen. An exposure of eight or more hours to levels of 10 parts per million of carbon monoxide will raise the level in non-smokers to over 2 percent. If air having such high CO concentration is breathed in for several hours, the oxygen–carrying capacity of the human body is decreased by nearly 20 percent; this is because CO forms a stable blood carboxyhemoglobin which reduces the oxygen carrying capacity of the blood leading to blood poisoning and death [19]. Concentration for a normal 8-hour workday to which humans may be exposed to CO without adverse effects is put at 25 parts per million (Table 5).

ii) Sulfur Dioxide (SO_2)

Sulfur dioxide (SO_2) is a colorless, irritant gas, generated primary by the burning (combustion) of sulphur containing fuels. This gas is a major contributor to acid rain and smog, accentuates disease symptoms in man, and causes death in patients with lung cancer and bronchitis. The signs and symptoms of intoxication include irritation of the eyes, nose, and throat [20][21]. Concentration of sulfur dioxide for a normal 8-hour workday to which humans be repeatedly exposed without adverse effects is 2 parts per million. Maximum concentration that should not be exceeded at any time during a 15-minute exposure period is 5 parts per million [20] (cf: Table 4). During or shortly after the four day London smog of December 1952 nearly 4000 deaths occurred. Peak daily concentration of contaminants during the episode reached 4000 microgram per cubic meter (1.34 parts per million) of sulfur dioxide and 6000 microgram per cubic meter (4.46 milligram per cubic meter) of smoke. The killer smog began on Thursday, December 4, 1952 as a high-pressure air mass and created a subsidence temperature inversion over Southern England. A white fog formed in the London area. As sulfur dioxide and suspended particulate levels built up, because of the extensive use of coal as fuel, the fog became black. The build-up of the pollutants combined with the fog resulted in zero visibility. And by Saturday, pollutants were sufficiently concentrated to cause deaths. In December 1962, a similar high level of sulfur dioxide occurred, the smoke levels were however markedly lower, so also were death tolls much lower than in 1952. It was generally presumed that the lower smoke levels were responsible for the low mortality. Similar episodes of deaths occurred in other smog incidences in Donora, Pennsylvania in October 1948 and New York in 1966.

iii) Oxides of Nitrogen

The oxides of nitrogen, nitric oxide (NO) and nitrogen dioxide (NO_2) can cause adverse health effects and photochemical smog. The most common, nitrogen oxide, a brownish irritant gas, forms nitric acid in the atmosphere and contributes to the formation of acid rain. Nitrogen dioxide has a stinging and suffocating odour and causes nasal and eye irritation at concentrations of 12 parts per million; pulmonary discomfort at 50 parts per million, bronchiolitis fibrosa (inflammation of the bronchioles) at 150 to 200 parts per million and bronchopneumonia at 500 parts per million. This gas not only enhances susceptibility to respiratory infections but it may lead to increased resistance to the flow of air through the respiratory tract, that is, it causes bronchoconstriction (increased respiratory resistance) and increases sensitivity to it in asthmatics and those with bronchitis. Chronic exposure to nitrogen dioxide levels of 10 to 40 ppm cause lung damage leading to fibrosis, and reduced breathing capacity. The effect of the oxides of nitrogen on vegetation is mainly through acid rain. Prolonged exposure of nitrogen dioxide concentrations of 470 to 1880 microgram per cubic meter (an equivalent of 0.25 to 1.0 parts per million) may suppress the growth of some plants like tomatoes and oranges [2].

Leaded Gasoline

The use of lead in gasoline has a long history. In 1922, auto manufacturers realized that adding lead to gasoline could boost its octane rating and produce more power. Although lead in gasoline represents only 2.2 % of the total global lead use, leaded gasoline remains by far the single largest source of lead exposure in urban areas. Approximately 90 percent of all lead emissions into the atmosphere are due to the use of leaded gasoline. In developing countries, transport fuels account for more than 55 % of the total oil consumption, which has grown by 50 % since 1980, while vehicles account for up to 95 percent of the lead emitted. Populations in at least 100 countries are still exposed to air polluted with lead from gasoline

Concerns over the environmental health effects of lead in gasoline surfaced in 1924, when, in the experimental laboratories of the Standard Oil Company, 5 out of 49 workers died and 35 experienced severe neurological symptoms from organic lead poisoning. Thereafter, the state of New York, the city of Philadelphia, and some other municipalities in the United States of America briefly banned the sale of leaded gasoline. However, immediately the furor subsided, the use of lead in gasoline resumed. The amount of lead additives increased quickly, rising to 375,000 metric tons annually by early 1970s. By the same 1970, however, concerns for tailpipe emissions led to the introduction of catalytic converters in the United States and Canada. Because leaded gasoline is incompatible with catalytic converters, cars with converters required unleaded gasoline. In 1985, the United States Environmental Protection Agency (U.S. EPA) decided to accelerate its gradual phase out of leaded gasoline by slashing the use of lead in gasoline. The public health benefits of the reductions (between 1976 and 1990) led to a dramatic decline of average blood lead levels in the U.S. population from 14.5 to 2.8 microgams per deciliter, parallel in the phase-out of leaded gasoline. By the end 1996, 14 countries had completely phased out the use of leaded gasoline. In many developing countries of Africa and Southeast Asia, unleaded gasoline is scarce, and the maximum allowed lead content of gasoline may reach or exceed 0.8 grams per liter; in Nigeria, for instance, the maximum allowed lead content of gasoline is 0.7 grams per liter.

Although fuel consumption in these countries is considerably lower than the rest of the world, lead emissions represent a serious health hazard because of increasing use of motor vehicles. Osuji and Avwiri [3] exemplified the rate of gasoline consumption with vehicle growth in these developing countries, using Nigeria as case reference Even in Latin America, increased gasoline consumption associated with urban growth and car ownership is nevertheless forcing large increases in the total amount of lead emissions. In most European countries, roughly one half of the cars use unleaded gasoline, while the other half uses gasoline containing 0.15 grams per liter.

The health toll of lead exposure is particularly high among poor populations of developed and developing countries alike, both because exposures are typically higher and because the populations may be more susceptible. In urban areas, for instance, the poor may live near major roadways where exposure to vehicle emissions is high. Besides, the poor and their children are more likely to walk on the roads than ride, thereby getting more exposed to lead emissions from vehicles. They also tend to live in older houses, where the risk of lead-based paint is greater. In addition, lead is believed to be absorbed from the stomach more completely when the stomach is empty and when the diet lacks essential trace elements, such as iron, calcium and zinc. Besides the immediate health risk posed through inhalation, vehicular lead emissions can also accumulate in soil, contaminate drinking water, and enter the food chain. Nriagu [22] has traced the historical evolution of lead production and emission from 1850 to 1990 (Figure 4)

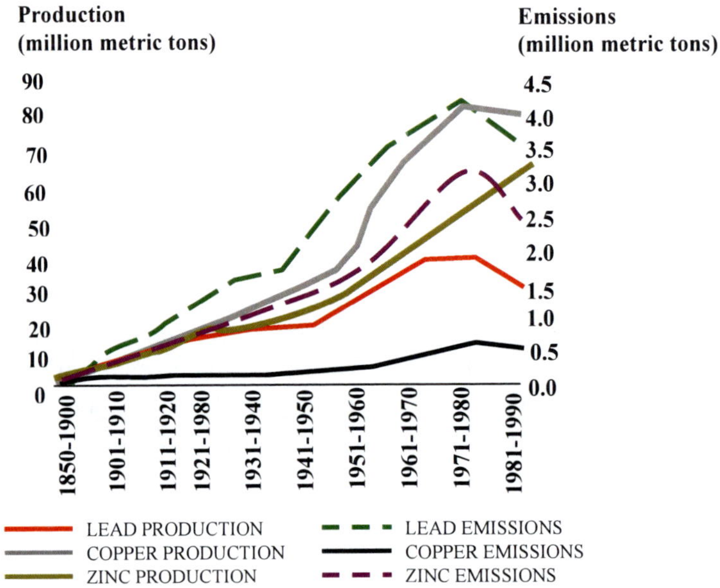

Figure 4. Heavy metal production and emission from 1850 to 1990 (after Nriagu [22]).

Trace Heavy Metals

Generally trace metals may be defined as metals occurring at 1000mg/kg or less in the earth's crust [1], but such metals may be classified as 'heavy' or 'light' with respect to density. Trace 'heavy' metals have densities greater than 5 grams per cubic centimeter (e.g., vanadium) whereas 'light' metals have densities less than 5 grams per cubic centimeter (e.g., beryllium). Some of the oldest diseases of humans can be traced to heavy metal poisoning. Many authors have worked on heavy metal contaminations and health implications. For instance, Nriagu [22] did an elaborate review of heavy metals and toxicity. Osuji and Onojake [23] investigated the ecological impact of some trace heavy metals associated with crude oil. However, even with the recognition of the hazards of heavy metals, the incidence of intoxication remains significant and the need for preventive strategies and effective therapy remains high [24][25]. A comprehensive data toxicity of al the trace metals from petroleum has been collected by Smith et al [26].

Biological Decay of Sewage and Refuse

Sewage and refuse are composed of methane (CH_4) hydrocarbon, odors, and hydrogen sulphide; effluents from these contaminants take the form of fumes and smoke. Garbage provides maternity wards and free lunch counters for flies and rats. Rubbish and trash provide the housing. High accumulations of such solid wastes are unsightly, deface the environment and produce unpleasant odors. Some of these substances find their way into water (streams, lagoons and rivers) with serious effects on human health [2].

Odors

Odors are gases, either single compounds or a mixture of complex compounds which stimulate olfactory sense organs (in the nostrils). The sensation may be sweet as in the case of odors from a perfumery or foul as in that from a sewage work or pungent as in ammonia gas. Odors are emitted petrol (gas) stations where motor spirit (gasoline) is sold from pumps ordinarily discharged or poured into containers. This is more common with poor quality gasolines (gasolines that were that were not desulfurized or deodorized); this is a common experience in and around petrol stations in Nigerian, where maximum sulfur content for motor gasolines should be 0.2 weight percent (Table 8). Also, .after a long periods of gas flare, water bodies around flare pits do emit unhealthy odors because of the depletion of oxygen induced by the elevated flare temperatures; this is usually the case even when the gas flared is sweet, that is, without the odorous hydrogen sulfide. Odors containing sulfur and nitrogen compounds (e.g., the rotten egg smell of hydrogen sulfide, the foul smell of methyl mercaptan and medicinal smell of phenols and carboxylic acid) are the most unpleasant or objectionable.

Odors are generally a nuisance and cause unhealthy conditions. Offensive odors affect the health and well-being of people by eliciting unpleasant sensations and objection to the smell. According to the US Board of Toxicology and Environmental Health Hazards, unfavorable responses to odours include nausea, vomiting, headaches; induction of shallow breathing and coughing, upsetting sleep, stomach and appetite; irritation of eyes, nose and throat; reduction in the enjoyment of home and the external environment; disturbance, annoyance, discomfort, depression, and decrease in heart rate.

Biomass / Wood Fuels

About half of the world's people cook all or some of their meals with biomass fuels. Until the twentieth century such fuels, mainly firewood, provided most of the world's energy. Today biomass in all its forms (wood, agricultural and forestry residues, and dung) meets about 14 percent of the world's energy demands. More than 80 percent of the world's energy demands. More than 80 percent is consumed I developing countries, where it still accounts for 35 percent of energy supplies. Biomass is used not only for cooking, but also in small-scale service industries, agricultural processing, and the manufacture of bricks, tiles, cement, and fertilizers. Such uses can be substantial, especially in and around towns and cities.

The use of biomass fuels for cooking gives rise to high indoor pollution. According to a 1992 world development report, indoor air pollution from burning wood, charcoal, and dung endangers the health of 400 to 700 million people in the world. The use of biomass fuel is also a source of ecological damage; the use of dung and crop residues depletes soil productivity, and deforestation often causes soil erosion. Finally, the poor thermal efficiency of biomass increases carbon dioxide and particulate emissions [11].

Spent Oil-based Drilling Fluid and Oily Drill Cuttings

Drilling fluids (also called drilling mud) are used for effective oil-well drilling. The ones formulated with diesel oil, and oily drill cuttings from th e bore hole usually contaminate the soil and aquatic environment when discharged in the course of drilling for oil and gas. For instance, an estimated 7 million cubic meter of drilling waste was produced in Nigeria between the beginning of oil prospecting in 1937 and 1995, an average bore hole giving rise to 2,500 cubic meters. When discharged under water, the fluids disperse sparingly and blankets parts of the sea beds with high concentrations of oily organic materials. Thus, the discharge increases turbidity of the water and sedimentation of solids, while the high organic matter content leads to increased bacterial activity and causes lower or even zero concentration in the waters (asphyxiation). Osuji and Uwakwe [4] reported the death of fishes as a result of the oxygen depletion of the water contaminated by spent oil-based drilling fluids. Similarly Otaigbe et al. [27] reported that a concentration of 621 parts per million (milligram per liter can kill 50 percent population of brackish water crustaceans within 96 hours.

The Greenhouse Gases and Global Warming

The moon is very cold because it is a long way from the sun, yet the earth, which is the same distance, is warm. This is largely because of the carbon dioxide (CO_2) in the earth's atmosphere. Heat from the sun passes through the carbon dioxide to the surface of the earth. But, when the earth radiates this heat back out into atmosphere, it is absorbed by the carbon dioxide which then radiates in all directions (some going into space and some going back to the earth). As a result the earth is warmer by 30 to 40 degrees Celsius. This process is called "Greenhouse Effect" because it works just like a greenhouse in cold climates. The glass panes of the greenhouse allow the short waves of the sun's heat to pass into the building, but the long-wave heat radiating from the warmed surfaces inside cannot pass out again. Thus the inside of the building becomes a lot warmer than the outside, so that plants can be grown in cold weather. Other greenhouse gases include water vapor, methane, nitrous oxide, and ozone. It follows therefore that if these gases are increasing in the atmosphere, then more heat will be trapped inside the biosphere and the earth's average temperatures will increase. As a result of the emission of these gases, global temperatures have increased by about 45 degrees Celsius since 1900, and mostly since 1940. this may seem insignificant, but it is enough to have caused the shrinking of Alpine and Polar ice caps, causing the a sea level rise of about 1.5 millimeter per year over the same period. These processes are complex, but they appear to be accelerating. Contributions to global emissions are oil (39 percent), solid fuels (42 percent), natural gas (11 percent), gas flaring and cement manufacturing (4 percent), and incineration (4 percent). Clearly these contributors are largely hydrocarbon-based. Table 7 gives an overview of global carbon dioxide emission.

Other Minor Sources of Hydrocarbon Contamination

Natural Seeps and Organic Matter

Although the majority of hydrocarbons in the soil environment are anthropogenic (man made) in nature, there are some natural sources of these materials. Included in this category are seeps from oil deposits and degradation of organic matter. There is a body of evidence to show that certain organisms, notably higher plants, are capable of synthesizing hydrocarbons and these too could find their way into the soil environment. These latter sources are however fairly minor and are unlikely to result in significant soil contamination.

Shale Oil and Condensates

Shale oil is a hydrocarbon liquid recovered only by strongly heating oil shales. Shale oil retorting plants provide another source of hydrocarbon contamination. Condensates are light liquid hydrocarbon mixtures are recovered from gas fields. These two oils are not yet in common usage, but are potential environmental hazards.

Conclusion and Recommendations

The growth in world demand for energy is the inevitable consequence of both an increasing population and growing expectation of an improved lifestyle. The more we seek to develop, the more we consume; and the more we are and consume, the more the production output, hence the greater the energy use, industrial wastes and resultant contamination. Judging from the complexity of hydrocarbon contaminants (crude oil, natural gases, lubricating oil, produced water, and so on), their chemical and physical attributes and economic implications, it is obvious that they can degrade their contact environment, impair human health and precipitate social disruptions as is the case in some oil bearing regions of the world. A preview of the levels to which humans can be exposed without adverse effects in comparison with the maximum levels that should not be exceeded, calls for a more expedient concern.

Two complementary approaches are recommendable for effective mitigation. The first should be the use of economic instruments and institutional reforms to encourage more efficient use of energy. Secondly, a technological development that can reduce the contaminating effects of conventional fuels or the development of less-polluting substitutes should be evolved. Developing countries should not compromise better quality fuels and automobiles in order to reduce vehicle emissions of the highly contaminable gases reviewed herein. Quality control and petroleum product specifications should be strictly and religiously adhered to; no motor gasoline entering through the borders of Nigeria, for instance, should contain more sulfur than the specified maximum of 0.2 weight percent, and more lead than 0.7 grams per liter.

Poverty alleviation schemes in developing countries should be strengthened where existent and introduced where non-existent. This would avoid the poverty-driven degradations of the environment that arise from lack of resources necessary for living within the minimum standard of human dignity and well-being. A careful husbandry of the scientific, political and socio-economic measures suggested herein will enhance the quality and stability of the environment.

REFERENCES

[1] Duffus, J. H. *Environmental toxicology*. Edward Arnold Publishers, London, 1994, pp. 23-55.

[2] Plaa, G. L. Introduction to toxicology: occupational and environmental. In: Katzung, B. G. (ed.), *Basic and clinical pharmacology*, McGraw-Hill, 2001, pp. 997-998.

[3] Osuji, L. C.; Avwiri, G. O. Flared gases and other pollutants associated with air quality in industrial areas of Nigeria: An overview. *Chem. Biodiv.* 2005, 2, 1277-1289.

[4] Osuji, L. C.; Uwakwe, A. A. Petroleum industry effluents and other oxygen demanding wastes in Niger Delta, Nigeria. *Chem. Biodiv*. 2006, 3, 705–717.

[5] Adesiyan, S. O. *Man and his environment*. University Press, Ibadan, 2005, pp. 68-82.

[6] Osuji, L. C.; Ukale, E. E. Post-oil-spill fires at Ugbomro (Niger Delta): A new vista in soil pollution studies. *Chem. Biodiv.* 2005, 2, 1079–1085.

[7] Manahan, S. E. *Environmental chemistry*. CRC Press, Inc. Florida, 1994, 811pp.

[8] Elsom, D. M. *Atmospheric pollution-a global problem*. Blackwell, Oxford, 1992, pp. 68-74.

[9] Folinsbee, L. J. Human health effects of air pollution. *Environ. Health Perspect.* 1992 100, 45-54.

[10] Faith, W. L.; Atkinson Jr, A. A. *Air pollution*. Wiley Interscience, New York, 1972 pp. 8–61.

[11] World Bank. *World development report*. Oxford University Press, 1992, pp. 134-169.

[12] Smith, K. Air pollution: assessing total exposure in developing countries. *Environ.* 1988, 30, 16.

[13] Ndes. *Niger Delta environmental survey phase -1 report*; ERML, Lagos, Nigeria, 1999, pp. 48–68.

[14] Osuji, L. C.; Inimfon, A. U.; Ogali, R. E. Attenuation of petroleum hydrocarbons by weathering: A case study. *Chem. Biodiv*. 2006, 3, 422-433.

[15] Gelder-Ottway, S. V.; Knight, M. A review of world oil spillages 1960-1975. In: Baker, J. M. (ed.), *Marine ecology and oil pollution*. Applied Science Publishers, Essex, England, 1976, pp. 483–520.

[16] Obute, G. C.; Osuji, L. C.; Kalio, C. Genotoxicity of petroleum refinery waste water in Nigeria. *Glob. Journ. Environ. Sci.* 2002, 3, 55-58.

[17] Yardley-Jones, A.; Anderson, D.; Parke, D. V. The toxicity of benzene and its metabolism and molecular pathology in human risk assessment. *Br. J. Indust. Med.* 1991, 48, 437.

[18] Odiete, W. O. *Environmental physiology of animals and pollution*, Diversified Resources Limited, Lagos, Nigeria, 1999, 261pp.

[19] Penney, D. G.; Howley, J. W. Is there a connection between carbon monoxide exposure and hypertension? *Environ. Health Perspect.* 1991, 95, 191-1996.

[20] Seinfeld, J. H. Urban air pollution: state of the science. *Science*. 1989, 243, 745-752.

[21] Forbes, V. E.; Forbes, T. L. *Ecotoxicology in theory and practice*. Chapman and Hall, London, 1994, pp. 104-118.

[22] Nriagu, J. O. History of global metal pollution. *Science*. 1996, 272, 223–224.

[23] Osuji, L. C.; Onojake, C. M. Field reconnaissance and estimation of petroleum hydrocarbon and heavy metal contents of soils affected by the Ebocha-8 oil spillage in Niger Delta, Nigeria. *Journ. Environ. Managt*. 2005, 20, 1–7.

[24] Friberg, L.; Nordberg, G. F.; Vouk, V. (Eds.), *Handbook on the toxicology of metals*. 2nd Edition, Elsevier, 1986, pp. 62-71.

[25] Kosnett, M. J. Heavy metal intoxication and chelation. In: Katzung, B. G. (ed.), *Basic and clinical pharmacology*, international edition, McGraw-Hill, 2001, pp. 999-1010.

[26] Smith, I. C.; Ferguson, T. L.; Carson, B. L. *Metals in new and used petroleum products and by-products: quantities and consequences*, Elservier, 1999, pp. 124–144.

[27] Otaigbe, J. O. E.; Osuji, L. C.; Azubike, A. N. Quantal response of *Paleomonetes africanus* in locally formulated drilling mud lubricants. *Toxicol. Environ. Chem.* 2006, 88.

In: Air Quality in the 21st Century ISBN 978-1-60456-793-9
Editors: G.C.Romano and A.G.Conti, pp. 287-310

Chapter 7

BIVARIATE STOCHASTIC VOLATILITY MODELS APPLIED TO MEXICO CITY OZONE POLLUTION DATA

***Jorge A. Achcar* and Henrique C. Zozolotto*†**
Faculdade de Medicina de Ribeirão Preto
Universidade de São Paulo
Ribeirão Preto – SP, Brasil
***Eliane R. Rodrigues*‡**
Instituto de Matemáticas – UNAM
Area de la Investigación Científica
Circuito Exterior, Ciudad Universitaria
México, D.F. 04510, México

Abstract

In this paper, we consider recently introduced bivariate stochastic volatility models commonly used to analyse financial time series, to study problems related to air pollution data. Such models are used here to estimate the volatility of weekly averaged ozone measurements taking into account two different sets of data provided by the monitoring network of Mexico City. A Bayesian analysis is developed using Markov Chain Monte Carlo (MCMC) methods to simulate samples from the joint posterior distributions and perform the estimates of interest.

1. Introduction

Air pollution is one of the main problems in several large cities throughout the world. Among the many existing pollutants, ozone is one that most affect these cities. In many of them, including Mexico City, environmental authorities have implemented measures aiming to reduce the level of pollutants in general. Such measures are very important because

*E-mail address: jorge@icmc.usp.br;
†E-mail address: henrique_ceretta@yahoo.com.br
‡E-mail address: eliane@math.unam.mx

when ozone concentration stays above a given threshold for a certain period of time, individuals exposed to the pollutant may experience serious health problems (see for example Wilson *et al.*, 1980; Loomis *et al.*, 1996; Bell *et al.*, 2004; Bell *et al.*, 2005; Bell *et al.*, 2007). In particular, it is well known that for ozone levels above 0.11 parts per million (0.11ppm) a very sensitive part of the population (such as the newborn and the elderly) experience a deterioration in their health. Therefore, being able to understand the long term behaviour of such pollutant is of great importance for policy makers. If there is a trend in the measurements that is decreasing with time and it is such that the variations are not so large, or otherwise if it is increasing or with large variations, then environmental authorities could have an idea if in the long run the preventive measures taken by them are producing a desirable result or not.

Several methods have been used in order to predict the violation of an air quality standard. Among them we may refer to Roberts (1979a, 1979b), Horowitz (1980) and Smith (1989) when the interest resides in applying extreme values theory to perform predictions. However, other techniques may also be used to study this type of problems. As examples we may quote, multivariate analysis (Guardani *et al.*, 2003), neural networks (Guardani *et al.*, 1999), Poisson models (Javits, 1980; Raftery, 1989; Leadbetter, 1991; Achcar *et al.*, 2007) and Markov chain models (Larsen *et al.*, 1990; Austin and Tran, 1999; Álvarez *et al.*, 2005).

Another possibility is to consider time series modeling the daily or the weekly averaged pollution data (see for example, Loomis *et al.*, 1996; Lanfredi and Macchiato, 1997; Zolghadri and Henry, 2004; Pan and Chen, 2007).

In this paper we use stochastic volatility (SV) models (see for instance Ghysels *et al.*, 1996; Kim *et al.*, 1998; Meyer and Yu, 2000) to study the behaviour of weekly averaged measurements of ozone provided by the monitoring network of Mexico City. The type of volatility models considered here have been extensively used to analyse financial time series (see for example Danielsson, 1994; Yu, 2002), as a powerful alternative for the usual existing ARCH (autoregressive conditional heteroscedastic) models introduced by Engle (1982) and the generalised autoregressive conditional heteroscedastic (GARCH) models introduced by Bollerslev (1986). The SV-type models have many advantages when they are used to analyse time series. The main reason being that they consider two processes to model the series: a process modelling the observations and another modelling the latent volatility.

A Bayesian inference approach using Markov Chain Monte Carlo (MCMC) methods (see for instance Gelfand and Smith, 1990; Smith and Roberts, 1993) has been performed to analyse the SV models considered here. The use of MCMC methods is necessary because we can have great difficulties when using standard classical inference approach. Some of the problems present are high dimensionality, likelihood function with no closed form and also a possible high computational cost.

This paper is organised as follows. In Section 2 the volatility models used here are introduced. Section 3 presents the Bayesian formulation of the problem. In Section 4 we establish the criterion for selecting the best model that fits the data used. An application to the data provided by the Mexico City monitoring network is given in Section 5. Finally, in Section 6 we conclude.

2. Bivariate Stochastic Volatility Models (BSV)

Different classes of multivariate stochastic volatility models are introduced in the literature (see for example Yu and Meyer, 2005). In the present paper we consider six bivariate models. In order to describe those models, we start by considering $N \geq 1$ a fixed integer number that will represent the size of the data set used in the analysis. Let $\mathbf{Z}(t) = (Z_1(t), Z_2(t))'$, $t = 1, 2, \ldots, N$ be the series recording the results of the same event performed in two different locations at the same time. (In here, for v a vector or a matrix we use v' to indicate the transpose of v.) In our case $\mathbf{Z}(t)$, $t = 1, 2, \ldots, N$ will represent the weekly averaged ozone measurements taken in two different regions of the Metropolitan Area of Mexico City. Consider a vector of latent variables $\mathbf{h}(t) = (h_1(t), h_2(t))$, $t = 1, 2, \ldots, N$ where $h_i(t)$ are defined by the following auto-regressive models AR(1)

$$\left\{ \begin{array}{rcl} h_1(1) & = & \mu_1 + \eta_1(1) \\ h_2(1) & = & \mu_2 + \eta_2(1) \end{array} \right. \tag{1}$$

and for $t = 2, 3, \ldots, N$

$$\left\{ \begin{array}{rcl} h_1(t) & = & \mu_1 + \phi_{11} \left[h_1(t-1) - \mu_1\right] + \eta_1(t) \\ h_2(t) & = & \mu_2 + \phi_{22} \left[h_2(t-1) - \mu_2\right] + \eta_2(t) \end{array} \right. \tag{2}$$

where $0 < \phi_{11}, \phi_{22} < 1$ and $\boldsymbol{\eta}(t) = (\eta_1(t), \eta_2(t))$ has a bivariate Normal distribution with mean vector $\mathbf{0} = (0, 0)'$ and variance-covariance matrix given by the 2×2 diagonal matrix $\text{diag}\left(\sigma^2_{\eta_1}, \sigma^2_{\eta_2}\right)$.

Consider $\mathbf{H}(t)$, $t = 1, 2, \ldots, N$ a 2×2 diagonal matrix with diagonal $e^{h_1(t)/2}$ and $e^{h_2(t)/2}$, i.e., $\mathbf{H}(t) = \text{diag}\left(e^{h_1(t)/2}, e^{h_2(t)/2}\right)$. Let $\mathbf{Y}(t) = (Y_1(t), Y_2(t))'$, $t = 1, 2, \ldots, N$ be modelled by

$$\mathbf{Y}(t) = \mathbf{H}(t)\boldsymbol{\epsilon}(t)$$

where $\boldsymbol{\epsilon}(t) = (\epsilon_1(t), \epsilon_2(t))'$ is the vector of error components having a bivariate Normal distribution with mean vector $\mathbf{0}$ and variance-covariance matrix Σ_ϵ given by

$$\Sigma_\epsilon = \begin{pmatrix} 1 & \rho_\epsilon \\ \rho_\epsilon & 1 \end{pmatrix}$$

with $\rho_\epsilon \geq 0$. Hence, $\mathbf{Y}(t)$, $t = 1, 2, \ldots, N$ is such that $Y_1(t) = e^{h_1(t)/2}\epsilon_1(t)$ and $Y_2(t) = e^{h_2(t)/2}\epsilon_2(t)$. (Usually, $Y_1(t)$ and $Y_2(t)$, $t = 1, 2 \ldots, N$ are the logarithms of the returns of $Z_1(t)$ and $Z_2(t)$ centred around their averages.)

Remark. By definition, we observe that $\text{E}[\mathbf{Y}(t)] = \mathbf{0}$ and the variance-covariance matrix for $\mathbf{Y}(t)$ is given by,

$$\begin{array}{rcl} \Sigma_Y & = & \text{Var}\left(\mathbf{Y}(t)\right) = \mathbf{H}'(t)\Sigma_\epsilon\mathbf{H}(t) \\ & = & \begin{pmatrix} e^{h_1(t)} & \rho_\epsilon e^{h_1(t)/2} e^{h_2(t)/2} \\ \rho_\epsilon e^{h_1(t)/2} e^{h_2(t)/2} & e^{h_2(t)} \end{pmatrix}, \end{array} \tag{3}$$

for $t = 1, 2, \ldots, N$. Furthermore, $\mathbf{Y}(t) = (Y_1(t), Y_2(t))'$, $t = 1, 2, \ldots, N$ has a bivariate Normal distribution with density

$$f(Y_1(t), Y_2(t) \mid h_1(t), h_2(t)) = \frac{1}{2\pi\sqrt{(1-\rho_\epsilon^2)\, e^{h_1(t)+h_2(t)}}} \exp\left\{-\frac{1}{2(1-\rho_\epsilon^2)}\left[\frac{Y_1^2(t)}{e^{h_1(t)}} + \frac{Y_2^2(t)}{e^{h_2(t)}} - \frac{2\rho_\epsilon Y_1(t)Y_2(t)}{e^{h_1(t)/2}e^{h_2(t)/2}}\right]\right\}$$

Also observe from (1), (2) and the definition of $\boldsymbol{\eta}(t)$, that the latent variables $\mathbf{h}(t) = (h_1(t), h_2(t))$ have Normal distributions. In fact, the density function of $h_i(1)$ is a Normal distribution $\mathrm{N}(\mu_i, \sigma_{\eta i}^2)$ and given $h_i(t-1)$ the variable $h_i(t)$ has density function $\mathrm{N}(\mu_i + \phi_{ii}\left[h_i(t-1) - \mu_i\right], \sigma_{\eta i}^2)$ for $i = 1, 2$ and $t = 2, \ldots, N$.

The different models considered in this paper are described as follows

2.1. Model I

In this model we assume that the error coordinates $\epsilon_i(t)$, $i = 1, 2$ are independent, i.e., $\rho_\epsilon = 0$. Hence, $\mathbf{Y}(t)$, $t = 1, 2, \ldots, N$ will have Normal distribution with mean vector $\mathbf{0}$ and with variance-covariance matrix a diagonal matrix given by $\Sigma_{\epsilon,0} = \mathrm{diag}(e^{h_1(t)}, e^{h_2(t)})$.

2.2. Model II

In this version of the model the covariance ρ_ϵ between $\epsilon_1(t)$ and $\epsilon_2(t)$ is considered to be an unknown quantity that ought to be estimated. Hence, $\mathbf{Y}(t)$, $t = 1, 2, \ldots, N$ will have Normal distribution with mean vector $\mathbf{0}$ and with variance-covariance matrix given by (3).

2.3. Model III

In a third version of the model we keep the assumption of Model II, except for the way we define the latent variable $h_2(t)$, $t = 1, 2, \ldots, N$. In this version, we assume the presence of the Granger causality when modelling $h_2(t)$, i.e., the latent variable $h_2(t)$ is now given by

$$h_2(t) = \mu_2 + \phi_{21}\left[h_1(t-1) - \mu_1\right] + \phi_{22}\left[h_2(t-1) - \mu_2\right] + \eta_2(t) \tag{4}$$

where $t = 2, 3, \ldots, N$ and $0 < \phi_{21}, \phi_{22} < 1$.

Remark. Note that from (4), that if $\phi_{21} \neq 0$, then the volatility of the second return is considered with the Granger causality induced by the volatility of the first return.

2.4. Model IV

In this version of the model we take the assumptions of Model II except that we consider the additional hypothesis on the correlation between $\epsilon_1(t)$ and $\epsilon_2(t)$. In this way, we assume that $\boldsymbol{\epsilon}(t) = (\epsilon_1(t), \epsilon_2(t))'$ has a bivariate Normal distribution with mean vector $\mathbf{0}$ and variance-covariance matrix $\Sigma_\epsilon(t)$ a 2×2 matrix given, for $t = 1, 2, \ldots, N$, by

$$\Sigma_\epsilon(t) = \begin{pmatrix} 1 & \rho_\epsilon(t) \\ \rho_\epsilon(t) & 1 \end{pmatrix},$$

where $\rho_\epsilon(t) = \left(e^{q(t)} - 1\right) / \left(e^{q(t)} + 1\right)$, $q(1) = \psi_0 + \sigma_\rho v(t)$ and for $t = 1, 2, \ldots, N$, we have $q(t) = \psi_0 + \psi_1 \left[q(t-1) - \psi_0\right] + \sigma_\rho v(t)$, with $v(t)$, $t = 1, 2, \ldots, N$, independent and identically distributed quantities with common distribution a Normal distribution $N(0, 1)$ (see for example Yu and Meyer, 2005). We further assume that $q(1)$ has a Normal distribution $N\left(\psi_0, \sigma_\rho^2\right)$ and for $t = 2, 3, \ldots, N$, we also assume that given $q(t-1)$, the quantity $q(t)$ has a Normal distribution $N(\psi_0 + \psi_1 \left[q(t-1) - \psi_0\right], \sigma_\rho^2)$.

Remark. In this stochastic volatility model with dynamic correlation between the error components we see that the correlation also change as well as the volatility. Also observe that we must have $-1 < \rho_\epsilon(t) < 1$ in order to have a well defined variance-covariance matrix $\Sigma_\epsilon(t)$.

2.5. Model V

Consider the same hypothesis of Model IV with the exception of the hypothesis made for $h_2(t)$. In the case of this latent variable we consider the assumption used in Model III. Therefore, we consider the presence of the Granger causality for the latent variable $h_2(t)$, i.e., $h_2(t)$ is given by (4).

2.6. Model VI

In this model we consider a setting similar to that of Model I. The difference being that now the error coordinates $\epsilon_i(t)$, $i = 1, 2$, have a bivariate Student distribution with ν degrees of freedom, $t = 1, 2, \ldots, N$. Hence, we assume that $\boldsymbol{\epsilon}(t)$ has a bivariate Student distribution with mean vector $\mathbf{0}$ and variance-covariance matrix $[\nu/(\nu-2)]\Sigma_{\epsilon,0}$, for $\nu > 2$. Therefore, the vector $\boldsymbol{\epsilon}(t) = (\epsilon_1(t), \epsilon_2(t))'$ has density function a $t\,(\mathbf{0}, \Sigma_{\epsilon,0}, \nu)$, where ν is the of degree of freedom, i.e., its density function is given by,

$$f\left(\boldsymbol{\epsilon}(t)\right) = \frac{\Gamma\left[(\nu+2)/2\right]}{\nu\pi\Gamma\left[\nu/2\right]} \left|\Sigma_{\epsilon,0}\right|^{-1/2} \left(1 + \frac{1}{\nu}\boldsymbol{\epsilon}'(t)\Sigma_{\epsilon,0}^{-1}\boldsymbol{\epsilon}(t)\right)^{-(\nu+2)/2} \tag{5}$$

where $\Gamma\left(x\right)$ denotes a Gamma function.

Remark. Using this heavy tail Student distribution, we could have the presence of extra Kurtosis for the return distributions. It is also interesting to observe that many other modifications for the BSV models could be considered.

3. Bayesian Analysis

In this section we are going to describe the Bayesian formulation of the models splitting them into three categories. Those in the so-called Class I, are the models that have the error vector $\boldsymbol{\epsilon}(t) = (\epsilon_1(t), \epsilon_2(t))'$ Normally distributed with mean vector $\mathbf{0}$ and a constant correlation for the error components model, i.e., the correlation either is zero or is a constant not depending on t. The models falling into this class are Models I, II and III. In Class II, we include models where $\boldsymbol{\epsilon}(t)$ is Normally distributed with mean vector $\mathbf{0}$, but the correlation between $\epsilon_1(t)$ and $\epsilon_2(t)$ also depends on time, i.e., Models IV and V. Finally, in Class III we have Model VI.

Bayesian inference will be performed through a sample taken from the posterior distribution of the parameters of the models. Therefore, we need to specify in each case which prior distributions we are considering and what likelihood functions are used. In all cases we assume prior independence among the parameters of the models.

3.1. Class I

If Model I is considered then the vector of parameters to be estimated is $\boldsymbol{\theta}_I = (\phi_{11}, \phi_{22}, \sigma^2_{\eta 1}, \sigma^2_{\eta 2}, \mu_1, \mu_2)$. We assume that ϕ_{ii}, $\sigma^2_{\eta i}$ and μ_i have as prior distributions a Beta, an Inverse Gamma and a Normal distributions, respectively, $i = 1, 2$, i.e., ϕ_{ii}, $\sigma^2_{\eta i}$ and μ_i have, respectively, prior distributions Beta(a_{ii}, b_{ii}), IG(c_i, d_i) and N(e_i, f_i), where the hyperparameters a_{ii}, b_{ii}, c_i, d_i, e_i and f_i are known, $i = 1, 2$. (In here, we are considering the Beta(a, b) and IG(c, d) as the Beta and the Inverse Gamma distributions with means $a/(a + b)$ and $d/(c - 1)$ and variances $ab/[(a + b)^2(a + b + 1)]$ and $d^2/[(c - 1)^2(c - 2)]$, $c > 2$, respectively.)

When Model II is considered the vector of parameters is $\boldsymbol{\theta}_{II} = (\boldsymbol{\theta}_I, \rho_\epsilon)$, where we assume that ρ_ϵ has as prior distribution an uniform distribution U(-1, 1) and use the same priors distributions for $\boldsymbol{\theta}_I$ and in Model I, with possibly different values for the hyperparameters.

In Models I and II, the joint density functions of the latent variables $\mathbf{h}(t) = (h_1(t), h_2(t))$ given the vector of parameters $\boldsymbol{\theta}$ are given by

$$\begin{cases} g\left(\mathbf{h}(1)\,|\,\boldsymbol{\theta}\right) \propto \prod\limits_{l=1}^{2} \left(\sigma^2_{\eta l}\right)^{-1/2} \exp\left[-\frac{1}{2\sigma^2_{\eta l}}\left(h_l(1) - \mu_l\right)^2\right], \, t = 1 \\ g\left(\mathbf{h}(t) \mid \mathbf{h}(t-1), \boldsymbol{\theta}\right) \propto \prod\limits_{l=1}^{2} \prod\limits_{t=2}^{N} \left(\sigma^2_{\eta l}\right)^{-1/2} \exp\left\{-\frac{1}{2\sigma^2_{\eta l}}\left[h_l(t) - \mu_l \right.\right. \\ \qquad \left.\left. -\phi_{ll}\left(h_l(t-1) - \mu_l\right)\right]^2\right\}, \, t = 2, 3, \ldots, N. \end{cases} \tag{6}$$

Let $\boldsymbol{\theta} = \boldsymbol{\theta}_{II}$ and take $\varphi = (\boldsymbol{\theta}, \mathbf{h})$, with $\mathbf{h} = (\mathbf{h}(1), \mathbf{h}(2), \ldots, \mathbf{h}(N))$. Hence, we have that the joint likelihood function of $\boldsymbol{\theta}$ and $\mathbf{h}$ for Model II is given, for $\mathbf{Y} =$

$(\mathbf{Y}(1), \mathbf{Y}(2), \ldots, \mathbf{Y}(N))$, by

$$
\begin{aligned}
L(\varphi \mid \mathbf{Y}) &\propto \prod_{t=1}^{N} p\left(\mathbf{Y}(t) \mid \mathbf{h}(t), \boldsymbol{\theta}\right) \\
&\propto \left(1-\rho_\epsilon^2\right)^{-N/2} \exp\left\{-\frac{1}{2}\left[\sum_{t=1}^{N} h_1(t) + \sum_{t=1}^{N} h_2(t)\right]\right\} \\
&\quad \exp\left\{-\frac{1}{2\left(1-\rho_\epsilon^2\right)}\left[\sum_{t=1}^{N} Y_1^2(t) e^{-h_1(t)}\right.\right. \\
&\quad \left.\left. + \sum_{t=1}^{N} Y_2^2(t) e^{-h_2(t)} - 2\rho_\epsilon \sum_{t=1}^{N} Y_1(t) Y_2(t) e^{-h_1(t)/2} e^{-h_2(t)/2}\right]\right\} \qquad (7)
\end{aligned}
$$

In the case of $\boldsymbol{\theta} = \boldsymbol{\theta}_I$ we just set $\rho_\epsilon = 0$ in (7).

Therefore, for $\varphi = (\boldsymbol{\theta}_I, \mathbf{h})$ or $\varphi = (\boldsymbol{\theta}_{II}, \mathbf{h})$ we have that the joint posterior distribution of the vector of parameters and $\mathbf{h}$ is given by

$$
\pi\left(\varphi \mid \mathbf{Y}\right) \propto \pi\left(\boldsymbol{\theta}\right) g\left(\mathbf{h}(1) \mid \boldsymbol{\theta}\right) \left(\prod_{t=2}^{N} g\left(\mathbf{h}(t) \mid \mathbf{h}(t-1), \boldsymbol{\theta}\right)\right) L(\varphi \mid \mathbf{Y}) \qquad (8)
$$

where $\pi(\boldsymbol{\theta})$ is the prior distribution of the vector of parameters with $\boldsymbol{\theta} = \boldsymbol{\theta}_I, \boldsymbol{\theta}_{II}$, $L(\varphi \mid \mathbf{Y})$ is the likelihood function of the model given by (7), and $g(\mathbf{h}(1) \mid \boldsymbol{\theta})$, $g(\mathbf{h}(t) \mid \mathbf{h}(t-1), \boldsymbol{\theta})$, $t = 2, 3, \ldots, N$ are given by the set of recursive functions (6).

When we assume the presence of the Granger causality for $h_2(t)$ and constant correlation for the error components, i.e., Model III, the vector of parameters is $\boldsymbol{\theta}_{III} = (\boldsymbol{\theta}_{II}, \phi_{21})$. We take the same prior distributions for $\boldsymbol{\theta}_{II}$ with possibly different hyperparameters. Additionally, for the variable ϕ_{21} we take as its prior distribution a Beta distribution Beta(a_{21}, b_{21}) with known hyperparameters a_{21} and b_{21}.

The likelihood function of Model III is given by (7). The joint posterior distribution of $\varphi = (\boldsymbol{\theta}, \mathbf{h})$ also has the same expression as in Models I and II, i.e., the expression is given by (8) but now taking $\boldsymbol{\theta} = \boldsymbol{\theta}_{III}$ and replacing the density function $g(\mathbf{h}(t) \mid \mathbf{h}(t-1), \boldsymbol{\theta})$, $t = 2, 3, \ldots, N$ by

$$
\begin{aligned}
& g\left(\mathbf{h}(t) \mid \mathbf{h}(t-1), \boldsymbol{\theta}\right) \propto \\
& \left(\prod_{t=2}^{N} \left(\sigma_{\eta_1}^2\right)^{-1/2} \exp\left\{-\frac{1}{2\sigma_{\eta_1}^2}\left[h_1(t) - \mu_1 - \phi_{11}\left(h_1(t-1) - \mu_1\right)\right]^2\right\}\right) \\
& \left(\prod_{t=2}^{N} \left(\sigma_{\eta_2}^2\right)^{-1/2} \exp\left\{-\frac{1}{2\sigma_{\eta_2}^2}\left[h_2(t) - \mu_2 - \phi_{21}\left(h_1(t-1) - \mu_1\right)\right.\right.\right. \\
& \qquad \left.\left.\left. -\phi_{22}\left(h_2(t-1) - \mu_2\right)\right]^2\right\}\right). \qquad (9)
\end{aligned}
$$

3.2. Class II

The models in this class have a stochastic volatility model with dynamic correlation for the error components. When Model IV is considered we have that the vector of parameters is $\boldsymbol{\theta}_{IV} = (\boldsymbol{\theta}_I, \psi_0, \psi_1, \sigma_\rho^2)$. The prior distribution for the components of $\boldsymbol{\theta}_I$ are taken to be the same as in Class I. The parameters ψ_0, ψ_1 and σ_ρ^2 have as prior distributions a Normal distribution $\mathrm{N}(0, f_3)$, a Beta distribution $\mathrm{Beta}(f, g)$ and a Inverse Gamma distribution $\mathrm{IG}(c_3, d_3)$, where the hyperparameters c_3, d_3, f_3, f, g are considered to be known.

Hence, the joint likelihood function of this class of models is given, for $\boldsymbol{\theta} = \boldsymbol{\theta}_{IV}$, by,

$$\begin{aligned} L(\boldsymbol{\theta}, \mathbf{h} \,|\, \mathbf{Y}) \;\propto\; & \prod_{t=1}^{N} p\left(\mathbf{Y}(t) \mid \mathbf{h}(t), \boldsymbol{\theta}\right) \propto \left\{ \prod_{t=1}^{N} \left(1 - \rho_\epsilon^2(t)\right)^{-1/2} \right\} \\ & \exp\left\{ -\frac{1}{2} \left[\sum_{t=1}^{N} h_1(t) + \sum_{t=1}^{N} h_2(t) \right] \right\} \\ & \exp\left\{ -\frac{1}{2} \left[\sum_{t=1}^{N} \frac{Y_1^2(t) e^{-h_1(t)}}{1 - \rho_\epsilon^2(t)} + \sum_{t=1}^{N} \frac{Y_2^2(t) e^{-h_2(t)}}{1 - \rho_\epsilon^2(t)} - \right.\right. \\ & \left.\left. -2 \sum_{t=1}^{N} \frac{\rho_\epsilon(t) Y_1(t) Y_2(t) e^{-h_1(t)/2} e^{-h_2(t)/2}}{1 - \rho_\epsilon^2(t)} \right] \right\}. \end{aligned} \tag{10}$$

The joint posterior distribution of $\varphi = (\boldsymbol{\theta}, \mathbf{h}, \mathbf{q})$, where $\boldsymbol{\theta} = \boldsymbol{\theta}_{IV}$ and $\mathbf{q} = (q(1), q(2), \ldots, q(N))$, is given by

$$\begin{aligned} \pi\left(\varphi \mid \mathbf{Y}\right) \;\propto\; & \pi\left(\boldsymbol{\theta}\right) g\left(\mathbf{h}(1) \,|\, \boldsymbol{\theta}\right) \left(\prod_{t=2}^{N} g\left(\mathbf{h}(t) \mid \mathbf{h}(t-1), \boldsymbol{\theta}\right) \right) \\ & f\left(\mathbf{q}(1) \,|\, \boldsymbol{\theta}\right) \left(\prod_{t=2}^{N} f\left(q(t) \mid q(t-1), \boldsymbol{\theta}\right) \right) L(\varphi \,|\, \mathbf{Y}) \end{aligned} \tag{11}$$

where $\pi\left(\boldsymbol{\theta}\right)$ is the joint prior distribution for $\boldsymbol{\theta}$; $g\left(\mathbf{h}(1) \,|\boldsymbol{\theta}\right)$ and $g(\mathbf{h}(t) \mid \mathbf{h}(t-1), \boldsymbol{\theta})$ are defined by (6), $f\left(q(1) \,|\, \boldsymbol{\theta}\right)$ and $f\left(q(t) \mid q(t-1), \boldsymbol{\theta}\right)$ are the Normal density functions of $q(1)$ and the conditional Normal density function of $q(t)$ given $q(t-1)$, respectively, and $L(\varphi \,|\, \mathbf{Y})$ is the likelihood function defined in (10).

When considering a stochastic volatility model with dynamic correlation for the error components and the presence of the Granger causality for $h_2(t)$, $t = 2, 3, \ldots, N$, i.e., Model V, the vector of parameters here is $\boldsymbol{\theta}_V = (\boldsymbol{\theta}_{IV}, \phi_{21})$. We take the same prior distribution for $\boldsymbol{\theta}_{IV}$ except for the parameters ψ_0 and ψ_1 which now will have as prior distributions, Gamma distributions with appropriate hyperparameters. We still take a $\mathrm{Beta}(a_{21}, b_{21})$ prior distribution for ϕ_{21}.

The likelihood of the model is also given by (10). The posterior distribution is similar to (11), the difference is that $g(\mathbf{h}(t) \,|\, \mathbf{h}(t-1), \boldsymbol{\theta})$, $t = 2, 3, \ldots, N$ are given by (9) with $\boldsymbol{\theta}$ replaced by $\boldsymbol{\theta}_V$.

3.3. Class III

In this class of models we assume that the error components have a bivariate Student distribution with degree of freedom $\nu > 2$. The latent variables $\mathbf{h}(t) = (h_1(t), h_2(t))$, $t = 1, 2, \ldots, N$ are defined as in Model I.

The likelihood function could be obtained directly from (5) or representing the bivariate Student distribution as a mixture of a bivariate Normal distribution with a Gamma distribution (see for example Bernardo and Smith, 1995). Following the latter approach, first we have that the distribution for $\boldsymbol{\epsilon}(t)$ is given by,

$$f_\epsilon\left(\boldsymbol{\epsilon}(t)\right) = \int f_{Normal}\left(\epsilon(t) \mid \boldsymbol{\mu}, \Sigma_Y z_t\right) f_{Gamma}\left(z_t \mid \alpha, \beta\right) dz_t \tag{12}$$

where $f_{Normal}\left(\boldsymbol{\epsilon}(t) \mid \boldsymbol{\mu}, \Sigma\right)$ denotes bivariate Normal density with mean vector $\boldsymbol{\mu}$ and variance-covariance matrix Σ and $f_{Gamma}\left(z \mid \alpha, \beta\right)$ is a Gamma density with mean α/β and variance α/β^2.

Therefore, take $z_t = 1/\sqrt{w_t}$, where w_t has as distribution a Gamma(λ_w, λ_w) distribution and is such that $\mathrm{E}\left(w_t\right) = 1$ and $\mathrm{Var}\left(w_t\right) = \lambda_w^{-1}$. Hence, the conditional distribution of $\mathbf{Y}(t)$ given Σ_Y and w_t is a bivariate Normal distribution with mean vector $\mathbf{0}$ and variance-covariance matrix $\frac{1}{\sqrt{w_t}}\Sigma_Y$, where Σ_Y is given by (3).

The vector of parameters for Model VI is $\boldsymbol{\theta}_{VI} = (\boldsymbol{\theta}_{II}, \lambda_w)$ and, for $\mathbf{w} = (w_1, w_2, \ldots, w_N)$, the likelihood function of the model is given, for $\varphi = (\boldsymbol{\theta}_{VI}, \mathbf{h}, \mathbf{w})$, by

$$\begin{aligned} L(\varphi, \mid \mathbf{Y}) \;\propto\; & \prod_{t=1}^{N} p\left(\mathbf{Y}(t) \mid \mathbf{h}(t), w_t\right) \propto \left(1-\rho_\epsilon^2\right)^{-N/2} \left\{\prod_{t=1}^{N} w_t^{1/2}\right\} \\ & \exp\left\{-\frac{1}{2}\left[\sum_{t=1}^{N} h_1(t) + \sum_{t=1}^{N} h_2(t)\right]\right\} \\ & \exp\left\{-\frac{1}{2\left(1-\rho_\epsilon^2\right)}\left[\sum_{t=1}^{N} Y_1^2(t) w_t^{1/2} e^{-h_1(t)} + \right.\right. \\ & \quad + \sum_{t=1}^{N} Y_2^2(t) w_t^{1/2} e^{-h_2(t)} \\ & \quad \left.\left. -2\rho_\epsilon \sum_{t=1}^{N} Y_1(t) Y_2(t) w_t^{1/2} e^{-h_1(t)/2} e^{-h_2(t)/2}\right]\right\}. \end{aligned} \tag{13}$$

The prior distributions of $\boldsymbol{\theta}_{II}$ are the same as used in Class I models (with possibly different hyperparameters) and we take a Gamma prior distribution Gamma(f_λ, g_λ) for λ_w. We also have that f_λ and g_λ are known hyperparameters.

Hence, the joint posterior distribution of $\varphi = (\boldsymbol{\theta}_{VI}, \mathbf{h}, \mathbf{w})$, is given by

$$\begin{aligned} \pi\left(\varphi \mid \mathbf{Y}\right) \;=\; & \pi\left(\boldsymbol{\theta}_{VI}\right) g\left(\mathbf{h}(1) \mid \boldsymbol{\theta}_{VI}\right) \left(\prod_{t=2}^{N} g\left(\mathbf{h}(t) \mid \mathbf{h}(t), \boldsymbol{\theta}_{VI}\right)\right) \\ & \left(\prod_{t=1}^{N} g_w\left(w_t \mid \boldsymbol{\theta}_{VI}\right)\right) L(\varphi \mid \mathbf{Y}), \end{aligned} \tag{14}$$

where $\pi\left(\boldsymbol{\theta}_{VI}\right)$ is the joint prior distribution of $\boldsymbol{\theta}_{VI}$, $g\left(\mathbf{h}(1) \mid \boldsymbol{\theta}_{VI}\right)$ and $g(\mathbf{h}(t) \mid \mathbf{h}(t-1), \boldsymbol{\theta}_{VI})$ are given by (6), and $\prod_{t=1}^{N} g_w(w_t \mid \boldsymbol{\theta}_{VI})$ is the product of Gamma(λ_w, λ_w) densities, that is,

$$\prod_{t=1}^{N} g_w\left(w_t \mid \boldsymbol{\theta}_{VI}\right) \propto \prod_{t=1}^{N} w_t^{\lambda_w - 1} e^{-\lambda_w w_t}$$

and $L(\varphi \mid \mathbf{Y})$ is the likelihood function (13).

Samples of the joint posterior distributions are generated using MCMC methods such as the Gibbs sampling algorithm or the Metropolis-Hastings algorithm (see for example Gelfand and Smith, 1990; Smith and Roberts, 1993). A great simplification in the simulation of the samples is obtained using the software WinBugs (Spiegelhalter *et al.*, 2003), where we only need to specify a distribution for the data and the prior distribution of the parameters.

4. Model Selection

In this paper we use the Deviance Information Criterion to select the model that best fit the data. The Deviance Information Criterion (DIC) (see Spiegelhalter *et al.*, 2002) is given by,

$$DIC = \widehat{D} + 2p_D \tag{15}$$

where $\widehat{D}$ is the deviance evaluated at the posterior mean and p_D is the effective number of parameters in the model which is given by $p_D = \overline{D} - \widehat{D}$, where $\overline{D}$ is the posterior mean deviance.

The smallest value of DIC indicates the best BSV model. Note that this value could be negative.

5. An Application to the Weekly Ozone Averages in Mexico City

In this section we present a description of the data to which the theory described earlier in this paper is applied. We also present the results obtained.

5.1. Description of the Data

The data was obtained from the monitoring network of the Metropolitan Area of Mexico City (www.sma.df.gob.mx/simat/). The Metropolitan Area is split into five regions or sectors corresponding to the Northeast (NE), Northwest (NW), Centre (CE), Southeast (SE) and Southwest (SW) and the ozone monitoring stations are placed throughout the city (see Álvarez *et al.*, 2005; Achcar *et al.*, 2007).

The primary data used in the analysis corresponds to sixteen years of the daily maximum ozone measurements, ranging from 01 January 1990 to 31 December 2005 (inclusive), of the daily maximum measurements in regions NE, CE and SW. The measurements are obtained minute by minute and the averaged hourly result is reported at each station. The daily maximum measurement for a given region is the maximum over all the maximum averaged values recorded hourly during a 24-hour period by each station placed in the region. The sixteen-year average measurements in regions NE, CE and SW are 0.1, 0.141 and 0.132, respectively, with standard deviations 0.042, 0.055 and 0.048. Regions NE, CE and SW were chosen because the wind direction in the Metropolitan Area of Mexico City is mainly from region NE to SW. Many of the ozone precursors are produced in regions NE and CE. In general, region SW is the one with frequently high levels of ozone.

We perform two analysis in separate. In the first analysis we consider jointly the data from regions NE and CE. In a second instance we consider the joint information provided by regions CE and SW. The data actually used in the analysis were produced by the weekly averaged ozone measurements for regions NE, CE and SW. Hence, for instance, if $X_1(t)$, $t = 1, 2, \ldots, T$, indicate the daily maximum measurements of ozone in region NE, then $Z_1(t)$ is the mean measurement in week t in region NE, i.e., $Z_1(1) = (1/7)\sum_{i=1}^{7} X_1(i)$, $Z_1(2) = (1/7)\sum_{i=8}^{14} X_1(i)$, and so on.

Inference was performed using a sample drawn from the joint posterior distributions after a burn-in period of size 5000. After the burn-in period, we selected 1000 Gibbs samples by keeping every 10th sample in order to have approximately uncorrelated values.

In Figure 1, we have the plots of the weekly ozone averages versus weeks for regions NE, CE and SW.

Observing Figure 1 it is possible to note a decreasing trend for the weekly ozone averages for these three regions of Mexico City, especially after the 400th week (close to the year 1998).

In Figure 2, we have the plots of the log-returns $Y_j(t) = \log(Z_j(t)/Z_j(t-1))$, $t = 1, 2, \ldots, N$, centred on their averages for regions NE, CE and SW. (In here we have $N = 834$.)

5.2. Application of the Models and Results

We are going to analyse each model and pair of data separately.

5.2.1. Model I

The hyperparameters of the prior distributions are given as follows. When either the pair NE and CE or the pair CE and SW is considered, we have that the hyperparameters of the prior distributions of ϕ_{ii} and $\sigma_{\eta i}$ are $a_{ii} = b_{ii} = 1$, $c_i = d_i = 1$, $i = 1, 2$. When the pair NE and CE is considered the hyperparameters of the prior distribution of μ_i are $e_i = 0$, $f_i = 100$, $i = 1, 2$. However, when the pair CE and SW is considered then we have $e_i = 0$, $f_i = 1$, $i = 1, 2$. In Table 1 we have the summary of the quantities estimated when the pair NE-CE is considered as well as when regions CE and SW are taken into account. (Note that in Table 1, and whenever suitable, we report the estimates of $\tau_{\eta i} = 1/\sigma_{\eta i}^2$ and the estimate of $\tau_\rho = 1/\sigma_\rho^2$ instead of reporting the estimates of $\sigma_{\eta i}^2$ and σ_ρ^2, $i = 1, 2$.)

In Figure 3 we have the plots of the estimated square roots of the volatility when regions NE and CE (top two plots) and when regions CE and SW (bottom two plots) are taken into account and when Model I is used.

5.2.2. Model II

In the case of Model II, the hyperparameters of the prior distributions are given as follows. In both cases, i.e., when regions NE and CE are considered and when the pair CE-SW is taken into account, we have that the hyperparameters of the prior distributions of

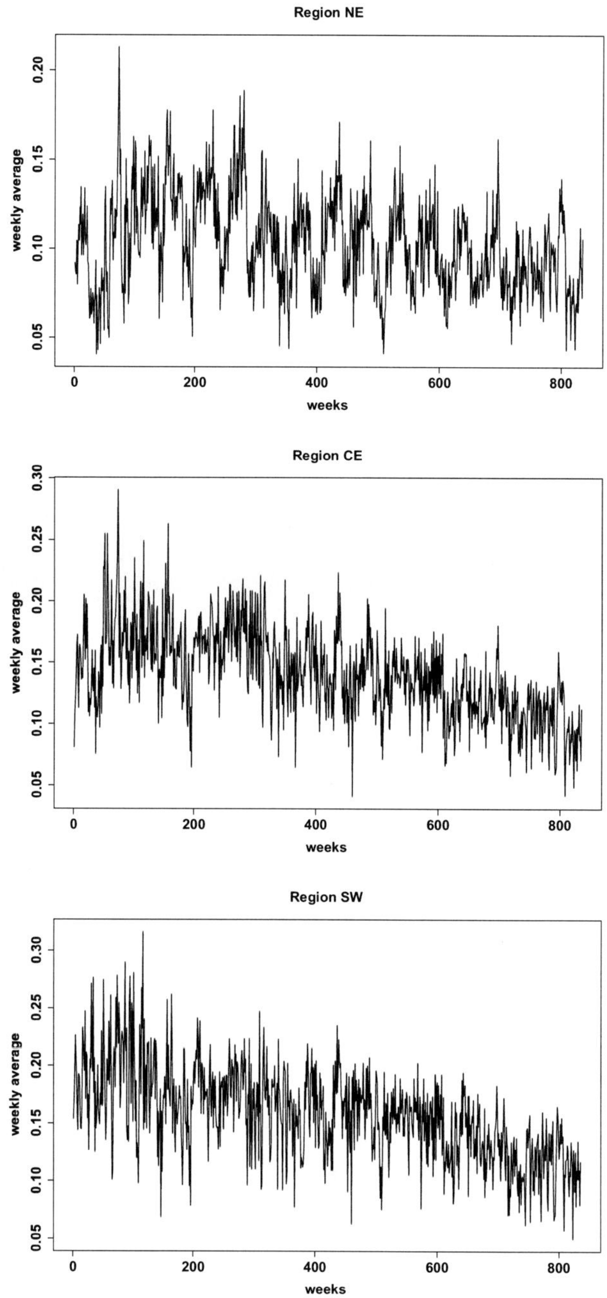

Figure 1. Weekly ozone averages for the regions NE, CE and SW from 01 January 1990 to 31 December 2005.

ϕ_{ii} and μ_i are $a_{ii} = b_{ii} = 1$ and $e_i = -3$, $f_i = 100$, $i = 1, 2$. In the case of the parameters $\sigma_{\eta i}$, $i = 1, 2$, we have that if we consider regions NE and CE then $c_1 = 5$, $c_2 = 4$ and $d_1 = d_2 = 1$, $i = 1, 2$. When data from regions CE and SW are used, we have that $c_1 = 3$, $c_2 = 4$ and $d_1 = d_2 = 1$. We would like to call attention to the fact that the changes in the

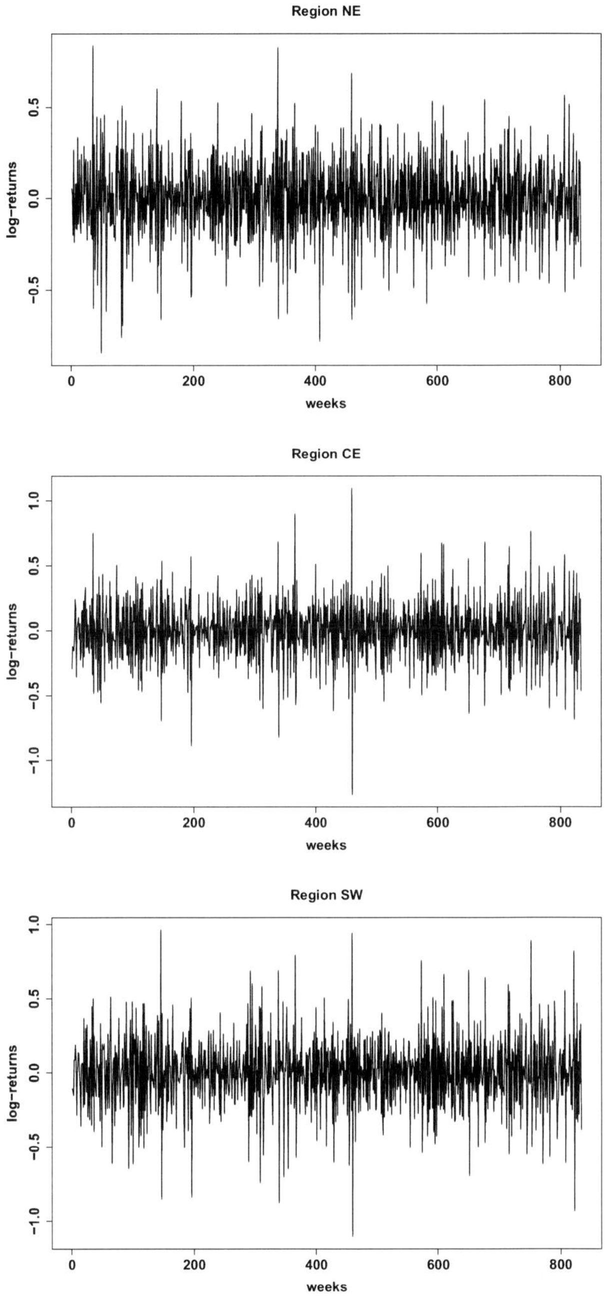

Figure 2. Log-returns centred in their averages for regions NE, CE and SW from 01 January 1990 to 31 December 2005.

parameters are because we are using information provided by the results given by Model I. Hence, we are using an empirical Bayes approach to analyse the problem (see Carlin and Louis, 2000). In Table 2 we present the summary of the quantities estimated.

Figure 4 shows similar plots to those of Figure 3, but now we use Model II instead of

Table 1. Posterior estimated mean, standard deviation (SD) and 95% Credible Interval for the quantities of interest when Model I is used and the pairs of regions NE and CE, and CE and SW are considered.

Regions	Parameters	Mean	SD	95% Credible Interval
NE – CE	τ_{η_1}	4.827	1.387	(2.85; 8.128)
	τ_{η_2}	3.957	1.088	(2.3930; 6.48)
	μ_1	-3.07	0.0633	(-3.19; -2.951)
	μ_2	-2.986	0.0763	(-3.13; -2.838)
	ϕ_{11}	0.4945	0.1491	(0.184; 0.7526)
	ϕ_{22}	0.6334	0.0855	(0.4595; 0.7791)
CE – SW	τ_{η_1}	3.291	0.9023	(1.929; 5.214)
	τ_{η_2}	4.058	1.185	(2.26; 6.954)
	μ_1	-2.936	0.07911	(-3.091; -2.781)
	μ_2	-2.969	0.07879	(-3.124; -2.82)
	ϕ_{11}	0.6277	0.0865	(0.4359; 0.7712)
	ϕ_{22}	0.6327	0.1012	(0.4128; 0.8007)

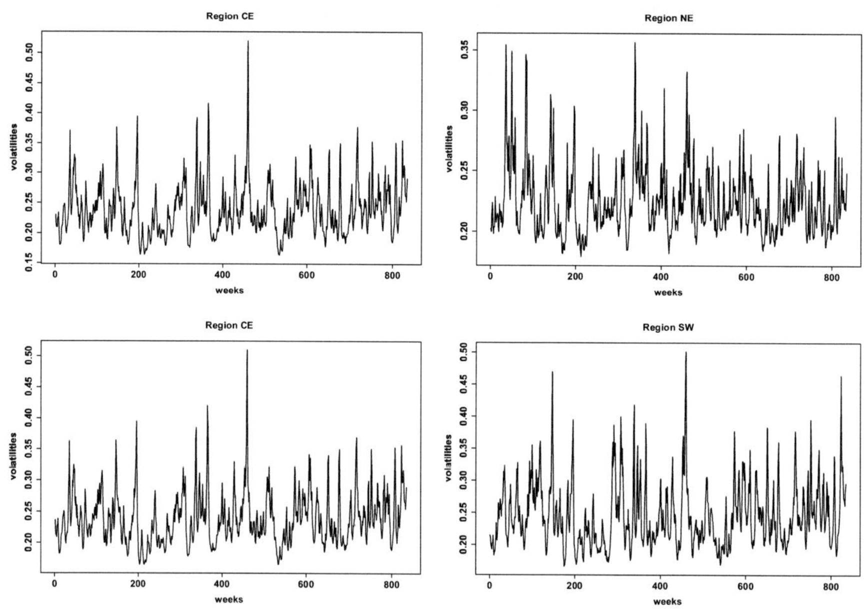

Figure 3. Square roots of the volatility for regions NE, CE and SW when Model I is used and when regions are paired in regions NE and CE (top row) and regions CE and SW (bottom row).

Model I.

5.2.3. Model III

Consider now Model III. The hyperparameters of the prior distributions ofϕ_{ii}, μ_i, $i = 1, 2$ and ρ_ϵ are the same as taken in Model II for both pair of regions NE and CE and CE

Table 2. Posterior estimated mean, standard deviation (SD) and 95% Credible Interval for the quantities of interest when Model II is used and the pairs of regions NE and CE, and CE and SW are considered.

Regions	Parameters	Mean	SD	95% Credible Interval
NE – CE	$\tau_{\eta 1}$	9.426	2.389	(5.3840; 14.65)
	$\tau_{\eta 2}$	7.436	1.86	(4.6590; 11.68)
	μ_1	-3.022	0.07	(-3.163; -2.888)
	μ_2	-2.916	0.0744	(-3.074; -2.772)
	ϕ_{11}	0.7307	0.0818	(0.535; 0.858)
	ϕ_{22}	0.6985	0.0807	(0.5263; 0.8335)
	ρ_ϵ	0.7546	0.0165	(0.7209; 0.7849)
CE – SW	$\tau_{\eta 1}$	7.96	1.49	(5.247; 10.99)
	$\tau_{\eta 2}$	9.628	2.116	(6.425; 14.38)
	μ_1	-2.853	0.06426	(-2.981; -2.731)
	μ_2	-2.892	0.0649	(-3.023; -2.766)
	ϕ_{11}	0.6924	0.07087	(0.5383; 0.8049)
	ϕ_{22}	0.7207	0.07001	(0.5629; 0.8338)
	ρ_ϵ	0.8825	0.008724	(0.865; 0.8989)

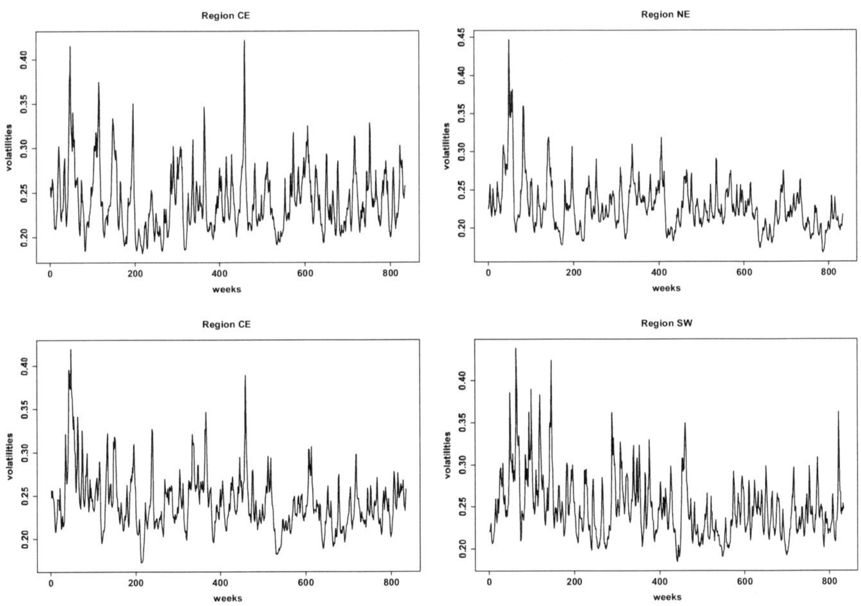

Figure 4. Square roots of the volatility for regions NE, CE and SW when Model II is used and when regions are paired in regions NE and CE (top row) and regions CE and SW (bottom row).

and SW. The hyperparameters of the prior distributions of $\sigma^2_{\eta 1}$ and $\sigma^2_{\eta 2}$ are $c_1 = 9$, $c_2 = 7$ and $d_1 = d_2 = 1$ when regions NE and CE are considered and are $c_1 = 8$, $c_2 = 10$ and $d_1 = d_2 = 1$ when regions CE and SW are taken into account. The Beta prior distribution of quantity ϕ_{21} has hyperparameters $a_{21} = b_{21} = 1$ when either regions NE and CE or regions CE and SW are considered. Table 3 gives the summary of the estimated quantities.

Table 3. Posterior estimated mean, standard deviation (SD) and 95% Credible Interval for the quantities of interest when Model III is used and the pairs of regions NE and CE, and CE and SW are considered.

Regions	Parameters	Mean	SD	95% Credible Interval
NE – CE	$\tau_{\eta 1}$	12.97	2.875	(8.1970; 19.5)
	$\tau_{\eta 2}$	9.233	2.308	(5.645; 14.32)
	μ_1	-3.04	0.0826	(-3.205; -2.882)
	μ_2	-2.917	0.0734	(-3.061; -2.777)
	ϕ_{11}	0.8346	0.0405	(0.7379; 0.9028)
	ϕ_{22}	0.4009	0.1997	(0.0408; 0.7466)
	ϕ_{21}	0.4851	0.2302	(0.1391; 0.9634)
	ρ_ϵ	0.7473	0.0177	(0.7131; 0.781)
CE – SW	$\tau_{\eta 1}$	13.83	2.755	(9.063; 19.34)
	$\tau_{\eta 2}$	13.39	2.694	(8.944; 19.41)
	μ_1	-2.838	0.09852	(-3.038; -2.639)
	μ_2	-2.903	0.08818	(-3.077; -2.733)
	ϕ_{11}	0.8848	0.02978	(0.823; 0.9338)
	ϕ_{22}	0.3309	0.1318	(0.06118; 0.5704)
	ϕ_{21}	0.5628	0.1278	(0.3077; 0.827)
	ρ_ϵ	0.8808	0.008337	(0.8627; 0.78962)

In Figure 5, we have the plots of the estimated square roots of the volatility when Model III is considered. The top two plots correspond to the case where regions NE and CE are jointly considered and the two plots at the bottom correspond to the case where regions CE and SW are used.

5.2.4. Model IV

When Model IV is used the hyperparameters of the prior distributions of$\sigma^2_{\eta i}$, $i = 1, 2$ are $c_1 = 13$, $c_2 = 9$ and $d_1 = d_2 = 1$ when regions NE and CE are considered and are $c_1 = 14$, $c_2 = 13$ and $d_1 = d_2 = 1$ when data from regions CE and SW are used. The hyperparameters of the prior distributions of the quantities ϕ_{ii}, $i = 1, 2$ are as in Model II and III in both sets of data, i.e., the one produced by regions NE and CE and the one produced by regions CE and SW. We also have that in both cases the hyperparameters of the prior distributions of μ_i are $e_i = -3$ and $d_i = 10$, $i = 1, 2$. The parameter ψ_0 has a Normal $N(0, 10)$ prior distribution and ψ_1 has a $Beta(20, 1)$ prior distribution and the hyperparameters of the prior distribution ofσ^2_ρ are $c_3 = d_3 = 1$ when either regions NE and CE or regions CE and SW are taken into account. In Table 4, the estimates of the quantities of interest are presented.

Figure 6 shows similar plots to the ones given by Figures 3, 4 and 5, but now we have used Model IV.

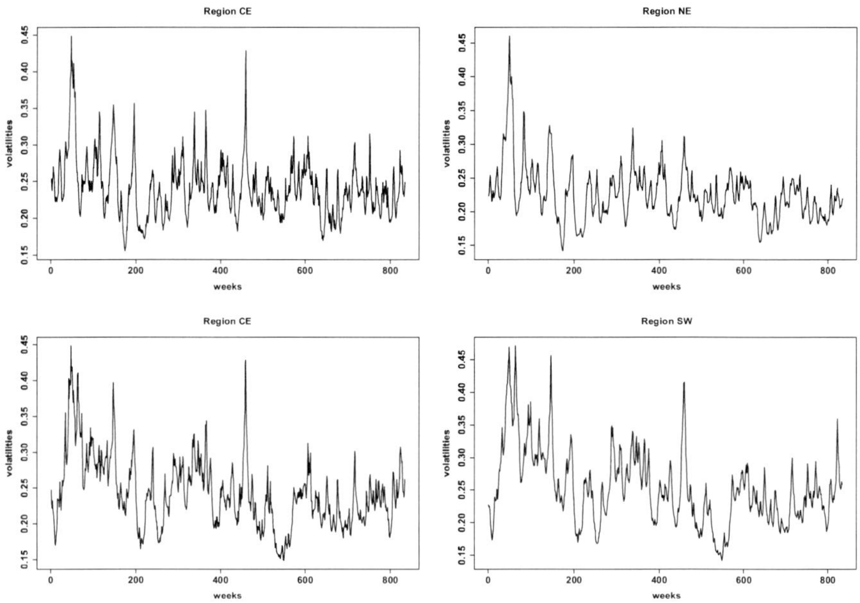

Figure 5. Square roots of the volatility for regions NE, CE and SW when Model III is used and when regions are paired in regions NE and CE (top row) and regions CE and SW (bottom row).

Table 4. Posterior estimated mean, standard deviation (SD) and 95% Credible Interval for the quantities of interest when Model IV is used and the pairs of regions NE and CE, and CE and SW are considered.

Regions	Parameters	Mean	SD	95% Credible Interval
NE – CE	τ_ρ	4.126	1.196	(2.3870; 7.19)
	$\tau_{\eta 1}$	17.05	3.605	(10.46; 24.39)
	$\tau_{\eta 2}$	11.61	2.891	(6.645; 18.94)
	μ_1	-3.017	0.0591	(-3.132; -2.898)
	μ_2	-2.891	0.0662	(-3.017; -2.763)
	ϕ_{11}	0.7238	0.1254	(0.3807; 0.8692)
	ϕ_{22}	0.7188	0.0782	(0.5527; 0.85)
	ψ_1	0.8492	0.0424	(0.7584; 0.9197)
	ψ_0	2.1280	0.1481	(1.838; 2.427)
CE – SW	τ_ρ	6.787	2.047	(3.161; 11.05)
	$\tau_{\eta 1}$	15.41	2.974	(10.66; 22.13)
	$\tau_{\eta 2}$	16.43	3.486	(10.66; 24.15)
	μ_1	-2.792	0.04915	(-2.888; -2.693)
	μ_2	-2.844	0.05712	(-2.961; -2.74)
	ϕ_{11}	0.5875	0.1198	(0.3019; 0.7707)
	ϕ_{22}	0.5518	0.1328	(0.25; 0.772)
	ψ_1	0.9281	0.02514	(0.8721; 0.9684)
	ψ_0	2.989	0.2412	(2.488; 3.423)

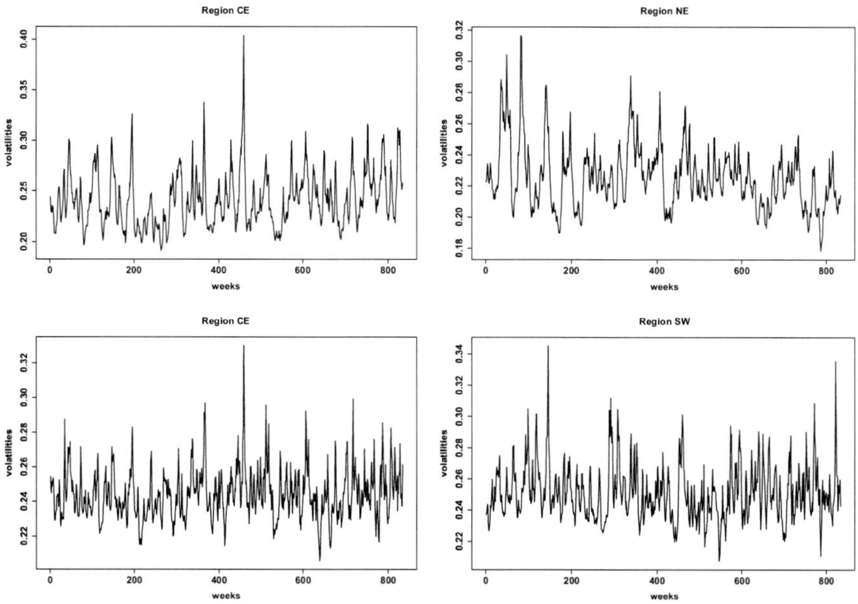

Figure 6. Square roots of the volatility for regions NE, CE and SW when Model IV is used and when regions are paired in regions NE and CE (top row) and regions CE and SW (bottom row).

5.2.5. Model V

In the case of Model V, i.e., when the Granger causality is present in $h_2(t)$, $t = 1, 2 \ldots, N$, we have that the hyperparameters of the prior distributions of the quantities ϕ_{ii}, μ_i, $i = 1, 2$ and σ_ρ^2 are the same as in Model IV and the hyperparameters of the prior distribution of ϕ_{21} are the same as in Model III for both sets of data, i.e., the one produced by regions NE and CE and the one produced by regions CE and SW. The hyperparameters of the prior distributions of $\sigma_{\eta 1}^2$ and $\sigma_{\eta 2}^2$ are $c_1 = 17$, $c_2 = 12$ and $d_1 = d_2 = 1$ in the case of regions NE and CE, and are $c_1 = 15$, $c_2 = 16$ and $d_1 = d_2 = 1$ in the case of regions CE and SW, ψ_0 has a Gamma prior distribution whose hyperparameters are (2,1) and (3,1) when regions NE and CE and regions CE and SW are considered, respectively. The quantity ψ_1 also has a Gamma prior distribution with hyperparameters (1,1) when either regions NE and CE or regions CE and SW are considered. (In here we are considering a Gamma(α, β) distribution with mean α/β and variance α/β^2.) Table 5 gives the summary of the quantities of interest.

The plots of the estimated volatility produced by Model V and using both data set (i.e., the one considering measurements from regions NE and CE and the one considering those from regions CE and SW) are given in Figure 7.

5.2.6. Model VI

When Model VI is taken into account, we have that the quantities ϕ_{ii}, μ_i, $i = 1, 2$ and ρ_ϵ have prior distributions with the same hyperparameters as those in Model IV in both cases,

Table 5. Posterior estimated mean, standard deviation (SD) and 95% Credible Interval for the quantities of interest when Model V is used and the pairs of regions NE and CE, and CE and SW are considered.

Regions	Parameters	Mean	SD	95% Credible Interval
NE – CE	τ_ρ	3.026	1.018	(1.592; 5.96)
	$\tau_{\eta 1}$	19.99	4.338	(12.26; 30.15)
	$\tau_{\eta 2}$	13.1	2.618	(8.77; 19.48)
	μ_1	-3.041	0.0722	(-3.177; -2.902)
	μ_2	-2.899	0.0654	(-3.032; -2.773)
	ϕ_{11}	0.8006	0.1086	(0.44; 0.906)
	ϕ_{22}	0.6795	0.1082	(0.4001; 0.8329)
	ϕ_{21}	0.2122	0.1471	(0.023; 0.6022)
	ψ_1	0.7811	0.0605	(0.6516; 0.8925)
	ψ_0	2.107	0.127	(1.861; 2.3640)
CE – SW	τ_ρ	6.241	2.303	(2.323; 11.18)
	$\tau_{\eta 1}$	16.72	3.248	(10.78; 24.01)
	$\tau_{\eta 2}$	18.96	3.744	(12.710; 27.17)
	μ_1	-2.835	0.07064	(-2.974; -2.693)
	μ_2	-2.883	0.06774	(-3.015; -2.738)
	ϕ_{11}	0.8215	0.0503	(0.7096; 0.9102)
	ϕ_{22}	0.4195	0.1152	(0.1955; 0.6503)
	ϕ_{21}	0.434	0.1126	(0.2492; 0.6855)
	ψ_1	0.9124	0.03554	(0.8235; 0.9644)
	ψ_0	2.953	0.205	(2.541; 3.342)

i.e., when regions NE and CE are considered and when region CE and SW are taken into account. The hyperparameters of the prior distributions of$\sigma^2_{\eta 1}$ and $\sigma^2_{\eta 2}$ are $c_1 = 20$, $c_2 = 13$ and $d_1 = d_2 = 1$ when regions NE and CE are considered and are $c_1 = 17$, $c_2 = 19$ and $d_1 = d_2 = 1$ when we use data from regions CE and SW. The hyperparameters of the prior distribution of quantity λ_w are $f_\lambda = g_\lambda = 1$ when regions NE and CE are considered and are $f_\lambda = 1$ and $g_\lambda = 2$ when the data from regions CE and SW is used. In Table 6 we present the summary of the quantities of interest.

In Figure 8, we have similar plots to those in Figures 3, 4, 5, 6 and 7, but now using Model VI.

5.2.7. Model Selection

The selection of the best model to fit the data was performed using the Deviance Information Criterion. Table 7 presents the estimated values of DIC for each model each set of data.

6. Conclusion

Observe that Model V (a stochastic volatility model with dynamic correlation and the presence of the Granger causality for $h_2(t)$, $t = 2, \ldots, N$) is the best model to explain the behaviour of the ozone pollution data of Mexico City when regions NE and CE are jointly

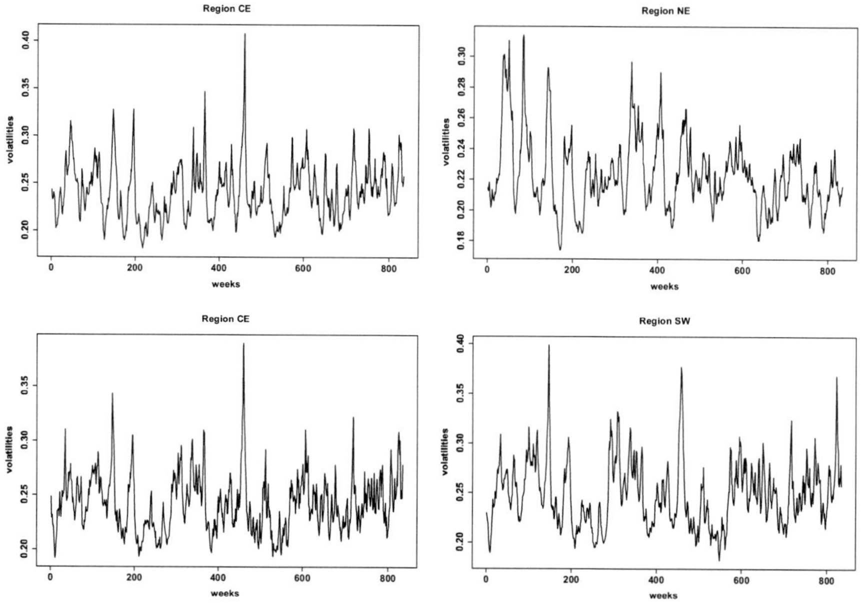

Figure 7. Square roots of the volatility for regions NE, CE and SW when Model V is used and when regions are paired in regions NE and CE (top row) and regions CE and SW (bottom row).

Table 6. Posterior estimated mean, standard deviation (SD) and 95% Credible Interval for the quantities of interest when Model VI is used and the pairs of regions NE and CE, and CE and SW are considered.

Regions	Parameters	Mean	SD	95% Credible Interval
NE – CE	λ_w	6.028	1.329	(3.918; 9.111)
	$\tau_{\eta 1}$	23.13	4.632	(14.53; 32.98)
	$\tau_{\eta 2}$	13.71	2.954	(8.616; 19.93)
	μ_1	-3.155	0.0726	(-3.283; -3.012)
	μ_2	-3.028	0.0688	(-3.167; -2.895)
	ϕ_{11}	0.7818	0.0848	(0.5817; 0.9136)
	ϕ_{22}	0.7097	0.1062	(0.3987; 0.8477)
	ρ_ϵ	0.7482	0.0169	(0.7143; 0.7816)
CE – SW	λ_w	4.385	0.8142	(3.104; 6.257
	$\tau_{\eta 1}$	19.41	3.721	(13.08; 27.61)
	$\tau_{\eta 2}$	21.45	4.869	(13.54; 33.2)
	μ_1	-2.993	0.07343	(-3.13; -2.855)
	μ_2	-3.035	0.06504	(-3.166; -2.912)
	ϕ_{11}	0.6176	0.1358	(0.2897; 0.8083)
	ϕ_{22}	0.5509	0.2054	(0.06438; 0.8096)
	ρ_ϵ	0.878	0.009271	(0.8582; 0.895)

considered, since it produces the smallest value of DIC given by - 882.200 and it is also the model producing the best fit when regions CE and SW are jointly considered having the

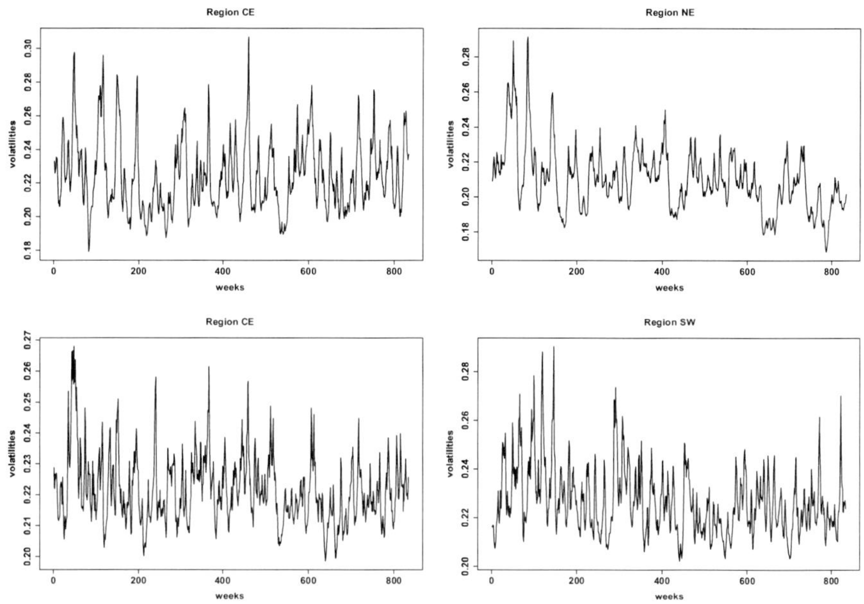

Figure 8. Square roots of the volatility for regions NE, CE and SW when Model VI is used and when regions are paired in regions NE and CE (top row) and regions CE and SW (bottom row).

Table 7. Deviance Information Criterion for Models I, II, III, IV, V and VI when the pairs of regions NE and CE, and CE and SW are considered.

	Deviance Information Criterion					
Regions	Model I	Model II	Model III	Model IV	Model V	Model VI
NE – CE	- 128.708	- 745.865	- 753.381	- 877.872	- 882.2	- 740.968
CE – SW	4.382	- 1090.84	- 1118.85	- 1251.66	- 1268.99	- 1080.25

smallest DIC value of -1268.99.

It is important to observe from Figures 3, 4, 5, 6, 7 and 8, that for both regions NE and CE assuming all six models, there is smaller and more stable volatility for the ozone weekly measurements after the 400th week (close to the year 1998). Similar conclusion can be reached when jointly considering regions CE and SW.

Note that even though the volatility decreases in all cases, we have a more stable plot, with rare exceptions, after week 600, specially when considering Model V. This week corresponds roughly to the end of year 2001 and beginning of year 2002. It is interesting to call attention to the fact that in the year 2000 we have the culmination of a series of measures that had been taken by the environmental authorities of Mexico since 1990 aiming to reduce the levels of ozone in big cities in the country and in particular in Mexico City. The series of measures may be roughly described as follows. In 1990, cars circulating in the Metropolitan Area of Mexico City would have to undergo periodic inspection of their mechanical condition. Additionally, restrictions were imposed on the circulation of car according to the ending number of their license plates. In 1997, further restrictions were implemented in terms of allowing clean cars to circulate freely. Additionally, in 1999, manufacturers were encouraged to produce cars with cleaner and modern technology. The production of such cars was made compulsory in 2001.

Therefore, we may see that some measures taken by the environmental authorities have produced a positive effect towards the aim of decreasing and stabilising the level of ozone measurements in Mexico City and in modifying its long term behaviour.

Finally, we would like to point out that besides their advantages when compared to the usual ARCH and GARCH models (mentioned earlier in this paper), stochastic volatility models also have an advantage when compared to non-homogeneous Poisson models (see Achcar *et al.*, 2007). The advantage is that besides indicating on average, the number of peaks that occur in a given period of time, stochastic volatility models also permit to analyse if the variability of the measurements stays stable. That is important since environmental authorities are interested not only in decreasing pollution levels but also in keeping the measurements stabilised in a low level.

Acknowledgements

The authors thank Guadalupe Tzintzun for providing the data from the monitoring network and also for providing information on the environmental issues. J.A.A. was partially funded by a CNPq grant number 300235/2005-4. H.C.Z. thanks to FAEPA from Hospital das Clínicas de Ribeirão Preto, FMRP-USP for financial support. E.R.R. thanks the Department of Statistics at the University of Oxford, where part of this work was developed, for all the support received during her stay at the Department. E.R.R was partially funded by DGAPA-UNAM grant number 968SFA/2007.

References

Achcar, J. A.; Fernández-Bremauntz, A. A.; Rodrigues, E. R.; Tzintzun, G. (2007). Estimating the number of ozone peaks in Mexico City using a non-homogeneous Poisson model. *Environmetrics* (http://www.interscience.wiley. com/10.1002/env.890).

Álvarez, L. J.; Fernández-Bremauntz, A. A.; Rodrigues, E. R.; Tzintzun, G. (2005). Maximum a posteriori estimation of the daily ozone peaks in Mexico City. *Journal of Agricultural, Biological, and Environmental Statistics*, **10**, 276–290.

Austin, J.; Tran, H. (1999). A characterization of the weekday-weekend behavior of ambient ozone concentrations in California. In: *Air Pollution* **VII**, 645–661. UK: WIT Press.

Bell, M. L.; McDermontt, A.; Zeger, S. L.; Samet, J. M.; Dominici, F. (2004). Ozone and short-term mortality in 95 US urban communities, 1987-2000. *Journal of the American Medical Society*, **292**, 2372–2378.

Bell, M. L.; Peng, R.; Dominici, F. (2005). The exposure-response curve for ozone and risk of mortality and the adequacy of current ozone regulations. *Environmental Health Perspectives*, **114**, 532-536.

Bell, M. L.; Goldberg, R.; Rogrefe, C.; Kinney, P. L.; Knowlton, K.; Lynn, B.; Rosenthal, J.; Rosenzweig, C.; Patz, J. A. (2007). Climate change, ambient ozone, and health in 50 US cities. *Climate Change*, **82**, 61-76.

Bernardo, J. M.; Smith, A. F. M. (1995). *Bayesian Theory*. New York: John Wiley and Sons.

Bollerslev, T. (1986). Generalized autoregressive Conditional heterocedasticity. *Journal of Econometrics*, **31**, 307–327.

Carlin, B. P.; Louis, T. A. (2000). *Bayes and Empirical Bayes methods for data analysis*. Second Edition. Boca Raton: Chapman and Hall Press.

Danielsson, J. (1994). Stochastic volatility in asset prices: estimation with simulated maximum likelihood. *Journal of Econometrics*, **64**, 375–400.

Engle, R. F. (1982). Autoregressive conditional heteroscedasticity with estimates of the variance of United Kingdon inflation. *Econometrica*, **50**, 987–1007.

Gelfand, A. E.; Smith, A. F. M. (1990). Sampling-based approaches to calculating marginal densities. *Journal of the American Statistical Association*, **85**, 398–409.

Ghysels, E.; Harvey, A. C.; Renault, E. (1996). A stochastic volatility. In: Rao, C.R.; Maddala, G.S., editors. Statistical Models in Finance. Amsterdam: North-Holland.

Guardani, R.; Nascimento, C. A. O.; Guardani, M. L. G.; Martins, M. H. R. B.; Romano, J. (1999). Study of atmospheric ozone formation by means of a neural network based model. *J. Air and Waste Management Assoc.*, **49**, 316–323.

Guardani, R.; Aguiar, J. L.; Nascimento, C. A. O.; Lacava, C. I. V.; Yanagi, Y. (2003). Ground-level ozone mapping in large urban areas using multivariate analysis: application to the São Paulo Metropolitan Area. *J. Air and Waste Management Assoc.*, **53**, 553–559.

Horowitz, J. (1980). Extreme values from a nonstationary stochastic process: an application to air quality analysis. *Technometrics*, **22**, 469–482.

Javits, J. S. (1980). Statistical interdependencies in the ozone national ambiente air quality standard. *J. Air Poll. Control Assoc.*, **30**, 58–59.

Kim, S.; Shepard, N.; Chib, S. (1998). Stochastic volatility: likelihood inference and comparison with ARCH models. *Review of Economic Studies*, **65**, 361–393.

Lanfredi, M.; Macchiato, M. (1997). Searching for low dimensionality in air pollution time series. *Europhysics Lett.*, **40**, 589-594.

Larsen, L. C.; Bradley, R. A.; Honcoop, G. L. (1990). A new method of characterizing the variability of air quality-related indicators. In: *Air and Waste Management Association's International Specialty Conference of Tropospheric Ozone and the Environment*. California. USA.

Leadbetter, M. R. (1991). On a basis for "peak over threshold" modeling. *Statistics and Probability Letters*, **12**, 357–362.

Loomis, D. P.; Borja-Arbuto, V. H.; Bangdiwala, S. I.; Shy, C. M. (1996). Ozone exposure and daily mortality in Mexico City: a time series analysis. *Health Effects Institute Research Report*, **75**, 1–46.

Meyer, R.; Yu, J. (2000). BUGS for a Bayesian analysis of stochastic volatility models. *Econometrics Journal*, **3,** 198–215.

Pan, J.-N.; Chen, S.-T. (2007). Monitoring long-memory air quality data using ARFIMA model. *Environmetrics* (http://www.interscience.wiley.com/10.1002 /env.882).

Raftery, A. E. (1989). Are ozone exceedance rate decreasing?, Comment of the paper "*Extreme value analysis of environmental time series: an application to trend detection in ground-level ozone*" by R. L. Smith. *Statistical Sciences*, **4,** 378–381.

Roberts, E. M. (1979a). Review of statistics extreme values with applications to air quality data. Part I. Review. *Journal of the Air Pollution Control Association*, **29**, 632–637.

Roberts, E. M. (1979b). Review of statistics extreme values with applications to air quality data. Part II. Applications. *Journal of the Air Pollution Control Association*, **29**, 733–740.

Smith, R. L. (1989). Extreme value analysis of environmental time series: an application to trend detection in ground-level ozone. *Statistical Sciences*, **4**, 367–393.

Smith, A. F. M.; Roberts, G. O. (1993). Bayesian Computation via the Gibbs Samples and related Markov Chain Monte Carlo Methods (with discussion). *Journal of the Royal Statistical Society Series B*, **55**, 3–23.

Spiegelhalter, D. J.; Best, N. G.; Carlin, B. P.; Van der Linde, A. (2002). Bayesian measures of model complexity and fit. *Journal of the Royal Statistical Society Series B*, **64**,583–639.

Spiegelhalter, D. J.; Thomas, A.; Best, N. G.; Lund, D. (2003). *Winbugs user manual.* Cambridge. United Kingdon: MRC Biostatistics Unit.

Wilson, R.; Colone, S. D.; Spengler, J. D.; Wilson, D. G. (1980). *Health effects in fossil fuel burning: assessment and mitigation.* Cambridge, USA:Ballenger.

Yu, J., (2002). Forecasting volatility in the New Zeland stock market. *Applied Financial Economics*, **12**,193–202.

Yu, J.; Meyer, R. (2005). Multivariate Stochastic Volatility models: Bayesian estimation and model comparison. To appear in *Econometrics Review.*

Zolghadri, A.; Henry, D. (2004). Minmax statistical models for air pollution time series. Application to ozone time series data measured in Bordeaux. *Environmental Monitoring and Assessment*, **98**, 275-294.

In: Air Quality in the 21st Century
Editors: G. C. Romano and A. G. Conti, pp. 311-338
ISBN 978-1-60456-793-9

Chapter 8

OPTIMIZATION APPROACHES FOR AIR QUALITY MONITORING NETWORK DESIGN

Esmaeil Fatehifar
Faculty of Chemical Engineering, Sahand University of technology, Tabriz, Iran
Ali Elkamel
Chemical Engineering Department, School of Engineering, University of Waterloo, Waterloo, ON. Canada
Mufreh Al-Rashidi
Coastal & Air Pollution Department, Environmental & Urban Development Division, Kuwait Institute for Scientific Research,
Po Box 24885, Safat 13109

ABSTRACT

Air pollution sampling site selection is one of the most important and yet most vexing of the problems faced by those responsible for regional and urban air quality management and for the attainment and maintenance of national ambient air quality standards. This chapter will present different optimization techniques for the design of optimal Air Quality Monitoring Networks (AQMN). The objective of the optimization is to provide maximum information about the presence and level of atmospheric contaminants in a given area and with a limited budget. A criteria for assessing the allocation of monitoring stations is developed by applying a utility function that can describe the spatial coverage of the network and its ability to detect violations of standards for multiple pollutants. The use of mathematical models based on the Multiple Cell Approach (MCA) to create spatial distributions for the concentrations of the pollutants emitted from different emission sources will be illustrated. The optimization techniques presented in the chapter are illustrated on a number of case studies and the number of monitoring stations and their locations for each case are obtained. In addition, the effect of the spatial correlation coefficient on total area coverage are discussed.

INTRODUCTION

Since one can not hope to monitor air quality at all locations at all times, selection of sites to give a reliable and realistic picture of air quality becomes a major issue and at the same time a difficult task. The location (configuration) and the number of stations may be based on many factors, some of which may depend on limited resources, country regulations and local conditions. The combination of these factors has made air quality surveys more complex, requiring comprehensive planning to ensure that the prescribed objectives can be attained in the shortest possible time and at the least cost. Furthermore, the choice and siting of the measuring network represents a factor of significant economic relevance for policymakers. In view of the fact that costs of equipment, maintenance and operating personnel are increasing dramatically, the possibility of optimizing monitoring design, is most attractive to the directors of air quality management programs.

AQMN-RESULTS

The result of the AQMN-design study will provide complete insight in all aspects needed to implement the network. This includes recommendations on:

- Minimum requirements for monitoring equipment and additional laboratory facilities.
- Lay-out and instrumentation of monitoring stations.
- Selection of monitoring sites (sites should be sensitive to sources and $\dfrac{\partial C}{\partial t} \Big/ \dfrac{\partial Q}{\partial t} = \dfrac{\partial C}{\partial Q}$ should be maximized)
- Data communication and data management.
- Data analysis and dispersion modeling.
- Network organization, staff requirements and training demands.
- Quality assurance and quality control.
- Maintenance and repair.
- Implementation schedule.

AIR QUALITY MONITORING SITING CRITERION

Table (1) gives the various criteria one would consider in attempting to meet the six objectives. It could be noted from the table that the representation of spatial-temporal patterns is one of the important objective of the monitoring networks as well as the violation of legal standards. Therefore, in the siting criterion the designer of an AQMN must taken into account the objectives listed in the table and the cost criteria that indicates the number of monitoring stations to be located.

METHODOLOGY DESCRIPTION

Usually, either concentration values coming from experimental campaigns of measures of the atmospheric pollution or approximations of such concentrations coming from a mathematical model are employed for designing a network. Normally, even when many real observations are available, the use of a mathematical model (i.e., Industrial Source Complex (ISC) or Multiple-cell Approach) is recommended for generating a more extensive database. In this way, an initial theoretical network as dense as desired can be generated. For the designed network to be effective, air quality data should be collected during a sufficiently extended period for the final data set to be representative of the area in which the network is to be located. One year is usually considered as minimum for reflecting the meteorological variability of the zone.

Starting from a set of potential sites in which a monitoring station can be located, the basic idea of the method consists of building up a utility function to meet the above siting principles and taking into account the coverage area and the legal violations detected of each station.

Table 1. Siting criteria for different objectives of the monitoring network (extracted and compiled from the WHO, 1977)

Monitoring Objective	Siting Criterion
1. Assess Compliance with air quality standards	Locate stations where concentrations are expected to be largest or locate stations where the spatial-temporal concentration distributions can be estimated most accurately
2. Assess long-term trends	Locate stations where concentrations are expected to be largest or locate stations where the spatial-temporal concentration distributions can be estimated most accurately
3. Provide data during episodes	Locate stations where concentrations are expected to be largest under conditions of stagnation or locate stations where the spatial-temporal concentration distributions can be estimated most accurately
4. Monitor source compliance with regulation	Locate stations at points where the sensitivity of concentration levels to source emission level changes is greatest or locate the stations at the high violation of the standard
5. Provide data to support enforcement actions	Locate stations at points where the sensitivity of concentration levels to source emission level changes is greatest or locate the stations at the high violation of the standard
6. Provide data for research	For the evaluation of diffusion models, locate stations where the spatial-temporal concentration distributions can be estimated most accurately

1. Representation of Spatial-Temporal Patterns

The spatial station coverage also called "Sphere of Influence (SOI)" is defined as the surrounding area over which the air quality data for a given station can be considered to be representative (Liu, 1986). The approach used in this study to calculate the SOI is based on the similarity between the information contained in a given station compared with the rest of them. To do this, the statistical properties of the spatial distributions of the pollutant concentrations are taken into account by the mean of the spatial correlation coefficient (r), calculated from the concentration values measured (or predicted) at each monitoring station. In this way, the spatial correlation coefficient provides an indication of the relationship between stations.

Leting $C_1 = (C_{11}, C_{12}, C_{13}, \ldots, C_{1n})$ and $C_2 = (C_{21}, C_{22}, C_{23}, \ldots, C_{2n})$ denote the pollutant concentrations in two different network locations measured at the same time, the spatial correlation coefficient for a sample size n can be expressed as :

$$r = \frac{\sum_{i=1}^{n}(C_{1i} - \bar{C}_1)(C_{2i} - \bar{C}_2)}{\sqrt{\sum_{i=1}^{n}(C_{1i} - \bar{C}_1)^2 \sum_{i=1}^{n}(C_{2i} - \bar{C}_2)^2}}, \tag{1}$$

Where $\bar{C}_1 = \frac{1}{n}\sum_{i=1}^{n} C_{1i}$ *and* $\bar{C}_2 = \frac{1}{n}\sum_{i=1}^{n} C_{2i}$ are the mean concentrations at location 1 and location 2, respectively.

The justification of the adopted approach is based on the fact that the correlation coefficient for concentration fluctuations is expected to decrease as the distance from the first station increases, as shown in Figure (1). This correlation coefficient can vary between 1 and -1.

Therefore, a cutoff distance S_c can be found so that the correlation coefficient is expected to be less than a certain value r_c. The assumptions implicit in this approach are:

- The data sets C_1 and C_2 are two correlated variables following a normal bivariate distribution.
- There are no significant temporal variations that could introduce spurious autocorrelation coefficients. Examples of such situations would be that of seasonal, weekly and diurnal cycles normally observed in air pollution concentrations.

After this, it can be said then that the sphere of influence of a station is the area surrounding it in which the spatial correlation coefficient of this station with the neighboring points is above a certain cutoff value. This means that the air quality data measured at this station can be considered representative and extrapolable with a certain degree of confidence to any point in this area.

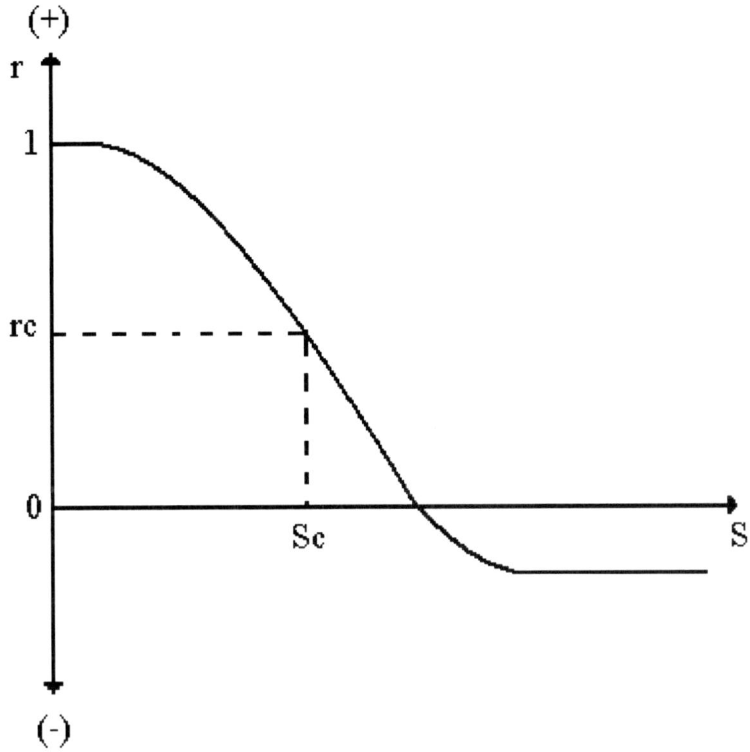

Figure 1. Correlation coefficient (r) versus distance (s) (Liu, 1986).

In fact, the value of r_c dose not imply a causal relationship between C_1 and C_2, but the existence of an association between both data sets, such that $100 r_c^2$ represents the percentage of concentration variations measured at one station explained by concentration variations measured at the other station (Ezekiel, 1941). The relation:

$$\text{Variance explained} = r_c^2 , \qquad (2)$$

is valid assuming a sufficient high sample size. Otherwise, it should be corrected as a function of the available number of samples.

In summary, the characterization procedure for the SOI of a station consists of:

1. Choosing the value of the explained variance.
2. Calculating the value of r_c by Equation (2).
3. In the case of having few samples, correcting the previous value with tables.

Once the SOI of a station has been characterized, the Coverage Area (CA) of this sphere is defined as the number of potential monitoring sites placed inside it. Herein, it is denoted by *"a pattern score* N_p^t".

The optimization problem could therefore be considered as the identification of the optimum number of monitoring stations and their configuration such that a variance of the order of r_c^2 is explained for a maximum area of the region at a minimum possible overlap. Before beginning with the optimization procedure, it is necessary to define terms such as *overlap region, coverage area, effective gain, effective coverage, pattern scores and coverage effectiveness* (Modak, 1985, a). Figure (2) shows the coverage areas and the overlap regions for 4 locations out of a mesh of candidate grid points. A *coverage area* or the SOI for a location is simply an area of which a variance of the order of r_c^2 is explained. An *overlap* for locations *i* and *j* is then defined as the common coverage area, i.e., the area common to the coverage area of locations *i* and *j*.

The coverage area has been quantified herein in terms of "*pattern scores*". A *pattern score* N_p^i for the *i*th candidate location is defined as the number of locations correlated to time series of concentrations (simulated at this location), above the specified cut off correlation coefficient r_c. Figure (2) shows the pattern scores for four locations arbitrarily selected. Based on the above definition of pattern score, the *coverage area* for location *i* could be defined as SOI_i where SOI is a set of locations correlated with *i* above the stipulated cut off correlation coefficient r_c.

In terms of set operation then, the *overlap* for locations *i* and *j* could be expressed as $(SOI_i \cap SOI_j)$. Furthermore, the *effective coverage* for both of these locations could be represented as $(SOI_i \cup SOI_j)$ and the *effective gain* on combining location *j* with *i* would be $(SOI_i \cup SOI_j - SOI_i \cap SOI_j)$. The *coverage effectiveness* for a combination of locations is then defined as the ratio of effective coverage (in terms of the pattern scores) and that of the number of candidate monitoring locations.

Since the interest of the optimization problem is to achieve a maximum coverage effectiveness at a minimum overlap, the first objective of the optimization problem could be formulated as, (in set notations)

$$\text{Maximize } (SOI^1 \cup SOI^2 \cup SOI^3 \cup \ldots \cup SOI^m). \tag{3}$$

Such that

A Partial Cover Problem is achieve

$$m = m^0 \tag{4}$$

or,

A Total Cover Problem is achieve

$$(SOI^1 \cup SOI^2 \cup SOI^3 \cup \ldots \cup SOI^m) = (SOI^1 \cup SOI^2 \cup SOI^3 \cup \ldots \cup SOI^N) \tag{5}$$

Where,
m = the number of monitors,
m^0 = the maximum number of monitors based on the budgetary constraint,
N = the number of candidate locations,
SOI^i = the set of N_p^i locations correlated with location *i*.

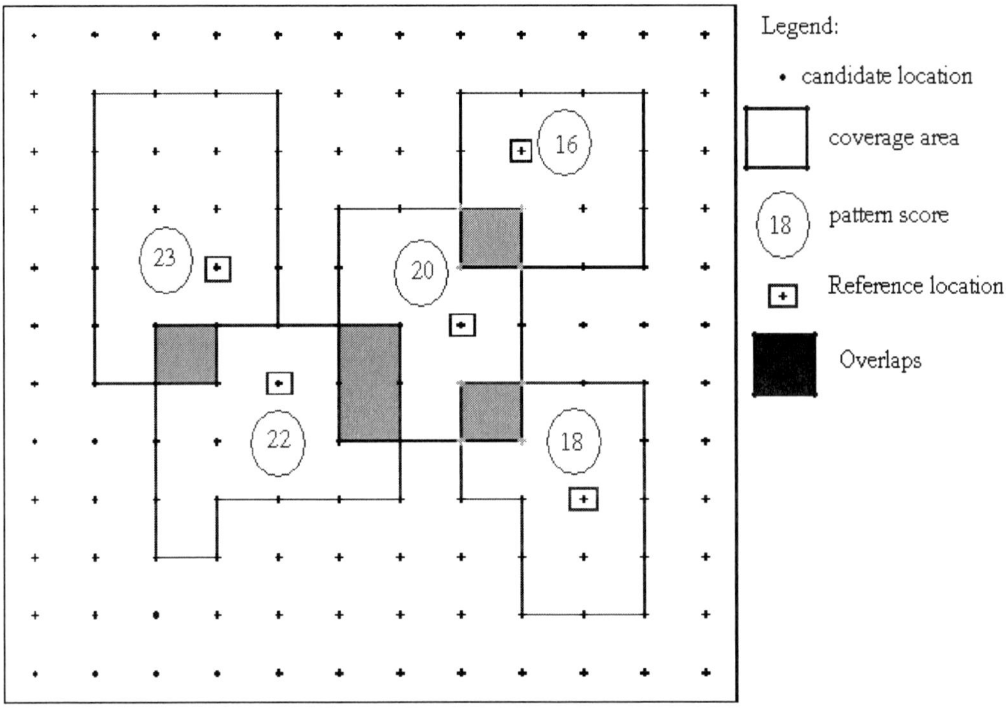

Figure 2. Representation of coverage area and overlaps for a typical grid of candidate locations.

At this stage the first objective of the optimization problem is formulated. The next section will present the second objective of the optimization problem based on the siting criteria listed in Table (1).

2. Detection of Violations over Ambient Standards

Several objective functions have been used in the past to measure the ability of a network to detect violations of standards. The approach taken here considers the potential of a monitoring site for the detection of violations in terms of violation scores. This choice was due to the need to incorporate in the multi-objective optimization algorithm the semantic distance (the relative significance) of the various alternative solutions, i.e. the alternative station locations in terms of meeting the objectives. A location with a high violation score is then considered to have a high potential for detection of violations. The computation of violation scores is essentially a weighted scoring corresponding to concentrations of SO_2, NO_2 and CO above prescribed thresholds. These scores are therefore dependent on,

1. The threshold levels.
2. The weighing factors between each threshold range and the weighing function.

A decision on thresholds and weighing factors (i.e. severity) is indeed pollutant-specific and further dependent upon the averaging time and populations concerned. Several weighing functions have been reported such as linear ones, segmented linear, non-linear, segmented non-linear etc. (Ott, 1978). In this chapter, a segmented nonlinear weighing function proposed by Modak et al. (1985, b) is used. The violation score for each candidate location is given by the equation,

$$N_v^i = \sum_{i=1}^{T} \sum_{k=1}^{N_t} \frac{(w_{k+1} - w_k)(x_i - x_k)X}{(x_{k+1} - x_k)} \tag{6}$$

Where,

N_v^i = the violation score for the ith candidate location,
w_k = the weighing factor corresponding to threshold x_k,
x_k = the k-th threshold,
$X = 0$ if $(x_i - x_k) \leq 0$,
$X = 1$ otherwise
N^t = the total number of thresholds,
T = the total number of simulated observations,

The weighing function, associated with the violation scores, is important because the severity of threshold violations reported by two monitoring stations may differ. In order to assist public authorities to manage and reduce health hazards and other risks from air pollutant, the US-EPA (40 CFR Part 58, Air Quality Index Reporting, Appendix-B) limit values for most of the common pollutants were used. Based on those guidelines, the values for thresholds of SO_2, NOx and CO were chosen as seen in Table (2). In addition, the air pollution index (weighing factor) ranging from 0.5 to 5 according to the severity of threshold exceedance is also given in the table. The dose-response function underlying the assignment of the weighing factors is fitted to a binomial. This mathematical form penalizes non-linearly the exceedance of higher threshold values with regard to exceedance of lower ones. In the absence of authoritative epidemiological data regarding air pollution effects on ecological vulnerability and public health, the network design adheres to the precautionary principle in environmental management.

Since the interest of the optimization problem is to achieve a maximum detection of violations over ambient air standard, the second objective of the optimization problem could be formulated as:

Table 2. Air Pollution Index assigned to weighing factors for the violation score

SO2 (μg/m3)	NOx (μg/m3)	CO (μg/m3)	Weighing factors
80	30	4000	0.5
120	80	6000	1
140	100	8000	2.5
160	130	13000	3
190	160	20000	5

$$\text{Maximize} \sum_{i=1}^{m} N_v^i \tag{7}$$

Such that
A Partial Cover Problem is achieve

$$m = m^0 \tag{8}$$

or,

A Total Cover Problem is achieve

$$(SOI^1 \cup SOI^2 \cup SOI^3 \cup \ldots \cup SOI^m) = (SOI^1 \cup SOI^2 \cup SOI^3 \cup \ldots \cup SOI^N) \tag{9}$$

Where,
N^i_v = potential of violation of the ith position,
m = the number of monitors,
m^0 = the maximum number of monitors based on the budgetary constraint,
N = the number of candidate locations,
SOI^i = the set of N^i_p locations correlated with location *i*.

At this point, the two objective functions constructed above can be used to build up the utility function in which the two designed parameters for each station are:

N_p, a decision variable associated with the first objective.
N_v, a decision variable associated with the second objective.

UTILITY FUNCTION APPROACH

In order to meet the two objectives described above, a utility function approach is used in this study. This approach is based on the establishment of a utility function (UF) that encompasses the multiple objectives to design the monitoring network. A general form of the UF could be described as:

$$UF = f(d1, d2, d3, \ldots, dn) \tag{10}$$

Where d1, d2, d3, … and dn are the decision variables related to each of the objectives. In this case, the UF could be expressed as:

$$UF = f(N_p, N_v) \tag{11}$$

where
N_p = the pattern score,
N_v = the violation score.

The two variables above are calculated separately for each grid-point in the area of interest, which is treated as a candidate location for placing a monitoring station.

1. Structure of the Utility Function

Several forms of UF could be used in order to combine the two mentioned variables (N_p, N_v). The suggested one by Modak at el. (1985, b) of the following form is used,

$$UF = N_p \text{ x } (N_v)^b \tag{12}$$

With $b \geq 0$,

The parameter b is a new degree of freedom that is used to weigh the relative importance given to each objective, for which an estimation procedure depending on the purpose of the network to be developed. In Perez Caseiras (1988) this new procedure is explained and it was shown that by taking values of b between 0.075 and 0.1 a good compromise is found between the two objectives.

The above structure of UF ensures that if any one of the decision variables becomes zero, the UF will also assume a zero value. This condition would imply that if a location has a high violation score, but zero effective number of pattern scores (i.e., it is already covered by the previous locations), then it is discarded in the optimal design. The optimization problem over utility could therefore be state as:

$$\text{Maximize} \sum_{i=1}^{m} N_p^i \tag{13}$$

while,

$$\text{Maximize} \sum_{i=1}^{m} N_v^i \tag{14}$$

Such that

$$(SOI^1 \cup SOI^2 \cup \ldots \cup SOI^m) = (SOI^1 \cup SOI^2 \cup \ldots \cup SOI^N)$$

or

$m = m^o$

Where,

N^i_v = potential of violation of the ith position,

m = the number of monitors,

m^0 = the maximum number of monitors based on the budgetary constraint,

N = the number of candidate locations,

SOI^i = the set of N^i_p locations correlated with location *i*.

2. Computational Algorithm of the Utility Function Based on Single Pollutant

To accomplish the above objectives through the use of the UF in this study, a selection methodology has been used on a sequential basis. This method could also be considred as a constructive method of heuristics, where the basic idea is to build up a single feasible solution, in a deterministic sequential fashion. This belongs to the general class of greedy algorithms. These do there best in each single decision step. A well known example of such an approach is the minimum spanning tree used in network analysis problems. The presented optimization algorithm has been therefore termed as Minimum Spanning Tree (MST) algorithm. A description of the optimization algorithm for the solution of the above problem is as follows:

20) Starting from the concentration data obtained through the mathematical model (i.e., Multiple Cell Approach or ISC) of the pollution phenomena, a network formed by a matrix of *M* x *N* cells where *M* is the number of observed (or predicted) concentrations at a potential locations numbered from 1 to *N*.
21) The correlation coefficient matrix $r_{i,j}$ is calculated, with $i = 1, 2, \ldots N; j = 1, 2, \ldots N$.
22) The SOI_i are calculated for $i = 1, 2, \ldots, N$ where the SOI_i is the set formed by locations *m* and correlated with location *i* in which the correlation coefficient $r_{i,m} \geq r_c$.
23) The violation score N^i_v are calculated for $i = 1, 2, \ldots, N$ using equation (3) and the coefficients shown in Table 2.
24) The coverage area in terms of pattern score N^i_p are formed for each SOI obtained and for $i = 1, 2, \ldots, N$.
25) The utility function sUF_i are calculated for $i = 1, 2, \ldots, N$ using equation (12) with a suitable value for *b*.
26) The *i*th location (station) of a maximum UF is chosen.
27) In order to avoid overlaps, the *i*th location(s) (stations) belonging to the SOI of the station being selected in step 7 are deleted from all sets SOI_i to which they belonged.
28) Back to step 5. The loop ends when the number of stations is adequate.

In this method, it is necessary to specify at the beginning the "adequate" or desired number of stations. This number depends on budgetary constraints, such that if an organization has \$ 10,000,000 to construct a monitoring network and each station cost \$ 1,000,000 they must build at most 5 monitoring stations. In this study and for illustration purposes five monitoring stations are considered for the area under study.

3. Computational Algorithm of the Utility Function Based on Multi Pollutants Network Design

The principle of Pareto optimality for multi objectives decisions is adopted in this study. The objective of Pareto optimal network is to look for something attractive for a group of pollutants as a whole rather than focusing on specific pollutants. The basic idea in such philosophy is to have maximum opportunity of measuring several pollutants at common sites. A design for such a compromise is described herein as the Pareto Optimal Design (POD). An approach of additive utility to accommodate a global interest of various sub-utilities (all of which are to be maximized) is normally categorized as a method of Pareto optimality.

A POD could be defined as follows: if the shape function U(u) is POD, then any other U(u) either the values of all the objective functions remain the same or at least one of them worsens compared with its value at U(u). A POD can be computed by maximizing or minimizing a single performance index obtained by combining the multiple objectives in a weighted sum as,

$$U^{i}(u) = \sum_{j=1}^{P} a_j\, u_j^{i} \tag{15}$$

$$\sum_{j=1}^{P} a_j = 1 \tag{16}$$

Where,

a_j = the importance associated with pollutant j;
u_j^i = the sub-utility for pollutant j at location i;
$U^i(u)$ = the cumulative utility for P pollutants at location i.

In the POD therefore, the interest of each pollutant is treated as a sub-utility and the optimization of the AQMN configuration is carried out for a single additive utility function. A POD formulation attains a general quasi-additive form, such as,

$$u_j^i = N_{pj}^i\,(N_{vj}^i)^b \text{ and } b \geq 0 \tag{17}$$

Where,

N_{pj}^i = the patterns score for pollutant j at location i;

N_{vj}^i = the violation score for pollutant j at location i.

The solution algorithm for the POD is easily derived as an extension of the basic utility approach. The various steps of this algorithm are similar to that of the algorithm for MST discussed above for the single pollutant case, except that the utilities are defined on a more comprehensive basis, i.e., in the interest of more than one pollutant as in equation (15). Figure

(3) shows the flowchart depicting the above described algorithm. This algorithm was implemented in MATLAB.

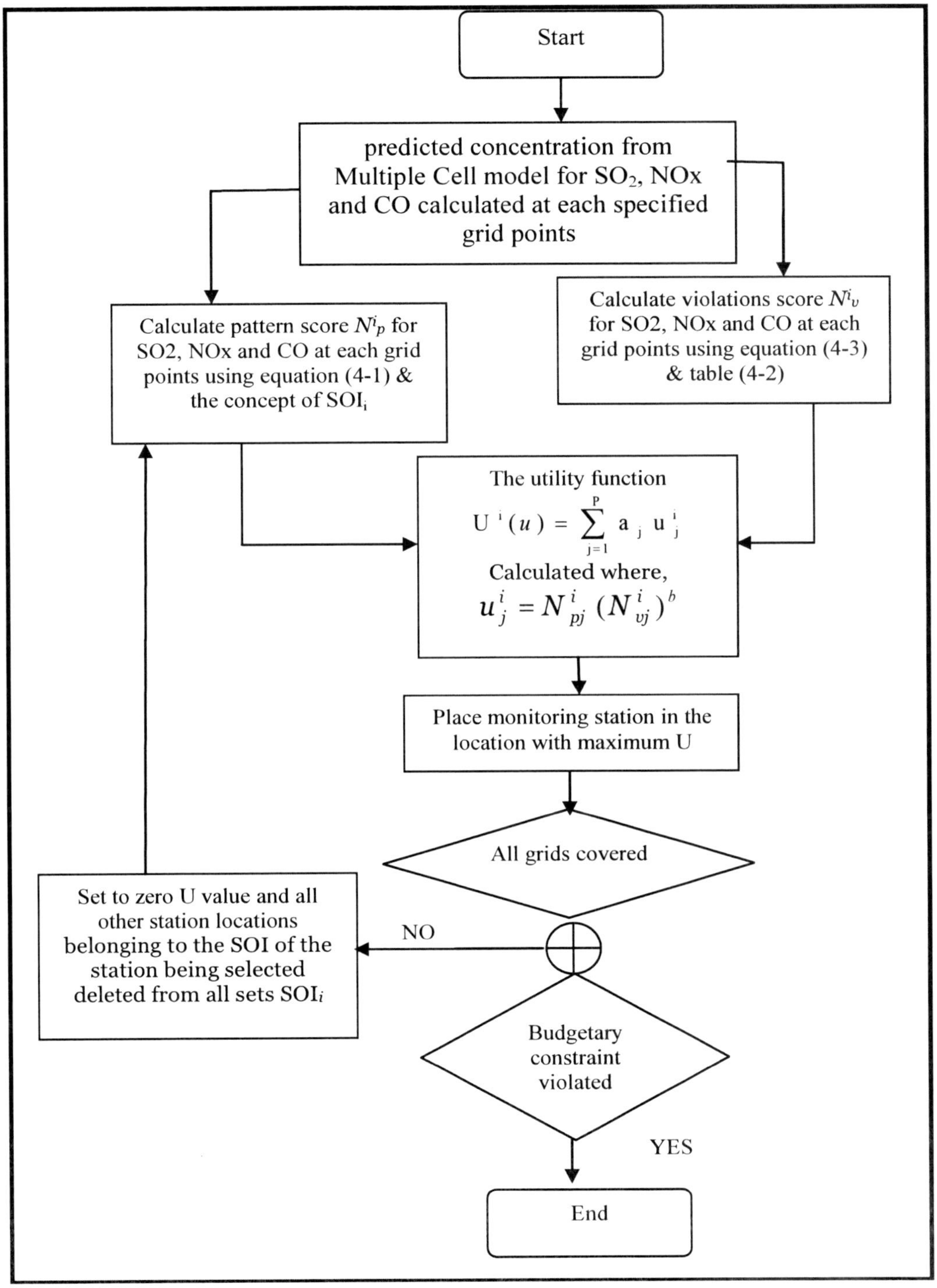

Figure 3. Flowchart of the procedure for allocating monitoring stations.

CASE STUDY

The optimization models discussed above were used to identify the optimal locations of the monitoring network within a 41 x 30 km of a typical pollutant oil field area in Kuwait and 10 x 5 km area of a Refinery in Iran.

STUDY AREA OPTIMIZATION MODELING RESULTS DISCUSSION

A matrix of 365 x 1280 for SO_2, NOx, and CO concentrations that has been generated at the specified 1280 candidate locations by the ISCST3 model was used as an input to the above algorithm. Figure (4) shows the emission sources (CG's) and the candidate locations (grid points), where each 4 connected points represent a 1 km^2 area to cover a 41 x 30 km of the vicinity under study. These candidate locations represent potential monitoring stations each with 365 daily predicted concentrations ready to be selected as optimal monitoring locations by the above algorithm.

Daily predicted concentrations of SO_2, NOx, and CO by the ISCST3 model within the typical pollutant oil field area were fairly low, and the violations of the daily Kuwait EPA ambient air quality standard (AAQS) did not exceed the EPA standards (Al-Rashidi, 2002). For the purpose of illustration therefore, the concentrations of SO_2 and NOx are scaled up by a factor of 10 whereas, the concentrations of CO are scaled up by a factor of 100 to justify the objective of compliance. It is to be noted that a constant multiplier such as 10 or 100 does not change the correlation coefficient matrix, so the analysis is not affected as far as the patterns are concerned.

The optimization models were applied to two different cutoff values in the correlation coefficient matrix. A value of 0.85 and 0.9 for rc were chosen to study the effect of rc on the percentage of coverage effectiveness of the monitoring networks. Figure (5) presents the effect of the rc value on the coverage effectiveness of the optimal selected locations for NOx as an illustration. It can be observed from the figure that as the cut-off correlation coefficient rc is increased from 0.85 to 0.9, the number of monitors required for the total representation of the pattern also increases. For a stipulated budget, the air quality monitoring organization (i.e., Kuwait EPA, US-EPA ...etc) could maintain either a high or a low rc value based network. A high rc based network may not necessarily cover the entire region, but the covered region will be well represented. A low rc based network on the other hand, would offer more coverage of the region, but the covered region may not be satisfactorily represented. The ultimate decision depends on is the air quality monitoring organization.

At the same time, the marginal effectiveness of the monitoring network decreases as the monitors are sequentially selected. In other words, the costs of additional monitors to cover the entire region, say beyond 20% (12 stations) to 25% (30 stations), are rather significant as compared to the accrued benefit. In many cases, therefore, it is possible that the organization would be interested in maintaining the AQMN only up to a gain of information of 20% for 12 stations.

Figures (5) to (11) show the concentrations distribution and the configuration of the optimal locations for an rc of 0.85 and 0.90 for SO_2, NOx, and CO, respectively for a maximum number of 10 stations. The number of stations is of course subject to the budgetary constraint. It could be noted that a network designed for a higher value of rc is not necessarily a simple extension, in terms of a mere addition of some more monitors, as compared to that of the network obtained at a lower rc value. It appears therefore, that the expansion of a network to a higher reliability of representation, may not only require an additional number of monitors but also a relocation of existing ones as well.

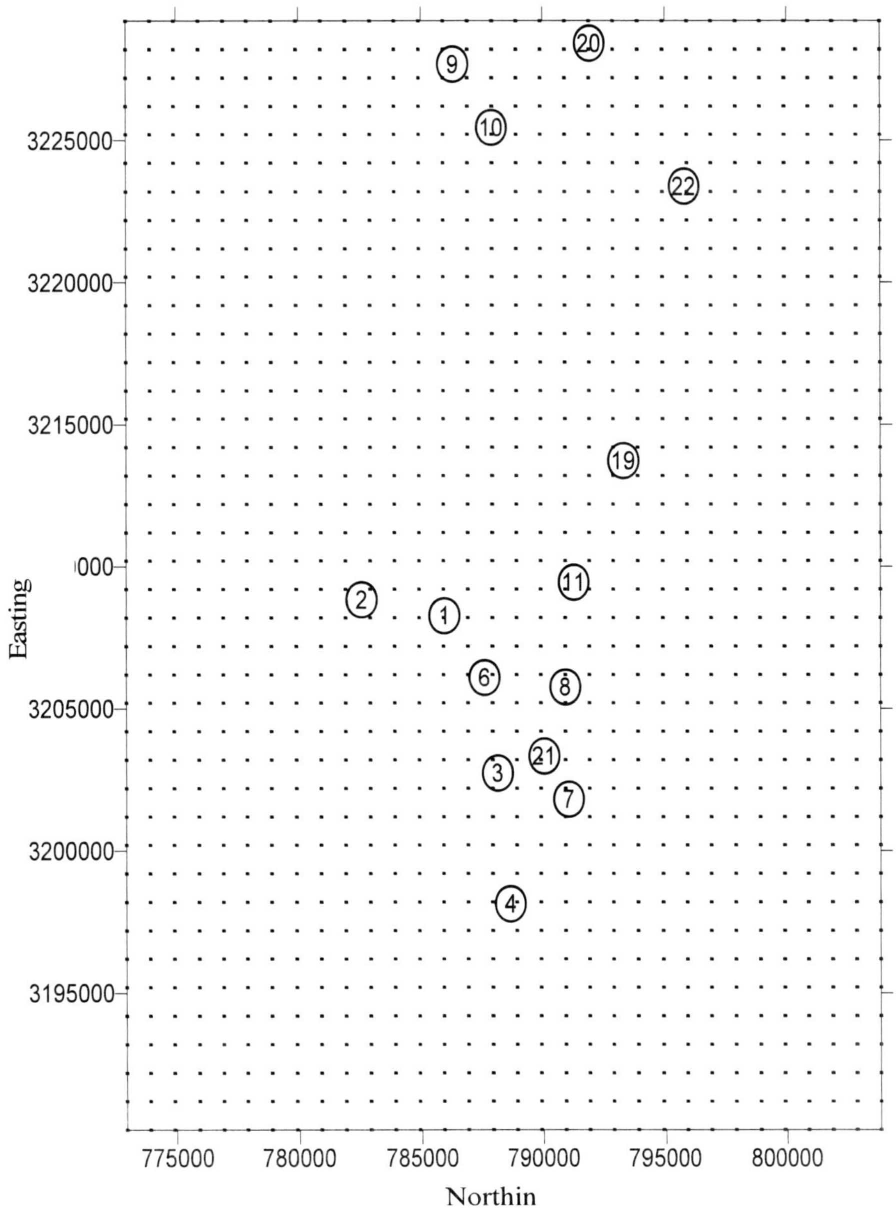

Figure 4. Emission Sources and Candidate locations of the study area.

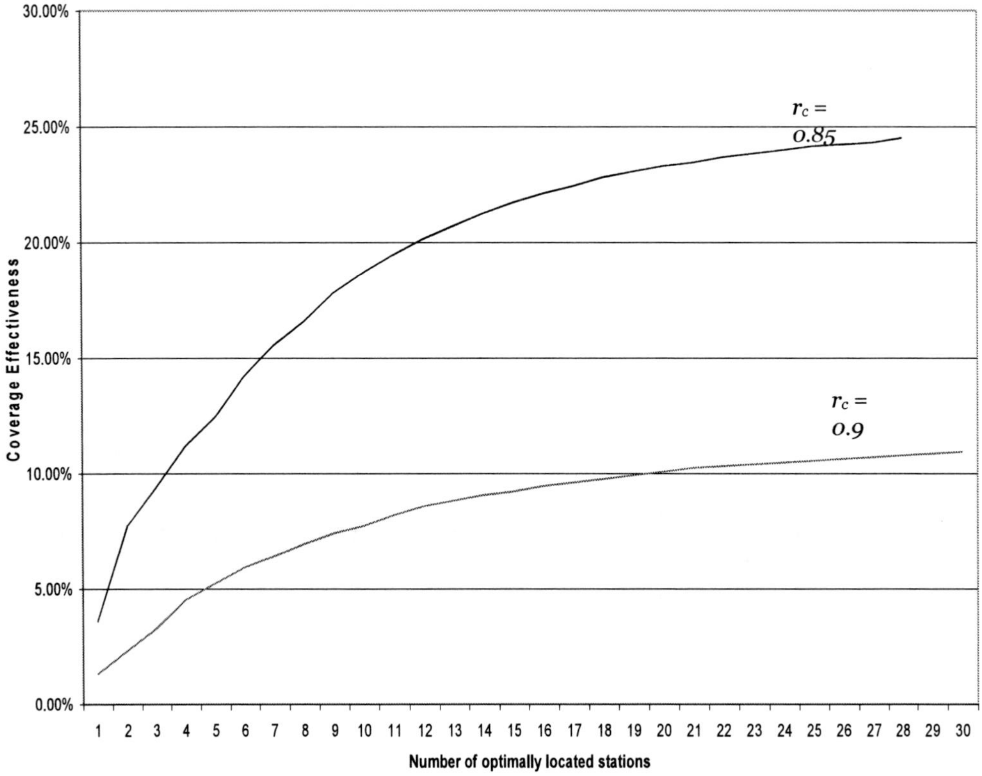

Figure 5. Relationship between cut-off correlation coefficient, number of monitors and the effectiveness of the optimally located stations.

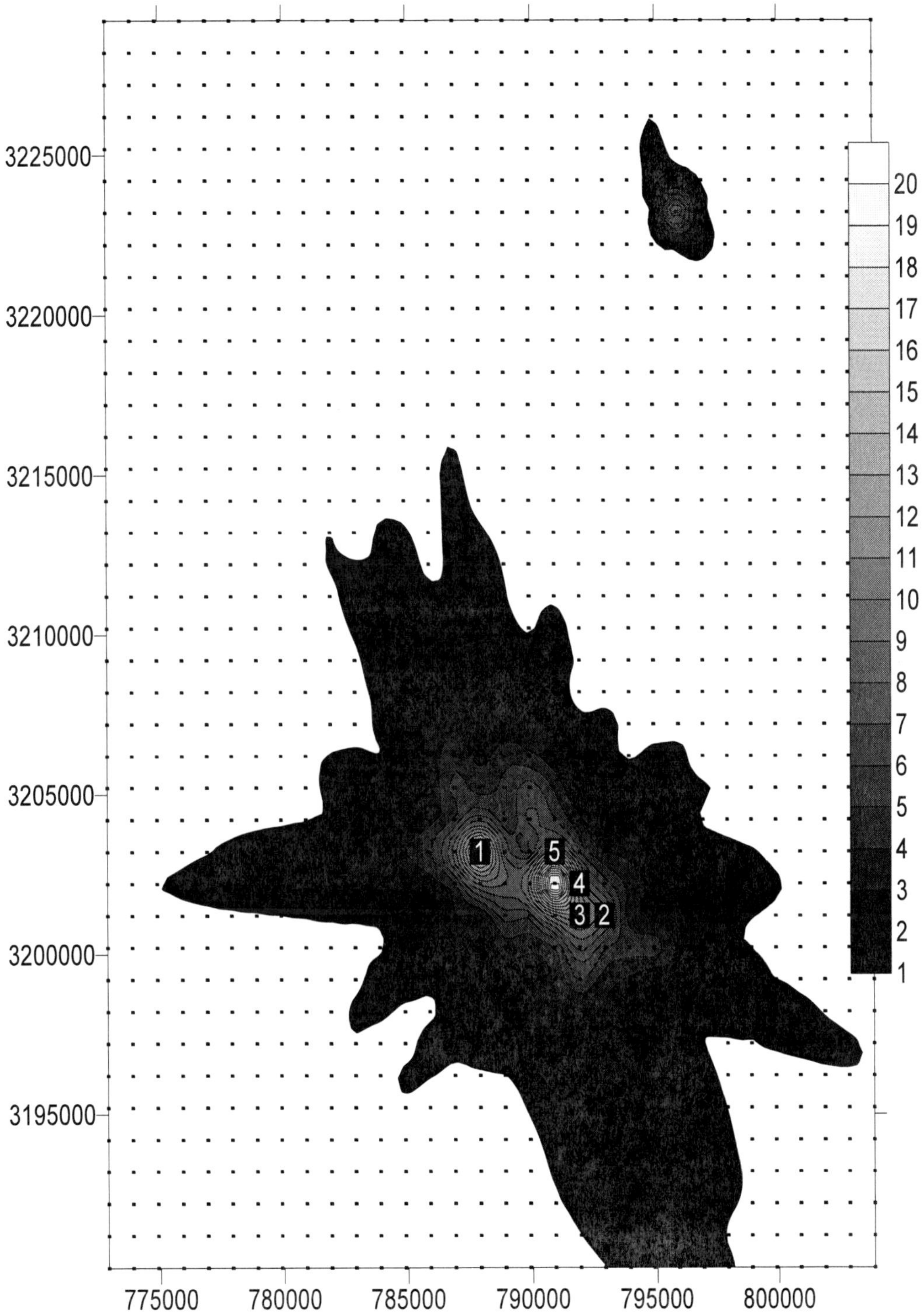
3225000
3220000
3215000
3210000
3205000
3200000
3195000
775000
780000
785000
790000
795000
800000
20
19
18
17
16
15
14
13
12
11
10
9
8
7
6
5
4
3
2
1
1
5
4
3
2

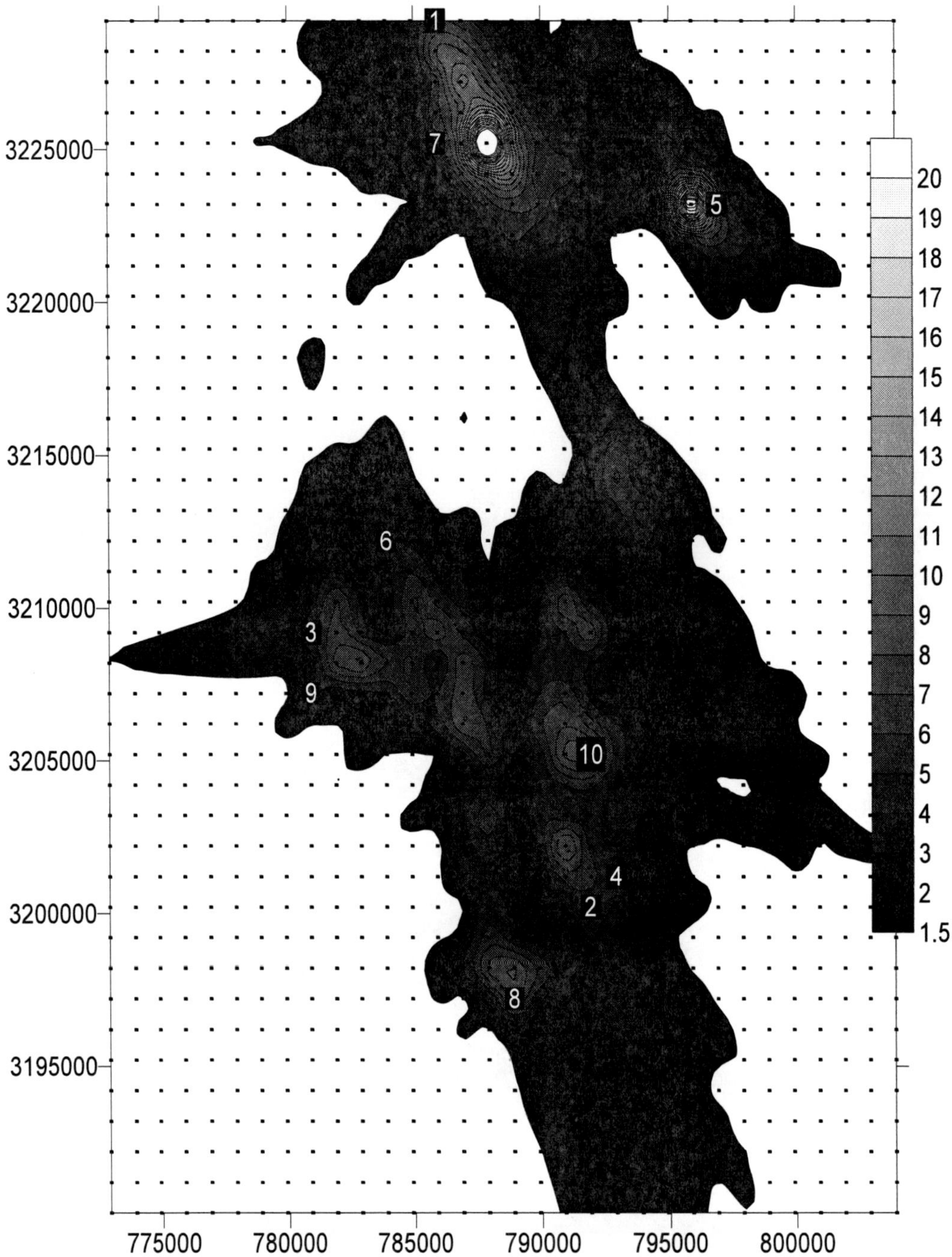
3225000
3220000
3215000
3210000
3205000
3200000
3195000
775000
780000
785000
790000
795000
800000
20
19
18
17
16
15
14
13
12
11
10
9
8
7
6
5
4
3
2
1.5
1
2
3
4
5
6
7
8
9
10

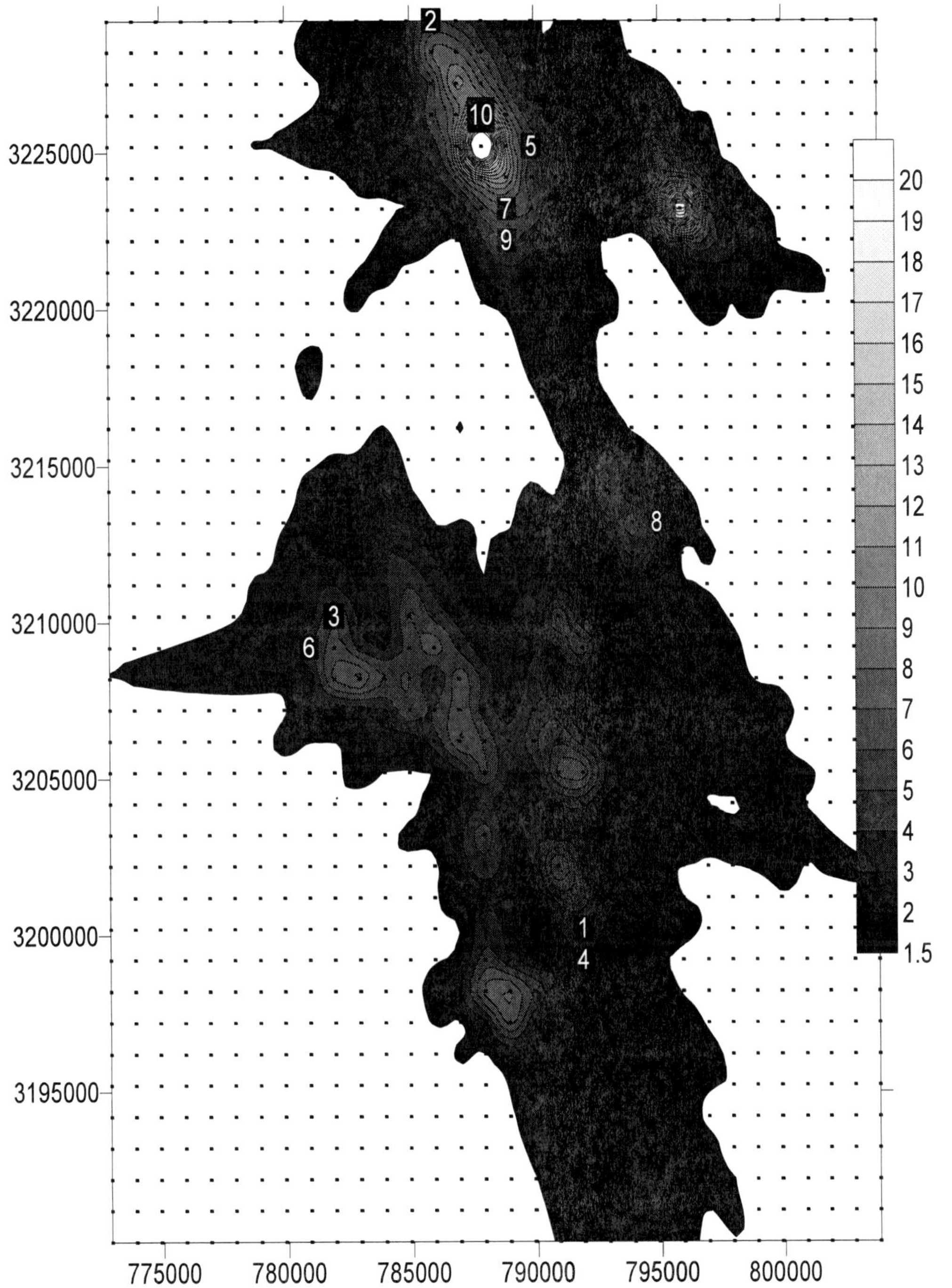
2
10
5
7
9
8
3
6
1
4
3225000
3220000
3215000
3210000
3205000
3200000
3195000
775000
780000
785000
790000
795000
800000
20
19
18
17
16
15
14
13
12
11
10
9
8
7
6
5
4
3
2
1.5

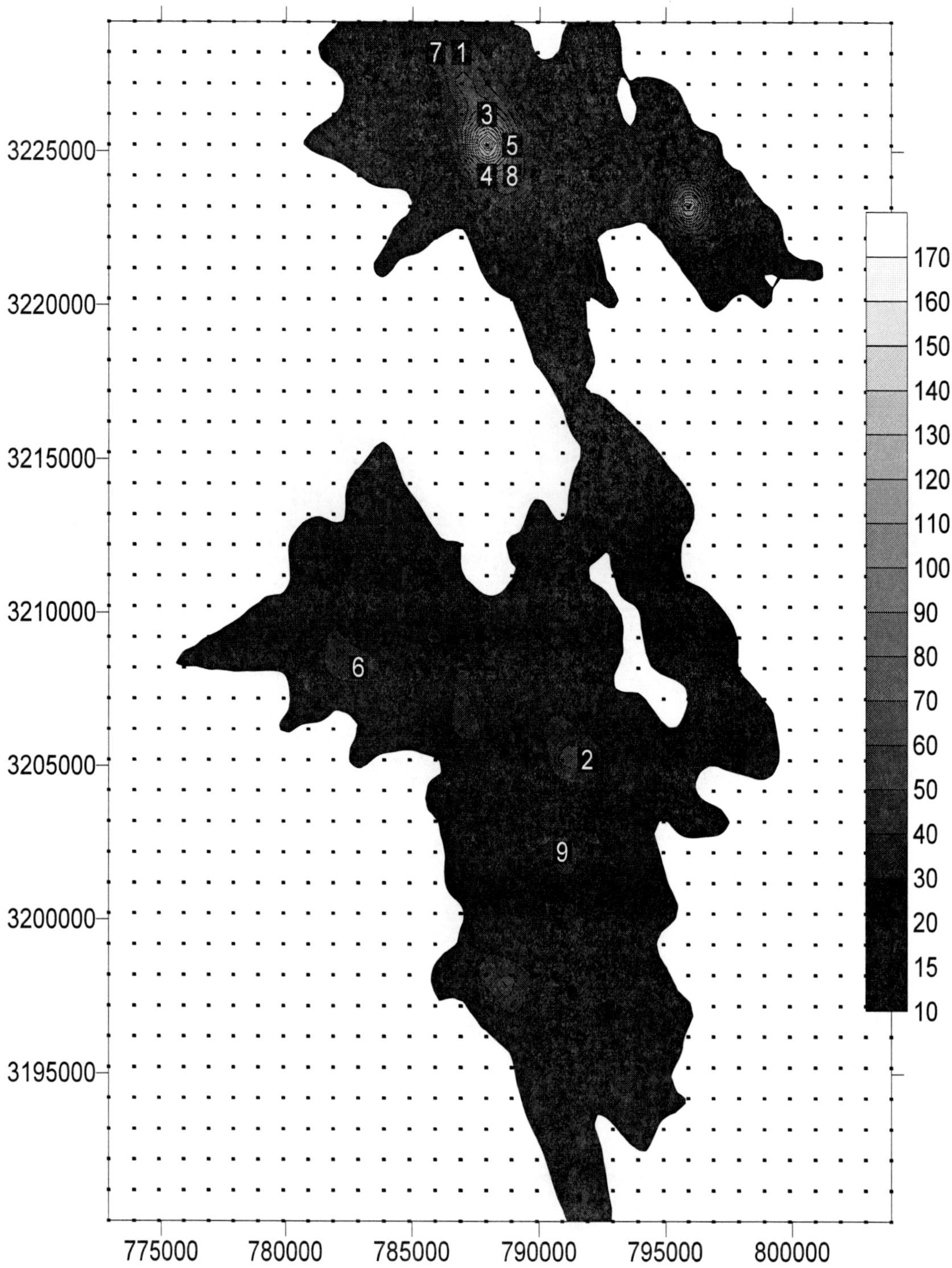
7
1
3
5
4
8
6
2
9
3225000
3220000
3215000
3210000
3205000
3200000
3195000
775000
780000
785000
790000
795000
800000
170
160
150
140
130
120
110
100
90
80
70
60
50
40
30
20
15
10

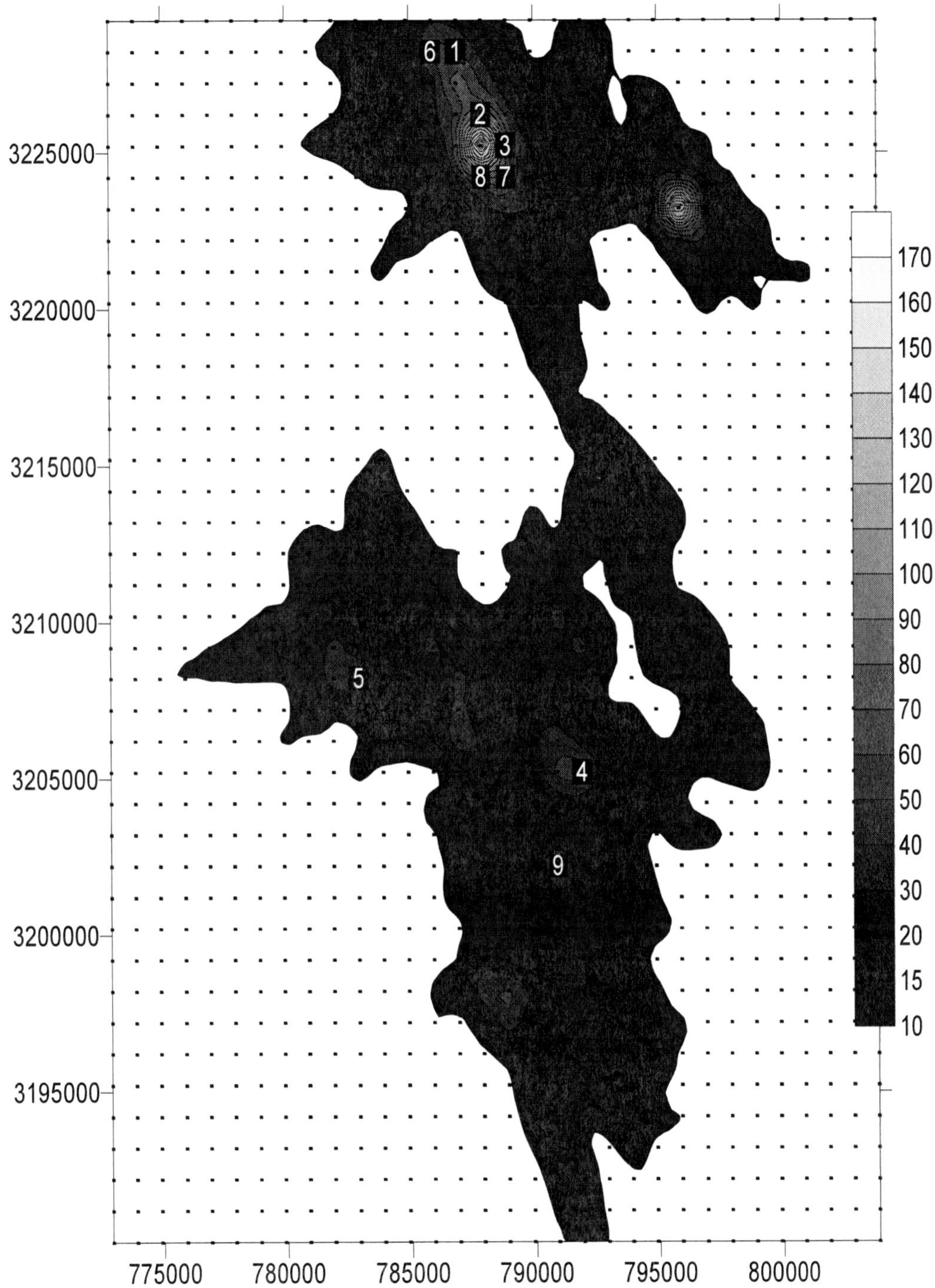
3225000
3220000
3215000
3210000
3205000
3200000
3195000
775000
780000
785000
790000
795000
800000
170
160
150
140
130
120
110
100
90
80
70
60
50
40
30
20
15
10
6
1
2
3
8
7
5
4
9

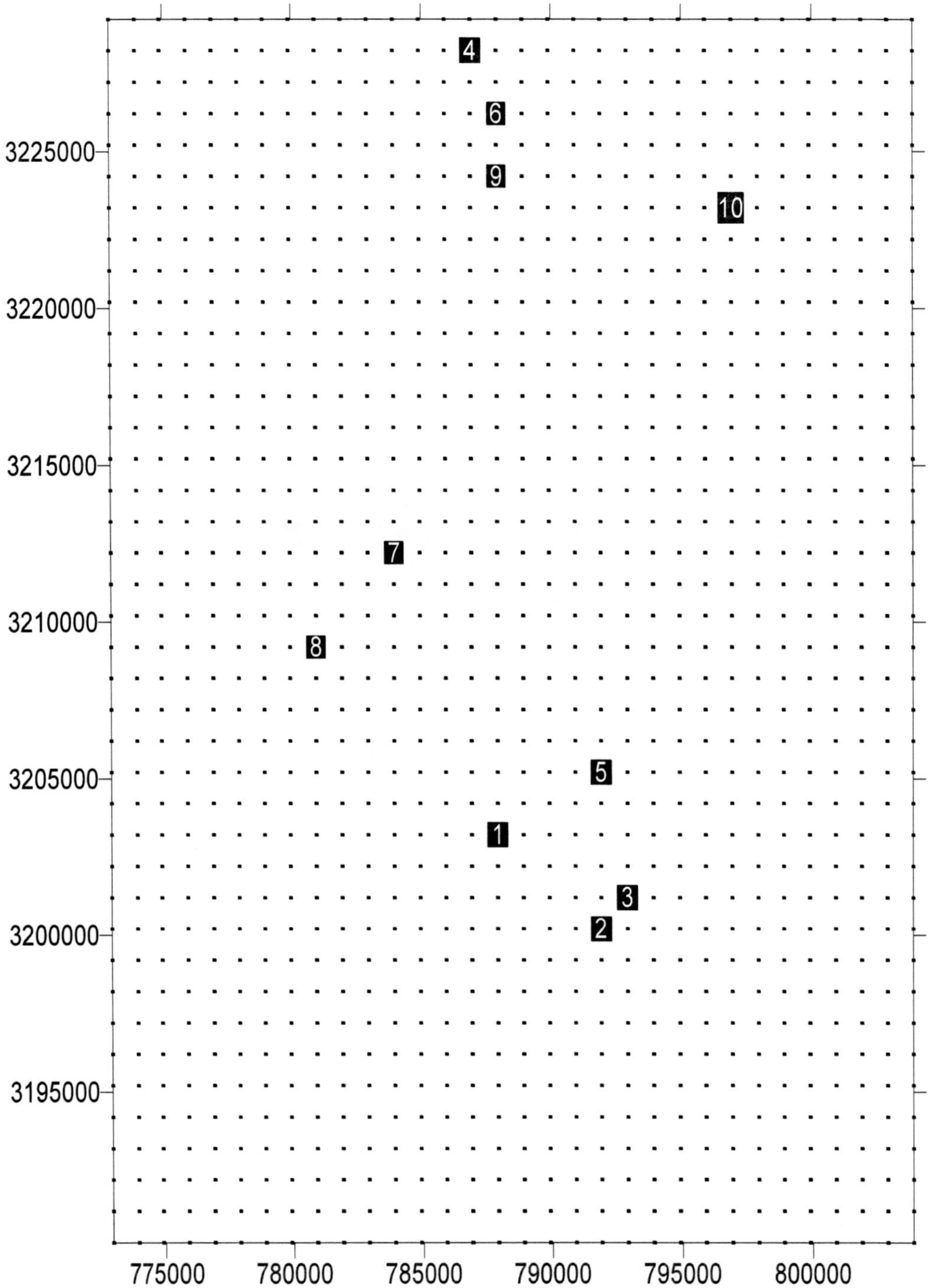
4
6
9
10
7
8
5
1
3
2
3225000
3220000
3215000
3210000
3205000
3200000
3195000
775000
780000
785000
790000
795000
800000

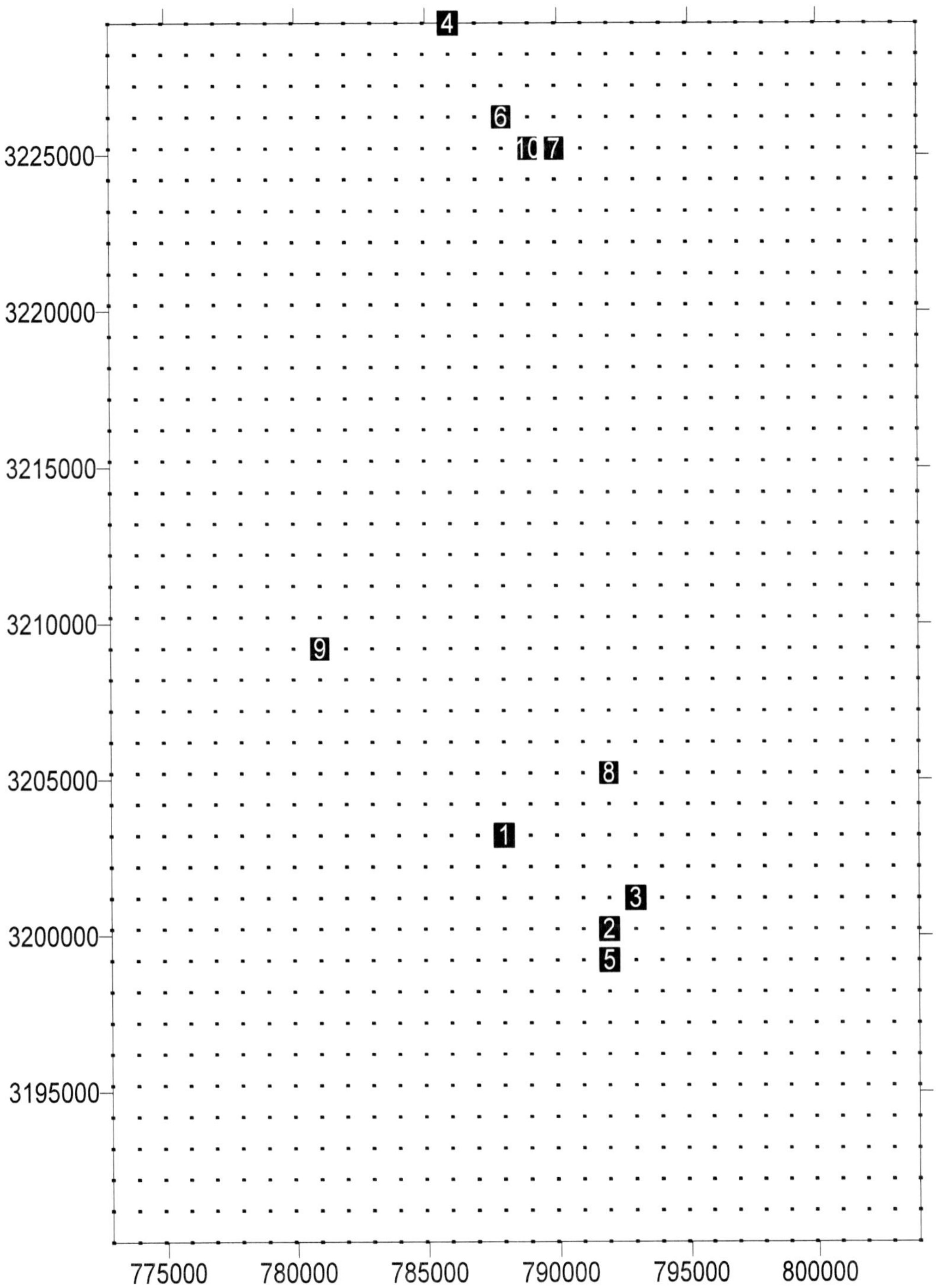
4
6
10
7
9
8
1
3
2
5
3225000
3220000
3215000
3210000
3205000
3200000
3195000
775000
780000
785000
790000
795000
800000

Table 3. Wind velocity distribution for different years

YEAR	JAN	FEB	MAR	APR	MAY	JUN	JUL	AGU	SEP	OCT	NOW	DEC
1995	2.1	2.15	2.15	4.05	3.95	3.6	4.45	4.15	3.3	3.15	2.45	2.75
1994	1.65	3.05	1.55	2.6	3.65	4.3	4.55	4.75	4.2	3.3	2.3	2.55
1993	2.25	2.75	2.25	5.65	4.1	3.75	4.6	4.35	4.3	4.15	3.5	2.3
1992	1.7	2.3	3.2	4.75	3.2	3.8	3.75	4.1	3.7	4.55	3.25	3.15
1991	1.6	1.9	4.05	4.9	4.55	4.85	4.75	4.4	3.75	3.55	2.95	2.2
1990	2	2.8	3.1	4.4	3.2	4.95	4.9	4.9	3.65	2.6	1.6	2.75

1. Single Pollutant Monitoring Network Optimization Modeling Results

The optimization model was used to find the optimal locations and the configuration of the monitoring networks for SO_2, NOx and CO respectively within the area under study. The 6 year (monthly) predicted concentrations of SO_2, NOx, and CO by the Multiple Cell Model (Fatehifar, 2005) within the Tabriz Refinery area were fairly low, and the violations of the EPA ambient air quality standard (AAQS) were never exceeded. For the purpose of illustration therefore, the concentrations of SO_2, CO and NOx were scaled up by a factors of 10000, 1000 and 100, respectively. It is to be noted that a constant multiplier such as 100 or 1000 does not change the correlation coefficient matrix, so the analysis is not affected as far as the patterns are concerned.

The optimization models were implemented for different cutoff rc values in the correlation coefficient matrix. Values in the range of 0.75 to 0.95 for rc were chosen to study the effect of rc on the percent of coverage effectiveness of the monitoring networks. Figure 12 presents the optimal monitoring network for CO with 0.75 and 0.95 rc values. The optimal locations for other rc values are shown in Table 4. It can be observed that as the cutoff correlation coefficient rc is increased from 0.75 to 0.95, the number of monitors required for the total representation of the pattern also increases. For a stipulated budget, the air quality monitoring organization could maintain either a high or a low value rc based network. A high rc based network may not necessarily cover the entire region, but the covered region will be well represented. A low rc based network on the other hand, would offer more coverage of the region, but the covered region may not be satisfactorily represented. The ultimate decision in such a case is of course left to the air quality monitoring organization.

Tables 5 and 6 show optimal locations of AQMN for NOx and SO_2 for a maximum number of 5 stations. The number of stations is of course subject to the budgetary constraint. It could be noted that a network designed for a higher value of rc is not necessarily a simple extension, in terms of a mere addition of some more monitors, as compared to that of the network obtained at a lower rc value. It appears therefore, that the expansion of a network to a higher reliability of representation, may not only require an additional number of monitors but perhaps also a relocation of existing ones as well.

Figures 13, 14 and 15 show the coverage efficiency of AQMN for different rc values. As shown in the figures, for this area, three stations for AQM are enough because, there is less difference when we use more than 3 stations. The coverage efficiency increases only by 2-3%.

Table 4. Optimal locations for AQMN for CO (x and y are in m)

Station	RC=0.75	RC=0.80	RC=0.85	RC=0.90	RC=0.95
1	X=800	800	800	800	800
	Y=320	320	320	320	320
2	7440	8480	7760	8240	7280
	320	320	320	320	320
3	2240	2320	2800	2800	9920
	320	320	160	160	160
4	9920	2960	2400	2480	2480
	640	160	320	320	320
5	3120	8320	8720	9920	2480
	160	640	160	160	160

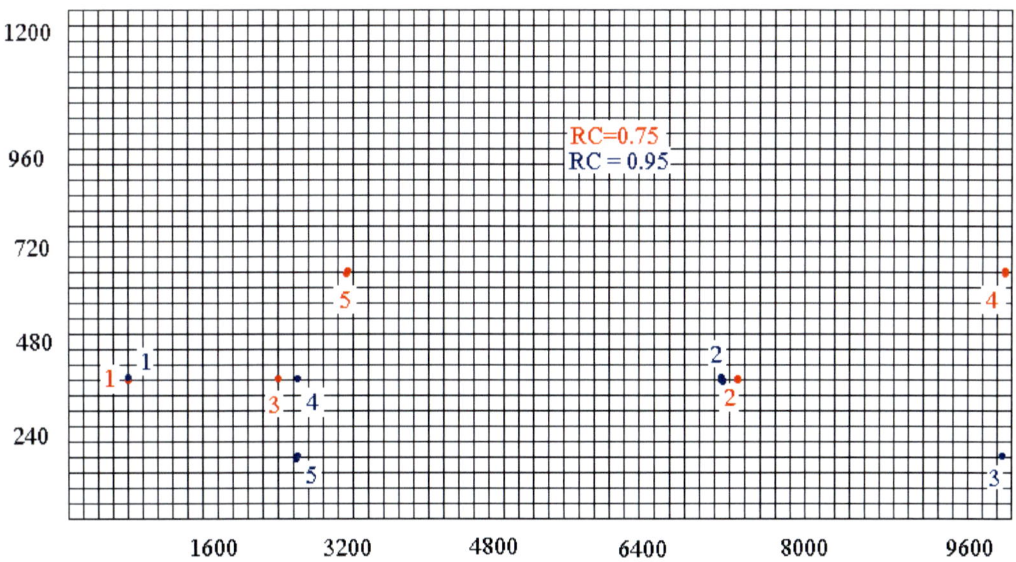

Figure 12. Optimal Location of AQMN for CO (meter).

Table 5. Optimal locations for AQMN for NOx (x and y are in m)

Station	RC=0.80	RC=0.85	RC=0.90	RC=0.95
1	4000	4240	4400	4320
	320	320	320	320
2	6400	4240	5360	6140
	480	480	480	480
3	6320	5680	6140	3120
	160	160	160	320
4	4240	4400	7680	4160
	480	480	480	480
5	5040	5280	5520	5600
	320	320	320	320

Table 6. Optimal locations for AQMN for SO_2 (x and y are in m)

Station	RC=0.90	RC=0.95
1	3200	3280
	160	160
2	8640	7600
	320	320
3	2480	9920
	320	160
4	4800	2480
	160	320
5	4880	4800
	800	800

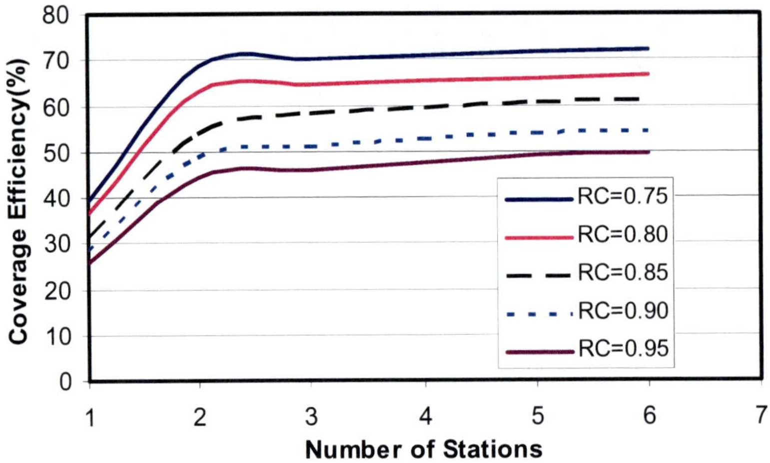

Figure 13. Coverage Efficiency Vs Number of stations for CO as a function of rc.

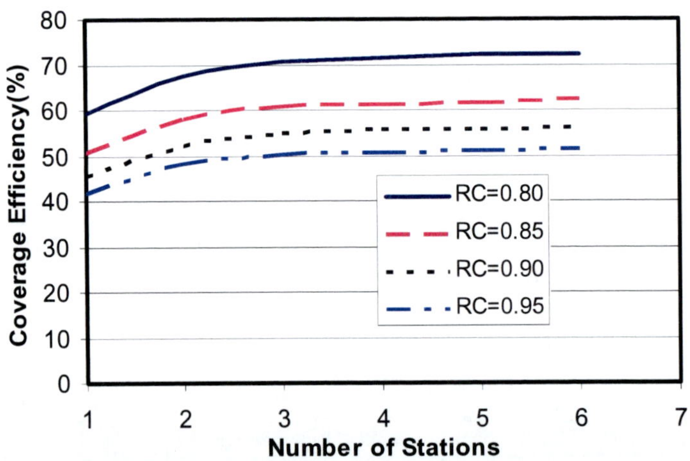

Figure 14. Coverage Efficiency Vs Number of stations for NOx as a function of rc.

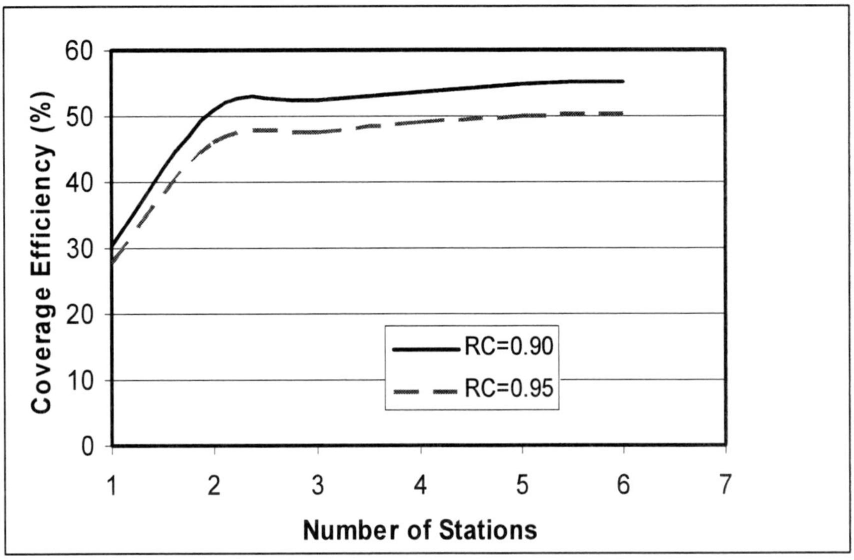

Figure 15. Coverage Efficiency Vs Number of stations for SO_2 as a function of rc.

2. Multi-Pollutant Monitoring Network Optimization Modeling Results

The optimization model was used to find the optimal locations and the configuration of the monitoring networks for monitoring simultaneously SO_2, NOx and CO within the area under study. For this purpose, the 6 year (monthly) predicted concentrations of SO_2, NOx, and CO by the Multiple Cell Model within Tabriz Refinery area were also used in this case. Tables 5 to 11 show the optimal locations of AQMN for a maximum number of 5 stations. Figure 16 shows the coverage efficiency of AQMN for different rc values. As shown in the figure, three stations for AQM are enough to also represent the case of multiple pollutants.

Table 7. Optimal locations for AQMN for multi-pollutant (x and y are in m)

Station	RC=0.75	RC=0.80	RC=0.95
1	2880	3120	3040
	160	160	160
2	8320	8560	7600
	320	320	320
3	2720	2640	7920
	480	480	160
4	3840	3920	2480
	640	640	480
5	9760	10000	5440
	640	1200	800

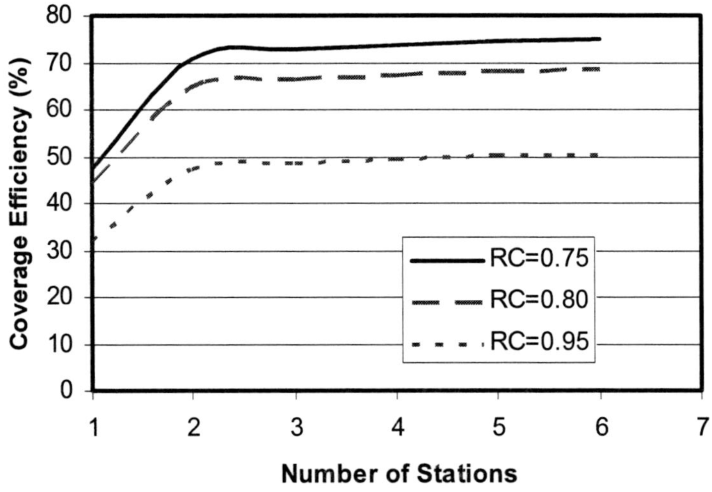

Figure 16. Coverage Efficiency Vs Number of stations for multi-pollutants as a function of rc.

REFERENCES

Al-Rashidi, M. (2002) *Optimization of multi-pollutant air quality monitoring network around typical pollutant oil field,* Department of Chemical Engineering, Kuwait University.

Fatehifar, E., (2005) Development of an optimal design of a multi-pollutant air quality monitoring network, Ph. D. Thesis, Department of Chemical Engineering, Shiraz University, Shiraz, Iran.

Liu, M. K. (1986) "Methodology for designing air quality monitoring networks: I. Theoretical aspects", *Environmental monitoring and Assessment*, 6, p. 1-11.

Modak, P. M. and Lohani, B. N., (1985a) "Optimization of ambient air quality monitoring networks: PartI", *Environmental Monitoring and Assessment*, 5, 1.

Modak, P. M. and Lohani, B. N., (1985b) "Optimization of ambient air quality monitoring networks: Part II", *Environmental Monitoring and Assessment*, 5,21.

Ott, W. J., (1977) "Development of Criteria for Siting Air Monitoring Stations", *Journal of the Air Pollution Control Association,* 27, 543-547.

World Health Organization (WHO) (1977), *'Air Monitoring Program Design for Urban and Industrial Areas', Global Environmental Monitoring System,* WHO Offset Publication No. 38.

In: Air Quality in the 21st Century
Editors: G. C. Romano and A. G. Conti, pp. 339-351
ISBN 978-1-60456-793-9

Chapter 9

SENSITIVITY OF LAND USE PARAMETERS AND POPULATION ON THE PREDICTION OF CONCENTRATION USING THE AERMOD MODEL FOR AN URBAN AREA

Ashok Kumar[1,*], Charanya Varadarajan[2], and Kanwar Bhardwaj[1]
[1]Department of Civil Engineering, University of Toledo,
2801 W. Bancroft St., Toledo, Ohio 43606, USA
[2]ENSR, 999 West Town and Country Rd, Orange, California 92606, USA

ABSTRACT

This paper examines the sensitivity of predicted concentrations for regulatory applications using air quality models for urban areas. The issue is explored in detail using the latest air quality model AERMOD promulgated by the US Environmental Protection Agency (EPA) and the meteorological and source data available from Toledo, Ohio for the years 1990, 1991, and 1992. The analysis suggests using the twelve sectors' land use parameter values around three kilometer radius of the meteorological site and the population around the major sources to obtain predicted concentrations. The use of this method resulted in lower predicted concentrations than the concentrations predicted from using the area method. The study also found that the AERMOD model was highly sensitive to variation in urban population for shorter averaging time periods and almost had negligible effect for the longer averaging time periods. This study demonstrates that the input of land use parameters and population affect the ground level concentrations for an urban area with multiple sources.

INTRODUCTION

Air quality models have been used to support air quality laws and/or regulations in industrialized nations. The U.S. EPA and the scientific community have conducted extensive

* Corresponding Author. Phone: (419) 530 8120, (419) 530 8136; Fax: (419) 530 8116. Email: akumar@utnet.utoledo.edu

studies to assess the performance of air quality models over the last 25 years. A detailed list of these studies is presented in Riswadkar and Kumar (1994). The recommended air quality models in the U.S. are given in the 'Guideline on Air Quality Models' (U.S. EPA, 2003).

The AERMOD model (Cimorelli et al. (2005)) was promulgated by the US EPA in December 2005 for studying the impact of industrial releases. AERMOD fully replaced the Industrial Source Complex (ISC3) model for regulatory applications since December 2006. Numerous studies have been conducted to evaluate the performance of the new regulatory AERMOD model (Paine et al., (1998, 2003), Hanna et al. (1999), Kumar et al. (2006)) using the statistical performance measures. These measures include measures of difference such as bias, variance of difference, gross variability of the difference, and measures of correlation such as time, space, and time and space combined. The performance measures are explained in detail by Kumar et al. (2006). The results of various studies carried out by the EPA on single tall stacks using the AERMOD model have been included in a paper by Perry et al. (2005).

Grosch and Lee (1999) conducted a study on the sensitivity of AERMOD on the selection of land use parameters by the model and it was concluded that modeled design concentrations could vary substantially due to normal ranges of variations in the albedo, Bowen ratio, and surface roughness length. Long et al. (2004) also studied the sensitivity of AERMOD to input parameters in the San Francisco Bay area and found AERMOD to be very sensitive to surface roughness in comparison to other input parameters such as solar radiation, urban population, albedo, and cloud cover for all the three source types (elevated, ground level, and volume). Solar radiation and cloud cover parameters also had an effect on the performance of the model. The sensitivity of AERMOD varied contrastingly among the three source types indicating the interaction of AERMOD with the input parameters a function of the source types. Note that both studies were based on hypothetical sources and no monitored concentrations were used for comparison purposes. In their hypothetical case study, Laffoon et al. (2005) evaluated the behavior of AERMOD in comparison to ISC, compared use of onsite meteorological data to the National Weather Service data, and the sensitivity of AERMOD to changes in input land use parameters. The study on the effect of surface parameters (albedo, Bowen ratio, and surface roughness) on modeled concentrations by Carper and Ottersburg (2004) revealed surface roughness to have the greatest impact when one surface characteristic was varied and the others were kept constant within AERMET for the entire modeling period. The study concluded that surface roughness also affected the plume dispersion, travel speed and distance. This observation is in line with the findings of Hanna et al. (1977) on the variation of dispersion coefficient in y-direction is affected by surface roughness, and of Kumar (1979) that the dispersion coefficients depend on surface roughness and release height.

The study by Kumar et al. (2006) focused on computing ambient air concentrations of sulfur dioxide (SO_2) for 1-hr, 3-hr, and 24-hr averaging periods using the emission inventory data for Lucas County, Ohio for the year 1990 using the AERMOD dispersion model. The estimated concentrations in this study were classified based on the stability parameter, Monin-Obukhov length (L), for the two monitoring stations located in the area. The data were divided into two atmospheric stability classes (stable and convective cases) as used in the AERMOD model. These categories were further grouped into five sub categories based on the value of L to learn the fine details of model performance. The study concluded that the performance of the AERMOD model did not meet the general criteria specified by Kumar et al. (1993) for air quality models for the 1-hr and 3-hr averaging periods. However, the model showed a better performance for the 24-hr concentrations as compared to the 1-hr and 3-hr averaging periods. The model also had a tendency to underpredict in all the cases thus showing that the multi-source regions were far more complex than single stack evaluations. The study also incorporated a comparative analysis of predicted SO_2 concentration using different land use types to understand the degree to which the results are affected by the choice of land use parameters. The study recommended that future work should focus on the role of land use parameters in predicting concentrations at the monitors and finding ways to quantify errors due to other factors.

The application of air dispersion models involves uncertainty in predicting ground level concentrations due to the specification of input data and their quality. This has implications on the confidence in model estimates of ground level concentration and on their use in the assessment of possible alternative scenarios in a regulatory framework. Hence, it is important to address the sensitivity due to the input parameters on the results of an air quality model. Currently the documentation available from the USEPA website on AERMOD provides a mixture of theory and practice, with specific guidance on technical issues but very little information on the inputs. The modeling guidance does not completely address the type, accuracy and location of additional measurements needed to apply AERMOD successfully. For example, specifying land use parameters, not directly measured at meteorological stations, are somewhat subjective, and there are no clear guidelines for determining values. Guidance does not clearly address the population of urban area in AERMOD. There is no study conducted which relates (if any) the simultaneous effect on the variation on urban population and surface characteristics on the ground level concentrations. Unlike the ISCST3 model that requires the simple selection of urban or rural default settings, the AERMOD dispersion model's meteorological pre-processor (AERMET) requires the input of site-specific land use parameters corresponding to land-use categories, including albedo, Bowen ratio, and surface roughness. It is important to know the sensitivity of model results to changes in various input parameters. This information is needed to gauge the accuracy required for these input parameters.

The purpose of this study is to present the results of limited sensitivity analysis of the AERMOD version 99351 from the Lake Environmental Consultants Inc., Waterloo, Canada for assisting environmental professionals involved in regulatory applications for the input of surface characteristics and population in a multiple source region. To the best of authors' knowledge, no study has been reported that use the observed concentration values to discuss with the predicted values for such an analysis using the AERMOD model.

It was assumed that the quality of input data used is acceptable for sensitivity analysis because the study had used data available from the regulatory agency or associated government agency.

ANALYSIS PROCEDURE

The requirements of the chosen modeling scenario involved detailed information of the database containing emission inventory data, the meteorological data, the monitoring data, and land use parameters. AERMOD's meteorological pre-processor AERMET, requires the input of site-specific land use parameters, such like albedo, Bowen ratio and surface roughness around the meteorological station. Additional geophysical input parameters such as population and a single value of urban surface roughness length (of source site) are required for the calculation of ground level concentration in AERMOD for urban areas. No removal of the pollutant was considered during modeling.

A detailed emission inventory for Lucas County was obtained from the Department of Public Utilities, City of Toledo, Ohio for all sources emitting sulfur dioxide (SO_2) for the years 1990, 1991 and 1992. The emission data had the measurement from stack monitors for three years. The point source report consisted information on facility, stack/discharge and air pollutant emissions. The information on facility included the Universal Transverse Mercator (UTM) coordinates and the UTM zone in which the stack was situated. The information on air pollutant emissions had the uncontrolled, actual and allowable annual emissions of pollutants. The monitoring data was obtained from the U.S. EPA's air quality system monitoring system (AIRS) for the two monitors in Lucas County located at 26 Main Street and 600 Collins Park for 1990, 1991, and 1992. The UTM coordinates of these two monitoring stations are (289304, 4613488) and (293889, 4615115). The monitoring data consisted of hourly observed values at the above two monitoring stations for three years and details of geographical location of the monitoring sites. The meteorological data for the AERMOD model was prepared for the year 1990, 1991 and 1992 using the surface data from the Toledo Airport and the upper air data from Flint, Michigan. Land use data for AERMET was prepared by Geographical Information System (GIS) application ARC Map. The Toledo data are explained by Kumar et al. (1999) and Bhardwaj (2005).

The land use parameters were examined around the meteorological site using land use/land cover map of Lucas County. GIS application was used for this purpose and the map was obtained from the Ohio Department of Natural Resources. The land use parameters were calculated by two different methods. An example of the land use values for summer is shown in Table 1. One was by analyzing the correct area of land cover type within a circular area extending approximately three kilometer radius (called the 'area method' hereafter) and the second was by calculating the weighted average of characteristics by surface area within a 30-degree sector in 12-pie shaped sectors (called 'sector method' henceforth) within the three kilometer circular area with the meteorological station (Toledo airport) as the center ((Högström and Högström (1978), Paine (1987)). The AERMOD model setup for all the analyses were the same for each issue. The surface characteristics within the meteorological pre-processor were only varied. The maximum concentrations for 1-hr, 3-hr, and 24-hr averaging periods were analyzed. The predicted concentration values from the above two methods were compared with the observed concentration values for both the monitoring stations for all three years. The input urban surface roughness length is taken as 0.722 m for these runs.

Three realistic scenarios were constructed to study the effect of population on the AERMOD results. The surface characteristic values input were based on the sector method and area method. The population was varied to analyze the variation of the ground level concentrations. The populations for the three different cases were based on: 1) population around the major sources emitting sulfur dioxide, 2) population which is intermediate of city population and the population around major sources, and 3) the overall population of city of Toledo. The population for case one was calculated using the population density of Toledo and the area covered by the major sources. The population of the city of Toledo was taken from the U.S. census. Input population values for the three cases were 40,000, 100,000, and 300,000. The concentrations predicted from these three cases were compared with the observed concentrations.

This study analyzed the 1st and 2nd highest concentration values for 1-hr, 3-hr, 24-hr, annual and monthly averaging periods. The objective was to assist environmental professionals in their choice of the input values of land use and population for regulatory applications. Downwash was neglected because the modeling accounted for the major and minor sources based on the concept of super stacks.

Table 1. Surface Characteristics for the Area and Sector Methods at the Meteorological Site for Summer Season

Sector	Surface Characteristic for Sector Analysis		
	Albedo	Bowen Ratio	Surface Roughness (m)
1	0.162	0.566	0.566
2	0.162	0.675	0.740
3	0.156	0.600	0.509
4	0.175	0.574	0.417
5	0.177	0.611	0.451
6	0.157	0.525	0.735
7	0.151	0.472	0.881
8	0.170	0.415	1.002
9	0.168	0.453	0.826
10	0.163	0.617	0.491
11	0.158	0.589	0.447
12	0.159	0.578	0.472
	Surface Characteristic for Area Analysis		
3 km Radius Circular Area	0.163	0.557	0.642

Results and Discussion

This section is divided in four parts. The results for the choice of land use parameters are given in Section a. Section b discusses the results for the population analysis, Section c gives the 99.5 percentile value analysis results, and Section d discusses and compares the concentration predictions for a multi-source region to the single source studies by the EPA.

a. Specification of Land Use Parameters

The purpose of this analysis was to help environmental professionals in selecting surface characteristics inputs for regulatory air modeling applications using AERMOD. Two different cases of the surface characteristics were used to gauge the realistic picture. Albedo, Bowen ratio, and surface roughness were calculated for all the cases. Both the approaches are mentioned in the U.S. EPA reports for different type of modeling.

For the analysis, highest, 2nd highest, and robust highest concentration values for 1-hr, 3-hr, and 24-hr averaging periods were considered. It is assumed that the analysis of these indicators will model typical regulatory applications. Tables 2 to 4 compare the predicted concentrations with the observed values using both the methods described earlier for calculating surface characteristics for different population at both monitoring stations. Care should be taken during the interpretation of the results because the emission rates used in modeling are based on average yearly values and are not a function of time. The results of this analysis showed the following:

1. The highest and 2nd highest predicted concentrations are closest to the observed concentration values for 1-hr averaging period most of the time when AERMET is run using the twelve sectors' land use parameter values around three kilometer radius at the meteorological site and the population is either 40,000 or 100,000. The predictions for the 300,000 population case are mixed.
2. The cases when highest and 2nd highest predicted concentrations are closest to the observed concentration values for 3-hr averaging period are mixed i.e., in many cases the area method produced better results than the sector method or vice versa.
3. The differences in predictions for the highest and 2nd highest 24-hr averaging time are small for both the methods. This is due to the fact that variations in concentration due to the use of different land use procedures are reduced to smaller degree over a 24-hour period.

The predicted values in Table 2 have been reported without the background concentrations in the area. The authors realize the importance of background concentration in reporting regulatory concentrations. Note that adding background concentration to the predicted concentration has little or no impact on the trends observed above if the background concentration is small for the modeling area.

The closeness of predicted concentrations to the observed concentrations depends on averaging time and urban population. The model predictions generally differ from the corresponding monitored values because it is not possible to incorporate all the variables in an air quality model that affect the concentration for a particular time and monitor. The differences in observed and predicted values are generally due to model formulation errors, stochastic variations, model inputs, and the observed concentrations. In this study, an attempt has been made to use the model inputs as they are used in a typical assessment in the US following the guidance from the EPA documents. The observed concentrations are obtained from the monitoring network maintained by local regulatory agency. The emission rates were obtained from the information submitted by the industries to Ohio EPA. It is quite possible that the scatter in predicted and observed values is largely due to the specification of emission rates.

Table 2. Comparison of the Concentrations using Two Methods for Calculating Surface Characteristics

Main Street Monitoring Station															
Year	Averaging Period	Highest Concentrations (ug/m^3)							2nd Highest Concentrations(ug/m^3)						
			Population							Population					
			40, 000		100, 000		300, 000			40, 000		100, 000		300, 000	
		Observed	Area	Sector	Area	Sector	Area	Sector	Observed	Area	Sector	Area	Sector	Area	Sector
1990	1-hr	363.1	688.9	600.6	456.5	448.0	293.9	300.6	357.9	564.2	525.6	378.6	359.1	267.0	254.4
	3-hr	346.6	229.8	224.3	152.2	149.4	109.1	100.3	309.1	188.0	175.2	136.2	127.0	98.1	97.9
	24-hr	147.8	54.9	53.7	42.9	41.3	42.1	34.9	91.6	43.5	37.63	41.5	35.6	36.6	34.7
1991	1-hr	386.6	370.3	374.5	271.5	276.6	240.4	249.6	271.7	291.9	305.9	218.5	227.7	208.3	206.1
	3-hr	198.5	211.7	238.6	185.1	210.8	174.9	184.7	192.4	184.9	200.5	151.8	166.1	120.9	133.1
	24-hr	59.96	66.3	71.9	56.6	61.8	50.3	52.9	58.3	47.5	48.19	47.6	48.1	45.5	45.7
1992	1-hr	394.4	495.1	463.1	343.4	331.8	243.9	233.6	321.3	459.3	437.9	320.1	323.0	224.2	224.5
	3-hr	202.9	165.1	155.9	125.4	132.4	119.4	116.4	191.6	153.1	154.5	120.0	118.8	117.5	112.6
	24-hr	85.29	43.5	46.49	38.6	42.0	38.5	41.2	73.59	43.4	45.17	38.4	41.6	37.8	38.4
Collins Park Monitoring Station															
1990	1-hr	389.2	1486.4	1120.3	947.8	893.9	589.1	900.4	389.2	897.8	845.4	575.1	552.3	363.9	544.6
	3-hr	348.3	502.9	474.4	327.6	310.3	217.4	206.3	200.3	299.6	282.1	191.9	184.4	153.8	137.4
	24-hr	96.76	72.1	68.05	47.1	44.6	34.8	38.5	78.36	55.4	55.44	41.1	42.7	31.78	32.1
1991	1-hr	428.4	988.3	850.6	653.6	608.2	427.1	620.7	368.3	674.7	647.0	483.5	459.8	359.4	486.6
	3-hr	268.2	333.7	289.0	223.2	210.8	150.8	238.6	244.7	225.4	216.5	161.8	166.1	149.4	200.5
	24-hr	59.9	65.1	59.6	47.4	53.8	42.3	52.2	58.26	54.9	55.20	45.4	46.9	36.7	39.6
1992	1-hr	684.4	718.6	697.9	508.4	481.5	368.6	359.3	365.7	699.0	682.4	505.0	476.5	344.4	337.9
	3-hr	379.6	288.4	271.3	249.6	222.9	196.8	179.8	250.8	239.8	232.9	169.7	181.1	177.8	171.3
	24-hr	78.6	50.2	48.28	44.1	40.7	35.7	34.2	75.30	42.1	41.06	31.9	33.6	35.7	34.0

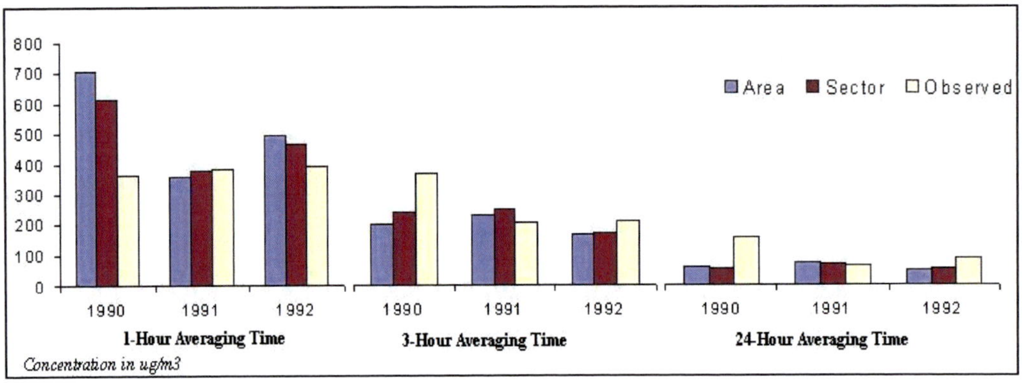

Figure 1. Comparison of the Highest Concentrations using Two Methods for Calculating Surface Characteristics at the Main Street Monitoring Station for population of 40, 000.

Therefore, it is difficult to conclude whether the sector or the area method is a more suitable approach for calculating the surface characteristics (albedo, Bowen ratio and surface roughness) as to obtain predicted concentrations closer to the observed values. However, based on the physics of atmospheric dispersion the sector method for calculating surface characteristics is more appealing than the area method.

b. Specification of Population Data

AERMOD uses the urban-rural temperature difference estimates based on the data collected by Oke (1998) for Canadian cities. It is possible that the current algorithm used in AERMOD may be of limited value in the American conditions. The attempt in this study is to document the sensitivity of using different population values during modeling.

The analysis of the input population data to concentration calculations revealed that the variation of the input population affected the concentration calculations for all averaging periods. The 1st and 2nd highest concentration values for 1-hr, 3-hr, 24-hr, and monthly averaging periods were analyzed. The fluctuations were more in shorter averaging periods as compared to longer averaging periods. In general, the maximum variation with respect to population was for the 1-hour averaging time period.

The results for the 1-hr averaging period did not yield to drawing a firm conclusion on the specification of population. The only conclusion which can be made is that the model performance was generally worst when the input population is taken of city of Toledo (i.e., 300, 000). For 3-hr and 24-hr averaging time periods, the results indicated that when the input population was around the major sources (40,000 or 100,000), the predicted concentration values were closest to the observed values. Figures 2 to 4 show the effect of population on the highest concentration for the monthly and annual averaging periods. The results indicated little variation in the predicted concentrations due to variation of input population for monthly and annual averaging periods.

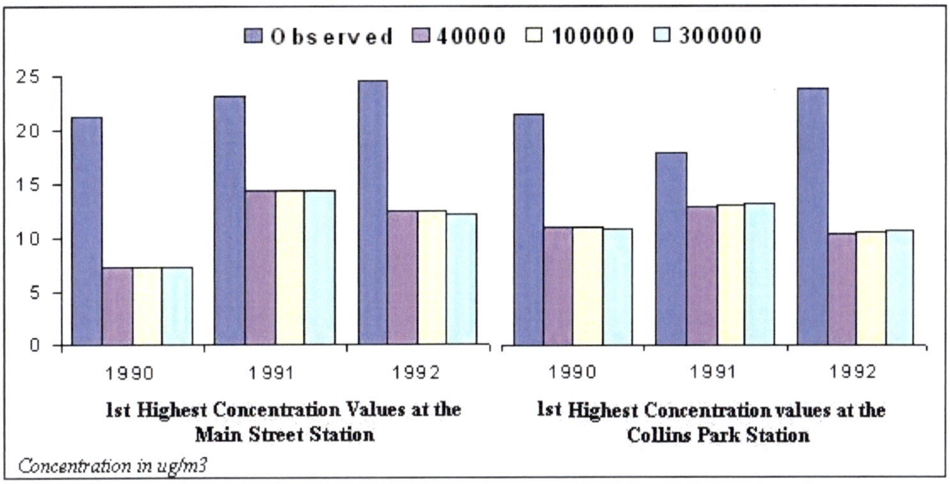

Figure 2. Effect of population on the highest predicted monthly concentrations for Sector method.

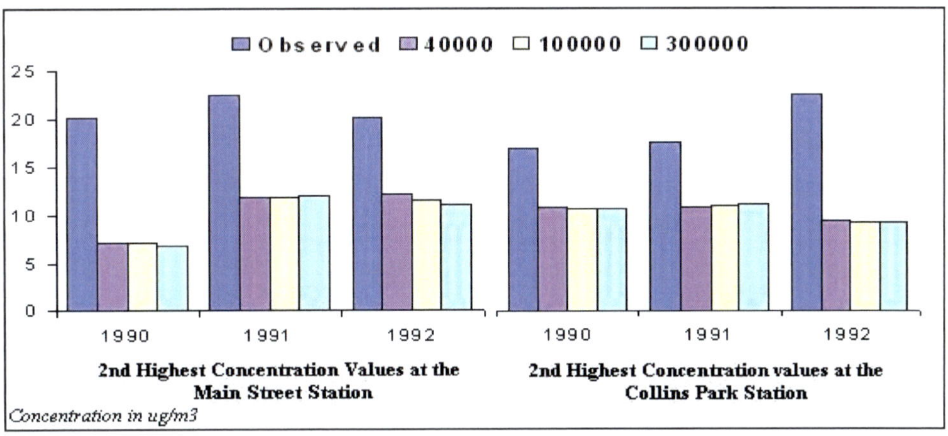

Figure 3. Effect of Population on the 2nd Highest Predicted Monthly Concentrations.

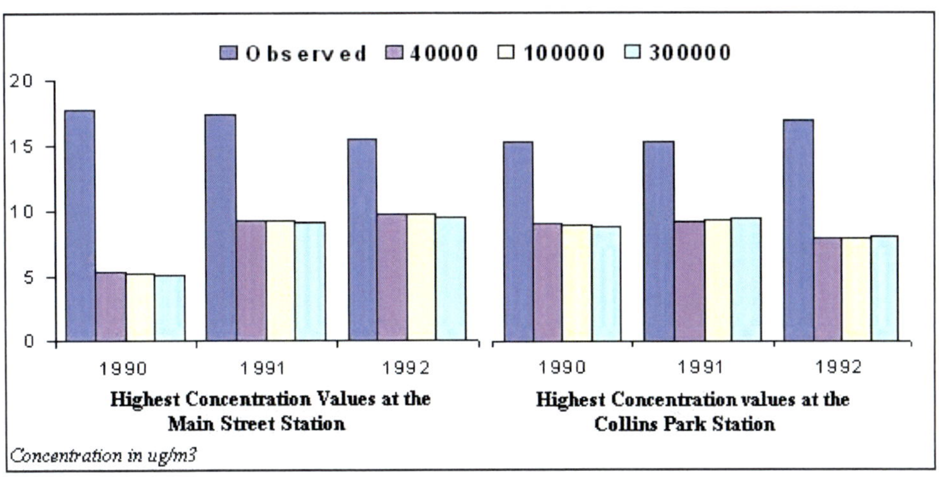

Figure 4. Effect of Population on Predicted Highest Annual Concentrations.

The analysis showed that, in general, overall model performance using input population near the major sources (40, 000) was better than the other two population cases. This may be due to the fact that the model computes the increase in temperature due to the urban population and readjusts the dispersion parameters. A lower value of population as compared to the population of the entire city appears to be better representative of this calculation. It is speculated that one may use the percent of urban area in the modeling area to calculate the population for the urban option.

The following are general conclusions from the population analysis:

- The population input is most sensitive to lower averaging times and least to annual average.
- The predicted results are close to the observed when the input population is taken around the area surrounding the major sources as compared to the entire study area. Our results may be biased because of the location of monitors.

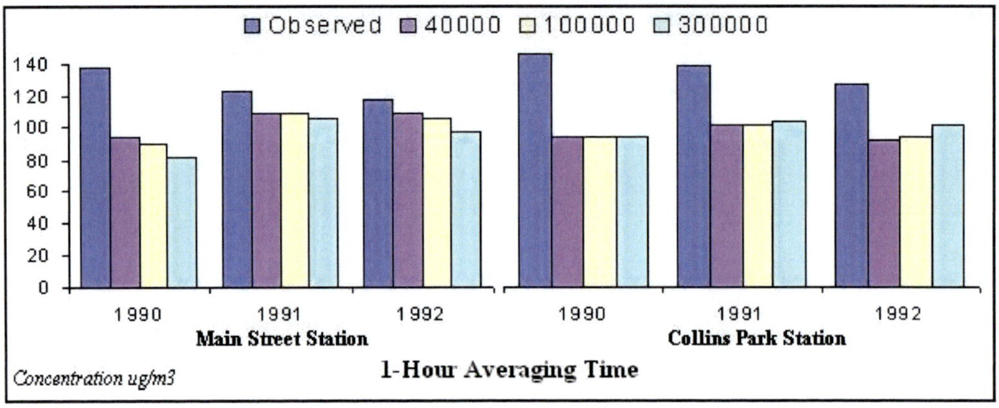

Figure 5. Impact of Population on 99.5 Percentile Values for 1-hr Averaging Period based on sector analysis.

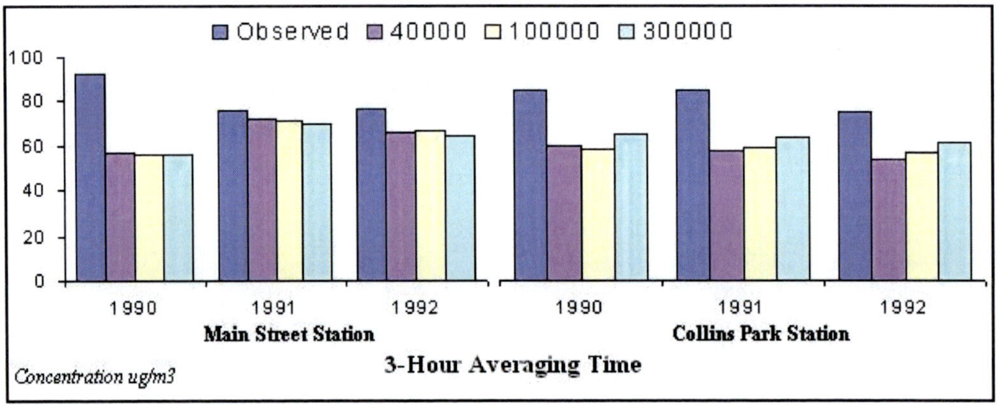

Figure 6. Impact of Population on 99.5 Percentile Values for 3-hr Averaging Period based on sector analysis.

c. 99.5 Percentile Values for 1-hr and 3-hr Averaging Periods

In some countries the regulations are based on percentile value of concentration. Therefore, the use of 99.5 percentile value was also examined in this study. The variation in the input of population had very little impact on the 99.5 percentile values for 1-hr and 3-hr averaging periods (refer Figures 5 and 6). This implies that if the regulations are based on this percentile value, then population is not a critical parameter for running the AERMOD model.

d. Discussion on Concentration Predictions for a Multi-source Region vs. Single Sources

The study on the performance of the AERMOD model for continuous releases from single sources by Perry et al. [(2005), Tables 3 and 4) showed that the air dispersion model performed well for the short-term as well as the long-term averaging periods. The work reported by Kumar et al. (2006) indicated that the AERMOD model tends to underpredict the regulatory concentrations for the multi-source region considered. The sensitivity of land use parameters and population found in this paper may explain some of the possible reasons.

CONCLUSIONS

The study provides the following useful results for the sensitivity of input values of land use parameters and population that would help environmental professionals in applying AERMOD to an urban area:

i. The predicted concentrations are sensitive to the specification of land use parameters. It is suggested that during AERMOD modeling the meteorological input is obtained using the twelve sectors' land use parameter values around three kilometer at the meteorological site. This approach is physically realistic.
ii. The results indicated that the input of population around the major sources are likely to provide predicted results closer to observed values for short-term maximum or near maximum ground level concentrations. However, it was difficult to determine the precise extent of the area for population calculations.
iii. The variation in the input of population has very little impact on the 99.5 percentile value for 1-hr and 3-hr averaging time period. This means that if the regulations are based on these percentiles values, population becomes an insignificant parameter for modeling urban areas.

Note that the above conclusions should be verified by studying more urban areas and monitoring points. We agree with the readers that our results and the resulting recommendations may be biased due to the dominance of elevated sources. One may find different results for small stacks and downwash cases.

This study demonstrated that the input of land use parameters and population affect the ground level concentrations for an urban area with multiple sources. More studies should be conducted on different source heights and other type of sources such as area sources and volume sources to properly gauge the sensitivity of all the input parameters. Future work should also consider the role of cloud cover, solar radiation, and height at which ambient temperature is measured in obtaining boundary layer parameters.

ACKNOWLEDGEMENTS

The authors would like to acknowledge the complimentary copy of the AERMOD model provided by the Lake Environmental Consultants Inc., Waterloo, Canada. This paper is a revised version of the papers titled 'Examination of Application Issues Related to Air Quality Models for Urban Areas' presented on January 7, 2006 at the International Interdisciplinary Conference on Sustainable Technologies for Environmental Protection, Coimbatore, India and 'Examination of Sensitivity of Land Use Parameters and Population on the Performance of the AERMOD Model for an Urban Area' presented at the 2006 A&WMA conference in New Orleans, USA.

REFERENCES

Bhardwaj, K., 2005. *Examination of Sensitivity of Land Use Parameters and Population on the Performance of the AERMOD Model for an Urban Area"* M.S. Thesis, University of Toledo.

Carper, E., and Ottersburg, E., 2004. *Sensitivity Analysis Study Considering the Selection of Appropriate Land-Use Parameters in AERMOD Modeling Analyses, Proceedings of the 97th A&WMA Annual Conference and Exhibition,* Paper # 167, Indianapolis, Indiana, June 22-25.

Cimorelli, A. J., Perry S. G, Venkatram A., Weil, J. C, Paine R., Wilson R. B., Lee, R. F., Peters, W. D., and Brode, R.W, 2005. AERMOD: A Dispersion Model for Industrial Source Applications. *Part I: General Model Formulation and Boundary Layer Characterization, Journal of Applied Meteorology,* Volume 44(5), pp. 682-693.

Grosch, G. T, and Lee, F. R, 1999. *Sensitivity of the AERMOD Air Quality Model to the Selection of Land Use Parameters,* (http://www.trinityconsultants.com/downloads/tp_wessex99.pdf, accessed 2008).

Hanna, S. R., Egan, B. A., Purdum, J., and Wagler, J., 1999. *Evaluation of ISC3, AERMOD, and ADMS Dispersion Models with Observations from Five Field Sites,* HC Report P020, API, 1220 L St. NW, Washington, DC 20005-4070.

Hanna, S. R., Briggs, G. A., Deardorff, J., Eagan, B. A., Gifford, F.A., and Pasquill, F., 1977. *AMS Workshop on Stability Classification Schemes and Sigma Curves-Summary of Recommendations, Bulletin of American Meteorology Society,* Volume 58, pp. 1305-1309.

Högström and Högström., 1978. A Practical Method for Determining Wind Frequency Distributions for the Lowest 200 m from Routine Data, *Journal of Applied Meteorology,* Volume 17, pp. 942-954.

Kumar, A., 1979. Estimation of Atmospheric Dispersion Coefficients for Elevated Releases, Preprint Volume, *Fourth Symposium on Turbulence, Diffusion, and Air Pollution* (A.M.S.), pp. 19-26, January.

Kumar, A., Bellam, N. K., and Sud, A., 1999. Performance of an Industrial Source Complex Model: *Predicting Long Term Concentrations in an Urban Area, Environmental Progress, Volume 18(2), pp. 93-100.*

Kumar, A., Luo, J., and Bennett, G., 1993. Statistical Evaluation of Lower Flammability Distance (LFD) *Using Four Hazardous Release Models, Process Safety Progress,* Volume 12(1), pp. 1-11.

Kumar, A., Dixit, S., Varadarajan, C., Vijayan, A., and Masuraha, A., 2006. *Evaluation of the AERMOD Dispersion Model as a Function of Atmospheric Stability for an Urban Area, Environmental Progress,* Volume 25(2), 141-151.

Laffoon, C., Rinaudo, J., Bowie, J., Soule, R., Meyers, C., Madura, R., and Pakunanya, S., 2005. Developing State-Wide Modeling Guidance for the Use of AERMOD-A Workgroup's Experience, *Proceedings of the 98th A&WMA Annual Conference and Exhibition,* Paper # 1302, Minneapolis, Minnesota, June 21-24.

Long, E. L., Cordova, F. J., and Tanrikulu, S., 2004. An Analysis of AERMOD Sensitivity to Input Parameters in the San Francisco Bay Area, *13th Conference on the Applications of Air Pollution Meteorology with the A &WMA, and American Meteorological Society,* Vancouver, BC, August 23-25.

Oke, T. R., 1998. An Algorithmic Scheme to Estimate Hourly Heat Island Magnitude. Preprints, 2nd Urban Environment Symposium, *American Meteorological Society,* Boston, MA, 80-83.

Paine R. J., Lee R. F., Brode R., Wilson R. B., Cimorelli A. J., Perry S. G., Weil J. C., Venkatram A., and Peters W. D., 1998. *Model Evaluation Results for AERMOD,* U.S. EPA, RTP. NC 27711.

Paine, R. J., Brode, R. W., Wilson, R.B., Cimorelli, A.J., Perry, G.S., Weil, J.C., Venkatram, A., Peters, W. D., and Lee, R.F., 2003. AERMOD: Latest Features and Evaluation Results, *Proceedings of the 96th A&WMA Annual Conference and Exhibition,* Paper # 69878, San Diego, California, June 22-26.

Paine, R. J., 1987. User's Guide to the CTDM Meteorological Preprocessor (METPRO) Program. EPA-600/8-88-004, U.S. *Environmental Protection Agency, Research Triangle* Park, NC. (NTIS No. BP 88-162102).

Perry S. G, Cimorelli, A. J., Paine, R. J., Brode, R. W., Weil, J.C., Venkatram, A., Wilson, R. B., Lee, R. F., and Peters, W. D., 2005. AERMOD: A Dispersion Model for Industrial Source Applications. *Part II: Model Performance against 17 Field Study Databases, Journal of Applied Meteorology,* Volume 44(5), pp.694-708.

Riswadkar, R. M., and Kumar, A., 1994. Evaluation of the ISC Short Term Model in a Large-Scale Multiple Source Region for Different Stability Classes, *Environmental Monitoring and Assessment,* 1-14.

In: Air Quality in the 21st Century
Editors: G. C. Romano and A. G. Conti, pp. 353-380
ISBN 978-1-60456-793-9

Chapter 10

SOLAR RADIATION AND CFD PHOTOCHEMICAL MODELING IN THE URBAN CANOPY

***Stamatis Zoras*[*,1], *Vasilis Evagelopoulos*[2], *Stelios Garas*[2], *Athanasios G. Triantafyllou*[2] *and Panagiotis Kosmopoulos*[1]**

[1]Laboratory of Environmental and Energy Design, Department of Environmental Engineering, Democritus University of Thrace, Xanthi, Greece
[2]Laboratory of Atmospheric Pollution and Environmental Physics, Department of Pollution Control Technologies, Technological Education Institute (TEI) of West Macedonia, Kozani, Greece

ABSTRACT

The combination of chemical substances and specific meteorological parameters cause the formation of the ground-level ozone layer. High temperature, sunlight intensity and increased surface pressure are the "obvious" conditions that mostly help ozone in its formation. In addition, local circulations (e.g. sea breeze, valley winds) with light or stronger winds may assist in ozone formation by creating adequate dilution conditions that accelerate photochemical reactions. Wind direction has also been considered as an important factor in the formation of ozone.
The strong spatial and temporal variability of traffic-related air pollution detected at roadside locations in large or medium-sized cities has raised the question of how representative the site and time period of air quality measurements actually can be. In general, measurements follow symmetric patterns and show their dependence on meteorological conditions. A CFD urban street canyon modeling approximation in local scale is presented in order to finalise a few issues that are observed in relation to ozone production, temperature and solar radiation.

INTRODUCTION

Traffic emissions into the atmosphere include volatile organic compounds (VOCs) and NO_X that in the presence of sunlight react photochemically forming ozone. Benzene is

* Corresponding author. Email: stamatis@airlab.teikoz.gr (Dr. Stamatis Zoras)

primarily emitted in vehicle exhaust as unburned fuel and as a product of combustion. Toluene is added in order to raise octane rating in gasoline that may also be emitted into the air. Formation of ozone depends on the contribution of VOCs to the conversion of NO to NO_2. In specific, hydrocarbons react with the hydroxyl radical producing peroxy radicals that react rapidly with NO to form NO_2.

The combination of chemical substances and specific meteorological parameters cause the formation of the ground-level ozone layer. The formation and accumulation of ozone at ground level is dangerous for people with respiratory diseases (Bernard et al., 2001) and damages crops, forests and materials (Friedfeld et al., 2002). High temperature, sunlight intensity and increased surface pressure have been proved to help in the formation of ozone (Chen et al., 2003). In addition, local circulations e.g. sea breeze (Petrakis et al., 2003), with light winds (Chiu et al., 2005) or stronger winds (Kelessis et al., 2006) may assist in ozone formation by creating adequate dilution conditions that accelerate photochemical reactions. Wind direction has also been considered as an important factor in the formation of ozone (Petrakis et al., 2003). Moreover, a petrochemical facility was responsible for ozone production for a distance of up to 26 km (Chiu et al., 2005).

Ground station measurements are influenced by small sources in their vicinity, surface dry deposition and small-scale wind effects. However, consideration of meteorological conditions may be required to, properly, assess the spatial concentration differences within the street canyon. The case that is presented here is a more complicated one in terms of source variety (i.e. coal fired power plants' operation, mining activities combined with urban sources), terrain complexity and meteorological conditions.

The medium-sized city of Kozani in northwestern Greece is a heavily industrialized area with power plants and opencast mining operations. The basin north of the city is governed by nocturnal stagnant conditions favoring ozone production and accumulation within the surface boundary layer during sunny mornings. Stack emissions may affect the city under specific meteorological conditions that in combination with urban sources may cause severe air pollution episodes (Triantafyllou et al., 2002).

In order to reveal the most important factors in ozone production and spatial distribution within the street canyon, pollutants' concentrations are modeled by a computational fluid dynamics (CFD) fast chemistry module. The module is integrated into a holistic CFD package (ANSYS-CFX) for the estimation of photolysis reaction rates and three dimensional flow dispersion.

Background

In previous CFD modeling studies of dispersion in urban street canyons, passive (non-reactive) pollutant (scalar) was exclusively considered and its concentration was calculated using a prognostic equation for the passive pollutant. In urban areas, a main pollutant source is automobiles and the pollutants emitted from automobiles, for example NO and NO_2, are chemically reactive. Complex photochemical processes in densely built-up urban areas with traffic often result in a serious air pollution problem. Therefore, to further enhance our understanding of street canyon dispersion, reactive pollutants need to be taken into account.

In this study a commercial general use CFD software package ANSYS-CFX was employed. The fidelity of simulation is linked directly to the choice of physical models available. The ANSYS - CFX software contains a large number of physical models to provide accurate simulation for a wide variety of industrial applications. It provides the user with common steps for the creation of the urban street canyon model starting with the geometry and meshing, then with the physical models definition and Solver settings.

THE OZONE PHOTOLYSIS REACTION

Ozone occurs in the troposphere, as a secondary pollutant by a chemical reaction with NOx and VOC, an extended class of gas compounds, mainly non-methane hydrocarbons, which are emitted by oil and its by-products and by biological materials, with concentrations higher than that of the natural background, commonly known as photochemical smog.

Ozone production by chemical reaction is increased by emissions of anthropogenic origin in the atmosphere. The increase is attributed to an increase in NOx emissions associated with the switch to fossil fuels during the industrial period and increases in biomass burning. Historical records imply a large anthropogenic contribution to the nowadays O_3 background concentration at northern mid latitudes. Ozone occurs in the troposphere for different reasons: 1. Intrusion from the stratosphere: The transportation downwards of stratospheric air across the tropopause into the troposphere in special meteorological situations (usually at extratropical latitudes) is one reason that ozone is found in the lower atmosphere. 2. Photochemical production by substances (precursors) such as NOx (Figure 1):

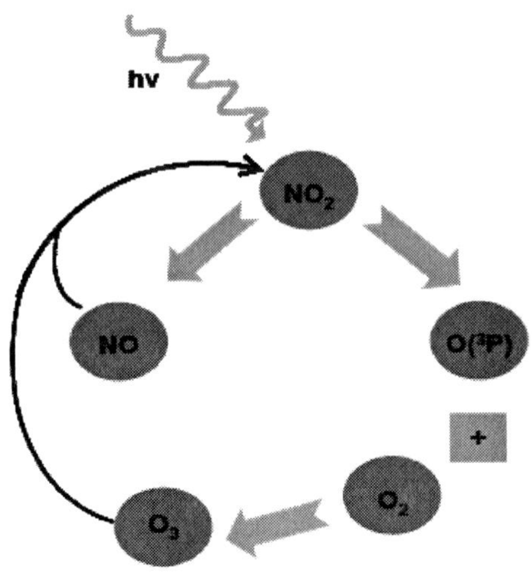

Figure 1. Ozone photolysis cycle.

NO_2 + sunlight (λ< 420 nm) $\rightarrow$ NO + (O^3P)

(O^3P) + O_2 + M $\rightarrow$ O_3 + M

$$NO + O_3 \rightarrow NO_2 + O_2 \Rightarrow \text{ fast reaction}$$

Where $O(^3P)$ represents the ground state for atomic oxygen. This is known as "*the null cycle*" of ozone, i.e. O_3 is destroyed about as fast as it is produced: However, ozone is not produced only by NOx but also VOCs play an important role in its formation. Therefore, the photolysis cycle is completed as in Figure 2.

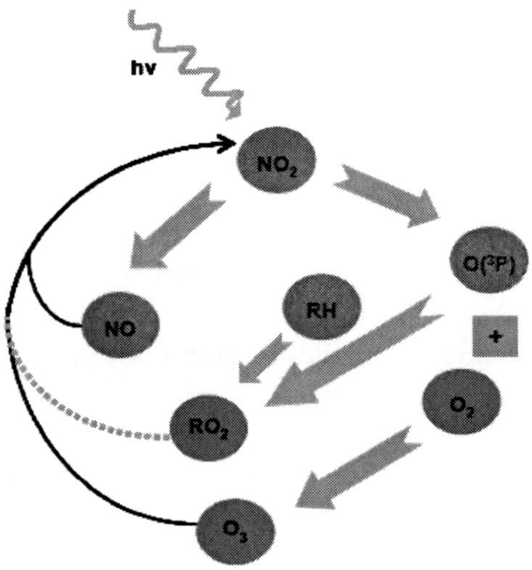

Figure 2. Ozone photolysis with VOC.

High levels of ground level ozone are found primarily in urban areas, particularly those with large industrial centers and high rates of traffic. The formation of ozone depends greatly on the topography of the land, the prevailing weather systems, and the level of pollutants being emitted. However, ground level ozone can also be found in rural areas far from its urban point of origin. While the transferal of ozone over long distances reduces its concentration, it can still pose a risk to plants, people, and wildlife.

The following are typical ozone levels or mixing ratios:

Natural background (pre-industrial):	10-20 ppb
Remote locations in the Northern Hemisphere:	20-40 ppb (varying by season and latitude)
Rural areas during region-wide pollution events:	80-100 ppb
Peak O3 in urban areas during pollution events:	120-200 ppb
Maximum urban O3 (Los Angeles, Mexico City):	490 ppb
Stratospheric ozone layer:	15000 ppb

WEATHER CONDITIONS FAVORING PHOTOCHEMICAL POLLUTION

Certain conditions are required for the formation of photochemical smog. So far the effects of the NOx and VOC cycles on the formation of O_3 were discussed, with the involvement of other species in the chemical cycle. In general, important "input" in the formation of photochemical smog over urban areas is the high concentration of NOx and VOCs that are associated with industrialization and transportation.

The time of the day is a very important factor in the amount of photochemical smog present. The severity of smog problem is closely related to weather conditions. Ideal weather conditions for its formation are:

- Sunlight.
- Stable weather conditions associated mainly with high-pressure systems.
- Temperature inversions.
- Calm winds. However, even if winds are strong problems may arise in distant areas that receive the pollution.
- On the other hand, precipitation can alleviate photochemical smog as the pollutants are washed out of the atmosphere with the rainfall.
- The photochemical smog mostly occurs in the summer (May to September) between noon and early evening, although it may be formed also in the winter with particulates being the main component.
- Topography is another important factor influencing how severe a smog event can become (e.g., valleys) as it can induce local airflow that may influence the transfer or trapping of smog. Cities situated in valleys are, therefore, more susceptible to photochemical smog. Hills and mountains surrounding them tend to reduce the airflow, allowing for pollutant concentrations to rise. Valleys are sensitive to photochemical smog because relatively strong temperature inversions can frequently develop in these areas. Normally, during the day the air near the surface is heated and as it warms it rises, carrying the pollutants with it to higher elevations. However, if a temperature inversion develops pollutants can be trapped near the Earth's surface. Temperature inversions cause the reduction of atmospheric mixing and therefore reduce the vertical dispersion of pollutants. Inversions can last from a few days to several weeks.
- At street scale, airflow highly depends, apart from street characteristics, on the coupling with the layer above the street canyon. A simplified model using only the NO-NO_2-O_3 fast cycle predicts high NO_x levels on the leeward side of a narrow street canyon, while O_3 is depleted within the street canyon (Figure 3).
- Diurnal variation of ozone. The time of day is a very important factor in the amount of photochemical smog present. Early morning traffic increases the emissions of both NOx and VOCs as people drive to work. NO is oxidized to form NO_2. Later in the morning, traffic dies down with the NOx and VOCs begin reacting forming NO_2, increasing its concentration. As the sunlight becomes more intense later in the day, NO_2 is broken down and its

formed by-products increase concentrations of O_3. At the same time, some of the NO_2 can react with VOCs to produce toxic chemicals such as PAN (peroxyacetyl nitrate) (see Figure 4). As the sun goes down, the production of O_3 is halted. The O_3 that remains in the atmosphere is then consumed by several different reactions.

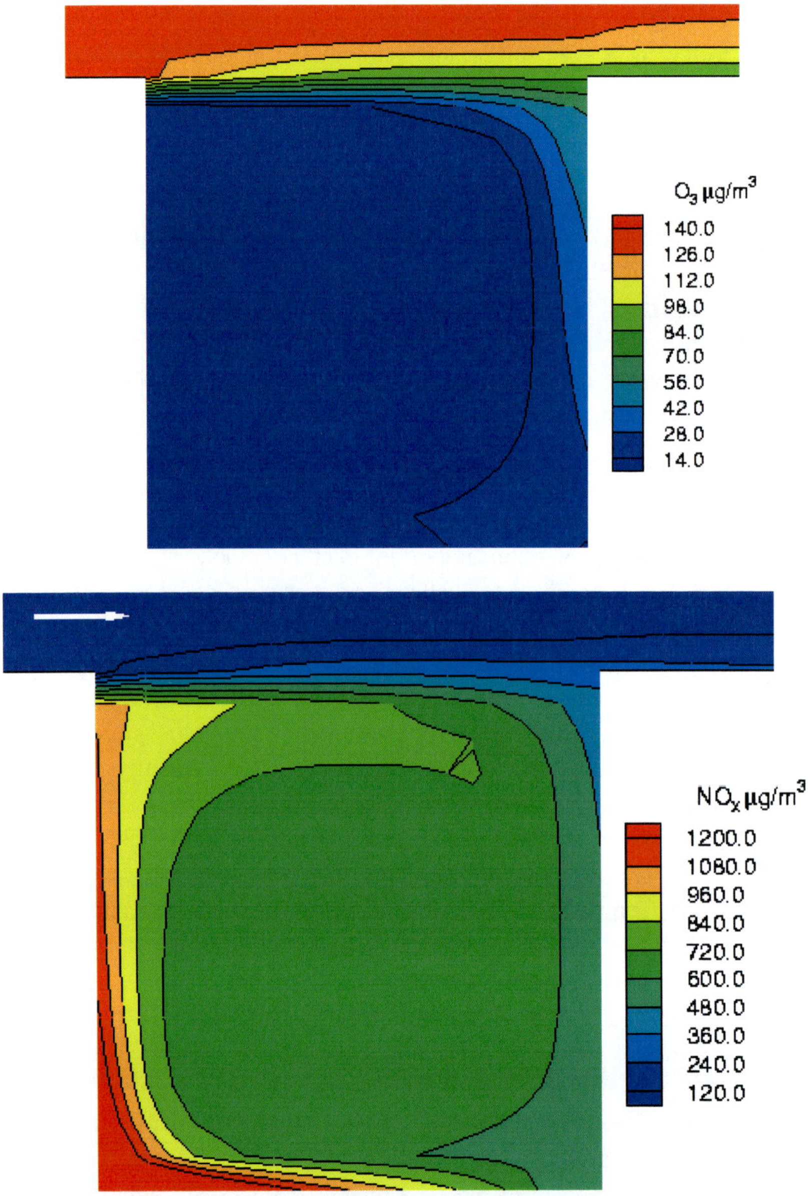

Figure 3. NOx (right) and O_3 (left) fields within a street canyon as predicted with the use of a module based on NO-NO_2-O_3 fast cycle (Theodoridis and Moussiopoulos, 2000).

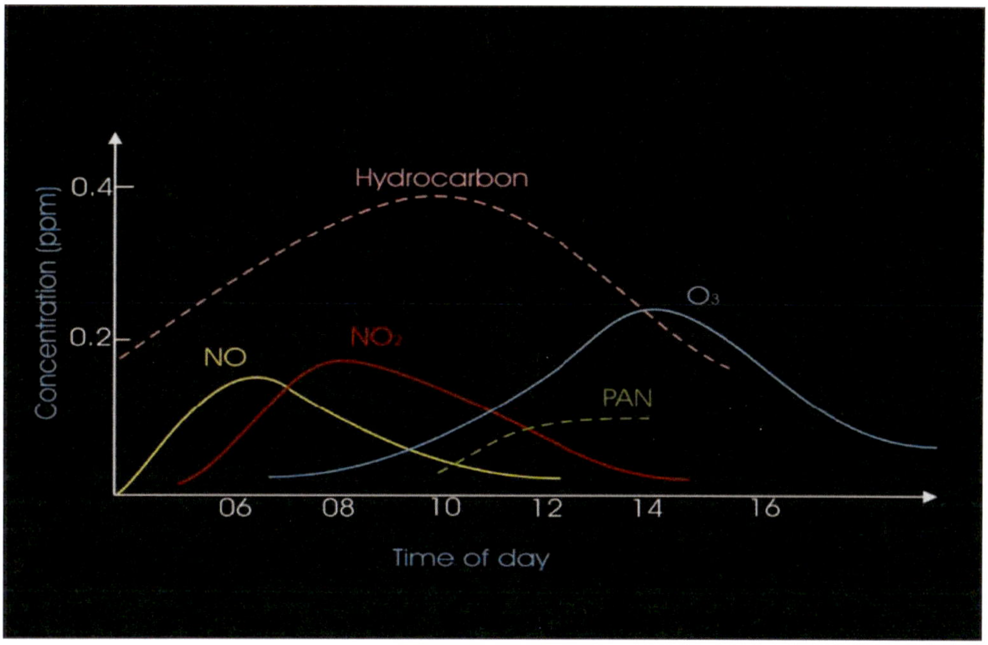

Figure 4. Diurnal cycle of O_3 and its precursors.

Street Canyons

Many studies have shown that relatively high wind speeds, a flat topography and low density of land use do not allow accumulation of pollutants during most of the days as the atmosphere is normally cleaned up during the night. However, in recent years, modern cities are becoming narrow, densely populated and compartmentalized and flanked by buildings or by concrete blocks in which air and noise pollution constitute an extended problem. Thus, it is of no surprise that air dispersion in street canyons are different than in open flat regions.

Many authors have referred the term *street canyon* as a relatively narrow street with buildings lined up continuously along both sides. However, the same term has been used previously to refer to larger streets, the so-called *avenue canyons*. In the real world, a wider definition has been applied, including urban streets that are not necessarily flanked by buildings continuously on both sides (Vardoulakis *et al*., 2003).

Within these urban canopies, wind vortices, low-pressure areas, channelling effects, and wind attenuation may be created under certain meteorological conditions and give rise to air pollution hotspots (Vardoulakis, *et al*., 2003). For example, many authors have also reported that high concentrations levels have been often been observed on the leeward side of the regular canyons under perpendicular wind conditions. Due to the increased concentrations of traffic emissions and reduced natural ventilation levels of pollutants are usually higher in these narrow urban street canyons (Vardoulakis *et al*., 2003).

Physical Dynamics within Street Canyons

The street canyon configuration has several impacts on several processes (like thermal modulation, pollution dispersion and concentration).

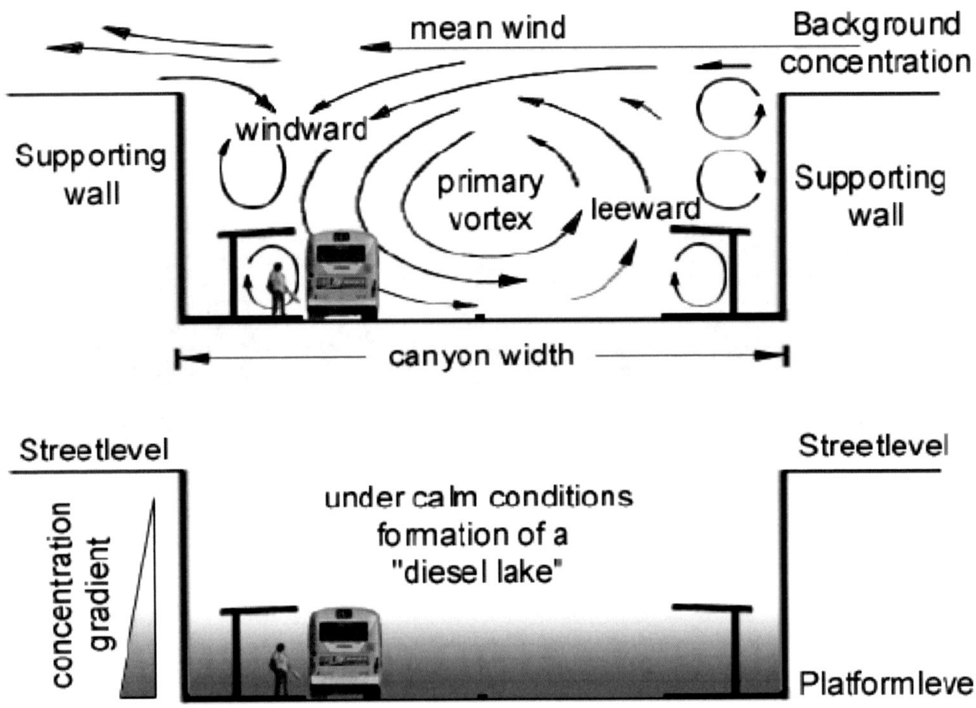

Figure 5. Pollutant dispersion in a regular street canyon under windy conditions (top) and calm & cloudy conditions (bottom) (modified after Vardoulakis *et al.*, 2003).

Pollution Concentration and Flow

Flow of air in a street canyon is affected by both the meteorological conditions and the geometry of the street canyon. For urban traffic emissions, when combined local and regional dispersion analysis with curbside dispersion (microscale effects) to account for the trapping of pollutants in street canyons. Previous measurements done by several authors showed that the wind speeds and directions, and building configurations have significant influence on the distribution of pollutant concentrations. The clockwise vortex circulation generated in the street canyon led to higher pollutant concentration levels on the leeward side than on the windward side of the buildings and decreased exponentially in the vertical direction of the leeward side of the upstream (Figure 5) building. However, in order to maintain just vortices, the ambient wind speed should exceed at certain critical value and it depends on ambient wind direction. This explains the importance of the local wind to the concentration of pollutants in street canyons, with emphasis on wind direction, which places an important role in determining vortex number and flow field in the street canyon. Typically, concentrations of pollutants are highest at the base of the leeward wall. Due to the smaller rotational velocity of the lower vortex circulation, it is difficult to remove the pollutant discharged from the line source out of the street canyon. Hence, it leads to direct impact on human health, particularly on drivers, pedestrians, and people working in the canyon.

Future Guidelines in Street Canyon Geometry

Many large and populated cities in the past and today have been constructed with no or little knowledge of pollution dispersion and its relationships to street canyon geometry. However, recent research studies placed importance to canyon flow to assess the air quality in urban areas and they can promote a better standard of canyon air and several rules can be made for the purpose of urban planning considerations. It has been found that the released pollutants from a street canyon become more diluted in the following cases: the lower the height of the street canyon, the higher the wind speed; the higher the height of the leeward building (compared to the windward), the better the ventilation effect. Wider canyon promotes better pollutant diffusion.

Urban areas cannot be considered as homogenous entities; the largest pollution levels occur in a street canyon, where the dilution of car exhausts gases is limited by the presence of buildings. Moreover, this study deals with pollution phenomena that take place in the immediate vicinity of the source, the car traffic. Therefore modeling of this pollution must account for micro-scale processes.

Chemical Processes in Street Canyons

Transport and dispersion processes are not the only factors determining relationships between emissions and ambient concentrations. Chemistry plays a crucial role in transformation of pollutants resulting in degradation of some species and formation of other. Considering transport of pollution on a large scale, when the transport times involved are of the order of hours or even days, often hundreds of chemical reactions must be considered in order to account for the chemical composition of air.

The situation is quite different when dealing with processes in street canyons. Due to the very short distances between the sources and receptors, only the fastest chemical reactions have a significant influence on the transformation processes in a street canyon's air. It means that most of pollutants emitted from traffic can be considered as inert compounds (e.g. CO and hydrocarbons).

The nitrogen oxide gases are, however, subject to fast chemical reactions. Of special interest for us are the reactions involving NO-NO_2-O_3. The time scales characterising these reactions are of the order of tens of a second and thus comparable with residence time of pollutants in a street canyon. Consequently, the chemical transformations and exchange of the street canyon air with the ambient air are of importance for processes leading to NO_2 formation. Taking this into account, the exchange rate between the concentrations of NO, NO_2, and O_3 in the street.

CFD Software Package

Computational fluid dynamics (CFD) has become an integral part of the engineering design and environmental analysis when needed to predict the performance of new designs or processes before they are manufactured or implemented. Understanding the motion of liquids and gases is crucial in many branches of engineering. Until recently, studies of fluids in motion were confined to the laboratory. But with the rapid growth in computer processing power, software applications now bring numerical analysis and solutions of flow problems to the desktop. In addition, the use of common interfaces and workflow processes makes fluid dynamics accessible to designers as well as analysts.

ANSYS - CFX Package Design Steps

The fidelity of simulation is linked directly to the choice of physical models available. The ANSYS - CFX software contains a large number of physical models to provide accurate simulation for a wide variety of industrial applications. Because almost all physical models interoperate with each other and in conjunction with all element types, across all grid interface connection types, using the coupled multigrid solver, in parallel, with accurate numerics, the ability to obtain an accurate solution is greatly enhanced by the following models:

- *Multiphase:* The Particle Transport model allows the solution of one or more discrete particles phases within a continuous phase. A general framework for Interphase Mass Transfer is also incorporated.
- *Turbulent:* A variety of well-established models are available such as k-ε and SST, which includes the scalable wall function model that ensures solution accuracy, is improved with mesh refinement. The advanced turbulence capabilities contain the first commercial implementation to model laminar to turbulent flow transition.
- *Heat Transfer:* Optimizing heat transfer between fluids and solids is critical in many types of industrial equipment. CFX solves fluid flows in 3-D domains including conjugate heat transfer for calculation of thermal conduction through solid materials.
- *Radiation:* Wide classes of radiative heat transfer models, from transparent to participating non-gray media, are available and include applications such as combustion, heating and ventilation and radiation through solid materials, among others.
- *Reaction definition:* All species are solved as a single coupled system, accelerating convergence especially for complex reaction mechanisms. Models include multi-step eddy break-up, finite rate chemistry and NOx and soot models.
- *Fluid Structure Interaction:* FSI approach preserves the individually validated specialized software components in the computational fluid dynamics and stress analysis disciplines, while at the same time permitting state-of-the-art interaction between the fluid and solid.

The steps that have been followed for the development urban street canyon model by the CFX software were the following:

- Geometry creation
- Meshing
- Pre-Processing, reaction definition and physical model selection
- Solver

Pollutant Concentration and Parameterization

Concentration: It depends on how many molecules of the particular gas (or the mass of the gas molecules) are in the sample volume of gas (es), and the total volume of the sample:

$$\text{Concentration } C = \frac{\text{Amount of a Substance}}{\text{Volume occupied by substance}}$$

Mixing ratio: It gives the ratio of a particular substance to the sum of all the other substances (e.g., "1 part in 10 parts", where there is, say, 10 grams of "A" and 100 grams of everything else):

$$\text{Mixing ratio} = \frac{\text{Amount of a substance in a mixture}}{\text{Amount of all substances in the mixture}}$$

Typical mixing ratios for pollutants result in extremely small fractions. For this reason, fractions such as parts per million (ppm) or parts per billion (ppb) are often used. Specifications of whether it is referred to mass (ppm_m) or volume (ppm_v) mixing ratios are also determined:

- ppm_v based on volume: ratio of moles of pollutant to 10^6 moles in total.
- ppm_m based on mass: ratio of grams of pollutant to 10^6g of total material.

Mole fraction of the gaseous species i can be written as

$$y_i = \frac{n_i}{n_t} = \frac{P_i}{P_t} = \frac{V_i}{V_t} \tag{1}$$

Where P_i and V_i are the partial pressure and partial volume, respectively.

By definition of the mole fraction $y_i = \frac{ppm}{10^6}$ ($ppm = \frac{1\,part}{10^6\,parts}$)

Mass fraction defined as:

$$f_i = \frac{m_i}{m_t} \tag{2}$$

Where m_i is the mass of gas i (pollutant), and m_t the mass of the total material (all substances).

The mole and mass fractions are related by the equation:

$$f_i = y_i \frac{M_i}{M_{avg}} \tag{3}$$

Where M_i is the molecular weight of gas (pollutant) i, and M_{avg} the average molecular weight, is defined by:

$$M_{avg} = \sum y_i M_i \tag{4}$$

(For the air $M_{air} = M_{avg} \approx 29$)

The ppm_v and ppm_m of gas i (pollutant) are related by:

$$ppm_m = ppm_V \frac{M_i}{M_{air}} \tag{5}$$

The transition from concentration unit $\mu g/m^3$ to mixing ratio unit *ppb* is performed through:

$$C(ppb) = \frac{83.14 \times T}{P \times M_i} C(\mu g / m^3) \tag{6}$$

Where T is the temperature in ^oK, P is the pressure in *mbar* and M_i is the molecular weight of the substance.

Real Case Study Simulation

The street canyon is located in the medium sized city of Kozani, in Northwestern Greece. It is a heavily industrialized area, due to lignite power stations that contribute about 70% of total electrical energy produced in Greece.

The one-way street simulated (Figure 6) by CFX, has an average width 10m and a length of 95m. It has about 19 building with heights varying between 3m and 18m. Details on the DOAS measurements in relation to meteorological conditions and the surrounding industrial pollution sources can be found elsewhere (Zoras et al., 2008).

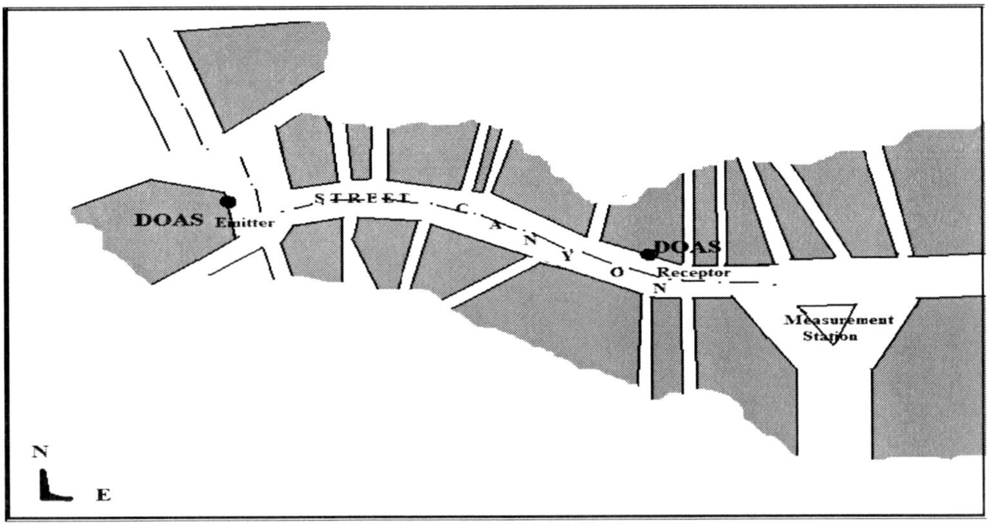

Figure 6. Map showing Street configurations with surrounded buildings.

Car Emissions

Traffic emissions in the street are calculated knowing the traffic flow (vehicle/hour) and emission factors (g/km). The basic formula for estimating emissions, using experimentally (The Core Inventory of Air Emissions in Europe CORINAIR, 3rd Edition 2003) obtained emission factors is:

Emissions per Period of Time [g] = Emission Factor [g/km] × Number of Vehicles [veh.] × Mileage per Vehicle per Period of Time [km/veh.]

The emissions factor depends on the vehicle speed for NOx. From CORINAIR $E = 0.5595 - 0.01047V + 10.8e^{-5} V^2$ [g/km] where V is the vehicle speed. Table 1 illustrates the emissions rates in kozani city, for an average speed of 40km/h:

Chemical Coupling of O_3, NO, and NO_2

The reactive pollutants concerned with in this study are nitrogen oxide NO and nitrogen dioxide NO_2, which are supposed to be emitted from automobiles within the street canyon in the presence of background ozone O_3. The fast chemistry module in the CFD model consisted of the following three reactions:

$$NO_2 + sunlight\ (\lambda < 420\ nm) \rightarrow NO + (O^{3p})\ (R_0)$$

$$(O^{3p}) + O_2 + M \rightarrow O_3 + M\ (R_1)$$

$$O_3 + NO \rightarrow NO_2 + O_2\ (R_2)$$

M represents a molecule (N_2 or O_2 or another third molecule).

Table 1. Emissions rate of cars in kozani city

Hour	Number of cars	NOx (g/s)
1	50	0.011
2	30	0.006
3	20	0.004
4	10	0.002
5	10	0.002
6	100	0.021
7	250	0.055
8	500	0.105
9	500	0.105
10	500	0.105
11	750	0.158
12	750	0.158
13	1000	0.211
14	1000	0.211
15	750	0.158
16	750	0.158
17	500	0.105
18	500	0.105
19	500	0.105
20	300	0.063
21	300	0.063
22	100	0.021
23	100	0.021
24	100	0.021

Equations of State: Ideal Gas Equation of State for NO and NO_2

For an ideal gas, the relationship by the ideal gas law is:

$$\rho = \frac{M.P}{R.T} \tag{7}$$

where M is the molecular weight of the gas, ρ is the density, R is the universal gas constant, T is the temperature, and P is the pressure.

Redlich-kwong Equation of State for Ozone

The Redlich-kwong equation (Soave, 1979) can handle various pure fluids, and includes ozone, oxygen, carbon monoxide, hydrogen, helium, argon, ect.

The Redlich-Kwong equation of state is considered one of the most accurate two parameter corresponding states equations. This equation of state is quite useful from an engineering point of view because it only requires that the user know the fluid critical temperature and pressure (T_c, p_c). The critical temperature, Tc, of a material is the temperature above which distinct liquid and gas phases do not exist. As the critical temperature is approached, the properties of the gas and liquid phases become the same resulting in only one phase: the supercritical fluid. Above the critical temperature a liquid cannot be formed by an increase in pressure, but with enough pressure a solid may be formed for materials other than water.

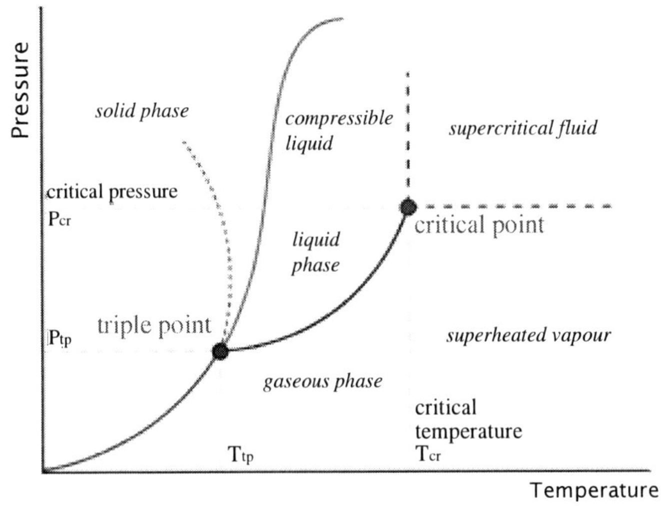

Figure 7. The critical point in a phase diagram is at the high-temperature extreme of the liquid-gas phase boundary (The dotted green line gives the anomalous behaviour of water).

The critical pressure is the vapor pressure at the critical temperature. On the diagram showing (Figure 7) the thermodynamic properties for a given substance, the point at critical temperature and critical pressure is called the critical point of the substance.

The Redlich-kwong equation of state provides much better accuracy near the critical point. The form of this equation of state is used by ANSYS CFX given by:

$$p = \frac{RT}{v - b + c} - \frac{a(T)}{v(v + b)} \tag{8}$$

This form differs form the original by the additional parameter c, which is added to improve the behaviour of isotherm conditions near the critical point. The parameters a, b, and c are given by:

$$a = a_0 \left(\frac{T}{T_c} \right)^{-n} \tag{9}$$

$$a_0 = \frac{0.42747 R^2 T_c^2}{p_c} \tag{10}$$

$$b = \frac{0.08664 R T_c}{p_c} \tag{11}$$

$$c = \frac{R T_c}{p_c + \frac{a_0}{v_c (v_c + b)}} + b - c \tag{12}$$

In the expression for a(T), optimum values of n depend on the pure substance. The best fit polynomial for n in terms of the acentric factor w is:

$$n = 0.4986 + 1.2735 w + 0.4754 w^2 \tag{13}$$

where:

$$w = -\log_{10} \left(\frac{p_v}{p_c} \right) - 1 \tag{14}$$

the vapour pressure p_v is calculated at $T = 0.7 T_c$.

Arrhenius Reaction Rate

Chemical kinetics characterize the rate at which chemical species appear or disappear. The kinetic rate constant (of reaction) is functions of temperature and given in the form:

$$k(T) = A T^{\beta} \exp\left(\frac{-E_a}{RT} \right) \tag{15}$$

This equation is called the Arrhenius equation (used by ANSYS CFX), where A is a pre-exponential factor, β is the temperature exponent, R is the universal gas constant, T is the temperature, and E_a is the activation energy, which represents the amount of energy a molecule requires before it can react. For simplicity, the term E_a/R is often combined into a single constant to eliminate confusion in the units of energy. Rate constants vary considerably in magnitude because some reactions are very fast while others are very slow. The units of the rate constants depend on reaction orders, first, second, or third order. And given by: $Time^{-1}$

$(mol\ m^{-3})^{1-n}$ where n is the sum of the number of reaction orders, (plus 1 for third body term, if present). The table below shows the rate constant parameters for the reactions **R_1** and **R_2**

Table 2. Reaction contants

Reaction	A E_a β
R_1	6.10^{-34} 0.0 -2.3
R_2	2.10^{-12} 2.782 0.0

where E_a in kcal/mol, and A in cm, molecule, s depends on the reaction orders
for reaction **R_1** (third order), A in cm^6 $molecule^{-2}$ s^{-1}
for reaction **R_2** (second order), A in cm^3 $molecule^{-1}$ s^{-1}.

FAST CHEMISTRY MODULE AND RADIATION

Material Creation and Fluid Domain

The Materials tab (in CFX) contains the Material Selector, which shows a list of all the available materials. It allows to create new fluids and to edit the properties; the new or modified materials can then be selected for use in simulation.

Create the Substances (O_3, O_2, NO, and NO_2)

Any material can consist of one or more materials. If a material contains only a single pure substance, then it is known as a pure substance (O_3, O_2, NO, and NO_2), if it contains more than one species, then it is known as a mixture (our fluid domain).

The pure substance option should be used to create a substance whose properties (thermodynamic, transport, and radiation) such as viscosity, density, or molar mass, are known. It must be specified the equation of state for each substance, equation of state is Ideal-Gas for NO and NO_2, and Redlich-kwong for O_2 and O_3.

Reaction Creation

For type of reactions **R_1**and **R_2** the list of reactants was specified (O_2 and O_3 for **R_1**,NO for **R_2**) with the ratio that they react each other and the order. A list of products was also set (O_3 and O_2 for **R_1**, NO_2 for **R_2**), with the ratio they are produced. Also, optional forward reaction rate and third body terms were applied.

The reactions **R_1** and **R_2** were used as multi-step reactions, which were set in our material fluid domain, with specified reacting mixture option.

As list of substances, the fluid domain involved NO, NO_2, O_3, O_2, and Air at 25°C. In the simulation model it was specified transport equation for NO, NO_2, O_3, and O_2; and Air at 25°C as constrain.

The Radiation Model

ANSYS - CFX includes several radiation modelling options the Rosseland model (or Diffusion Approximation model), the P-1 model (also known as the Gibb's model or Spherical Harmonics model), the Discrete Transfer model and the Monte Carlo model.

Many fluid flows of practical interest occur in situations where the fluid and/or the enclosing boundaries are hot. In such situations, the effect of radiant heat transfer may become important. A typical environment where radiation plays a significant role is a furnace or other such as a combustion chamber. The goal of radiation modelling is to solve the radiation transport equation, obtain the source term, S, for the energy equation, and the radiative heat flux at walls, among other quantities of interest.

The Monte Carlo Model was used in our case. The Monte Carlo method simulates the underlying processes which govern the system of interest, i.e. the physical interactions between photons and their environment. A photon is selected from a photon source and tracked through the system until its weight falls below a minimum where it 'dies.' When using the Monte Carlo radiation model, can be set Non-thermal radiation sources (fluxes at boundaries) that are divided into 2 groups: The Directional Radiation Source and Directional Radiation Flux. The Directional Radiation Source allows the specification of the source strength and its direction. The direction can be specified by either Cartesian Components or Cylindrical Components using a local axis. The Directional Radiation Flux Allows the specification of collimated non-thermal radiation flux at boundaries. The direction can be set by using Cartesian Components, Cylindrical Components using a local axis, or Normal to Boundary.

CFD Equations Solved

Computational fluid dynamics (CFD) modelling is based on the governing fluid flow and dispersion equations, which are derived from basic conservation and transport principles:

- The mass conservation (continuity) equation.
- The three momentum conservation (Navier-Stokes) equations in x, y, z.
- The transport equation for pollution concentration.

The air within the street canyon can be regarded as an incompressible turbulent flow, and the air and pollutants densities are assumed to be constant. These assumptions are reasonable for lower atmosphere environment as described by Sini (1996).

The turbulence production due to the buoyancy effect is not included because the thermal effect in the street canyon is not taken into consideration in the present study.

The continuity equation:

$$\frac{\partial U_i}{\partial x_i} = 0 \tag{16}$$

The momentum equation:

$$\frac{\partial U_i}{\partial t} + U_j \frac{\partial U_i}{\partial x_i} = -\frac{1}{\rho}\frac{\partial \bar{p}}{\partial x_i} + \frac{\partial}{\partial x_j}\left(\upsilon \frac{\partial U_i}{\partial x_j} - \overline{u_i' u_j'} \right) \tag{17}$$

k and ε transport equations in the standard k - ε model:

$$\frac{\partial k}{\partial t} + \vec{V} gradk = div(\frac{\upsilon_t}{\sigma_k} gradk) + P - \varepsilon \tag{18}$$

$$\frac{\partial \varepsilon}{\partial t} + \vec{V} grad\varepsilon = div\left(\frac{\upsilon_t}{\sigma_\varepsilon} grad\varepsilon \right) + \frac{\varepsilon}{k}\left(C_{\varepsilon_1} P - C_{\varepsilon_2} \varepsilon \right) \tag{19}$$

k is the turbulent kinetic energy, ε denotes the turbulent dissipation rate.
P: production of k , ε: dissipation of k .

Where: $\upsilon_t = C_\mu \frac{k^2}{\varepsilon}$; $P = 2\upsilon_t S_{ij} S_{ij}$; $\overline{u_i' u_j'} = -2\upsilon_t S_{ij} + \frac{2}{3} k\delta_{ij}$;

$$S_{ij} = \frac{1}{2}\left[\frac{\partial U_i}{\partial x_j} + \frac{\partial U_j}{\partial x_i} \right].$$

The constants for k -ε turbulence model are:

C_μ	σ_k	σ_ε	C_{ε_1}	C_{ε_2}
0.09	1	1.3	1.44	1.92

Pollutant concentration is calculated with the convective-diffusion equation:

$$\frac{\partial C_i}{\partial t} + U_j \frac{\partial C_i}{\partial x_j} = \frac{\partial}{\partial x_j}\left(K_t \frac{\partial C_i}{\partial x_j} \right) + S_i \tag{20}$$

Where C_i denotes the pollutant concentration, K_t is the eddy diffusivity coefficient; and S_i represents all sources and sink terms.

3-D Model

Figure 8 illustrates the computational domain with the building configuration. This study models reactive pollutant dispersion in a long street when the wind direction (x-direction) is parallel to the street direction. The origin of the coordinate system is located at the left bottom corner of the street in the computational domain. The domain size is 95m in the x-direction, 16m in the y-direction, 21m in z-direction.

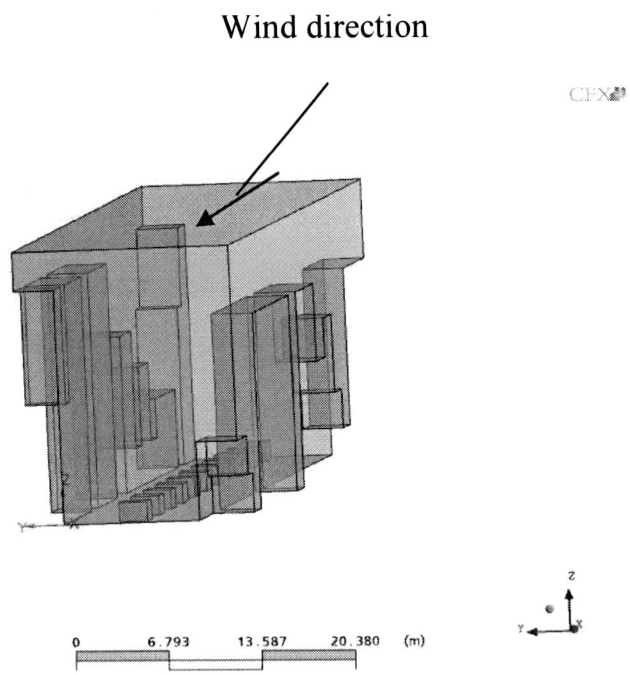

Figure 8.Computational domain and buildings configuration.

The finite element mesh for the domain is shown in Figure 9. Tetrahedrales are accounted at 902585 elements.

The meshing parameters are summarizing as:

- Type Face spacing constant (Constant Edge length 1m and Expansion Factor 1.2).
- Number of inflated layers 5 (prisms).
- Number of Nodes 284282.
- Max Edge length ratio 59.1084.
- Volume of domain 25935 m^3.
- Volume of sub-domain (car) 10.36 m^3.

INITIALIZATION

Emission Sources

The emission sources considered in this study are sub-domains (volume sources), created along the street in x-direction (11 sub-domains or cars), with size of 3.5m×2m×1.48m (each car), and the distance between two cars is 5m (Figure 9). The vehicles were assumed to emit NO (90% of NO_x), NO_2 (10% of NO_x). It was estimated the emission rate for each car, as NO emission rate 16.5 μg/m^{3}s, and 2.5 μg/m^{3}s for NO_2. A background ozone concentration of 70 μg/m^3 was then set for the entire domain gathered from experimental data in the area.

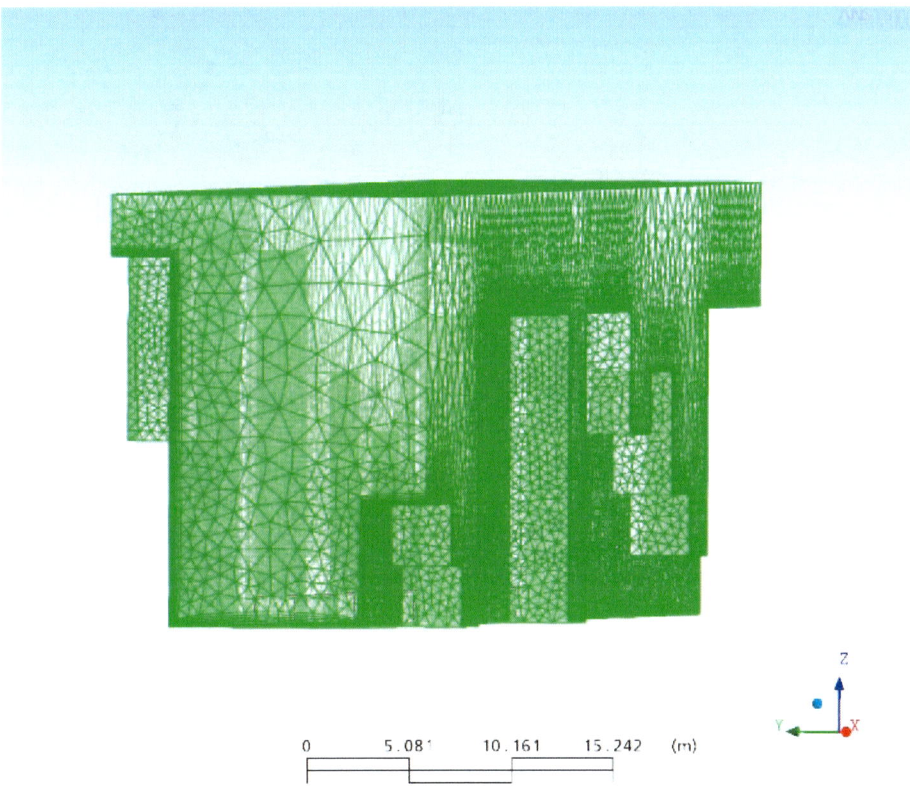

Figure 9. Mesh plots for the complete computational domain.

Mass Fractions

Initial mass fractions were considered the values measured by DOAS (average values) in the 23rd of June 2006 (at 15:00), as following:

- Ozone mass fraction was f_{O3}= 5.8×10^{-8} (70 μg/m^3).
- NO_2 mass fraction was f_{NO2}= 2.025×10^{-8} (24 μg/m^3).
- NO mass fraction was f_{NO}=1.02×10^{-8} (12 μg/m^3).

At the inflow boundary, wind observed to blow along the canyon from east to west; with a speed of 1.5 m/s. The pressure and temperature were set at 1 atm and 25 ^{0}C, respectively.

The radiation flux at the boundaries (Monte Carlo model) was set at an intensity of 700W/m^2 acquired from a pyranometer's measurements at TEI (Technological Education Institute), and the direction set by using Cartesian Components. Note that, the meteorological parameters of temperature, surface pressure and wind speed and direction were obtained from the meteorological station in the canyon.

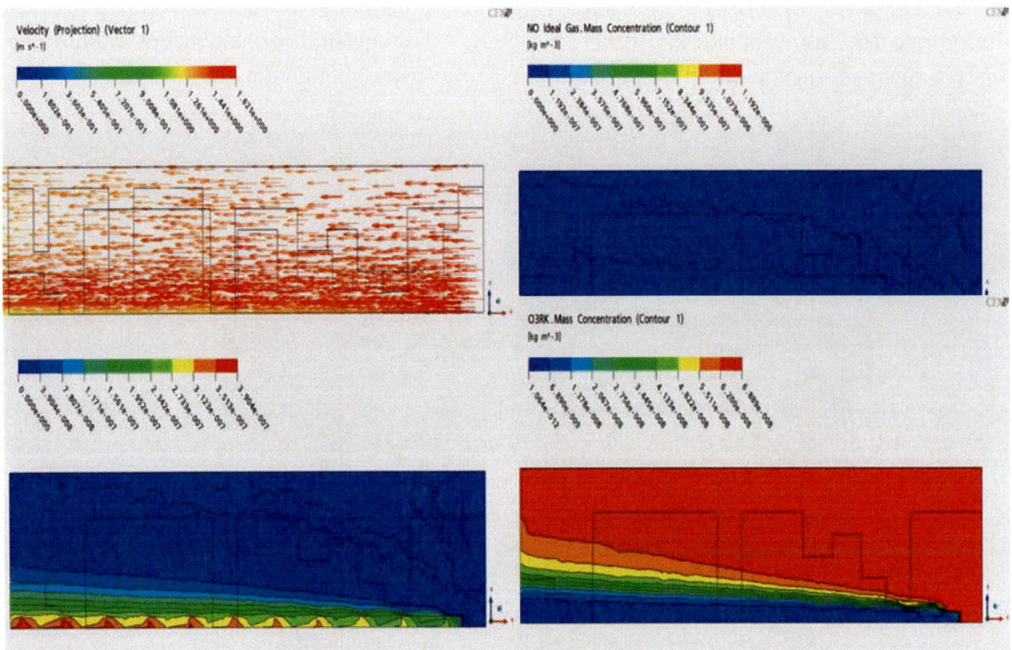

Figure 10. Velocity vectors; NO, NO_2, and O_3 concentration contours on x-z plane at y = -5 (middle of street).

RESULTS AND DISCUSSION

As shown in Figure 10 the effect of wind direction was crucial and a Key factor determining the dispersion of pollutants. It was very interesting to see how the concentration distribution is behaving with respect to the flow field.

Wind parallel to the street direction resulted in higher O_3 concentration levels in the middle upper area of the street. The accumulation of pollutants (NO, NO_2) along the canyon axis dominates close to the vehicular emission sources. Ozone (O_3) concentration is high apart from the upper region of the street and at the inlet. This is explained by ozone rich air aloft being entrained into the canyon, followed by dispersion and reaction. A very interesting symmetric level has been occurred between ozone O_3 and NO_2 which follow opposite trends. A steady rise in ozone level was observed with decreased NO_2 concentration.

In Figure 11, it is illustrated clearly that concentration of NO and NO_2 in the areas near the emission source-traffic road, are significantly higher than that of any other areas and very low close to the buildings. The spatial distribution of NO_2 is larger than NO, because NO is quickly transformed to NO_2. Furthermore the concentrations of NO and NO_2 are much lower in the inflow region than the rest of the canyon. This is in contrast with O_3 that is high at the inlet. This is explained by the consummation of ozone O_3 by NO in the street canyon to varying degrees. And the flow speed in the left side near the ground level is weak and hence the pollutants have sufficient time spent in the mixing and reaction process of the chemical species. On the other hand, there is a formation of NO_2 due to the reactions between O_3 and NO and thus, NO_2 dispersed more significantly than NO.

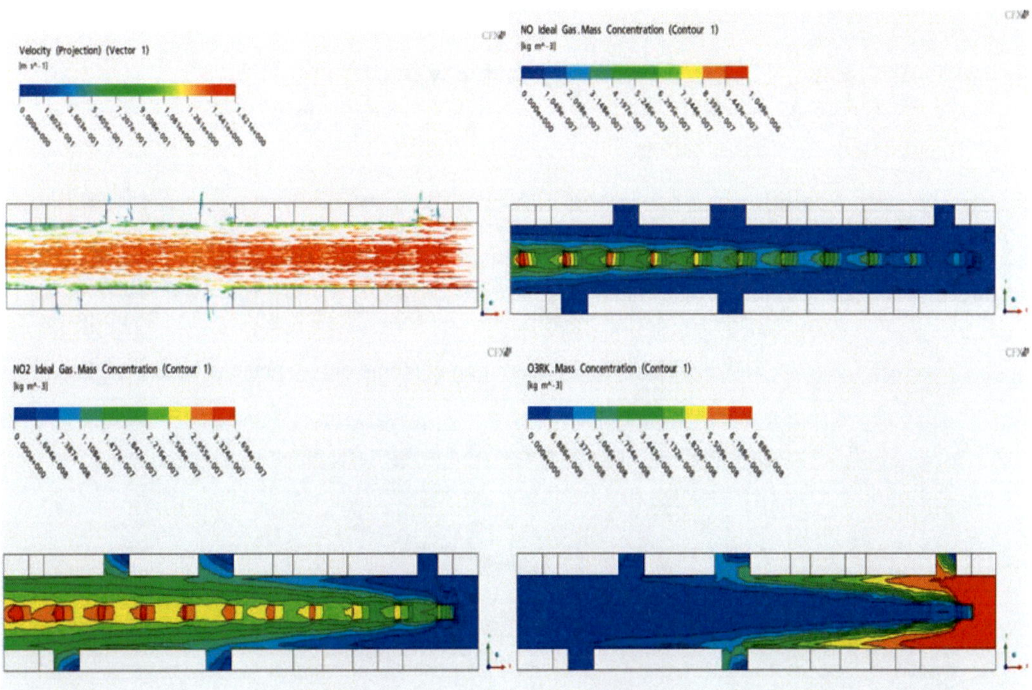

Figure 11. Velocity vectors; NO and NO_2, and O_3 concentration contours on x-y plane at level z= 1m (near the ground).

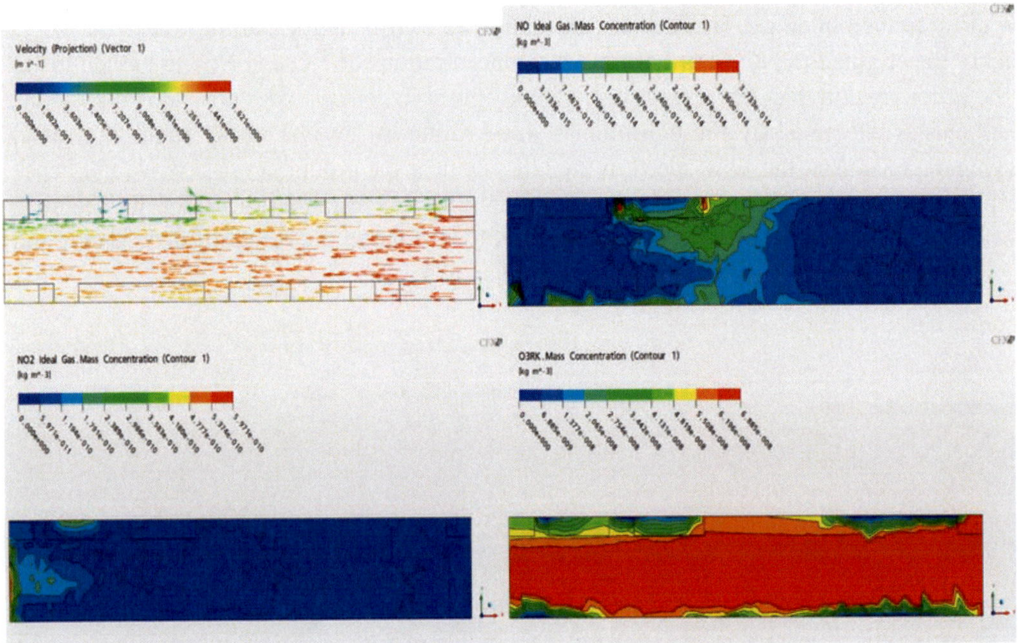

Figure 12. Velocity vectors; NO, NO_2, and O_3 concentration contours on x-y plane at level z= 18m.

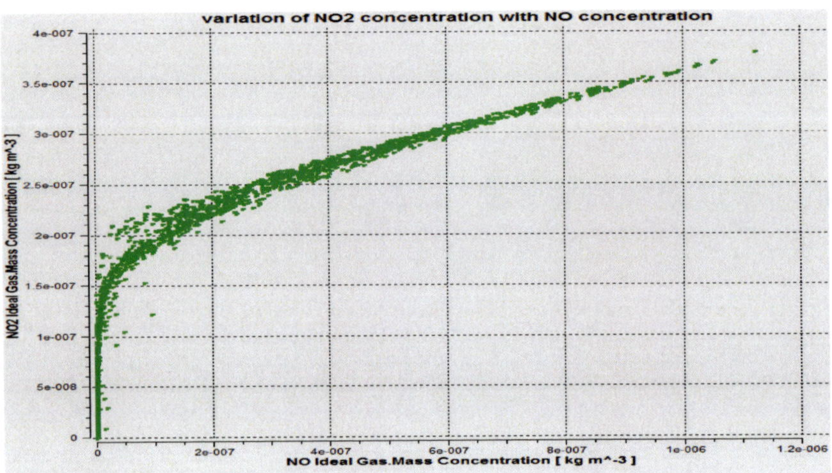

a)

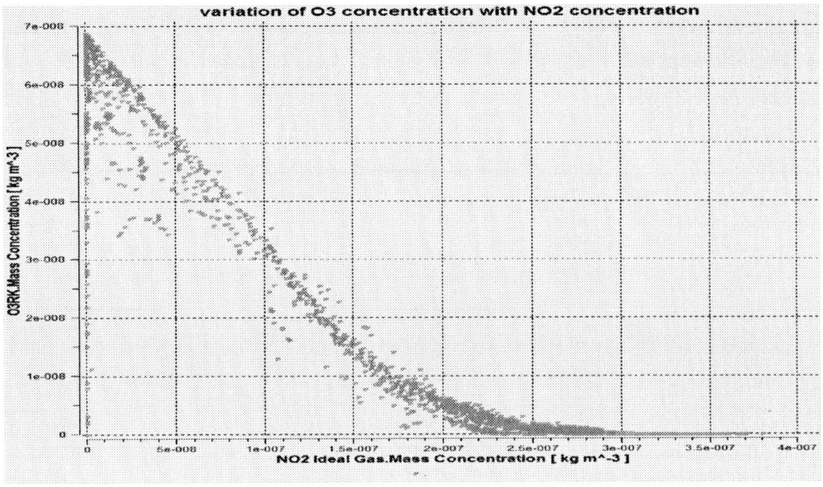

b)

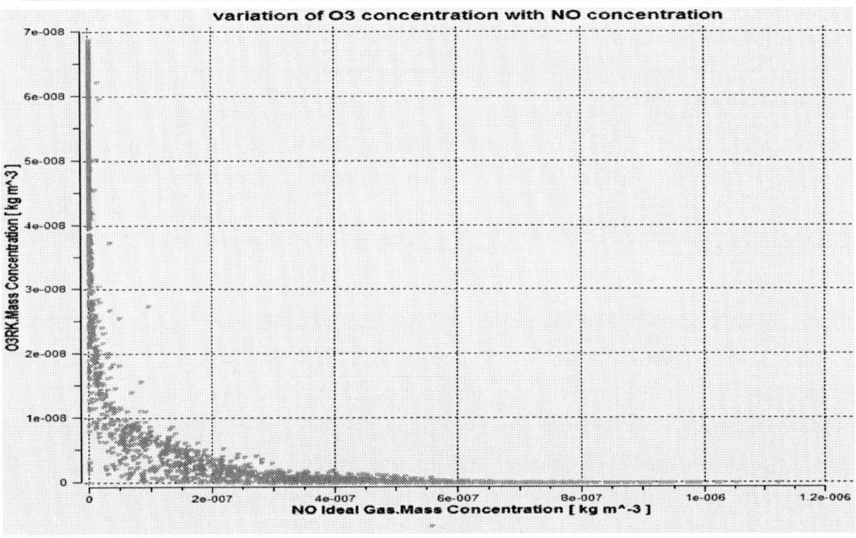

c)

Figure 13. Spatial variation of concentrations in the domain a) variation of NO_2 concentration with NO b) variation of O_3 concentration with NO_2 c) variation of O_3 concentration with NO concentration.

When compared the concentration levels of NO and NO2 at ground and upper level, are reduced by 12% and 5.7%, respectively. For ozone concentration 6.88×10-8kg/m3 (68.8μg/m3) as shown in Figures 4 and 5 was in a good agreement with the DOAS reading of 70μg/m3 at the roof level.

At z = 18m level, NO and NO_2 concentrations were low with values of 3.73×10^{-14}, 5.95×10^{-10} kg/m^3, respectively. On the contrary ozone (O_3) concentration was significantly higher. This can be explained by the stronger wind speed, that transfers background ozone into the canyon from inflow (inlet), and in particular in the middle of the street where the wind speed becomes stronger. Also ozone increases with wind speed because fresh oxygen is brought into the reaction sector $O+O_2+M\rightarrow O_3+M$.

In addition the low concentration of NO means that the chemical reaction process between NO and O_3 is negligible there, and led to higher O_3 concentration. Except the layer above the buildings' top, a significant decrease of O_3 concentration has been observed, due to the low wind speed, in areas depended on the different heights of buildings, which obstruct and decrease the wind speed.

As expected, NO and NO_2 concentration levels decrease with height while that of ozone generally increases.

The spatial variation of each reactive pollutant's concentration, against the other in Figure 13, help to interpret performance analysis results.

Figure 13 shows the variation of NO_2 concentration with NO concentration. The increasing NO_2 concentration with increasing NO concentration is displayed. This variation is approximately linear above 1.5×10^{-7}kg/m^3 of NO_2 concentration, until the maximum values of NO and NO2 concentrations, which are ~1.19×10^{-6} and ~3.9×10^{-7} kg/m^3, respectively. This case reflects the high concentration levels of NO and NO_2 near the ground, where their emission source is located. Note that, the values of NO concentration is greater than NO_2 concentration, due to the rate of emission source that was assumed with a ratio of one-tenth (1/10), of NO_2 to NO.

Except in the region below 1.5×10^{-7}kg/m^3 where NO_2 concentration varied between zero to 1.5×10^{-7}kg/m^3 with nearly zero NO. This was because NO_2 dispersed more than NO corresponding to areas at high levels because NO is quickly transformed to NO_2.

The corresponding NO_2 and O_3 concentration levels are displayed in Figure 13b, that shows decreasing O_3 concentration with increasing NO_2 concentration. This indicates the oxidation of NO by ozone O_3 that leads to an increase in the NO_2 concentration and linear decrease of ozone levels.

It is interesting in Figure 13c to observe the difference between NO and O_3 concentrations. As it shown the increase of NO concentration (for 2×10^{-7}kg/m^3 until a maximum value) with low O_3 concentration, implies that the ozone is consumed in high degree by NO due to the reaction with it near the emission source region.

A similar behaviour is observed for O_3 concentration (more than 10^{-8}kg/m^3) with low NO. Because O_3 concentration is higher in the upper region of the street and in the inlet boundary where NO concentration was negligible in these regions.

In addition there are some areas (Figure 13c) where low O_3 concentration (less than 10^{-8}kg/m^3), the NO concentration was low too (less than 2×10^{-7}kg/m^3). This result corresponds to areas close to building surfaces, at all levels, where O_3 and NO were both considerably low, due to low wind speed and weak dispersion ability of NO. Also note that NO_2 concentration is low in these areas.

CONCLUSION

This study was carried in order to show what the steps of a modeling fast chemistry approach in a street canyon should be. It examined reactive pollutant dispersion in a 3D urban street canyon in steady state. A CFD general code was used (ANSYS-CFX), with a standard k-ε turbulence model using transport equations for the mean concentration and concentration variance of the scalars, incorporating simple NO-NO_2-O_3 photochemistry. Vehicular emission sources of NO and NO_2 were considered in the presence of background O_3. Experimental meteorological data of temperature, wind speed and direction, solar radiation and NO, NO_2 pollutants (from DOAS) concentrations were used as initialization.

It was found that the wind direction (parallel to street) and its distribution are playing a significant role in determining pollutant levels. High NO, NO_2 concentrations can be found at low heights. In specific, close to emission sources peak concentration values were reached at ground level. It was also evident that the reactive gases NO and NO_2 emitted into the canyon by traffic against a background of ozone. This implied that background ozone that has been transported into the canyon from inflow region, is destroyed by the NO emissions from the motor vehicles, in particular at low levels.The fast reaction of the emitted NO with O_3 increases NO_2 concentration (because NO is quickly transformed to NO_2). In addition, it was observed that NO_2 dispersed more significantly than NO to higher levels, probably because it comes from both primary combustion sources within the street canyon (direct emission from cars) and secondary formation from the NO+O_3 reaction. This resulted to higher O_3 concentration at upper levels and in the region where ambient O_3 enters into the canyon, while NO and NO_2 are depleted in a high degree in this areas. The highest ozone O_3 concentration occurred under high wind speed due to increased mixing under the sunlight's presence. The results indicated that there is a strong influence of the street geometry on the wind field and consequently on the pollutant dispersion around the buildings' surfaces where ozone was found low.

It can be concluded that the use of a general CFD software could be employed on a laboratory or design basis to efficiently assess the urban air quality due to photochemical pollution. The addition of experimental surface temperatures could lead to even more detailed ozone distributions over the urban canopy.

REFERENCES

Bernard SM, Samet JM, Grambsch A, Ebi KL, Romieu I. The potential impacts of climate variability and change on air pollution-related health effects in the United States. *Environmental Health Perspectives* 2001;109: 199-209.

Chiu KH, Sree U, Tseng SH, Wu C, Lo J. Differential optical absorption spectrometer measurement of NO_2, SO_2, O_3, HCHO and aromatic volatile organics in ambient air of Kaohsiung petroleum refinery in Taiwan. *Atmospheric Environment* 2005;39: 941-955.

Friedfeld S, Fraser M, Ensor K, Tribble S, Rehle D, Leleux D, Tittel F. Statistical analysis of primary and secondary atmospheric formaldehyde. *Atmospheric Environment* 2002;36: 4767-4775.

Kelessis AG, Petrakakis MJ, Zoumakis NM. Determination of benzene, toluene, ethylbenzene and xylenes in urban air of Thessaloniki, Greece. *Environmental Toxicology* 2006;21: 440-443.

Petrakis M, Psiloglou B, Kassomenos PA, Cartalis C. Summertime measurements of benzene and toluene in Athens using a differential optical absorption spectroscopy system. *Journal of Air and Waste Management Association* 2003;53: 1052-1064.

Sini JF, Anquetin S, Mestayer G. Pollutant dispersion and thermal effects in urban street canyons. *Atmospheric Environment* 1996;30: 2659–2677.

Soave GS. Application of the Redlich-Kwong-Soave equation of state to solid-liquid equilibria calculations. *Chemical Engineering Science* 1979;34(2):225-299.

The Core Inventory of Air Emissions in Europe, CORINAIR, 3rd Edition 2003. Group 7, Road Transport.

Triantafyllou AG, Kiros ES, Evagelopoulos VG. Respirable particulate matter at an urban and nearby industrial location: Concentration and variability and synoptic weather conditions during high pollution episodes. *Journal of Air and Waste Management Association* 2002;52: 287-296.

Vardoulakis S, Fisher BEA, Pericleous K, Gonzalez-Flesca N. Modelling air quality in street canyons: a review. *Atmospheric Environment* 2003;37(2): 155-182.

Walton A, Cheng AYS, Yeung WC. Large-eddy simulation of pollution dispersion in an urban street canyon—Part I: comparison with field data. *Atmospheric Environment* 2002;36(22): 3601-3613.

Zoras S, Triantafyllou AG, Evagelopoulos V. Aspects of year-long differential optical absorption spectroscopy and ground station measurements in an urban street canyon nearby Industrial pollution sources. *Atmospheric Environment,* 2008;42: 4293-4303.

INDEX

A

B

C

D

E

F

G

H

I

J

K

L

M

N

Q

W

Y

Z